The Marine and Freshwater Fishes of Ceylon

The Marine and Freshwater Fishes of Ceylon

by

IAN S.R. MUNRO

BIOTECH

2018

BIOTECH BOOKS®

ISBN 978-81-7622-044-6

Published by: **BIOTECH BOOKS®**
4762-63/23, Ansari Road, Darya Ganj,
New Delhi - 110 002
Phone: +91-011-23262132
E-mail: biotechbooks@yahoo.co.in

PRINTED IN INDIA

FOREWORD

For people of the coastal areas and islands of south and south-east Asia, fishing is an ancient and honourable occupation. It has for many centuries been an important factor in the economic life of the people.

Although fisheries do not occupy such a permanent place in our Australian economy, we have developed techniques and research directed towards the development, conservation, and regulation of our fisheries.

When the Government of Ceylon asked whether Australia could assist in formulating a fisheries programme, I welcomed the opportunity of making our skills and experience available. Mr. I. S. R. Munro, of the Fisheries Division of the Commonwealth Scientific and Industrial Research Organization, was invited to go to Ceylon as the first Australian expert to be sent overseas under the Colombo Plan.

A specialist in marine biology, Mr. Munro had had considerable experience of fisheries in northern Australian and New Guinea waters. His task in Ceylon was to design and instigate a programme of biological research in relation to Ceylon's onshore and estuarine fisheries, and to train the newly appointed graduate research officers in general fisheries research methods and techniques.

The present volume is an additional result of Mr. Munro's visit. In listing and classifying all the known fish fauna of Ceylon, I believe he has made a useful contribution on which future researches and investigations can be built. I hope that the Handbook will be useful to our friends in Ceylon.

R. G. Casey

Minister for External Affairs

CONTENTS

INTRODUCTION

Under the Colombo Plan of technical aid to south and south-east Asia, the author spent several months during 1951 in Ceylon as adviser on fisheries research and taxonomic problems to the Ministry of Industries, Industrial Research and Fisheries. It was found that one of the initial requirements in the development of a fisheries research programme was that of the production of a comprehensive guide for identification of the fish species. As a direct result, the Australian Government, through the Minister for External Affairs, has made available funds and facilities for the production of this handbook.

Descriptions of Ceylonese fishes are scattered throughout a wide range of scientific publications which are not readily available to the average reader or student. Much is included in the works of the early ichthyologists of last century and many contributions have appeared in numbers of the *Ceylon Journal of Science* and in various reports of the Marine Biologists appointed by the Ceylonese Government. Other reports have been published from time to time by various foreign ichthyologists. Dr. P. E. P. Deraniyagala, Director of the Ceylon National Museums, has contributed probably more than any other individual, especially as regards the fresh water fish fauna, and in the production recently of a coloured atlas which deals with a portion of the known fish fauna. Amongst early attempts at complete treatment are the classical treatises of Bennett (1828) who published descriptions and coloured plates of 30 species, and Day (1875-78) who included many Ceylonese species in his well-known *Fishes of India*.

Eight hundred and fifty-six species are at present listed from the marine, brackish, and fresh waters of Ceylon. They include several introductions to fresh waters and some rare deep-sea species from the Gulf of Manaar and other surrounding waters. The author does not believe that this constitutes a complete list and sincerely hopes that this handbook will stimulate a greater interest in the fishes of this country, and that in so doing, will bring to record many other forms.

This present handbook aims at bringing under a single cover descriptions, illustrations and identification keys to all the marine, brackish and fresh water species of fish that have hitherto been recorded from Ceylon and the adjacent waters of the Gulf of Manaar with its pearl banks or paars, and the trawling grounds which have been investigated experimentally or fished commercially. The contents are based on a check-list prepared by the author in Colombo at the Ceylon Fisheries Research Station, and completed under his direction in Australia by Miss Marcia Whitehurst at the Marine Biological Laboratory of the Division of Fisheries, Commonwealth Scientific and Industrial Research Organization. The index is the work of Mrs. A. Temple, and the photographic copying used in preparation of plates 1-56 was carried out by Mr. R. Breach, also of the above named organization. Text-figures 1-19 are by the author.

Descriptions: A brief diagnosis is provided for each species. These descriptions have been compiled primarily from the more reliable standard reference works on fishes and from original published descriptions. Characters given under orders and families are not repeated in the descriptions of species. Likewise, characters employed in the keys are largely not repeated in the descriptions.

Illustrations: A figure is provided for the vast majority of the species included in the handbook. The plates (1-56) which illustrate this volume are compiled from reproductions of the best available published figures. The authority in each case is that quoted immediately preceding the plate reference given with each description. Each figure is identified on its plate by a reference number which is identical with that which accompanies the description of that particular species. The number of the plate on which the figure occurs is indicated in brackets at the end of the references quoted at the beginning of each such description.

References: The references to scientific literature are restricted to a citation of the original description and to the source of the figure employed. The principal synonyms and name combinations employed by authors dealing with Ceylonese fishes (see list, pp. 291-95) are also included but, for the sake of compactness, authors and references are omitted.

Vernacular names: In most cases no common names in the English language are in general use in Ceylon. The majority of names herein provided have been adopted from other countries or derived from roots of the trivial names or from diagnostic characters. Sinhalese and Tamil names also appear at the end of specific descriptions and are marked (S) and (T) respectively. They are mostly after Deraniyagala (1933), (1952) and are spelt in phonetic English by Mr. A. S. Mendis, Research Officer of the Ceylon Fisheries Research Station.

Keys: The orders and suborders of the cartilaginous fishes are distinguished in the key on p. 1, and those of the bony fishes in the key on pp. 18-20. In both classes the orders are dealt with separately and under each (see page references included in the keys on the pages quoted above) is provided a key and page reference to constituent families. For each family are given characters and a key to species. While it should be possible to identify any fish by systematically working through the keys to orders, families and species in that sequence, there may be circumstances where a key fails. In such cases supplementary guidance will be obtained from the plates which have been arranged with related forms grouped as closely together as possible.

Classification: The nomenclature and arrangement of families is, with certain modifications, that of L. S. Berg, "Classification of fishes both recent and fossil", *Trav. Inst. Zool. Acad. Sci. U.R.S.S.*, 1940, reprinted in U.S.A., 1947.

Terminology: For the guidance of readers who are unaccustomed to characters and terminology of structures used in the description of fishes, the following glossary and explanatory diagrams (text-figs. 1-19) are included.

HOW TO USE THIS BOOK

HOW TO IDENTIFY A FISH BY USING THE TEXT

1. A fish will belong to either of two classes. Decide which by comparing the characters on p. 1 and p. 18. Make use of the sketches on pp. 1, 2, 9 and 20.

   If it has five separate gill-openings and is either spindle-shaped or in the form of a flattened disc, it will belong to the Elasmobranchii and will be located in Part 1 (see pp. 1-17 and pls. 1-3).

   If it has only one gill-opening on each side and has either overlapping scales or a casing of bony plates, rings or sharp prickles, it will belong to the Teleostomi and will be located in Part 2 (see pp. 18-290 and pls. 4-56).

2. When the fish belongs to the Class Elasmobranchii, consult the key on p. 1 and determine which of the three constituent orders contains it.

   When the fish belongs to the Class Teleostomi, consult the key on pp. 18-20 and determine which of the 25 orders contains it. One of these, the Order Perciformes is very large and has been divided into 13 suborders.

3. Turn to the page reference obtained from the above key and check the characters of the order, and then determine from the key now before you, which of the constituent families contains your fish.

4. Now turn to the page reference obtained from the key to families and check the characters of the family, and then determine the identity of your fish from the key to species on the same page.

5. Having now obtained the scientific name of the species, consult the characters in the description that accompanies that name in the text which immediately follows the key. Each species description is accompanied by a serial number which identifies it amongst the 818 illustrations grouped in the 56 plates at the end of the book. The plate on which this particular figure is located is the plate number quoted in parenthesis at the end of the references which precede the description of that species.

HOW TO IDENTIFY A FISH BY USING THE ILLUSTRATIONS

The illustrations can also be used as a starting point when it is desired to identify a fish. Proceed by turning over the plates at the end of this book until something exactly matching or closely resembling your specimen is located. Now refer to the serial number in the text which corresponds to that of the illustration. The description of that species should then be consulted. It is always advisable to refer back to the key dealing with all the species included in the particular family concerned. In this way you can make certain that the correct species, and not one that is closely allied, has been selected.

HOW TO OPERATE THE KEYS

The keys throughout employ couplets (or occasionally triplets) of alternative characters. They are used simply by consulting the alternatives under the couplet number 1, which will indicate on the right either an identification or reference to another number, e.g. 2. The couplet under 2 on the left should then be consulted, and the procedure continued until an identification is obtained. In the case of species, the diagnosis and figure should then be consulted to confirm the identification.

GLOSSARY OF TECHNICAL TERMS

Abdomen—Belly
Acute—Sharp pointed
Adipose—Fatty
Adnate—One part adherent to another
Anastomosing—Interjoining
Anterior—Relating to the front
Anus—External opening of the intestine
Apical—At the tip or apex
Asperites—Rough bony excrescences
Auxiliary scales—Small scales superimposed on larger ones
Axilla—Upper angle of pectoral fin

Barbel—An elongate fleshy projection, usually on the head
Bifurcate—Forked
Bilobate—Having two lobes
Branchial—Pertaining to the gills
Buccal—Pertaining to the mouth cavity
Buckler—A bony shield

Canine teeth—Elongate conical teeth
Carapace—A bony covering encasing the body
Cardiform—Coarse and sharp like wool cards
Carinate—Having a keel or ridge along the mid-line
Catadromous—Travelling to the sea to spawn
Caudal—Pertaining to the tail
Cephalic—Pertaining to the head region
Ciliated—Fringed with fine, hair-like projections
Cirri—Fringes
Compressed—Flattened from side to side
Confluent—Joined together
Continuous—In actual contact or closely adjoining
Cranial—Pertaining to the skull
Crenulate—With the edge slightly scalloped
Ctenoid— With a rough edge
Cuirass—Armour consisting of back-plate and breast-plate
Cultrate—With a knife edge
Cuneate, cuneiform—Wedge-shaped
Cutaneous—Pertaining to the skin
Cycloid—With a smooth edge

Deciduous—Easily shed
Dendritic—Tree-like
Dentate, denticulate, dentigerous—With tooth-like projections
Depressed—Flattened from top to bottom
Dermal—Pertaining to the skin
Dimorphous—With two distinct forms
Distal—Remote from the point of attachment
Dorsal—Pertaining to the back

Edentulous—Without teeth
Emarginate—With the margin slightly hollowed
Epibranchial—A bone forming part of the gill-arch

Falcate, falciform—Scythe-shaped
Filiform—Thread-like
Fimbriate—Fringed at the margin
Fontanel—An opening for the discharge of secretions
Fossa—A pit or depression
Fusiform—Spindle-shaped

Gill-arch—The skeleton which supports the branchiae or gills
Gill-membranes—Membranes covering the gill-openings, attached to the branchiostegals
Gill-rakers—A series of bony projections along the front edge of the gill-arches
Gular plate—A plate covering the upper part of the throat

Hypural—The last and modified vertebra which joints the caudal fin

Incisors—Flattened front cutting teeth
Interorbital—The space above and between the eyes
Illicium—Modified first dorsal fin ray consisting of a stalk and fringed lobe forming a lure in Angler-fish
Iridocyte—An iris-like pigment cell
Isthmus—Fleshy projection of body separating the gill-openings

Jugular—Pertaining to the throat

Labial—Pertaining to the lips
Laminae—Thin layers or plates of bone, skin or other tissue
Lanceolate—Spear-shaped
Lateral—At or towards the side
Lobate—Having rounded divisions

Mandible—Lower jaw
Maxilla, maxillary—Upper jaw
Median, mesial—Pertaining to the middle
Melanophore—A cell containing melanin or black pigment
Molar or molariform teeth—Blunt and rounded grinding teeth
Monomorphous—Having essentially similar form or structure
Multiserial—Arranged in many series

Nape—The part of the neck adjoining the skull
Narial, nasal—Pertaining to the nostrils
Nasoral—Between the nostrils and the mouth
Nictitating membrane—A third or inner eyelid
Nuchal—Pertaining to the nape

Occipital, occiput—Pertaining to the hind part of the skull
Ocellus—An eye-like spot
Ocular—Pertaining to the eyes
Oculonarial—Between the eyes and nostrils
Osseous—Bony
Oviparous—Producing eggs that hatch after exclusion from the body

Palate—The roof of the mouth
Palmate—Having the shape of the hand
Papilla—A small fleshy projection
Papillate, papillose—Covered with papillae
Pectinate—Shaped like a comb
Pectoral—Pertaining to the breast
Pedicel, pedicle—A short supporting stalk or stem
Peduncle—The stalk or basal support of the tail
Pelvic—Relating to the hind paired fins or the bony girdle which supports them
Peripheral—Pertaining to external boundary or superficial parts
Peritoneum—The layer of tissue which lines the abdominal cavity
Pharynx—The cavity between the mouth and oesophagus
Photophore—Light-emitting organ or luminous spot
Pluriserial—Consisting of several rows
Postocular, postorbital—Behind the eyes
Procumbent—Leaning forwards
Procurrent—Projecting forwards
Protractile—Capable of being extended forwards
Pseudobranchiae—Small gills on the inner side of the gill-cover
Pungent—Pricking and producing a sting
Pyriform—Pear-shaped

Ramus—One branch or half of the jaw
Ray, radial—A jointed rod which supports a fin
Rectilinear—Formed or bounded by straight lines
Reticulated—Divided by veins into a network
Retrorse—Turned backwards
Rivulated—Marked by narrow irregular streaks of colour
Rostral—Pertaining to the snout
Rostrum—A projecting snout or beak

Scapular—Pertaining to the shoulder region
Scute—An external horny or bony plate
Septum—A partition
Serrae—Saw-like teeth or notches
Setiform—In the form of bristles
Spinous, spiniform, spinigerous—Bearing spines or sharp projections
Spiracle—An opening behind the eye in sharks and rays
Stellate—Star-shaped
Suborbicular—Not quite but nearly circular
Suprabranchial—Situated above the gills
Symphysis—Point of junction of the two rami of the jaw
Swim-bladder—A sac filled with gas lying beneath the backbone

Thoracic—Pertaining to the chest
Tricuspid—With three points
Truncate—Cut off square
Tubiform—In the form of a tube

Undulated—Waved
Urogenital—Pertaining to the kidneys and reproductive organs

Vent—External opening of the alimentary canal
Venulose—Full of small veins and their branches
Vermiculating—Forming a pattern of wavy worm-like lines
Vertical fins—The unpaired fins on the median line of the body
Vestigial—Pertaining to a remnant; rudimentary
Villi—Minute finger-like processes
Villiform teeth—Slender and small teeth forming velvety bands
Viviparous—Producing living young from within the body

PART ONE

SHARKS, SKATES AND RAYS

Class ELASMOBRANCHII

Subclass SELACHII

Skeleton composed of cartilage which is sometimes partly calcified. Skin covered with small denticles instead of overlapping scales, but smooth in Electric Rays and Eagle Rays. A single nostril on each side. Five separate gill-openings and no bony operculum covering the gill-arches. Never more than one spine in each fin. Caudal fin asymmetrical, upper lobe longer than lower. Males with a pair of claspers alongside pelvic fins.

1. Gill-openings on the sides; body cigar-shaped .... *Order* LAMNIFORMES (SHARKS), p. 2
   Gill-openings on lower surface; body flattened, discoid 2
2. Electric organs absent .... *Order* RAJIFORMES (SKATES, RAYS), p. 9
   Electric organs present .... *Order* TORPEDINIFORMES (ELECTRIC RAYS), p. 17

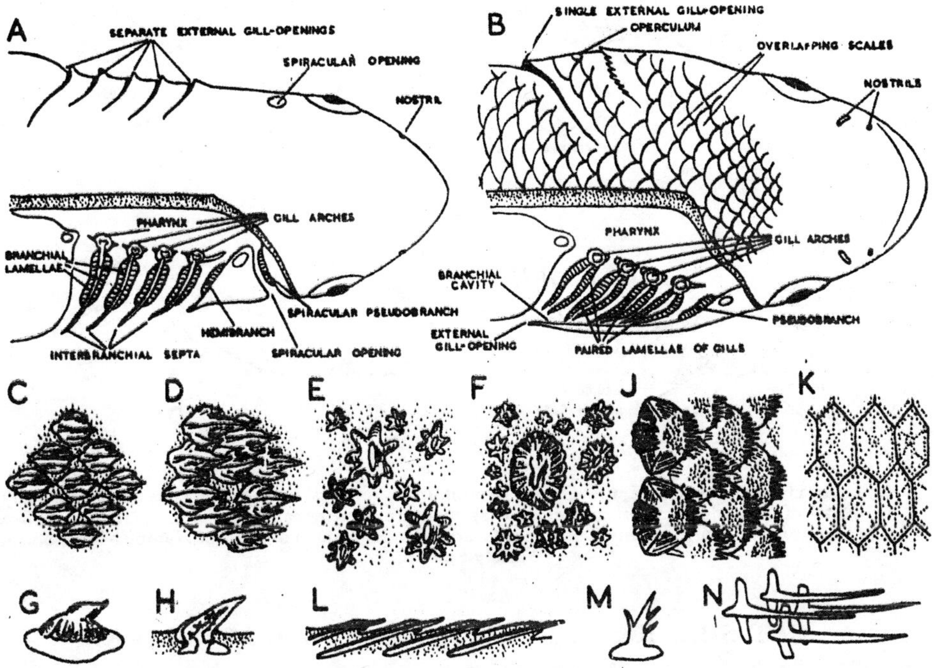

FIG. 1.

Explanation of differences between Class Elasmobranchii and Class Teleostomi. a.—Dissection of head of an elasmobranch showing gill structure. b.—Similar dissection of a teleost. c-d.—Denticles of sharks. e-f.—Asperites and bucklers of rays and skates. g.—Lateral view of a buckler. h.—Section of a denticle. j.—Overlapping scales of a typical teleost. k.—Plate-like scales of Ostracioidei. l.—Section through scales of a normal teleost. m.—Spine-like scale of Balistoidei. n.—Spine-like scales of Tetradontoidei.

SHARKS

Order LAMNIFORMES

Fusiform, cartilaginous fishes with 5 gill-openings, 2 dorsal fins and an anal fin.

1. Nictitating membrane absent .... (*Suborder* LAMNOIDEI) 2
   Nictitating membrane present .... (*Suborder* SCYLIORHINOIDEI) 3
2. Nasoral grooves present .... *Family* ORECTOLOBIDAE, p. 2
   Nasoral grooves absent .... *Family* LAMNIDAE, p. 4
3. Head with oculonarial expansions .... *Family* SPHYRNIDAE, p. 7
   Head without oculonarial expansions 4
4. Caudal pit absent; first dorsal fin behind or before pelvics .... *Family* SCYLIORHINIDAE, p. 4
   Caudal pit present (except *Myrmillo*); first dorsal fin before pelvics ... *Family* CARCHARINIDAE, p. 4

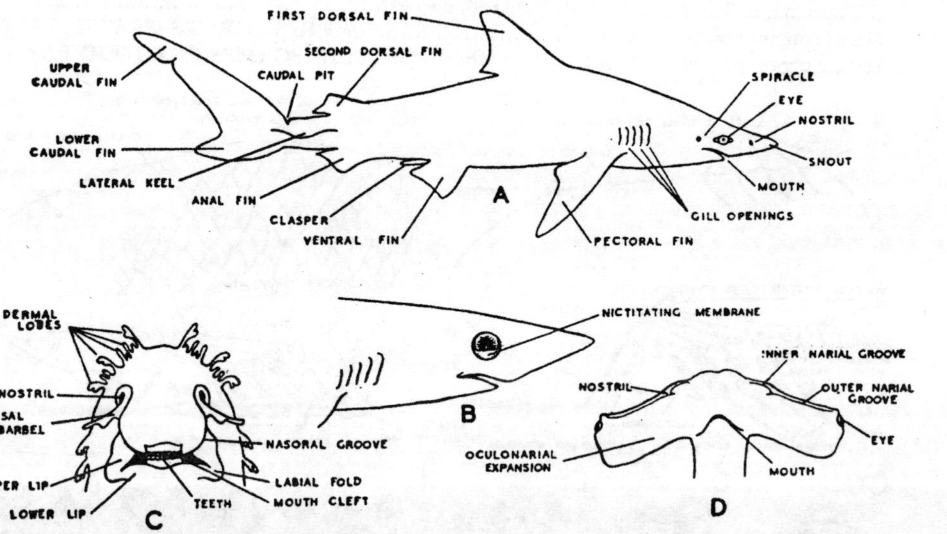

FIG. 2.

Explanation of characters of the Order Lamniformes (Sharks). a.—General external anatomy. b.—Head of Suborder Scyliorhinoidei showing nictitating membrane over eye. c.—Under surface of head of Carpet Shark, Family Orectolobidae. d.—Under surface of head of Hammer-head Shark, Family Sphyrnidae.

CAT SHARKS, CARPET SHARKS, WHALE SHARKS

Family ORECTOLOBIDAE

Body robust in *Rhincodon* but rather short in other genera, with long tail. Folds on both lips. Eyes small, without nictitating membrane. Includes very small and extremely large species. Littoral and deep water in habit. Viviparous and oviparous.

1. Nasal cirri present; body not massive; tail without lateral keels; caudal pit absent 2
Nasal cirri absent; body massive; tail with lateral keels; caudal pit present .... *Rhincodon typus*
2. First dorsal fin behind pelvics; three dermal ridges on the back .... *Chiloscyllium indicum*
First dorsal fin opposite pelvics; no dermal ridges on the back 3
3. Caudal fin long .... *Stegostoma fasciatum*
Caudal fin short 4
4. Teeth in 3 rows; second dorsal fin longer than anal fin .... *Nebrius concolor*
Teeth in more than 3 rows; second dorsal fin smaller than anal fin .... *Ginglymostoma ferrugineum*

1. CHILOSCYLLIUM INDICUM (Gmelin) RIDGE-BACK CATSHARK

Squalus indicus Gmelin, Syst. Nat. Linnaeus, i, 1789, p. 1503. *Chiloscyllium phymatodes* or *plagiosum*, *Hemiscyllium indicum* or *colax*. *Chiloscyllium indicum* Day, Fishes India, 1875-78, Pl. clxxxviii, fig. 3 (plate 1).

Slender ground shark. About 18 inches. Tail longer than trunk. Three rough ridges on back at front of first dorsal fin. Mottled or banded with red-brown. Oviparous.
Kana mora (S). Corungan schura, Karikkan schura (T).

2. STEGOSTOMA FASCIATUM (Hermann) ZEBRA SHARK

Squalus fasciatus Hermann, Tab. Affin., 1783, p. 302. *Squalus tigrinus*. *Stegostoma tigrinum* Day, Fishes India, 1875-78, Pl. clxxxvii, fig. 4 (plate 1).

Slender ground shark. Attains 5-6 feet. Caudal fin very elongate. Second dorsal fin very small. A ridge along the back. Colour pattern varies with age from broad cross-bars to black or brown spots. Young more distinctly banded than adults. Oviparous. Feeds on shellfish.

3. NEBRIUS CONCOLOR Rüppell TAWNY SHARK

Nebrius concolor Rüppell, Neue Wirbelth., Fische, 1835, p. 62. *Ginglymostoma concolor*. *Nebrodes concolor* Whitley, Mem. Q'land. Mus. x (4), fig. 1, redrawn (plate 1).

Small, elongate, spindle-shaped pelagic shark. 2 feet. The 4th and 5th gill-openings closer together than the others. Anal fin originates below middle of second dorsal fin. Teeth with sub-equal cusps. Uniform brownish colour.

4. GINGLYMOSTOMA FERRUGINEUM (Lesson) RUSTY SHARK

Scyllium ferrugineum Lesson, Voy. Coquille, Zool., ii (1), 1830, p. 95. *Ginglymostoma concolor* Müller and Henle, Syst. Beschr. Plagiostomen, 1841, Pl. vi (plate 1).

Elongate, spindle-shaped pelagic shark. Attains 8 feet. The 4th and 5th gill-openings closer together than the others. Anal fin inserted slightly behind second dorsal. Teeth with strong median cusp and 3-5 smaller lateral cusps. Rusty brown colour.
Be kal mora (S).

5. RHINCODON TYPUS Smith WHALE SHARK

Rhincodon typus Smith, Zool. Journ., iv, 1829, p. 443. *Rhinodon typicus*, *Rhineodon typus*. *Rhincodon typus* Smith, Zool. S. Africa, Fishes, 1849, Pl. xxvi (plate 2).

Largest shark in the world. Attains 50 feet. Body massive. Head broad and flattened. Mouth very wide. Teeth very small, numerous and in bands on both jaws. Gill-openings very wide. Gill-rakers long and numerous, used as strainers. Ground colour sandy brown with numerous pale spots on head, and series of pale spots alternating with pale vertical stripes on sides. Plankton feeders, gregarious, timid and bask on surface at sea.
Mini mutu mora (S). Pulli udumba schura, Usman schura, Makara schura (T).

MAN-EATING SHARKS, THRESHER SHARKS

Family LAMNIDAE

In Ceylon represented only by the Subfamily ALOPINAE. Easily recognised by upper lobe of caudal fin greatly enlarged and as long as the body. Body torpedo-shaped. Caudal base with notch. No nictitating membrane. Teeth triangular and smooth. A single species.

6. ALOPIAS VULPINUS (Bonnaterre) THRESHER OR FOX SHARK

Squalus vulpinus Bonnaterre, Tab. encyclop. Ichth., 1788, p. 9. *Alopias vulpes. Vulpecula marina* Garman, Mem. Mus. Comp. Zool. Harvard, xxxvi, 1913, Pl. vii, fig. 1 (plate 2).

Robust, torpedo-shaped body. Tail as long as body and used to whip the water and frighten together schools of small fish upon which it feeds. Second dorsal and anal fins very small. Attains 15 feet. Colour uniform green or deep blue. Occasionally leaps high out of the water. Not regarded as harmful to man.

LAZY SHARKS, MARBLED CATSHARKS

Family SCYLIORHINIDAE

Body elongate. Trunk slightly shorter than tail. Eyes large with nictitating membranes. Mouth angles with long labial folds. Anal fin before second dorsal fin. A single species.

7. ATELOMYCTERUS MARMORATUS (Bennett) MARBLED CATSHARK

Scyllium marmoratum Bennett, Life of Raffles, 1830, p. 693; Day, Fish. India, 1875-78, Pl. cxc, fig. 2 (plate 1).

Body elongate and slender. Eyes large. Labial folds well developed. Base of anal fin equal to base of second dorsal. Teeth small, tricuspid, median cusp longest. Attains 20 inches. Young with about 12 transverse brown bands, separated by pairs of white spots. Adults irregularly blotched with brown. Oviparous.

MAN-EATING SHARKS, TRUE SHARKS

Family CARCHARINIDAE

Body elongate. Head depressed. Nictitating membrane present. Spiracle small or absent. First dorsal fin before pelvics. No keel on caudal peduncle.

1.	Spiracle present	2
	Spiracle absent	4
2.	Caudal pit absent	*Myrmillo manazo*
	Caudal pit present	3
3.	Teeth monomorphous; lower serrated on both edges	*Galeocerda cuvieri*
	Teeth dimorphous; lower smooth	*Hemigaleus balfouri*
4.	Teeth polyserial, in bands	*Triaenodon obesus*
	Teeth monoserial, not in bands	5
5.	Teeth with smooth edges	6
	Teeth with serrated edges	8
6.	Anal base length more than twice second dorsal base length	*Scoliodon sorrakowah*
	Anal base length not more than twice second dorsal base	7
7.	Labial fold extending to upper jaw	*Scoliodon walbeehmi*
	Labial fold not extending to upper jaw	*Scoliodon palasorrah*
8.	In upper jaw teeth serrated on bases and cusps	9
	In upper jaw teeth serrated at bases only	*Hypoprion macloti*
9.	Second dorsal fin larger than anal	*Eulamia ellioti*
	Second dorsal and anal equal in size	10
10.	Second dorsal origin before anal origin	*Eulamia melanoptera*
	Second dorsal and anal origins opposite	11
11.	Snout extended; pectoral twice as wide as long	*Eulamia spallanzani*
	Snout bluntly rounded; pectoral less than twice as wide as long	*Eulamia dussumieri*

8. MYRMILLO MANAZO (Bleeker) GUMMY SHARK

Mustellus manazo Bleeker, Verh. Batav. Genootsch., xxvi, 1857, p. 42; Day, Fish. India, 1875-78, Pl. clxxxvi, fig. 3 (plate 1).

Snout markedly pointed, long. 45-50 rows of rounded rather flattened teeth in a pavement in upper, about 40 similar in lower jaw. Pupil of eye rounded. Sometimes white spots. Attains 5 feet. Very sluggish.

9. GALEOCERDA CUVIERI (Le Sueur) TIGER SHARK

Squalus cuvier Le Sueur, Journ. Acad. Nat. Sci. Philadelphia, ii, 1822, p. 351. *Galeocerda tigrinus* or *arcticus*. *Galeocerda rayneri* MacDonald and Barron, Proc. Zool. Soc., 1868, Pl. xxxii (plate 1).

Snout short and rounded. 18-20 strongly serrated powerful teeth in each jaw. Adults usually not marked. Young conspicuously striped. Attains 15 feet. A fierce shark known to attack man, but mostly a voracious scavenger, entering shallow water. Viviparous.

Koti mora, Thalagoi mora (S). Pulli schura (T).

10. HEMIGALEUS BALFOURI Day BALFOUR'S SHARK

Hemigaleus balfouri Day, Fish. India, 1875-78, p. 717, Pl. clxxxv, fig. 4 (plate 1).

Body slender, elongate. Trunk shorter than tail. Snout pointed. Teeth dimorphous; upper inclined, with denticles on basal part of outer edge; lower erect and smooth. Dark brown, fins grey, second dorsal with dark summit. Attains about 2½ feet.

11. TRIAENODON OBESUS (Rüppell) LIGHT TIP SHARK

Carcharias obesus Rüppell, Neue Wirbelth., Fische, 1835, p. 64. *Triaenodon obtusus* Day, Fish. India, 1875-78, Pl. clxxxix, fig. 3 (plate 1).

Body elongate, trunk longer than tail. Snout short and rounded. Teeth minute, numerous in both jaws, with central and lateral cusps. Light cinnamon brown. Fins more or less dark terminally but tips of first dorsal and caudal paler than rest of fin.

12. SCOLIODON SORRAKOWAH (Cuvier) YELLOW DOG SHARK

Carcharias sorrakowah Cuvier, Regne Animal, ed. 2, ii, 1829, p. 388. *Scoliodon laticaudus. Carcharias laticaudis* Day, Fish. India, 1875-78, Pl. clxxxviii, fig. 1 (plate 1).

Body elongate. Snout sharply pointed. Labial fold not extending to upper jaw. Teeth in 22 rows in upper jaw and 23 rows in lower, cusps directed obliquely outward with smooth edges and without swollen bases. Upper surface of body dull naples yellow, lower surface cream. Fins dull brown.

Hakuru mora (S). Karnpathi schura (T).

13. SCOLIODON WALBEEHMI (Bleeker)

WALBEEHM'S SHARP-NOSED SHARK

Carcharias (Scoliodon) walbeehmii Bleeker, Nat. Tijds. Ned. Ind., x, 1856, p. 353. *Carcharias walbeehmi* Day, Fish. India, 1875-78, Pl. clxxxv, fig. 2 (plate 1).

Snout pointed. Head depressed. 24-28 outwardly inclined triangular teeth in each jaw. Labial fold extends to upper jaw. Grey to brownish above with first dorsal fin greyish. Attains 4 feet.

14. SCOLIODON PALASORRAH (Cuvier) GREY DOG SHARK

Carcharias palasorrah Cuvier, Regne Animal, ed. 2, ii, 1829, p. 388. *Scoliodon* or *Charcharinus acutus. Carcharias acutus* Day, Fish. India, 1875-78, Pl. clxxxiv, fig. 3 (plate 1).

Snout long and pointed. Head depressed. 25-26 outwardly inclined triangular teeth in each jaw. Outer edges of teeth very finely serrate, other edges smooth. Pectorals and pelvics with light margins. Attains 4 feet. Grey to brownish above.

Kiri mora (S). Pal schura (T).

15. HYPOPRION MACLOTI (Müller and Henle) MACLOT'S SHARK

Carcharias (Hypoprion) macloti Müller and Henle, Syst. Beschr. Plagiostomen, 1841, p. 34, Pl. x (plate 1).

Body elongate. Snout acute. Labial folds behind each mouth angle, not extending on jaws. Teeth in 27 rows in both jaws, slightly inclined. Both sides of bases of upper teeth serrated. Colour uniform greyish, whiter below. Attains 3 feet.

Hudja mora (S). Muthra schura (T).

16. EULAMIA ELLIOTI (Day) ELLIOT'S GREY SHARK

Carcharias ellioti Day, Fish. India, 1875-78, p. 716, Pl. clxxxix, fig. 2 (plate 1).

Fusiform, trunk longer than tail. Snout rounded. Teeth in 24-26 rows above and 30-34 below. Upper teeth triangular without notch or basal enlargement but coarsely serrated. Lower teeth obliquely erect, awl-like. Second dorsal fin larger than anal. Greyish above, white beneath. Attains 5-6 feet.

Kachipudangia, Patchepadanguva (S). Adukupal schura, Padancan, Puducan (T).

17. EULAMIA MELANOPTERA (Quoy and Gaimard) BLACK SHARK

Carcharias melanopterus Quoy and Gaimard, Voy. Uranie, Zool. pts. 5-6, 1824, p. 194; Jordan and Evermann, Bull. U.S. Fish. Comm., xxiii (1), 1905, Pl. i (plate 1).

Fusiform. Snout moderately pointed in profile, rounded from above. Teeth in 26 to 28 rows. Upper teeth attenuated distally with inner edge obliquely straight, outer deeply notched. Lower teeth narrower and more erect. Grey-black above, tips of fins black. Attains 10 feet.

Kunda mora, Perunthalaya (S). Perunthalai schura (T).

18. EULAMIA SPALLANZANI (Le Sueur) BLACK-TIP SHARK

Squalus spallanzani Le Sueur, Journ. Acad. Nat. Sci. Philadelphia, ii (2), 1822, p. 351. *Lamna spallanzani. Mapolamia spallanzani* Whitley, Fish. Australia, i, 1940, fig. 89 (plate 1).

Snout broadly rounded. Labial folds very short. Teeth finely serrated, in 25 rows above, in 23 rows below. Upper teeth oblique with inner edge straight, outer notched. Lower teeth narrow, nearly erect, on broad bases. Brown. Fins black-tipped. Rarely exceeds 5 feet. Inhabits shallow water.

Mini muthu mora (S). Ganumu schura (T.)

19. EULAMIA DUSSUMIERI (Müller and Henle) WHITE-CHEEKED SHARK

Carcharias (Prionodon) dussumieri Müller and Henle, Syst. Beschr. Plagiostomen, 1841, p. 47. *Carcharias dussumieri* Day, Fish. India, 1875-78, Pl. clxxxvii, fig. 2 (plate 1).

Fairly slender body. Snout rather long, rounded from above. 24-25 rows of serrated triangular teeth in both jaws. Upper teeth with attenuated cusps and outer edge notched. Lower teeth erect, slender on wide bases. Attains 7 feet. Upper surface grey or fawn, underneath whitish. Fins greyish with white edges. Sides of head white.

HAMMERHEAD SHARKS

Family SPHYRNIDAE

Very ferocious sharks with head flattened in front and expanded sideways to resemble a hammer. Eyes at extremities of expansions.

1. Eyes and nostrils widely separated; oculonarial expansions long .... *Sphyrna blochii*
   Eyes and nostrils close together; oculonarial expansions short 2
2. Inner narial groove extending more than half way to middle of snout .... *Sphyrna zygaena*
   Inward extension of narial groove absent .... *Sphyrna tudes*

20. SPHYRNA BLOCHII (Cuvier) ARROW-HEADED HAMMERHEAD SHARK

Zygaena blochii Cuvier, Regne Animal, ii, 1817, p. 127. *Eusphyra blochii. Zygaena blochii* Day, Fish. India, 1875-78, Pl. clxxxiv, fig. 4 (plate 2).

Outer narial grooves greatly extended along much produced lateral expansions of head, the eye distant from the nostril by many times its diameter. Head expansions about twice as long as wide. First dorsal fin much larger than pectoral. Base of second dorsal much shorter than that of anal. Dusky or grey-brown above, lighter underneath.

21. SPHYRNA ZYGAENA (Linnaeus) ROUND-HEADED HAMMERHEAD SHARK

Squalus zygaena Linnaeus, Syst. Nat., ed. 10, i, 1758, p. 234. *Sphyrna malleus. Zygaena malleus* Day, Fish. India, 1875-78, Pl. clxxxvi, fig. 4 (plate 2).

Outer narial grooves short, inner extending more than half way to snout. Eye not more than twice its diameter from nostril. Front of snout viewed from above, convex, without median indentation. Oculonarial expansions 1·5 times longer than wide. Base of anal fin much shorter than that of pectoral, about equal to base of second dorsal. Grey or grey-brown above, whitish below. Fins black terminally. Attains over 20 feet.

Udhalu mora (S). Komben schura, Mammoti schura (T).

22. SPHYRNA TUDES (Valenciennes) SQUAT-HEADED HAMMERHEAD SHARK

Zygaena tudes Valenciennes, Mem. Mus. Hist. Nat. Paris, ix, 1822, p. 225; Day, Fish India, 1875-78, Pl. clxxxviii, fig. 4 (plate 2).

Inward extension of narial groove absent. Front margin of snout double-convex, with distinct median indentation. Oculonarial expansions shorter than their width. Eye distant from nostril by less than its diameter. First dorsal fin much larger than pectoral; base of second dorsal a little shorter than that of anal. Grey-brown above, whitish below.

SAW-FISHES, SKATES, AND RAYS

Order RAJIFORMES

Dorso-ventrally flattened cartilaginous fishes with 5 gill-openings on the under surface of the body.

1. Disc narrow and elongate 2
   Disc broad and expanded 3
2. Rostrum very much produced and saw-like .... *Family* PRISTIDAE, p. 11
   Rostrum very short and not saw-like .... *Family* RHINOBATIDAE, p. 9
3. Tail whip-like; dorsals reduced to spines 4
   Tail not whip-like, without spines; two small dorsals .... *Family* RAJIDAE, p. 11
4. Head distinct from disc, with a prominent snout 5
   Head not distinct from disc, without prominent snout .... *Family* TRYGONIDAE, p. 12
5. With horn-like cephalic flippers .... *Family* MOBULIDAE, p. 16
   Without horn-like cephalic flippers 6
6. Head bilobed or notched between the rostral fins; rostral fins separate ... *Family* RHINOPTERIDAE, p. 15
   Head neither bilobed nor notched between the rostral fins; rostral fins united into one lobe .... *Family* MYLIOBATIDAE, p. 15

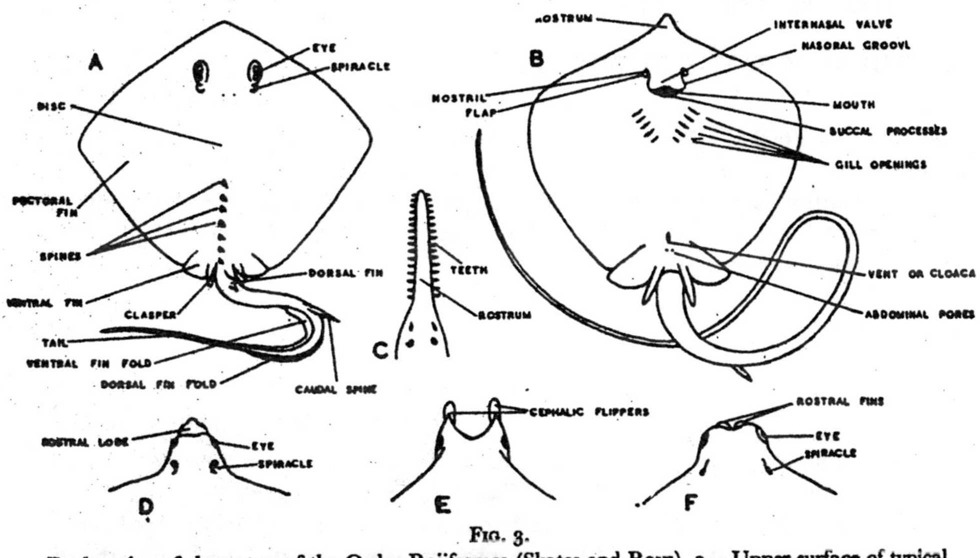

FIG. 3.

Explanation of characters of the Order Rajiformes (Skates and Rays). a.—Upper surface of typical ray of Families Rhinobatidae, Rajidae or Trygonidae. b.—Under surface of same. c.—Rostrum of Sawfish, Family Pristidae. d.—Head of Eagle Ray, Family Myliobatidae. e.—Head of Devil Ray, Family Mobulidae. f.—Head of Cow Ray, Family Rhinopteridae.

SHOVELNOSE SHARKS, FIDDLE SHARKS

Family RHINOBATIDAE

Depressed body with a tapering snout and elongate caudal portion. Teeth numerous in pavement-like bands. Spiracles large, close to eyes. 2 dorsal fins. Caudal well developed. Viviparous rays of shallow water. Feeding on bottom organisms.

1. Snout triangularly pointed; spiracles with folds on hind edge 2
Snout blunt, broad and rounded; spiracles without folds on hind edge .... *Rhina ancylostoma*
2. Origin of first dorsal distinctly nearer to tip of snout than to tip of caudal........... *Rhynchobatus djiddensis*
Origin of first dorsal distinctly nearer to tip of caudal than to tip of snout 3
3. Snout long and pointed; preorbital length more than 3½ times distance between spiracles
.... *Rhinobatos granulatus*
Snout short and blunt; preorbital length less than 2½ times distance between spiracles
.... *Rhinobatos annandalei*

23. RHINA ANCYLOSTOMA Bloch and Schneider

BOW-MOUTHED ANGEL FISH

Rhina ancylostomus Bloch and Schneider, Syst. Ichth., 1801, p. 352. *Rhynchobatus ancylostomus* Day, Fish. India, 1875-78, Pl. cxciii, fig. 3 (plate 2).

Disc sub-triangular, obtusely rounded in front. About 75 transverse rows of 22-25 rounded teeth. Spiracles large, without posterior folds and 1½ diameters behind eyes. Brownish, sometimes with light spots. Attains 8 feet.

Thitha mora, Pulaman oliya (S). Pulaman uluvai (T).

24. RHYNCHOBATUS DJIDDENSIS (Forskål)

WHITE-SPOTTED SHOVELNOSE RAY.

Raja djiddensis Forskål, Descript. Animal., 1775, p. 18. *Rhynchobatus laevis*. *Rhynchobatus djeddensis* Day, Fish. India, 1875-78, Pl. cxcii, fig. 1 (plate 2).

Snout narrow, produced, acute. Disc longer than wide. First dorsal opposite pelvics. Fairly large tubercles on back. Teeth blunt, about 40 rows in each jaw. Spiracle close to eye, 2 folds on hind margin. Sandy brown above, sides and pectorals with scattered white ocelli. Round black spot posteriorly on scapular region. Attains 10 feet.

Kiri uluva (S). Pal uluvai (T).

25. RHINOBATOS GRANULATUS Cuvier

GRANULATED SHOVELNOSE RAY

Rhinobatus granulatus Cuvier, Regne Animal, ed. 2, ii, 1829, p. 396. *Rhinobatus halavi* or *acutus*. *Rhinobatus granulatus* Day, Fish. India, 1875-78, Pl. cxcii, fig. 2 (plate 2).

Disc and snout triangularly pointed. Snout long and pointed. Length of nostril equal to internarial space and twice width of mouth. Preorbital length 3½-3⅜ times the distance between the spiracles. Brown above, coarsely tuberculate, rough to touch. Each side of rostral cartilages buff. Dorsals and caudal greyish. Attains 6 feet.

Bal oliya, Gal uluva (S). Kal uluvai (T).

26. RHINOBATOS ANNANDALEI Norman

ANNANDALE'S SHOVELNOSE RAY

Rhinobatus annandalei Norman, Proc. Zool. Soc., 1926, p. 960. *Rhinobatus columnae*.

Snout blunt, margins slightly concave. Preorbital length 2⅓-2½ times distance between spiracles. Length of nostril greater than internarial space and less than twice width of mouth. Skin covered with minute denticles, rather smooth to touch.

SAW-FISHES

Family PRISTIDAE

Skates with gill-slits on under surface with snout produced into a long rostrum armed with teeth, in the form of a saw. Viviparous.

1. First dorsal originates in front of pelvics; 17-20 teeth on each side of rostrum .... *Pristis microdon*
   First dorsal originates behind pelvics; 23-35 teeth on each side of rostrum 2
2. Upper margins of dorsals deeply concave with posterior lobes produced; subcaudal well developed .... *Pristis cuspidatus*
   Upper margins of dorsals slightly concave with posterior lobes not produced; subcaudal poorly developed .... *Pristis zijsron*

27. PRISTIS MICRODON Latham — SMALL-TOOTHED SAW-FISH

Pristis microdon Latham, Trans. Linn. Soc., ii, 1794, p. 280. *Pristis perotetti* Day, Fish. India, 1875-78, Pl. cxci, fig. 1 (plate 2).

17-20 rostral teeth in double series fairly wide-set. About 70 series of small, blunt teeth in each jaw. Greyish to brownish above. Attains 15 feet. Enters fresh water.
Illipara vellava (S). Vaal schura, Vela schura (T).

28. PRISTIS CUSPIDATUS Latham — POINTED SAW-FISH

Pristis cuspidatus Latham, Trans. Linn. Soc., ii, 1794, p. 279; Day, Fish. India, 1875-78, Pl. cxci, fig. 3 (plate 2).

23-24 rostral teeth, narrowly triangular, with inner subbasal barb, not extending on basal fourth of rostrum. About 62 series of teeth in jaws. Skin smooth. Dark drab or neutral grey above. Borders of fins and lateral fold of tail whitish.
Dhathi mora (S). Vela schura (T).

29. PRISTIS ZIJSRON Bleeker — GREEN SAW-FISH

Pristis zijsron Bleeker, Nat. Tijds. Ned. Ind., ii, 1851, p. 442. *Pristis pectinatus*. *Pristis zysron* Day, Fish. India, 1875-78, Pl. cxci, fig. 2 (plate 2).

25 or more rostral teeth. Enlarged denticles along median keels of back and along fins. Subcaudal without lobe. Greenish grey, becoming yellowish on sides and around eyes. Ventral surface whitish. Dorsals yellow-grey. Attains 24 feet.

SKATES

Family RAJIDAE

Body and head greatly depressed, united with pectorals to form a rhomboidal disc. Tail ending in blunt tip, without caudal fin. Two dorsal fins posteriorly without spines or sting. Oviparous. One species only.

30. RAJA MAMILLIDENS Alcock PRICKLY SKATE

Raja mamillidens Alcock, Ann. Mag. Nat. Hist. (6), iv, 1889, p. 380; Alcock, Illust. Zool. Investigator, 1894, Pl. viii, fig. 1 (plate 2).

Snout 3-3·5 times the interorbital distance. Dorsals very close together. Teeth in 24 rows in upper jaw, 18 rows in lower. Whole upper surface including ventrals and tail covered with sharp prickles. A single row of prominent spines on tail. Uniform jet black.

STINGRAYS, STINGAREES

Family TRYGONIDAE

Disc kite-shaped, flattened. Tail distinct from disc, narrow and tapering, usually with serrated spine. No dorsal fin. Teeth small, in pavement. Shallow water species. Viviparous.

1. Tail shorter than the length of the disc; disc twice as broad as long 2
   Tail as long as or much longer than length of disc; disc not twice as broad as long 3
2. Tail almost as long as disc .... *Gymnura poecilura*
   Tail less than half length of disc .... *Gymnura micrura*
3. Serrated caudal spine present; body not profusely covered with tubercles 4
   No serrated caudal spine; body profusely covered with tubercles .... *Urogymnus africanus*
4. Disc oval 5
   Disc quadrangular 6
5. Mouth straight; 5 buccal processes .... *Taeniura meyeni*
   Mouth curved; 2 buccal processes .... *Taeniura lymma*
6. Cutaneous folds on tail 7
   No cutaneous folds on tail 11
7. With only ventral cutaneous fold .... *Pastinachus sephen*
   With both dorsal and ventral cutaneous folds 8
8. Without buccal processes .... *Amphotistius zugei*
   With 2 buccal processes 9
9. Tail scarcely as long as disc .... *Amphotistius imbricatus*
   Tail exceeding length of disc 10
10. Blue ocelli on dorsal surface of disc .... *Amphotistius kuhlii*
    No ocelli on dorsal surface of disc .... *Amphotistius marginatus*
11. With 2 buccal processes .... *Himantura bleekeri*
    With 4 buccal processes .... *Himantura uarnak*

31. GYMNURA POECILURA (Shaw) LONG-TAILED BUTTERFLY RAY

Raja poecilura Shaw, General Zoology, v, 1804, p. 291. *Pteroplatea poecilura. Pteroplatea micrura* Day, Fish. India, 1875-78, Pl. cxciv, fig. 2 (plate 2).

Disc width 1·9 times length. No dorsal fin. Tail almost as long as disc. Brown above. Tail whitish with 9 broad blackish bands much wider than pale interspaces.

32. GYMNURA MICRURA (Bloch and Schneider)

SHORT-TAILED BUTTERFLY RAY

Raja micrura Bloch and Schneider, Syst. Ichth., 1801, p. 360. *Pteroplatea micrura* Garman, Mem. Mus. Comp. Zool. Harvard, xxxvi, 1913, Pl. xxxiii, fig. 3 (plate 2).

Disc width $1\frac{7}{8}$ times length. Small rudimentary dorsal as long as eye. Back uniform blackish brown. Tail with end broadly blackish and 2 blackish saddles above, below whitish.

33. UROGYMNUS AFRICANUS (Bloch and Schneider) THORNY RAY

Raja africana Bloch and Schneider, Syst. Ichth., 1801, p. 367. *Raja* or *Urogymnus asperrimus* Day, Fish. India, 1875-78, Pl. cxcv, fig. 1 (plate 3).

Tail long, tapering, with or without narrow fold below according to age. No caudal spine. Back covered with osseous tubercles and spinigerous bucklers, making an excessively rough skin. Body oval. Skin dark brown or bluish, with tubercles yellowish or whitish.
Eramudhu maduva (S). Mullam thirukai, Kallu thirukai (T).

34. TAENIURA MEYENI Müller and Henle MEYEN'S RAY

Taeniura meyeni Müller and Henle, Syst. Beschr. Plagiostomen, 1841, p. 172, Pl. lv (plate 2).

Disc rounded. Spiracle close behind and slightly less than eye. Mouth straight. 5 oral papillae. Skin smooth. Blackish-brown above, whitish below. Pectorals and ventrals with blackish edges.

35. TAENIURA LYMMA (Forskål) BLUE-SPOTTED FANTAIL RAY

Raja lymma Forskål, Descript. Animal., 1775, p. 8. *Taeniura melanospilos. Taeniura lymma* McCulloch, Rec. Aust. Mus. xiii (2), 1920, Pl. x (plate 3).

Disc oval. Spiracle large, deep, larger than eye. Mouth curved. 2 oral papillae. Skin smooth except for tubercles along mid-line of back and over shoulders. Tail slightly longer than body, with deep fold ventrally. Tan brown with large blue spots on disc. A broad blue stripe along each side of tail. Underneath light. Attains 8 feet.
Ali maduva, Ath maduva (S).

36. PASTINACHUS SEPHEN (Forskål) COWTAIL RAY

Raja sephen Forskål, Descript. Animal., 1775, p. 17. *Dasyatis* or *Dasybatus sephen. Trygon sephen* Day, Fish. India, 1875-78, Pl. cxcv, fig. 2 (plate 3).

Disc granular with one or two central tubercles. Dental surface of upper jaw angular. Tail long, with barb. No dorsal or caudal fins. Tail with deep cutaneous fold ventrally, not extending to anywhere near whip-like tip.
Pol kolle maduva (S). Ada thirukai (T).

37. AMPHOTISTIUS ZUGEI (Müller and Henle) PALE-EDGED STINGRAY

Trygon zugei Müller and Henle, Syst. Beschr. Plagiostomen, 1841, p. 165; Day, Fish. India, 1875-78, Pl. cxc, fig. 3 (plate 3).

Disc as long as wide. Spiracle much larger than eye. Skin smooth. Ventral cutaneous fold greater than dorsal. Large tubercles in median series, smaller ones on tail. Yellowish to reddish brown, rather paler marginally on disc. Cutaneous folds of tail dusky brown.

38. AMPHOTISTIUS IMBRICATUS (Bloch and Schneider) SCALY STINGRAY

Raja imbricata Bloch and Schneider, Syst. Ichth., 1801, p. 366. *Trygon imbricata* or *polylepis*, *Dasyatis imbricatus*, *Dasybatus walga*. *Trygon walga* Day, Fish. India, 1875-78, Pl. cxciv, fig. 3 (plate 3).

Spiracle as large as eye. Interorbital, cranium and middle of back broadly covered with minute rough plates. Spines from middle of disc in median line to tail. Tail without membranes. Disc nearly as wide as long. Reddish brown, yellow spotted. Whitish below.

39. AMPHOTISTIUS KUHLII (Müller and Henle) BLUE-SPOTTED STINGRAY

Trygon kuhlii Müller and Henle, Syst. Beschr. Plagiostomen, 1841, p. 164. *Dasyatis* or *Dasybatus kuhlii*. *Trygon kuhlii* Day, Fish. India, 1875-78, Pl. cxciii, fig. 2 (plate 3).

Spiracle greatly larger than eye. Kite-shaped. Scapular region smooth or with few median spines. Tail about as long as disc. Tail with short dorsal cutaneous fold behind barb and long ventral fold. Grey or brown spotted with black or rusty red and white. Tail tip black, preceded by two obscure cross bands.

40. AMPHOTISTIUS MARGINATUS (Blyth) BLACK-EDGED STINGRAY

Trygon marginatus Blyth, Journ. Asiatic Soc. Bengal, xxix, 1860, p. 38. *Dasyatis marginatus*. *Trygon marginatus* Annandale, Mem. Indian Mus., ii, 1909, fig. 5 (plate 3).

Spiracle nearly twice as large as eye. Head and centre of back with close-set, rounded, nearly flat denticles. Tail with similar denticles intermixed with stellate spines. Tail without cutaneous folds and with single large spine. Grey above, male with blue marks around disc in from margin. Ventral surface white with broad blackish margins. Tail blackish.

41. HIMANTURA BLEEKERI (Blyth) WHIP-TAIL STINGRAY

Trygon bleekeri Blyth, Journ. Asiatic Soc. Bengal, xxix, 1860, p. 41. *Dasyatis bleekeri*. *Trygon bleekeri* Day, Fish. India, 1875-78, Pl. cxcv, fig. 3 (plate 3).

Two long fingerlike processes on floor of mouth. Tail long, whiplike. Large round tubercle in middle of back, and usually three smaller ones before and three more behind. Others along upper surface of tail to caudal spine. No cutaneous fold. Uniform dark-brown above. Ventral surface white with broad dark-brown margin, increasing in area with age.

42. HIMANTURA UARNAK (Forskål) BANDED WHIP-TAIL STINGRAY

Raja uarnak Forskål, Descript. Animal., 1775, p. 18. *Dasyatis uarnak*. *Trygon uarnak* Annandale, Mem. Indian Mus., ii, 1909, Pl. ii, fig. 1, a (young) (plate 3).

Disc barely wider than long. Adult develops a few tubercles in middle of back. 4-7 skinny flaps on floor of mouth. Tail very long, thin, several times length of body, with no cutaneous folds. Varies from light sandy to dark-brown, sometimes with dark spots or meshwork pattern on back. Tail white with about 30-35 blackish-blue bands. Attains 5 feet across disc.

Vali maduva (S). Manal thirukai (T).

EAGLE RAYS, BULL RAYS

Family MYLIOBATIDAE

Rays with wide angular wings, whip-tails with spine. Head raised so that eyes look sideways. Mouth with a scroll of flat grinding teeth.

1. Teeth in single broad series; caudal spine present .... *Aetobatus narinari*
   Teeth in several rows of which lateral ones are narrower than the central; caudal spine absent 2
2. Origin of dorsal fin behind ends of pelvic bases .... *Aetomylaeus maculatus*
   Original of dorsal fin opposite ends of pelvic bases .... *Aetomylaeus nichofii*

43. AETOBATUS NARINARI (Euphrasen) SPOTTED EAGLE RAY

Raia narinari Euphrasen, Kon. Vet. Acad. Nya Handl. Stockholm, xi, 1790, p. 217. *Aetobatus narinari. Stoasodon narinari* Jordan and Evermann, Bull. U.S. Fish. Comm., xxiii (1), 1905, fig. 7 (plate 3).

Teeth in a single series of wide, flat, chevron-shaped grinders. A duck-billed appearance about snout. Wide pointed wings. Tail more than 4 times body length. Body purple brown with spots. Attains 11 feet across disc. Viviparous. Leaps from water.

Vavoul maduva (S). Valval thirukai, Kurivi thirukai (T).

44. AETOMYLAEUS MACULATUS (Gray) MOTTLED EAGLE RAY

Myliobatis maculatus Gray, Illustr. Indian Zool., Hardwicke, ii, 1834, pl. 101. *Aetomylaeus maculatus* Fowler, Proc. Acad. Nat. Sci. Philadel., lxxxvii, 1935, fig. 7 (plate 3).

Teeth with 3 narrow lateral rows each side of wide median row. Dorsal origin behind ends of ventral bases. Back with median small spines and tubercles. Back brown with dark edged rounded spots of white posteriorly. Tail indistinctly banded brown and darker.

Panjadiya (S). Panjadi thirukai, Neduvalai thirukai (T).

45. AETOMYLAEUS NICHOFII (Bloch and Schneider)
NIEUHOF'S EAGLE RAY

Raja nichofii Bloch and Schneider, Syst. Ichth., 1801, p. 364. *Myliobatis nieuhofii.* "Mookarah tenkee", Russell, Fish. Coromandel, i, 1803, Pl. 7 (plate 3).

Teeth with 3 narrow lateral rows each side of wide median row. Dorsal origin opposite ends of ventral bases. Disc twice as wide as long. Skin smooth in young with some asperites on disc with age. Back dark brown anteriorly with 3-5 transverse or horizontal grey bands; posteriorly with large, close-set rounded greyish blotches.

COW RAYS

Family RHINOPTERIDAE

Head divided into two rounded lobes by deep median notch. Teeth grinder-like. Head prominent. Pair of rostral fins, not joined with front of skull and not continuous at sides of head with pectorals. Tail long, slender, whiplike.

Teeth in 7 rows in upper jaw .... *Rhinoptera javanica*
Teeth in 9 rows in upper jaw .... *Rhinoptera adspersa*

46. RHINOPTERA JAVANICA Müller and Henle JAVANESE COW RAY

Rhinoptera javanica Müller and Henle, Syst. Beschr. Plagiostomen, 1841, p. 182, Pl. lviii (plate 3).

Teeth in flattened pavement, 7 rows in each jaw. Spiracles large, close behind eye. Skin smooth. Caudal spine serrate. Liver brown above. Attains 3 feet in width.

Valuvadi thirukai (T).

47. RHINOPTERA ADSPERSA Müller and Henle ROUGH COW RAY

Rhinoptera adspersa Müller and Henle, Syst. Beschr. Plagiostomen, 1841, p. 183.

Upper teeth in 9 rows, lower teeth in 7 rows. Back rough with small stellate based spines. Greenish brown above, lighter below.

Hudja maduva (S). Muthra thirukai, Mundekanmi thirukai (T).

DEVIL RAYS

Family MOBULIDAE

Disc wider than long, rhomboidal, pectoral tips falcate. Mouth well below the head. A pair of curling horn-like flippers in front of the head. One species in Ceylon.

48. MOBULA DIABOLUS (Shaw) LESSER DEVIL RAY

Raja diabolus Shaw, General Zoology, v, 1804, p. 291. *Mobula eregoodootenkee. Dicerobatis eregoodoo* Day, Fish. India, 1875-78, Pl. cxciii, fig. 1 (plate 3).

No spine at base of tail. In each jaw 30-80 oblique rows of about 10 teeth. Skin smooth. Grey to grey-brown above. Attains only about 4 feet across disc.

Anga maduva, Ath maduva (S). Kombu thirukai (T).

ELECTRIC RAYS

Order TORPEDINIFORMES

Somewhat like the rays but possess electric organs and a well-developed tail fin (without spine) and a soft naked skin. A single family.

ELECTRIC RAYS, NUMBFISH

Family TORPEDINIDAE

1. Two dorsal fins 2
   One dorsal fin .... *Narke dipterygia*
2. Uniform brown above .... *Narcine brunnea*
   Brown-spotted above .... *Narcine timlei*

49. NARKE DIPTERYGIA (Bloch and Schneider) NUMBFISH

Rhinobatus dipterygia Bloch and Schneider, Syst. Ichth., 1801, p. 359. *Astrape dipterygia* Day, Fish. India, 1875-78, Pl. cxcii, fig. 4 (plate 2).

One dorsal fin. Disc almost circular. Tail short. Skin smooth. Eyes small, about half length of spiracle which is continuous with them. Brown above with diffuse blackish cloudings medially. Dorsal and caudal brown.

50. NARCINE BRUNNEA Annandale BROWN ELECTRIC RAY

Narcine brunnea Annandale, Mem. Indian Mus., ii, 1909, p. 45. *Narcine firma. Narcine timlei* Day, Fish. India, 1875-78, Pl. cxcii, fig. 3 (plate 2).

Disc shorter than tail. Two dorsal fins. First dorsal entirely behind ventral. Chocolate brown, without spots. Narrow cream edges around disc. Dorsals and caudal edges with grey-white. Attains about 8 inches.

51. NARCINE TIMLEI (Bloch and Schneider) SPOTTED ELECTRIC RAY

Raja timlei Bloch and Schneider, Syst. Ichth., 1801, p. 359. *Narcina timlei* Jordan and Seale, Proc. Davenport Acad. Sci., v (10), 1905, Pl. i (plate 2).

Two dorsal fins. First dorsal origin behind hind basal end of ventral. Light brown above, with large round brown spots, larger than the interspaces. Spots sometimes purplish with pale edges. Dorsal and caudal edges white.

Hiri maduva, Thitha maduva, Thimbiriya (S). Thimili (T).

PART TWO

BONY FISH

Class TELEOSTOMI

Subclass ACTINOPTERYGII

Skeleton composed of true bone. Skin normally covered with overlapping scales which in some families may be obsolete, and in others modified by calcification into a hard covering, ossified dermal plates or a complete bony casing. Gills covered by a bony operculum and have only one external opening on each side. Caudal fin nearly symmetrical.

1. Body asymmetrical with eyes both on same side of head ...... *Order* PLEURONECTIFORMES, p. 256
   Body bilaterally symmetrical with eyes on opposite sides of head 2
2. Ventral fins absent 3
   Ventral fins present (if greatly reduced or absent, the snout is tube-like and the body encased in bony rings) 9
3. Body extremely long, eel-like or ribbon-like 4
   Body short, not eel-like or ribbon-like 7
4. Body cylindrical; caudal fin present 5
   Body flattened and ribbon-like; caudal fin absent
   .... *Order* PERCIFORMES (TRICHIUROIDEI), p. 216
5. Dorsal and anal fins without spines 6
   Dorsal and anal fins with spines .... *Order* MASTACEMBELIFORMES, p. 267
6. Two gill-openings, lateral in position .... *Order* ANGUILLIFORMES, p. 57
   One gill-opening on ventral surface .... *Order* SYMBRANCHIFORMES, p. 102
7. Scales modified into minute sharp granules, needle-like spines, calcareous plates or fused into a bony casing .... *Order* TETRADONTIFORMES, p. 269
   Scales when present are normal overlapping type 8
8. Large sub-cylindrical fishes with upper jaw prolonged into a sword
   .... *Order* PERCIFORMES (SCOMBROIDEI, part), p. 218
   Small oval fishes with upper jaw not prolonged into a sword
   .... *Order* PERCIFORMES (STROMATEOIDEI), p. 223
9. Ventral fins abdominal 10
   Ventral fins thoracic or jugular 20
10. Body naked or with bony plates or rings 11
    Body with normal overlapping scales 13
11. Snout produced into a tube; barbels absent 12
    Snout normal; several pairs of long barbels .... *Order* CYPRINIFORMES (SILUROIDEI), p. 38
12. Pelvics with 1-3 enlarged feeler-like rays .... *Order* PEGASIFORMES, p. 290
    Pelvics with 3-7 small normal rays .... *Order* SYNGNATHIFORMES, p. 79
13. One spineless dorsal fin 14
    Two distinct dorsal fins 18
14. Gill-membranes broadly united with isthmus; jaws without teeth
    .... *Order* CYPRINIFORMES (CYPRINOIDEI), p. 38
    Gill-membranes free from isthmus; jaws usually with teeth 15
15. Lateral line when present, situated in upper half of sides, never forming a raised ridge 16
    Lateral line extremely low on sides, forming a raised ridge 17
16. Head scaly and depressed; ventrals moderately large .... *Order* CYPRINODONTIFORMES, p. 85
    Head naked and compressed; ventrals small .... *Order* CLUPEIFORMES, p. 21

17. Tail tapering to a point; ventrals with 9 rays .... *Order* HALOSAURIFORMES, p. 70
Tail forked, truncate or rounded; ventrals with 6 rays .... *Order* BELONIFORMES, p. 71

18. First dorsal with soft rays only; second dorsal adipose .... *Order* SCOPELIFORMES, p. 36
First dorsal with spines only; second dorsal with soft rays 19

19. Pectoral with lowermost rays detached and filamentous .... *Order* POLYNEMIFORMES, p. 96
Pectoral fin entire, with no free rays .... *Order* MUGILIFORMES, p. 90

20. Ventral fins with 1 spine and 5 rays 30
Ventral fins with other than 1 spine and 5 rays 21

21. Fins without true spines (sometimes 1 or 2 ossified rays in dorsal only) 22
Fins with true spines 25

22. Two dorsal fins, first short usually with 1-2 ossified rays 23
A single dorsal fin with no ossified rays 24

23. Free caudal fin present .... *Order* GADIFORMES, p. 76
Tail tapers to sharp point without free caudal fin .... *Order* MACRURIFORMES, p. 77

24. Fresh water species with accessory air-breathing apparatus; depressed head covered with large plate-like scales; ventrals when present, thoracic with 6 rays ... *Order* OPHIOCEPHALIFORMES, p. 99
Marine species without accessory air-breathing apparatus; head naked or with small scales only; ventrals when present, jugular or mental with 1 or 2 rays
.... *Order* PERCIFORMES (OPHIDIOIDEI), p. 206

25. Scales present 26
Scales absent (or never wholly scaly) 28

26. Scales large and strongly serrated or spinous; head with mucous cavities
.... *Order* BERYCIFORMES, p. 86
Scales small and smooth; head without mucous cavities 27

27. Anal with 7 spines .... *Order* PERCIFORMES (SIGANOIDEI), p. 208
Anal with 2 or 3 spines .... *Order* PERCIFORMES (ACANTHUROIDEI), p. 210

28. Anterior dorsal spines modified as a movable lure equipped with a terminal organ
.... *Order* LOPHIIFORMES, p. 286
Anterior dorsal spines not modified 29

29. First dorsal with 2-4 short stout spines ... . *Order* BATRACHOIDIFORMES, p. 285
First dorsal with more than 4 spines .... *Order* PERCIFORMES (BLENNIOIDEI), p. 200

30. First dorsal fin modified to a sucking disc on upper surface of head ... *Order* ECHENEIFORMES, p. 268
Dorsal fins normal 31

31. Pectorals very large and divided into 2 portions .... *Order* DACTYLOPTERIFORMES, p. 255
Pectorals moderate and normal 32

32. A short anterior anal fin of 3 spines; first vertebra firmly attached to skull ... *Order* ZEIFORMES, p. 89
Anal spines not forming a separate anterior fin; first vertebra free from skull 33

33. Dorsal and anal fins followed by 5 or more detached finlets
.... *Order* PERCIFORMES (SCOMBROIDEI), p. 218
Dorsal and anal fins without series of detached finlets (except where lateral scutes are present) 34

34. Suprabranchial organ for accessory air breathing
.... *Order* PERCIFORMES (ANABANTOIDEI), p. 225
No accessory air breathing organ 35

35. Mouth much reduced; gill-openings reduced to small pore
.... *Order* PERCIFORMES (CALLIONYMOIDEI), p. 207
Mouth with moderate to large gape; gill-openings normally wide 36

36. Males with a characteristic denticulated crest on the supraoccipital
.... *Order* PERCIFORMES (KURTOIDEI), p. 228
No such crest in either sex 37

37. Bony ridge across cheek, its end articulating with the front edge of the opercle; head armoured with bony plates and many spiny projections .... *Order* PERCIFORMES (COTTOIDEI), p. 243
Head not armoured in such fashion 38
38. Ventral fins close together or united, forming a sucking disc
.... *Orde.* PERCIFORMES (GOBIOIDEI), p. 228
Ventral fins moderately far apart, never forming a sucking disc
.... *Order* PERCIFORMES (PERCOIDEI), p. 103

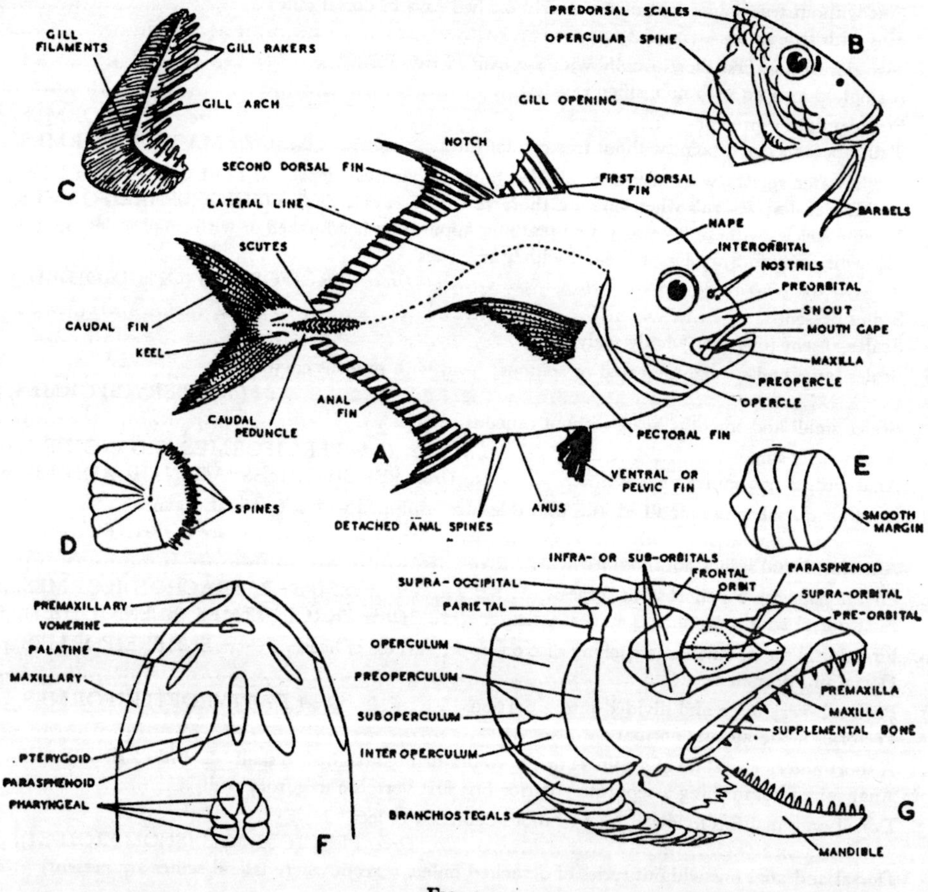

Fig. 4.

Explanation of characters of the bony fishes, Class Teleostomi. a.—General external anatomy. b.—Additional characters of the head. c.—Structures of a gill component. d.—Typical ctenoid scale. e.—Typical cycloid scale. f.—Inside of roof of mouth showing bones which may bear teeth. g.—Principal bones which form the skull and gill-coverings.

HERRING-LIKE FISHES

Order CLUPEIFORMES

Dorsal and anal fins without spines. Scales well developed, cycloid. One dorsal fin. Ventral fins usually small, sometimes absent. Maxillary and premaxillary bones border the mouth. Marine, estuarine and fresh water.

1. Dorsal far back, opposite anal; teeth large 2
   Dorsal median; teeth not enlarged 4
2. Photophores absent .... (*Suborder* CHIROCENTROIDEI) *Family* CHIROCENTRIDAE, p. 33
   Photophores present .... (*Suborder* STOMIATOIDEI) 3
3. Adipose fin present; dorsal between pectorals and ventrals ...... *Family* CHAULIODONTIDAE, p. 34
   Adipose fin absent; origin of dorsal and anal fins opposite .... *Family* STOMIATIDAE, p. 34
4. Lateral line present 5
   Lateral line absent .... (*Suborder* CLUPEOIDEI) 9
5. Branchiostegals 4 .... (*Suborder* CHANOIDEI) *Family* CHANIDAE, p. 34
   More than 4 branchiostegals .... (*Suborder* CLUPEOIDEI) 6
6. Deep water fishes with large mouth and feeble teeth; less than 9 branchiostegals .... *Family* ALEPOCEPHALIDAE, p. 33
   Shallow water fishes with villiform teeth; more than 20 branchiostegals 7
7. No gular plate on chin .... *Family* ALBULIDAE, p. 23
   Gular plate between each ramus of lower jaw 8
8. Body deep, oblong; scales large; last dorsal ray ending in filament ...... *Family* MEGALOPIDAE, p. 22
   Body moderately slender; scales small; fins without filaments .... *Family* ELOPIDAE, p. 21
9. Stomach gizzard-like; mouth small, inferior, toothless; maxillary simple ... *Family* DOROSOMIDAE, p. 29
   Stomach not gizzard-like 10
10. Mouth moderate, terminal; maxillary in 3 pieces 11
    Mouth very large, usually inferior below tapering pig-like snout; maxillary very long .... *Family* ENGRAULIDAE, p. 30
11. Belly rounded, covered with ordinary scales; supplemental maxillary very narrow .... *Family* DUSSUMIERIDAE, p. 28
    Belly compressed, with bony serrae; supplemental maxillary broad .... *Family* CLUPEIDAE, p. 23

GIANT HERRINGS, MILK FISH

Family ELOPIDAE

Branchiostegals 27 to 34. Scales small and thin. Anal fin with scaly basal sheath. Ventral rays 17-18. Last dorsal ray not elongate filament.

52. ELOPS MACHNATA (Forskål) GIANT HERRING

Argentina machnata Forskål, Descript. Animal., 1775, p. 68. *Elops saurus* or *indicus*. *Elops machnata* Richardson, Voy. Erebus & Terror, 1846, Pl. xxxvi, fig. 3 (plate 6).

D. 20-25. A. 13-17. L. lat. 100-120. Tr. 12-17. G.R. 16-17. Adipose eyelids present. Depth 5-6 in length. Pectorals low down. Dorsally olive, sides and belly silvery. Attains 30 inches.

Mannava, Ranava (S). Manna (T).

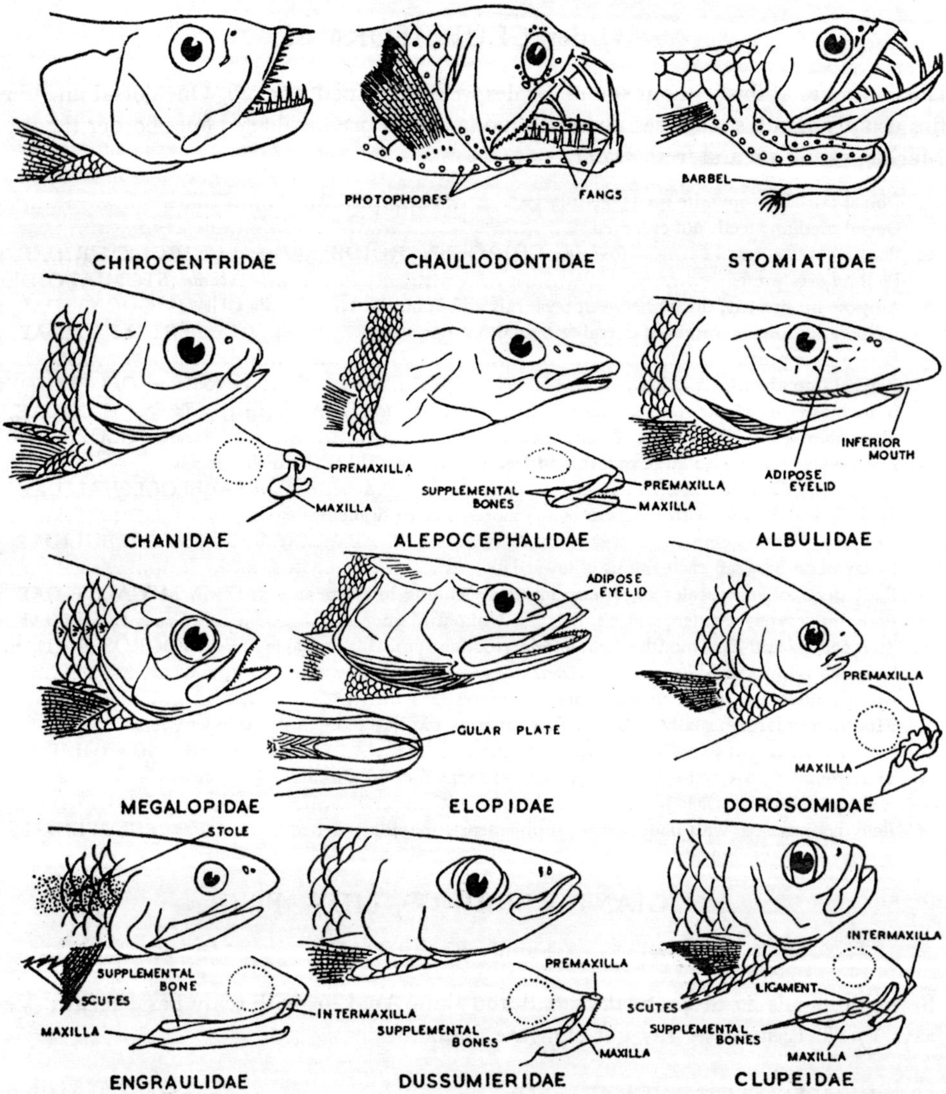

FIG. 5.

Principal differences between the twelve component families of the Order Clupeiformes (Herring-like fishes), showing characteristic structures within the group and detail of the bones bordering the upper jaw.

TARPON, OX-EYE HERRING

Family MEGALOPIDAE

Branchiostegals 23 to 27. Scales large, tough. Anal fin without scaly basal sheath. Ventral rays 10-11. Body laterally compressed. Last ray of dorsal fin filamentous.

53. MEGALOPS CYPRINOIDES (Broussonet) TARPON

Clupea cyprinoides Broussonet, Ichth., 1782, pl. 9. *Megalops kundinga. Megalops filamentosus* Bleeker, Atlas Ichth. Ind. Neerl., vi, 1872, Pl. cclxxiii, fig. 1 (plate 6).

D. 17-20, last ray elongate. A. 24-31. L. lat. 36-40. Tr. 6+8. G.R. (15-17)+(30-31). Adipose eyelids present. Dorsally olive, sides and belly silvery. Pectorals and unpaired fins yellowish. 40 inches. Estuaries.

Illeya, Mareva (S). Marua (T).

LADY FISH, BONE FISH, BANANA FISH

Family ALBULIDAE

Body elongate. Snout bluntly conical, mouth inferior with minute teeth. Ventrals 10-14. No scales on head.

54. ALBULA VULPES (Linnaeus) LADY FISH

Esox vulpes Linnaeus, Syst. Nat., edit. 10, i, 1758, p. 313. *Albula conorhynchus. Conorhynchus glossodon* Bleeker, Atlas Ichth. Ind. Neerl., vi, 1872, Pl. cclxx, fig. 1 (plate 6).

D. 15-17. A. 8. L. lat. 65-75. Tr. 8+12. G.R. 8-10. Body sub-cylindrical. Adipose eyelids present. Dorsally olive, sides and belly silvery. Snout and fins yellowish. 40 inches. Coastal waters and estuaries.

Vauva, Miya (S).

HERRINGS, SARDINES

Family CLUPEIDAE

Oblong or elongate fishes, mostly compressed. Belly keeled with scutes along ventral edge. Snout not overshot. Mouth large, jaws equal. Branchiostegals 6-15. No lateral line. Scales thin, deciduous.

1. Anal moderate, 15-25 rays; jaws equal; ventrals well developed 2
   Anal very long, more than 30 rays; lower jaw prominent; ventrals small or absent 13
2. Dorsal origin behind ventral origin .... *Kowala coval*
   Dorsal origin before ventral origin 3
3. Upper jaw with distinct median notch 4
   Upper jaw without median notch 6
4. Parietal ridges expanded and striated; lower gill-rakers 95-130 .... *Macrura kelee*
   Parietal ridges narrow and smooth; lower gill-rakers less than 95 or more than 120 5
5. Lower gill-rakers 70-95; caudal longer than head .... *Tenualosa sinensis*
   Lower gill-rakers 120-220; caudal as long as head .... *Tenualosa ilisha*
6. Last two rays of anal equal, not enlarged .... *Harengula ovalis*
   Last two rays of anal enlarged 7
7. Ventral scutes sharp, keeled, exposed 8
   Ventral scutes little evident, scarcely protrude from groove containing them 12
8. Ventral rays 9; dark spot at upper edge of opercle .... *Sardinella longiceps*
   Ventral rays 8; dark spot at bases of front dorsal rays 9
9. Depth $2\frac{1}{2}$ to $2\frac{7}{8}$ 10
   Depth 3 to 4 11

10. Lower gill-rakers 55-65; eye $3\frac{1}{8}$ in head .... *Sardinella abella*
    Lower gill-rakers 48-55; eye $3\frac{1}{8}$ to $3\frac{3}{8}$ .... *Sardinella perforata*
11. Lower gill-rakers 38-44 .... *Sardinella melanura*
    Lower gill-rakers 50-55 .... *Sardinella jussieu*
    Lower gill-rakers 70-75 .... *Sardinella fimbriata*
12. Depth $4\frac{3}{8}$ to $5\frac{1}{8}$; lower gill-rakers 36-40 .... *Amblygaster sirm*
    Depth $4\frac{1}{8}$ to $4\frac{3}{8}$; lower gill-rakers 32-33 .... *Amblygaster clupeoides*
13. Toothed bone between terminal end of premaxillary and middle of maxillary .... *Pellona ditchela*
    Ligament between terminal end of premaxillary and middle of maxillary 14
14. Ventral fins present 15
    No ventral fins 17
15. Scales less than 46 in medial lateral series; preventral scutes 16-20 .... *Euplatygaster indica*
    Scales more than 46 in medial lateral series; preventral scutes 20-28 16
16. Anal 44-52; depth more than $3\frac{1}{2}$ .... *Ilisha elongata*
    Anal 45-50; depth less than $3\frac{1}{2}$ .... *Ilisha filigera*
17. Dorsal fin present; maxillary broadly rounded behind, not extending beyond centre of eye .... *Opisthopterus tardoore*
    Dorsal fin absent; maxillary tapers behind, extends to gill-opening or beyond .... *Raconda russelliana*

55. KOWALA COVAL (Cuvier) — WHITE SARDINE

Clupea coval Cuvier, Regne Animal, ed. 2, ii, 1829, p. 318. *Clupeoides lile*, *Rogenia argyrotaenia*, *Kowal thoracata*. *Clupea lile* Day, Fish. India, 1875-78, Pl. clxii, fig. 1 (plate 4).

D. 13-15. A. 17-18. Lateral scales 38-41. Tr. 10-11. Predorsal scutes 16. Ventral scutes 18+10. G.R. 32-34. Head silvery with greenish yellow vertex. Body white, translucent with median lateral silver streak. Snout, top of head, back, caudal and anal base dotted with black. $2\frac{1}{2}$ inches. Coastal waters.

Silinda, Sudu sudaya (S). Vellai schudai (T).

56. MACRURA KELEE (Cuvier) — FIVE-SPOT HERRING

Clupea kelee Cuvier, Regne Animal, ed. 2, ii, 1829, p. 320. *Alosa kanagurta*. *Clupea kanagurta* Day, Fish. India, 1875-78, Pl. clxii, fig. 4 (plate 4).

D. 16-18. A. 18-22. Lateral scales 42-44. Tr. 13-14. Predorsal scales 10-15. Ventral scutes 18+13. G.R. 50+(95-130). Body light yellow dorsally, silvery on sides and abdomen. Black shoulder blotch about size of pupil, followed by row of 3-5 similar blotches in young specimens. Dorsal fin yellow with black tip. Caudal yellow. 8 inches. Coastal waters.

Seriya, Koiya, Katu massa (S). Koi meen (T).

57. TENUALOSA SINENSIS (Linnaeus) — CHINESE HERRING

Clupea sinensis Linnaeus, Syst. Nat., ed. 10, i, 1758, p. 319. *Alosa hilsa* or *Harengula toli*. *Clupea toli* Day, Fish. India, 1875-78, Pl. clxii, fig. 2 (plate 4).

D. 17-19. A. 18-21. Lateral scales 40. Tr. 14-15. Predorsal scales 16. Ventral scutes (17-18)+(11-13). G.R. 70-95 lower. Purplish green dorsally, six longitudinal rows of small diffuse brown scale marks. Sides silvery. Fins yellow. Dark diffuse edge to caudal. Dark shoulder mark in young. 13 inches. Coastal waters.

Ulla, Seriya (S). Ullam (T).

58. TENUALOSA ILISHA (Hamilton-Buchanan) RIVER SHAD

Clupanodon ilisha Hamilton-Buchanan, Fishes of Ganges, 1822, pp. 243-382. *Hilsa ilisha, Alausa palasah. Clupea ilisha* Day, Fish. India, 1875-78, Pl. clxii, fig. 3 (plate 4).

D. 18-20. A. 18-21. Lateral scales 45-48. Tr. 17-20. Ventral scutes (17-19)+(11-13). G.R. 70-95 lower. Parietal ridges covered with smooth skin. Silvery. Coastal waters.

59. HARENGULA OVALIS (Bennett) SPOTTED HERRING

Clupea ovalis Bennett, Mem. Life of Raffles, 1830, p. 690. *Harengula punctata, venenosa,* or *moluccensis, Clupea kunzei, Sardinella lineolatus* or *ovalis. Clupea* (*Harengula*) *molluccensis* Bleeker, Atlas Ichth. Ind. Neerl., vi, 1872, Pl. cclxiii, fig. 2 (plate 4).

D. 17-19. A. 17-19. Lateral scales 42-45. Tr. 10-11. Predorsal scales 14. Ventral scutes (16-20)+(10-14). G.R. 14+(31-32). Dark green dorsally, separated from silvery sides and belly by yellow lateral stripe. Two bright orange spots behind operculum, sometimes 10-12 spots along back. Snout brown. Dorsal fin tip yellow. Other fins hyaline. 5 inches. Coastal waters.

Kolamura salaya, Korrumburua, Ehalamurua (S).

60. SARDINELLA LONGICEPS Valenciennes OIL SARDINE

Sardinella longiceps Valenciennes, Hist. Nat. Poiss., xx, 1847, p. 273. *Harengula longiceps, Sardinella neohowii. Clupea longiceps* Day, Fish. India, 1875-78, Pl. clxi, fig. 2 (plate 4).

D. 16-17. A. 14-15. Lateral scales 46-47. Tr. 12-13. Predorsal scales 14-16. Ventral scutes 18+13. G.R. 158+200. Sides and belly silvery, darker above. Yellowish patch on shoulder. Back greenish yellow. Light yellow lateral stripe. Dorsal fin yellow, caudal dusted with black, others hyaline. 6 inches. Coastal waters.

Pesalaya (S). Pesalai (T).

61. SARDINELLA ALBELLA (Valenciennes) SHORT-BODIED SARDINE

Kowala albella Valenciennes, Hist. Nat. Poiss., xx, 1847, p. 362. *Sardinella, Clupea, Harengula, Harengus* or *Pellona brachysoma. Clupea* (*Harengula*) *hypelosoma* Bleeker, Atlas Ichth. Ind. Neerl., vi, 1872, Pl. cclxvii, fig. 2 (plate 4).

D. 18. A. 15-16. Lateral scales 40-45. Tr. 11-13. Predorsal scales 14-18. Ventral scutes (17-18)+(12-14). G.R. 30+65, or up to 121. Purplish green dorsally, sides silvery. Dark patch at base of first three dorsal rays. 5½ inches. Coastal waters.

Karalau, Hadalla, Sudaya (S). Schudai, Sudai (T).

62. SARDINELLA PERFORATA (Cantor) PERFORATED-SCALE SARDINE

Clupeonia perforata Cantor, Journ. Asiatic Soc. Bengal, xviii (2), 1849, p. 1276. *Clupea* (*Spratella*) *perforata* Bleeker, Atlas Ichth. Ind. Neerl., vi, 1872, Pl. cclxviii, fig. 2 (plate 4).

D. 18-19. A. 20. Lateral scales 37-41. Tr. 12. Predorsal scales 14. Ventral scutes (17-18)+14. G.R. 30+(58-60). Scales pitted apically. Sides and belly silvery, darker above. Coastal waters.

63. SARDINELLA MELANURA (Cuvier) BLACK-TIPPED SARDINE

Clupea melanura Cuvier, Regne Animal, ed. 2, ii, 1829, p. 318. *Harengus atricauda, Clupea* or *Harengula melanura. Clupea* (*Harengula*) *atricauda* Bleeker, Atlas Ichth. Ind. Neerl., vi, 1872, Pl. cclxviii, fig. 5 (plate 4).

D. 16-18. A. 18-21. Lateral scales 39-43. Tr. 10-12. Predorsal scales 11-16. Ventral scutes (17-18)+(11-13). G.R. (16-18)+(35-44). Back bluish-violet. Sides silvery. Tips of caudal lobes deep black. Other fins hyaline. 4 inches. Coastal waters.

Salaya (S). Salai (T).

64. SARDINELLA JUSSIEU (Lacépede) TEMBANG

Clupanodon jussieu Lacépède, Hist. Nat. Poiss., v, 1803, p. 469. *Clupea tembang, Sardinella gibbosa. Clupea* (*Spratella*) *gibbosa* Bleeker, Atlas Ichth. Ind. Neerl., vi, 1872, Pl. cclxvi, fig. 6 (plate 4).

D. 17-20. A. 16-19. Lateral scales 44-48. Tr. 11-12. Predorsal scales 13. Ventral scutes (17-19)+(12-14). G.R. 28+65. No teeth. Mainly silvery. Muzzle dusky. Dark spot at base of dorsal fin. Coastal waters.

65. SARDINELLA FIMBRIATA (Valenciennes) FRINGE-SCALE SARDINE

Spratella fimbriata Valenciennes, Hist. Nat. Poiss., xx, 1847, p. 359. *Harengula* or *Harengus fimbriata. Clupea* (*Spratella*) *fimbriata* Bleeker, Atlas Ichth. Ind. Neerl., vi, 1872, Pl. cclxxi, fig. 5 (plate 4).

D. 17-20. A. 16-19. Lateral scales 42-45. Tr. 10-12. Predorsal scales 12-15. Ventral scutes (17-19)+(12-15). G.R. 43+80. Borders of scales pectinate and pitted. Dark blue-green dorsally with yellow lateral stripe. Sides and belly silvery. Fins pale green. Dark blotch at base of first 5 dorsal rays. 5½ inches. Coastal waters.

Gal salaya (S). Chilai, Salai (T).

66. AMBLYGASTER SIRM (Walbaum) TRENCHED SARDINE

Clupea sirm Walbaum, Artedi Pisc., iii, 1792, p. 38. *Clupea* (*Sardinella*) *leiogaster* Bleeker, Atlas Ichth. Ind. Neerl., vi, 1872, Pl. cclxii, fig. 6 (plate 4).

D. 15-18. A. 18-20. Lateral scales 42-45. Tr. 11-12. Predorsal scales 15-16. Vertral scutes (16-18)+(13-16). G.R. 21+43. Ventral scutes in groove. Back greenish. Belly silvery with bluish tint. Lateral stripe pinkish with about 15 dark spots. Fins yellowish. 8 inches. Coastal waters.

Hurulla (S). Kirimeen chalai (T).

67. AMBLYGASTER CLUPEOIDES Bleeker

SHARP-NOSE TRENCHED SARDINE

Amblygaster clupeoides Bleeker, Journ. Indian Archip., iii, 1849, p. 73. *Clupea* (*Sardinella*) *clupeoides* Bleeker, Atlas Ichth. Ind. Neerl., vi, 1872, Pl. cclxxii, fig. 1 (plate 4).

D. 17-19. A. 16-17. Lateral scales 40-45. Tr. 11-12. Predorsal scales 14-18. Ventral scutes 16+(12-14). G.R. (17-22)+(32-33). Ventral scutes in groove. Nose sharper than

in preceding species. Blue-green dorsally with silvery sides. Fins pale greenish yellow. 8 inches. Coastal waters.
Gal hurulla (S).

68. PELLONA DITCHELA Valenciennes TOOTHED SHAD

Pellona ditchela Valenciennes, Hist. Nat. Poiss., xx, 1847, p. 314. *Neosteus ditchela, Pellona hoevenii. Ilisha hoevenii* Bleeker, Atlas Ichth. Ind. Neerl., vi, 1872, Pl. cclxix, fig. 2 (plate 4).

D. 17-18. A. 32-38. Lateral scales 40-45. Tr. 11-13. Predorsal scales 14-16. Ventral scutes (16-20)+(8-9). G.R. (12-13)+(21-27), fine but short. Adipose eyelids present. Fine teeth in jaws. Depth 3 in body length, laterally compressed. Mainly silvery. Coastal waters.

69. EUPLATYGASTER INDICA (Swainson) INDIAN SHAD

Platygaster indicus Swainson, Nat. Hist. Animals, ii, 1839, p. 294. *Pellona ditchoa. Ilisha indica* Bleeker, Atlas Ichth. Ind. Neerl., vi, 1872, Pl. cclix, fig. 4 (plate 4).

D. 13-16. A. 36-39. Lateral scales 42-44. Tr. 13. Predorsal scales 14-16. Ventral scutes (18-20)+(9-10). G.R. 12+(22-24). Back greenish, sides and belly silvery. Median streak along back. Dorsal and caudal fins dusted with dark pigment. Coastal waters.
As bokka, Venganawa (S).

70. ILISHA ELONGATA (Bennett) SLENDER SHAD

Alosa elongata Bennett, Life of Raffles, 1830, p. 691. *Pellona elongata. Ilisha elongata* Bleeker, Atlas Ichth. Ind. Neerl., vi, 1872, Pl. cclix, fig. 3 (plate 6).

D. 13. A. 38. Lateral scales 48-49. Tr. 7. Predorsal scales 17. Ventral scutes (19-23)+(8-11). G.R. 8+20. Fine teeth in narrow bands. Body and fins silvery. Coastal waters.
Puvalaya (S). Puvali (T).

71. ILISHA FILIGERA (Valenciennes) JEWELLED SHAD

Pellona filigera Valenciennes, Hist. Nat. Poiss., xx, 1847, p. 322; Day, Fish. India, 1875-78, Pl. clxv, fig. 4 (plate 6).

D. 18-20. A. 44-51. Lateral scales 50. Tr. 15-16. Ventral scutes (22-23)+(10-12). G.R. 18-19 lower. Jaws with teeth. Coppery tinge along back. Sides silvery with mother-of-pearl reflections. Dorsal and pectoral tipped with black. Coastal waters.

72. OPISTHOPTERUS TARDOORE (Cuvier) LONG-FINNED HERRING

Pristigaster tardoore Cuvier, Regne Animal, ed. 2, ii, 1829, p. 321. *Opisthopterus macrognathus* or *indicus. Opisthopterus tartoor* Day, Fish. India, 1875-78, Pl. clxiii, fig. 5 (plate 6).

D. 16-17. A. 58-60. Lateral scales 46. Tr. 13. Ventral scutes 30-33. G.R. (9-10)+(22-25). Upper surface of head concave, mouth pointing upwards. Ventral profile strongly convex. Ventral fins absent. Anal fin long and low. Dorsally greenish brown. Sides silvery. Dark patch on top corner of opercle. Dorsal and caudal fins yellow. 8 inches. Coastal waters.
Maisthre lagga, Panniki lagga, Thottawa (S). Thota (T).

73. RACONDA RUSSELLIANA Gray RUSSELL'S SMOOTH-BACK HERRING

Raconda russeliana Gray, Zool. Miscellany, i, 1831, p. 9; Day, Fish. India, 1875-78, Pl. clxiii, fig. 4 (plate 6).

Dorsal fin absent. A. 83-92, originating nearer mandible tip than caudal base. Lateral scales 60-64. Tr. 16. Ventral scutes 33-37. G.R. 22-26 lower. Narrow blackish band along back. Sides yellowish, silvery below. Dark spot on shoulder, sometimes continued on opercle. Palk Straits.

ROUND HERRINGS, SPRATS

Family DUSSUMIERIDAE

Body elongate, belly rounded. Scales deciduous. No lateral line. Jaws equal or subequal. Maxillary with supplemental bones. Teeth small. Branchiostegals 6 to 15. No scutes on belly. Fins small.

1. Maxillary with one supplemental bone .... *Ehirava fluviatilis*
   Maxillary with 2 supplemental bones 2
2. Dorsal origin nearer snout tip than caudal; anal rays 11-13 .... *Spratelloides japonicus*
   Dorsal origin nearer caudal than snout tip; anal rays 15-17 3
3. Lateral scales 40-46 .... *Dussumieria acuta*
   Lateral scales 52-56 .... *Dussumieria hasseltii*

74. EHIRAVA FLUVIATILIS Deraniyagala ESTUARINE SPRAT

Ehirava fluviatilis Deraniyagala, Spolia Zeylanica, xv, 1929, p. 35, Pl. xiv (plate 5).

D. 13. A. 15. Lateral scales 35-38. Tr. 6-7. Predorsal scales 15. Preventral scales 17. G.R. 25-27. Dorsal inserted behind ventrals, closer to caudal than to snout tip. Branchiostegals 6. Top of head with about 6 large melanophores arranged in ring in interorbital space and V-pattern on parietals. Body iridescent, pearly pink. Inhabits rivers and estuaries. 2 inches.
Ahirava, Gan ahirava (S).

75. SPRATELLOIDES JAPONICUS (Houttuyn) BLUE SPRAT

Atherina japonica Houttuyn, Verh. Holl. Maatsch. Haarlem, xx, 1782, p. 340. *Stolephorus japonicus*. *Spratelloides gracilis* Bleeker, Atlas Ichth. Ind. Neerl., vi, 1872, Pl. cclxvi, fig. 2 (plate 5).

D. 13-14. A. 13-14. Lateral scales 40. Tr. 8. Predorsal scales 13. G.R. 18+28. Branchiostegals 6. Ventral fins nearer anal origin than pectorals. Dark bluish above. Silvery below. Sometimes with a silvery lateral band. Coastal waters.

76. DUSSUMIERIA ACUTA Valenciennes COMMON SPRAT

Dussumieria acuta Valenciennes, Hist. Nat. Poiss., xx, 1874, p. 467; Bleeker, Atlas Ichth. Ind. Neerl., vi, 1872, Pl. cclxxi, fig. 1 (plate 5).

D. 18-20. A. 15-18. Lateral scales 42-46. Tr. 11-12. Predorsal scales 23. Branchiostegals 14-15. G.R. 11+21. Base of dorsal slightly longer than base of anal. Bluish green dorsally. Sides and belly silvery. Golden lateral stripe along sides. 5 inches. Coastal waters.
Tondaya (S). Tondai (T).

77. DUSSUMIERIA HASSELTII Bleeker VAN HASSELT'S SPRAT

Dussumieria hasseltii Bleeker, Nat. Tijds. Ned. Ind., i, 1850, p. 422; Bleeker, Atlas Ichth. Ind. Neerl., vi, 1872, Pl. cclxxi, fig. 2 (plate 5).

D. 18-19. A. 16. Lateral scales 52-56. Tr. 12-13. Predorsal scales 24. G.R. 12+26. Branchiostegals 15. Upper half dark greenish. Silvery on sides and belly. Golden lateral band along side. Edge of caudal black. Dark spot on first ray of dorsal and pectorals. Coastal waters.

GIZZARD SHADS, BONY BREAM

Family DOROSOMIDAE

Oval, deep, laterally compressed body. Scutes along ventral border of abdomen. Stomach gizzard-like. No teeth. Branchiostegals 6. No lateral line. Head naked. Dorsal fin median, usually behind ventrals and last ray often prolonged. Feed on mud. Estuarine.

1. Last dorsal ray prolonged into a filament .... *Nematalosa nasus*
   Last dorsal ray not prolonged or filamentous 2
2. Maxillary slender, terminally slightly expanded and curved downwards .... *Gonialosa manminna*
   Maxillary straight, thin, transversely expanded plate, tapering terminally .... *Anodontostoma chacunda*

78. NEMATALOSA NASUS (Bloch) LONG-RAY BONY BREAM

Clupea nasus Bloch, Naturg. Ausländ. Fische, ix, 1795, p. 116. *Chatoessus nasus, Clupanodon thrissa. Dorosoma nasus* Bleeker, Atlas Ichth. Ind. Neerl., vi, 1872, Pl. cclx, fig. 4 (plate 6).

D. 15-17. A. 20-24. Lateral scales 45-50. Tr. 15-19. Ventral scutes (16-19)+(10-12). Predorsal scales 20. G.R. 226-240, in young 130 or more. Last dorsal ray prolonged and filamentous. Body silvery with yellow fins. Occasionally a diffuse pink lateral band. Black shoulder spot present. Seven longitudinal rows of dark dots along scales of upper series. 9 inches. Bays and lagoons.
Suthara koiya, Katu goiya (S). Koi meen (T).

79. GONIALOSA MANMINNA (Hamilton-Buchanan) GANGES GIZZARD SHAD

Clupanodon manminna Hamilton-Buchanan, Fishes of Ganges, 1822, p. 247. *Chatoessus manminna* Day, Fish. India, 1875-78, Pl. clx, fig. 2 (plate 6).

D. 14-17. A. 22-26. Lateral scales 55-65. Tr. 21-25. Ventral scutes (16-19)+(10-13). Snout shorter than eye. Ventral below or in advance of dorsal fin. Dorsal fin without last ray prolonged. Silvery with sometimes a dark shoulder spot. Coastal waters.

80. ANODONTOSTOMA CHACUNDA (Hamilton-Buchanan)

SHORT-NOSE GIZZARD SHAD

Clupanodon chacunda Hamilton-Buchanan, Fishes of Ganges, 1822, p. 246. *Chatoessus chacunda. Dorosoma chacunda* Bleeker, Atlas Ichth. Ind. Neerl., vi, 1872, Pl. cclxi, fig. 6 (plate 6).

D. 17-18. A. 19-20. Lateral scales 37-40. Tr. 12-13. Predorsal scales 10-11. Ventral scutes (17-18)+(11-12). G.R. 65+83. Snout shorter than eye. Last ray of dorsal fin not prolonged. Body silvery, upper half brownish with indistinct longitudinal lines along the scale rows. Fins hyaline. Prominent black spot on shoulder. Coastal waters.

ANCHOVIES

Family ENGRAULIDAE

Body elongate and compressed. Snout bluntly rounded. Mouth inferior with maxilla sometimes extremely long. No lateral line. Eyes completely covered with skin. Belly with numerous keeled abdominal scutes. Small gregarious fishes, mostly translucent. Prefer sandy shores.

1. Body greatly elongate, tail long and tapering; caudal not deeply forked and continuous with long anal; upper pectoral rays produced as long filaments .... *Coilia quadragesimalis*
   Body only moderately long; anal free from deeply forked caudal; upper pectoral rays normal 2
2. A prominent silvery lateral stripe 3
   No silvery lateral stripe 4
3. Abdominal scutes 6-7 between pectorals and ventrals; maxillary reaches gill-opening .... *Anchoviella commersonii*
   Abdominal scutes 4-5 between pectorals and ventrals; maxillary reaches front edge of opercle .... *Anchoviella indica*
4. Upper pectoral ray produced; anal rays 51-60; anal fin 2-2½ in body without caudal ...... *Setipinna taty*
   Upper pectoral ray not produced; anal rays 30-50; anal fin 3-5 in length without caudal 5
5. No abdominal scutes in front of ventral .... *Thrissina baelama*
   Abdominal scutes extend from isthmus to vent 6
6. Maxillary not reaching beyond opercle 7
   Maxillary reaches at least to pectorals 9
7. Anal origin beneath end of dorsal; lower gill-rakers 13 .... *Thrissocles hamiltonii*
   Anal origin well behind end of dorsal; lower gill-rakers 20 or more 8
8. Abdominal scutes 27; lower gill-rakers 21-25; snout ⅔ in eye .... *Thrissocles malabarica*
   Abdominal scutes 23; lower gill-rakers 27; snout equals eye .... *Thrissocles kammalensis*
9. Maxillary reaches pectoral origin .... *Thrissocles mystax*
   Maxillary reaches beyond pectorals 10
10. Maxillary surpasses ventrals; lower gill-rakers 12; ventral scutes 28 .... *Thrissocles setirostris*
    Maxillary reaches or nearly reaches ventrals; lower gill-rakers 16-20; abdominal scutes 22-24 .... *Thrissocles dussumieri*

81. COILIA QUADRAGESIMALIS Valenciennes TAPER-TAIL ANCHOVY

Coilia quadragesimalis Valenciennes, Hist. Nat. Poiss., xxii, 1848, p. 83. *Demicoilia margaritifera.*

D. 14-15. A. 35-42, originating behind last dorsal ray and attached to caudal. Upper rays of pectoral free and filamentous. Lateral scales 35. Abdominal scutes 5+8, none before pectoral. Lower G.R. 23. Silvery, brownish on sides and back. Two longitudinal rows of pearl coloured spots below median line, upper row with 18-19, lower with 16-18, some golden yellow. On sides of belly third irregular row of 8-9 spots; 4-5 on thorax; 5 on lower jaw. Harbours.

82. ANCHOVIELLA COMMERSONII (Lacépède) COMMERSON'S ANCHOVY

Stolephorus commersonii Lacépède, Hist. Nat. Poiss., v, 1803, p. 381. *Engraulis* or *Anchoviella commersonii*, (?) *Engraulis brownii*. *Engraulis commersonianus* Day, Fish. India, 1875-78, Pl. clviii, fig. 1 (plate 5).

D. 15-16. A. 20-22. Lateral scales 36-40. Tr. 9. Predorsal 18-21. Ventral scutes 6-8 before pelvics, none behind. Gill-rakers (17-19)+(21-23). Maxilla reaches margin of gill cover. Dusky brown above, sides silvery with brilliant metallic lateral stripe. Fins hyaline, caudal powdered with darker colour. Attains 5 inches. Coastal waters.

Hal massa (S). Netholi (T).

83. ANCHOVIELLA INDICA (Van Hasselt) INDIAN ANCHOVY

Engraulis indica Van Hasselt, Algemein Konst- en Letterbode, 1823, p. 329. *Anchoviella indica*. *Stolephorus indicus* Bleeker, Atlas Ichth. Ind. Neerl., vi, 1872, Pl. cclix, fig. 2 (plate 5).

D. 16. A. 19-21. Lateral scales 37-42. Tr. 8-9. Predorsal scales 20-22. Ventral scutes 4-5 before pelvics. G.R. 20+(21-25). Maxilla reaches to margin of preopercle. Dusky on back, sides silvery with brilliant metallic lateral stripe. Fins hyaline, caudal powdered with black. Attains 6 inches. Coastal waters.

Hal massa (S). Netholi (T).

84. SETIPINNA TATY (Valenciennes) HAIR-FIN ANCHOVY

Engraulis taty Valenciennes, Hist. Nat. Poiss., xxi, 1848, p. 60. *Stolephorus* (*Setipinna*) *taty* Bleeker, Atlas Ichth. Ind. Neerl., vi, 1872, Pl. cclx, fig. 7 (plate 5).

D. 14-16. A. 51-60, its length $2\frac{1}{2}$-3 in body without caudal. Lateral scales 40-46. Tr. 12. Abdominal scutes 18+8. G.R. (12-13)+(15-16). Maxilla reaches preopercle angle. Dorsal fin high. First ray of pectoral fin prolonged. Yellowish with silvery hue. Fins hyaline. Coastal waters.

85. THRISSINA BAELAMA (Forskål) SHORT-JAW ANCHOVY

Clupea baelama Forskål, Descript. Animal., 1775, p. 72. *Anchoviella baelama* or *encrasicholoides*. *Stolephorus* (*Engraulis*) *boelama* Bleeker, Atlas Ichth. Ind. Neerl., vi, 1872, Pl. cclx, fig. 1 (plate 5).

D. 16. A. 27-30. Lateral scales 36-38. Tr. 9-11. Predorsal scales 14-20. Gill-rakers 15+19. Abdominal scutes (5-6)+(9-10), none before pectorals. Maxillary short, reaching not quite to gill-opening. Silvery, back dark. Generally dark patch on shoulder. No silvery lateral band. Dorsal and caudal greyish. Other fins white. Coastal waters.

Bilee lagga, Hal massa (S). Netholi (T).

86. THRISSOCLES HAMILTONII (Gray) HAMILTON'S ANCHOVY

Thrissa hamiltonii Gray, Illust. Indian Zool. Hardwicke, ii, 1834, p. 92. *Engraulis, Scutengraulis* or *Anchoviella hamiltoni, Engraulis grayi. Stolephorus (Thrissa) hamiltoni* Bleeker, Atlas Ichth. Ind. Neerl., vi, 1872, Pl. cclix, fig. 5 (plate 5).

D. 15-16. A. 36-40. Lateral scales 45-47. Tr. 12-13. Predorsal scales 18-22. Abdominal scutes (16-18)+(10-11). Opercles, cheeks, occipital and scapular regions venulose. G.R. (9-11)+(12-16). Maxilla extends nearly to pectoral origin. Silvery, back dark. Scapular region with black venules. Lower part of head powdered with black. Fins hyaline, caudal with dark margin. Coastal waters.

Pothu lagga (S).

87. THRISSOCLES MALABARICA (Bloch) MALABAR ANCHOVY

Clupea malabaricus Bloch, Naturg. Ausländ. Fische, ix, 1795, p. 115. *Spratelloides malabaricus, Anchoviella malabarica. Engraulis malabaricus* Day, Fish. India, 1875-78, Pl. clvii, fig. 5 (plate 5).

D. 13-15. A. 40-43. Lateral scales 37-40. Tr. 11-12. Abdominal scutes (17-18)+(9-10). Lower G.R. 21-25. Maxilla reaches barely beyond pectoral base. Silvery, shot with gold and purple. Black venules on shoulder. Fins yellow. Dorsal and end of caudal edged with black. Pectorals sometimes black. Coastal waters.

88. THRISSOCLES KAMMALENSIS (Bleeker) MADURA ANCHOVY

Engraulis kammalensis Bleeker, Verh. Batav. Genootsch., xxii, 1849, p. 13; Day, Fish. India, 1875-78, Pl. clvii, fig. 1 (plate 5).

D. 14-15. A. 32-35. Lateral scales 35-38. Tr. 9-10. Abdominal scutes 15+8. Lower G.R. 27. Maxillary reaches gill-opening, expanded above mandibular joint, tapers behind. Silvery, darker above. Across nape usually blackish spot, reaching downwards to scapula. Fins hyaline, dorsal and caudal usually blackish along margin. Coastal waters.

89. THRISSOCLES MYSTAX (Bloch and Schneider) MUSTACHED ANCHOVY

Clupea mystax Bloch and Schneider, Syst. Ichth., 1801, p. 426. *Engraulis* or *Anchoviella mystax. Stolephorus (Thrissa) mystax* Bleeker, Atlas Ichth. Ind. Neerl., vi, 1872, Pl. cclxi, fig. 2 (plate 5).

D. 15-16. A. 36-38. Lateral scales 40-45. Tr. 11-12. Predorsal scales 20-24. Abdominal scutes (17-18)+(10-12). G.R. (9-11)+(16-17). Maxillary reaches beyond gill-opening or to pectoral origin. Silvery, back dark. Scapular region with black venules. Lower part of head powdered with black. Dorsal and caudal brownish, others yellowish. Coastal waters.

Ata lagga, Hal massa (S). Netholi (T).

90. THRISSOCLES SETIROSTRIS (Broussonet) LONG-JAW ANCHOVY

Clupea setirostris Broussonet, Ichth., i, 1782, Pl. 2. *Anchoviella setirostris. Stolephorus (Thrissa) setirostris* Bleeker, Atlas Ichth. Ind. Neerl., vi, 1872, Pl. cclxi, fig. 1 (plate 5).

D. 14-17. A. 33-38. Lateral scales 40-44. Tr. 11-12. Predorsal scales 18-19. Abdominal scutes (16-18)+(8-10). G.R. 5+12. Maxillary extremely long, reaching to or beyond pelvics. Silvery, dark above. Scapular region with dark venules. Fins hyaline. Coastal waters.
Ravul lagga (S).

91. THRISSOCLES DUSSUMIERI (Valenciennes) DUSSUMIER'S ANCHOVY

Engraulis dussumieri Valenciennes, Hist. Nat. Poiss., xxi, 1848, p. 69. *Stolephorus (Thrissa) dussumieri* Bleeker, Atlas Ichth. Ind. Neerl., vi, 1872, Pl. cclx, fig. 2 (plate 5).

D. 15-16. A. 34-38. Lateral scales 40-42. Tr. 10-11. Predorsal scales 16-17. Abdominal scutes (15-16)+(7-8). G.R. (16-17)+(16-20). Maxillary extremely long, reaching origin of pelvics. Yellowish silvery, back dark. Blackish spot across nape to scapula. Fins hyaline, caudal edged with black. Coastal waters.

NAKED HEADS

Family ALEPOCEPHALIDAE

Deep sea fishes with compressed body, pointed snout and fairly large mouth set with feeble teeth. Fins without spines. No adipose fin. One species known from deep water in the Gulf of Manaar.

92. ALEPOCEPHALUS BLANFORDI Alcock BLANFORD'S NAKED HEAD

Alepocephalus blanfordi Alcock, Ann. Mag. Nat. Hist. (6), x, 1892, p. 357; Alcock, Illust. Zool. Investigator, ii, 1894, Pl. ix, fig. 1 (plate 5).

D. 16. A. 17. L. lat. 70. Body scales thick, deciduous. Head naked. Dorsal and anal fins low, opposite, in posterior third of body. Gill-rakers numerous, long and close set. Brown. Head and paired fins black. Gulf of Manaar in 902 fathoms.

WOLF HERRINGS

Family CHIROCENTRIDAE

Elongate and compressed body. Thin deciduous scales. Large mouth set with large fangs. No lateral line. Lower abdominal margin sharp.

Depth $6\frac{1}{2}$ to $7\frac{1}{2}$ in length .... *Chirocentrus dorab*
Depth 5 to $5\frac{1}{2}$ in length .... *Chirocentrus nudus*

93. CHIROCENTRUS DORAB (Forskål) DORAB; WOLF HERRING

Clupea dorab Forskål, Descript. Animal., 1775, p. 7. *Chirocentrus dorab*, Bleeker, Atlas Ichth. Ind. Neerl., vi, 1872, Pl. cclxxi, fig. 3 (plate 6).

D. 16-18. A. 33-35. Head $5\frac{1}{3}$ to 6 in body length. Maxillary not reaching opercle. G.R. 3+(14-15). Back blue. Sides silvery. Fins hyaline. Caudal yellowish. Attains 12 feet. Coastal waters.
Podi katuvalla, Panu katuvalla, Pat katuvalla (S).
Kallaku vallai, Tuppu vallai, Mullu vallai, Kuru vallai (T).

94. CHIROCENTRUS NUDUS Swainson SMOOTH WOLF HERRING

Chirocentrus nudus Swainson, Nat. Hist. Animals, Fishes, ii, 1839, p. 294. *Chirocentrus hypselosoma* Bleeker, Atlas Ichth. Ind. Neerl., vi, 1872, Pl. cclxix, fig. 3 (plate 6).

D. 17-19. A. 31-33. Head $3\frac{3}{4}$ to 4 in body length. Maxillary reaching beyond opercle. G.R. 5+(16-18). Scales minute, feeble. Back blue. Sides silvery. Fins hyaline. Attains 18 inches. Coastal waters.

Gal katuvalla (S). Kat karuvallai (T).

MILKFISH, BANDANG, SALMON HERRING

Family CHANIDAE

Compressed, torpedo-shaped body. Scales small, regular, grooved. Mouth small with no teeth. An accessory gill-structure present. 10-11 soft pelvic rays.

95. CHANOS CHANOS (Forskål) MILKFISH; SALMON HERRING

Mugil chanos Forskål, Descript. Animal., 1775, p. 74. *Leuciscus zeylonicus*, *Chanos salmoneus*. *Chanos chanos* Jordan and Evermann, Bull. U.S. Fish. Comm., xxiii, (1), 1905, fig. 10 (plate 6).

D. 14-16. A. 10-11. L. lat. 75-80. Head naked. Axillary scales above and below pectorals and ventrals. Maxillaries broad without supplemental bone. No teeth. Dorsal fin opposite ventrals. Caudal deeply forked. Greenish grey above, sides silvery. Anal and caudal margins dusky. Attains 6 feet. Coastal waters including lagoons, Gulf of Manaar.

Vaikka (S). Pal meen (T).

BOA-FISH

Family STOMIATIDAE

Elongate fish with short head and enormous fang-lined mouth. Long barbel below chin. Photophores present. Degenerate, black, snake-like fishes of deep water.

96. STOMIAS NEBULOSUS Alcock DUSKY BOA-FISH

Stomias nebulosus Alcock, Ann. Mag. Nat. Hist. (6), iv, 1889, p. 451; Alcock, Illust. Zool. Investigator, i, 1892, Pl. vii, fig. 1 (plate 10).

D. 16-17. A. 21. L. lat. 62. Barbel nearly as long as head. Origin of dorsal slightly behind that of anal. Large unequal curved fangs in both jaws. 16-17 photophores between branchiostegal rays and long series along ventral border of body. Uniform black with metallic gloss. Barbel and fins white. Deep water species from 500 fathoms in Gulf of Manaar.

FANG-FISH

Family CHAULIODONTIDAE

Elongate fish with large head and large mouth with great fangs. Large thin angular scales. Photophores in 2 ventral rows. Deep water fishes rising to surface at night.

97. CHAULIODUS SLOANI Bloch and Schneider SLOAN'S FANG-FISH

Chauliodus sloani Bloch and Schneider, Syst. Ichth., 1801, p. 430. *Chauliodus sloanii* Goode and Bean, Oceanic Ichthyol., 1895, Pl. xxx, fig. 115 (plate 10).

D. 6-7, placed far forward. Adipose dorsal above anal. A. 11-13. L. lat. 56-61. Tr. 7. G.R. 14-15. No chin barbel. 2 rows of small photophores along ventral border. Deep water species from 600 fathoms in Gulf of Manaar.

DEEP-SEA LANTERN FISHES, LIZARD-FISHES

Order SCOPELIFORMES

Dorsal and anal fins without spines. Adipose fin present. Deep-sea species with photophores.

1. No phosphorescent organs 2
   Photophores present .... *Family* SCOPELIDAE, p. 37
2. Maxilla narrow; less than 20 anal rays .... *Family* SYNODONTIDAE, p. 36
   Maxilla broad behind; more than 20 anal rays .... *Family* SUDIDAE, p. 37

LIZARD-FISHES

Family SYNODONTIDAE

Shallow water fishes with cylindrical bodies. Mouth large with numerous sharp teeth. Head lizard-like.

1. Inner pelvic ray about twice as long as outer 2
   Inner pelvic ray subequal to outer 3
2. Snout short and blunt .... *Trachinocephalus myops*
   Snout fairly pointed .... *Synodus indicus*
3. Pectoral with 12-13 rays .... *Saurida gracilis*
   Pectoral with 14-15 rays .... *Saurida tumbil*

98. TRACHINOCEPHALUS MYOPS (Bloch and Schneider)
SHORT-HEADED LIZARD-FISH

Salmo myops Bloch and Schneider, Syst. Ichth., 1801, p. 421. *Saurus myops* Bleeker, Atlas Ichth. Ind. Neerl., vi, 1872, Pl. cclxxviii, fig. 3 (plate 10).

D. 11-14. A. 15-16. L. lat. 54-58. Tr. $3\frac{1}{2}+(6\text{-}7)$. P. 12-13. Interorbital less than eye. Silvery yellow below, dark above, with faint grey irregular longitudinal stripes. Head and dorsal with irregular ocelli. Black shoulder spot. Fins light yellow. Attains 12 inches. Coastal waters and Gulf of Manaar.

99. SYNODUS INDICUS (Day) INDIAN LIZARD-FISH

Saurus indicus Day, Jour. Linn. Soc., Zool., xi, 1873, p. 526; Day, Fish. India, 1875-78, Pl. cxvii, fig. 4 (plate 10).

D. 12-13. A. 9-10. L. lat. 55-57. Tr. $3\frac{1}{2}$ above L. lat. P. 12-14. Silvery below, olive-brown above, mottled. Eye reddish. Fins pinkish, dorsal and lower caudal tips olive. Attains 10 inches. Coastal waters, trawling grounds and Gulf of Manaar.

100. SAURIDA GRACILIS (Quoy and Gaimard) SLENDER LIZARD-FISH

Saurus gracilis Quoy and Gaimard, Voyage Uranie, Zoologie, 1824, p. 224. *Saurida nebulosa* Bleeker, Atlas Ichth. Ind. Neerl. vi, 1872, Pl. cclxxvii, fig. 1 (plate 10).

D. 10-11. A. 9-10. L. lat. 45-49. Tr. $3\frac{1}{2}+6$. P. 12-13. Interorbital equal to eye. Brownish yellow, lighter below, with obscure mottlings and sometimes cross-bars on sides, also brownish stripes or patches on fins. Attains 12 inches. Coastal waters.

101. SAURIDA TUMBIL (Bloch) GREATER LIZARD-FISH

Salmo tumbil Bloch, Ausländ. Fische, ix, 1795, p. 112. *Saurida tumbil* Bleeker, Atlas Ichth. Ind. Neerl., vi, 1872, Pl. cclxxvii, fig. 4 (plate 10).

D. 11-13. A. 10-11. L. lat. 50-53. Tr. 4½-5 above L. lat. P. 14-15. Eye less than interorbital. Depth 6-7 in body length. Brownish, lighter below, mottled, with traces of cross-bars. Attains 18 inches. Coastal waters, trawling grounds and Gulf of Manaar.
Mudhu balla (S). Nai meen, Puna (T).

DEEP-SEA LIZARD-FISH

Family SUDIDAE

Deep water fishes usually with elongate body and short tail regions. Paired fins small and anal much longer than dorsal and with about twice as many rays. No phosphorescent organs present.

101a. STEMONOSUDIS ELONGATUS (Ege) DEEP-SEA LIZARD-FISH

Macroparalepis elongatus Ege, Vidensk. Medd. Dansk. Nat. Foren., xciv, 1933, p. 233.

D. 11. A. 50. P. 13. V. 9. Body region with 31-32 pigmented sections of peritoneum showing through body wall. Ventrals inserted below the interval between 6th and 7th sections, reckoned from behind. A rare species known only from an adolescent specimen taken in 500 fathoms to the SE. of Ceylon. 5 inches.

LANTERN FISHES

Family SCOPELIDAE

Deep water fishes with large mouth and feeble teeth. Body compressed. Scales deciduous. Photophores present, in definite groups, one on each scale.

102. DIAPHUS SPLENDIDUS (Brauer) LANTERN FISH

Myctophum (Nyctophus) splendidum Brauer, Zool. Anz., xxviii, 1904, p. 399. *Diaphus splendidus* Deraniyagala, Colour. Atlas Vert. Ceylon, i, 1952, Pl. iii (plate 10).

D. 13-15. A. 15-16. L. lat. 38-39. Luminous organs divided by a black septum. A large autorbital luminous organ, no suborbital photophore. First pectoral photophore and two suprapectoral photophores form obliquely ascending line to base of pectoral. No patches of luminous scales. A deep water species, coming to surface after dark. Colour bluish.

CARPS, BARBS, LOACHES AND CAT-FISHES

Order CYPRINIFORMES

Chiefly fresh water fishes representing two major groups which have in common a so-called Weberian apparatus, a bony structure which connects the ear with the swim bladder.

1. Skin covered with cycloid scales; mouth protractile; jaws toothless; no adipose fin .... (*Suborder* CYPRINOIDEI) 2

   Skin naked; mouth not protractile; intermaxillary with teeth; adipose fin generally present .... (*Suborder* SILUROIDEI) 3

2. No adipose fold on caudal peduncle; no supraorbital spine; eye free; barbels 4 or less .... *Family* CYPRINIDAE, p. 39

   Adipose fold on caudal peduncle; supraorbital spine present or absent; eye sub-cutaneous or free; barbels 6 or more .... *Family* COBITIIDAE, p. 48

3. Dorsal spineless. Anal very long 4

   Dorsal with a pungent spine 6

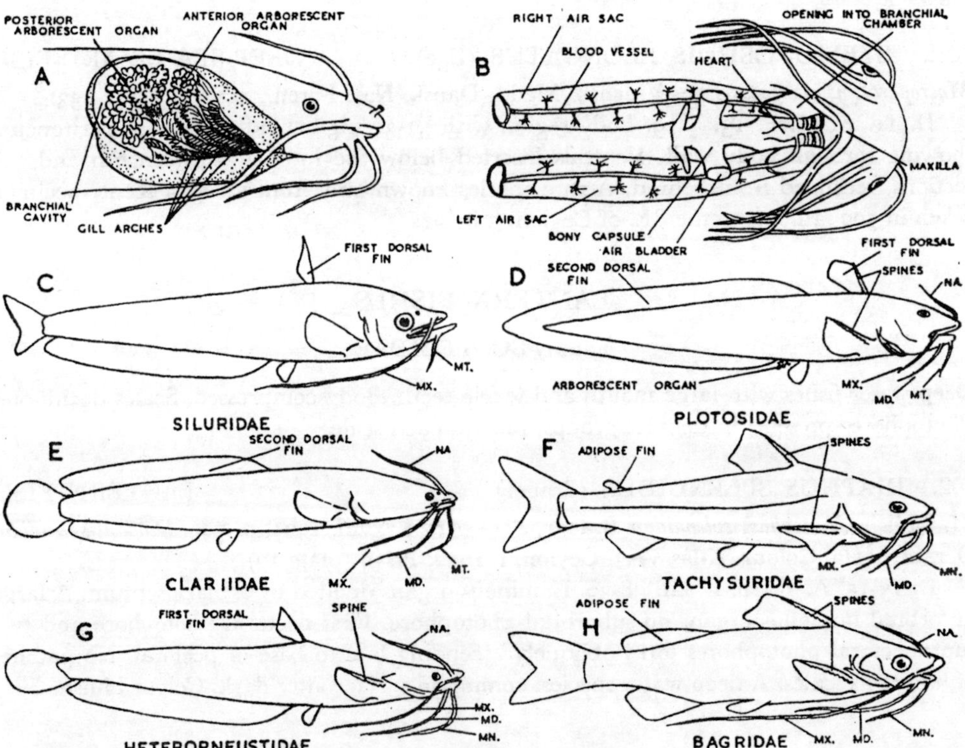

Fig. 6.

Explanation of characters of the Suborder Siluroidei (Cat-fishes). a.—Dissection of gill-chamber of *Clarias* showing accessory dendritic organ for aerial breathing. b.—Dissection of *Heteropneustes* from underneath showing lung-sacs used for aerial breathing. c-h.—Principal differences between the six component families of the Suborder.

4. Dorsal very short, rudimentary or absent; 2 pairs of barbels .... *Family* SILURIDAE, p. 49
   Dorsal with many rays; 4 pairs of barbels 5
5. Dorsal with 70 rays; branchial cavity with dendritic accessory respiratory organ
   .... *Family* CLARIIDAE, p. 49
   Dorsal with 7-8 rays; a long air sac extends back from branchial cavity
   .... *Family* HETEROPNEUSTIDAE, p. 50
6. Caudal pointed, united with procurrent caudodorsal .... *Family* PLOTOSIDAE, p. 50
   Caudal forked or truncate, without procurrent part 7
7. Nostrils close together, separated by valve without nasal barbel; gill-membranes connected with isthmus
   .... *Family* TACHYSURIDAE, p. 51
   Nostrils remote from each other, the posterior with a nasal barbel; gill-membranes free from isthmus
   .... *Family* BAGRIDAE, p. 55

CARPS, CARPLETS, BARBS, LABEOS, STONE-SUCKERS

Family CYPRINIDAE

Compressed scaly fishes. Head naked. Lateral line usually present. Mouth with upper border of intermaxillaries only. Anal and dorsal fins each with 4 spines. Fresh water fishes inhabiting streams, pools and tanks.

1. Abdomen compressed, ventral edge cultrate (*Sub-Family* ABRAMINAE) .... *Chela laubuca*
   Abdomen rounded 2
2. Lateral line in middle of tail (*Sub-Family* CYPRININAE)* 8
   Lateral line close to ventral edge (*Sub-Family* RASBORINAE)† 3
3. Barbels absent 4
   Barbels present 7
4. Lateral line absent .... *Horadandia atukorali*
   Lateral line present 5
5. Predorsal scales 11-17 6
   Predorsal scales 28-30 .... *Amblypharyngodon melettinus*
6. Lateral line reduced to 3-4 perforated scales; 25 scales in lateral series .... *Rasbora vaterifloris*
   Lateral line complete; 30-34 scales in lateral series .... *Rasbora daniconius*
7. No symphysial knob; 5 anal rays .... *Esomus danrica thermoicos*
   Symphysial knob present; 13-16 anal rays .... *Danio aequipinnatus*
8. Mouth terminal or subterminal 9
   Mouth markedly inferior 24
9. Postlabial groove continuous .... *Tor khudree longispinis*
   Postlabial groove interrupted in the middle 10
10. Dorsal with 17-21 rays 11
    Dorsal with 7-9 rays 13
11. Head with 4 barbels .... *Cyprinus carpio*
    Head without barbels 12
12. Pharyngeal teeth in 3 rows; L.lat. 40-43 .... *Catla catla*
    Pharyngeal teeth in single series; L.lat. 28-35 .... *Carassius carassius*

* Other fresh water Cyprinids introduced recently to Ceylon include *Ctenopharyngodon idellus* (Valenciennes) (Grass Carp), *Hypophthalmichthys molitrix* (Valenciennes) (Silver Carp), *Aristichthys nobilis* (Richardson) (Big Head), and *Puntius javanicus* (Bleeker) (Tawes).

† *Barilius bendelisis* (Ham.-Buch.) and *Barilius vagra* (Ham.-Buch.) are recorded from Ceylon but their occurrence requires confirmation.

13. Lateral line complete 14
    Lateral line incomplete 21
14. Dorsal spine smooth 15
    Dorsal spine serrate 19
15. Maxillary barbels absent; 3 bands across body .... *Puntius melanampyx sinhala*
    Maxillary barbels present 16
16. Dorsal spine strong; Tr. $4\frac{1}{2}+2\frac{1}{2}$ .... *Puntius dorsalis*
    Dorsal spine slender 17
17. Black spot over base of anal fin .... *Puntius filamentosus*
    Black spot behind base of anal fin or absent 18
18. Tr. $5\frac{1}{2}+3\frac{1}{2}$ .... *Puntius chola*
    Tr. $4\frac{1}{2}+2\frac{1}{2}$ .... *Puntius amphibius*
    Tr. $3\frac{1}{2}+(2\frac{1}{2}\text{-}3\frac{1}{2})$ .... *Puntius bimaculatus*
19. Barbels absent; 3 bands across body .... *Puntius nigrofasciatus*
    Rostral and maxillary barbels present 20
20. Lateral band from eye to caudal .... *Puntius pleurotaenia*
    Black spot on caudal peduncle .... *Puntius sarana*
21. Dorsal spine smooth; band on dorsal .... *Puntius vittatus*
    Dorsal spine serrate 22
22. Barbels absent 23
    Barbels present on maxillary .... *Puntius titteya*
23. Two bands across body .... *Puntius cumingi*
    Spot on shoulder and on caudal peduncle .... *Puntius ticto*
24. Chin without disc 25
    Chin with disc 27
25. Lateral rostral lobes present 26
    Lateral rostral lobes absent .... *Labeo dussumieri*
26. Origin of dorsal in middle of back .... *Labeo fisheri*
    Origin of dorsal nearer snout-tip than caudal .... *Labeo porcellus lankae*
27. Rostral fold with ventral median groove .... *Garra ceylonensis ceylonensis*
    Rostral fold without groove .... *Garra ceylonensis phillipsi*

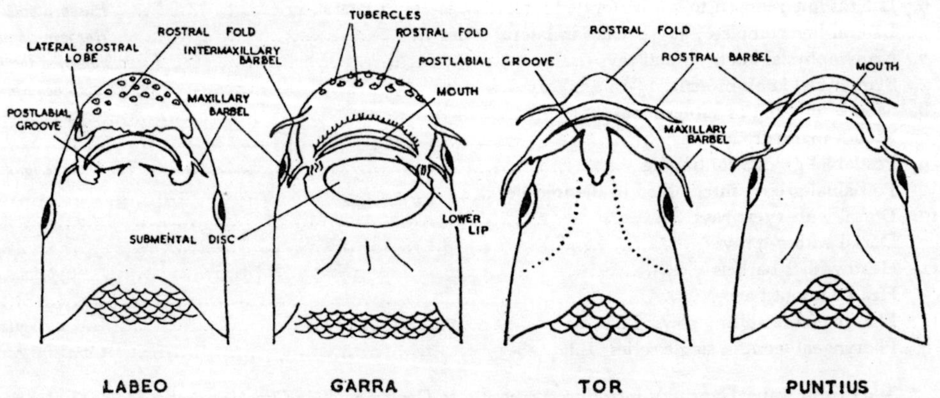

FIG. 7.

Under surfaces of heads of representative genera of Carp-like fishes, Family Cyprinidae, showing arrangement of barbels and specialized structures associated with the snout and mouth.

103. CHELA LAUBUCA (Hamilton-Buchanan) WINGED RASBORA

Cyprinus laubuca Hamilton-Buchanan, Fish. Ganges, 1822, p. 260. *Laubuca laubuca, Eüstira* or *Perilampus ceylonensis. Perilampus laubuca* Day, Fish. India, 1875-78, Pl. cli, fig. 5 (plate 7).

D. II, 7-8. A. III, 17-19. P. I, 9-11. L. lat. 33-35. Tr. ($6\frac{1}{2}$-$7\frac{1}{2}$)+($1\frac{1}{2}$-$3\frac{1}{2}$). Predorsals 17-19. Preventrals 22. Rhomboid compressed body with convex carinate belly. Bright metallic blue with hyaline fins. Dark shoulder spot; in young specimens a black spot on caudal peduncle. Attains $2\frac{1}{4}$ inches. Inhabits ponds, tanks, rivers in dry parts of low-country. Kara adaya, Tatu dandiya (S).

104. RASBORA VATERIFLORIS Deraniyagala VATERIA FLOWER RASBORA

Rasbora vaterifloris Deraniyagala, Spolia Zeylanica, xvi, 1930, p. 19, Pl. iii, fig. 2 (plate 7).

D. II, 8. A. II, 6. P. I, 11. L. lat. 25, only 3-4 perforated. Tr. 6. Predorsals 11-12. Preventrals 14. Oblong, compressed body increasing in depth with age. Characterized by incomplete lateral line. Greenish dorsally, blue on belly, head yellowish. Dorsal and caudal fins orange. Other fins hyaline. Attains $1\frac{1}{2}$ inches. Inhabits streams in Ratnapura district. Hal mal dandiya, Hal mal titteya (S).

105. RASBORA DANICONIUS (Hamilton-Buchanan) COMMON RASBORA

Cyprinus daniconius Hamilton-Buchanan, Fish. Ganges, 1822, p. 327. *Leuciscus* or *Rasbora dandia, Rasbora cephalotaenia*, or (?) *Rasbora buchanani. Rasbora daniconius* Day, Fish. India, 1875-78, Pl. cxlvi, fig. 2 (plate 7).

D. II, 6-7. A. II, 5. P. I, 13-14. Lateral line complete, 30-34. Tr. ($4\frac{1}{2}$-$6\frac{1}{2}$)+($1\frac{1}{2}$-$2\frac{1}{2}$). Predorsals 13-17. Preventrals 15-16. Elongate, rather rounded, compressed body. Dorsal profile more convex than ventral. Dorsally greenish-yellow. Sides silvery with blue-black lateral stripe, narrowly edged with gold. Fins pale orange. Attains 4 inches. Inhabits ponds, tanks, streams.
Kehel dandiya, Kiri dandiya, Kudu massa, Dandiya, Petiya (S).

106. AMBLYPHARYNGODON MELETTINUS (Valenciennes)
ATTENTIVE CARPLET

Leuciscus melettinus Cuvier and Valenciennes, Hist. Nat. Poiss., xvii, 1844, p. 304. *Amblypharyngodon grandisquammis* Jordan and Starks, Ann. Carnegie Mus., xi, 1917, Pl. xliv (plate 7).

D. II, 7. A. III, 5. P. I, 13-14. L. lat. incomplete ending after 11-18 scales. L. lat. 45-50. Tr. ($8\frac{1}{2}$-$10\frac{1}{2}$)+($4\frac{1}{2}$-$5\frac{1}{2}$). Predorsals 28-30. Body somewhat compressed. Mouth with knob in mandibular symphysis which develops prominent fleshy flaps during breeding season. Sides silvery, back brownish yellow, vertex bright green. Attains 3 inches. Inhabits ponds and streams.
Soreya (S).

107. HORADANDIA ATUKORALI Deraniyagala GREEN CARPLET

Horadandia atukorali Deraniyagala, J. Royal Asiatic Soc., xxxv, 1943, p. 158, fig. 1, redrawn (plate 7).

D. II, 6. A. III, 5. P. 10. L. lat. 22, no lateral line. Tr. 6-7. G.R. 12-14. No barbels. Pale yellowish green, sides silvery with pink reflections. Green mid-lateral line. Belly silvery. Fins hyaline. Attains 1 inch. Lives amongst weeds in fresh water ponds of coastal plains.
Horadandiya (S).

108. ESOMUS DANRICA THERMOICOS (Valenciennes) FLYING BARB

Nuria thermoicos Cuvier and Valenciennes, Hist. Nat. Poiss., xvi, 1842, p. 238. *Esomus danrica danio. Nuria danrica* Day, Fish. India, 1875-78, Pl. cxlv, fig. 8 (plate 7).

D. II, 6. A. III, 5. P. I, 10-12. Dorsal and anal spines smooth, weak and soft. L. lat. 31-34, complete. Tr. $5\frac{1}{2}+1\frac{1}{2}$. Predorsals 17-20. Preventrals 22-23. Barbels 2 on premaxillary, 0·75 in eye; 2 on maxillary reaching to tip of pectorals or beyond. Pale olive dorsally with greenish yellow lateral stripe above a steel blue one. Yellow or brown nuchal band. Belly and sides silvery. Paired fins hyaline, others pale yellow. Attains $2\frac{1}{2}$ inches. Inhabits ponds, tanks, ditches and can tolerate very muddy water during droughts.
Ravul dandiya (S).

109. DANIO AEQUIPINNATUS (McClelland) GIANT DANIO

Perilampus aequipinnatus McClelland, Indian Cyprinidae, Asiat. Res., xix (2), 1839, p. 393. *Perilampus malabaricus, Danio lineolatus,* or *micronema. Danio malabaricus* Day, Fish. India, 1875-78, Pl. cl, fig. 7 (plate 7).

D. II, 10-12. A. III, 13-16. P. I, 11. L. lat. 37-40. Tr. $(7\frac{1}{2}\text{-}8\frac{1}{2})+(1\frac{1}{2}\text{-}2\frac{1}{2})$. Predorsals 16-17. Preventrals 20-24. Rostral and maxillary pairs of barbels. Former 0·5 of eye, latter rudimentary or absent. Compressed oblong body. Bright metallic blue, greenish dorsally with 2-3 parallel lateral lines of yellow which break into spots anteriorly. Fins vary from yellow to deep orange-red. Dark shoulder patch. Attains 3 inches. Inhabits small streams ponds, tanks.
Dan kola salaya, Ratha kailaya, Salaya, Suddara (S).

110. CYPRINUS CARPIO Linnaeus MIRROR CARP

Cyprinus carpio Linnaeus, Syst. Nat., ed. 10, 1758, p. 320. *Cyprinus carpio* var. *specularis* Jones and Sarojini, Jour. Bombay Nat. Hist. Soc., l, 1952, fig. 4 (plate 7).

D. III, 17-22. A. III, 5. P. I, 15. L. lat. 35-39. Tr. $5+7\frac{1}{2}$. Mouth terminal with 4 barbels. Dorsal long, last simple ray ossified and ossified behind. Scales large. Lateral line complete. Body proportions and fins very variable. Olivaceous with silvery or gold sides. Introduced into fresh waters of Ceylon.

111. CARASSIUS CARASSIUS (Linnaeus)
GOLDEN-, CRUCIAN- OR PRUSSIAN CARP

Cyprinus carassius Linnaeus, Syst. Nat., ed. 10, 1758, p. 322. *Carassius vulgaris* Day, Fish. Gt. Brit. and Ireland, ii, 1884, Pl. cxxx, fig. 1 (plate 7).

D. 15-25. A. 8-9. V. 9-10. P. 13-18. L. lat. 28-35. Tr. (6-7)+9. Mouth terminal without barbels. Dorsal long, last ray simple, ossified and serrated behind. Scales large. Lateral line complete. Olive to brown above, yellow-grey on sides and white below. Fins dark with bluish or reddish tinge. Closely related to the ornamental goldfish, *C. auratus* (Linnaeus). 18 inches. Introduced to Ceylon in 1930.

112. CATLA CATLA (Hamilton-Buchanan) CATLA

Cyprinus catla Hamilton-Buchanan, Fish. Ganges, 1822, p. 287. *Catla buchanani* Day, Fish. India, 1875-78, Pl. cxxxiv, fig. 5 (plate 7).

D. III-IV, 14-16. A. III, 5. P. 21. L. lat. 40-43. Tr. $7\frac{1}{2}+9$. Mouth terminal without barbels. Greyish above, silvery on sides. Fins dark, sometimes nearly black. Attains 6 feet. Suitable for culture in tanks. Introduced into fresh waters of Ceylon.

113. TOR KHUDREE LONGISPINIS (Günther) MAHSIER

Barbus longispinis Günther, Cat. Fish., vii, 1868, p. 132. *Cyprinus, Barbus* or *Labeobarbus tor, Labeobarbus progenius. Tor tor* Deraniyagala, Spolia Zeyl., xvi, 1930, Pl. iv (plate 7).

D. IV, 9. A. III, 5. P. I, 15-17. L. lat. 23-24. Tr. $3\frac{1}{2}+2\frac{1}{2}$. Predorsals 8-9. Preventrals 13-14. Lateral line complete. Rostral barbels extend to eye. Maxillary barbels longer. Males smaller than females. Males have larger labial flap. Head and back olivaceous with blue and bronze reflections. Lighter below. Dorsal and caudal fins oliveaceous, other fins orange. In Ceylon attains 24 lb. Common in hill streams. Popular game fish for sportsmen.
Hora paleya, Lehella (S). Pu meen (T).

114. PUNTIUS MELANAMPYX SINHALA (Duncker) BLACK-BANDED BARB

Barbus singhala Duncker, Mitteil. Naturhist. Mus. Hamburg, xxix, 1912, p. 263. *Puntius singhala* Deraniyagala, Spolia Zeyl., xvi, 1930, Pl. ii, fig. 1 (plate 8).

D. III, 8. A. III, 5. P. I, 14-16. L. lat. complete, 20-22. Tr. $4\frac{1}{2}+2\frac{1}{2}$. Barbels absent. Elongate body with convex dorsal profile. Mouth subterminal, small. Light red, brownish dorsally. Three dark brown or black vertical bands across body, below dorsal, above anal and at base of caudal. Dorsal and caudal fins reddish with black tips. Attains $1\frac{1}{8}$ inches. Inhabits streams and ponds.
Goma titteya, Iri titteya, Pulli titteya, Thirelliya, Titha kayada (S).

115. PUNTIUS DORSALIS (Jerdon) LONG-SNOUTED BARB

Systomus dorsalis Jerdon, Madras Journ. Lit. and Sci., xv, 1849, p. 314. *Puntius innominatus, Leuciscus binotatus, Barbus tetraspilus,* or (?) *burmanicus. Labeo dorsalis. Barbus dorsalis* Day, Fish. India, 1875-78, Pl. cxlii, fig. 2 (plate 8).

D. IV, 8. A. III, 5. P. I, 11-13. L. lat. complete, 23-25. Tr. $4\frac{1}{2}+2\frac{1}{2}$. Predorsals 8-10, preventrals 12-14. Short pair of maxillary barbels. Body rounded elongate. Mouth subterminal. Snout long, pointed. Olive dorsally, sides silvery, belly white. Fins orange. Black blotch on tail and sometimes a black band along lateral line. Attains $9\frac{1}{2}$ inches. Inhabits rivers, streams, ponds and lakes.
Bin gattaya, Honda petiya, Kattu kuriya, Kattu petiya, Kendhaya, Kureya (S).
Kendhai (T).

116. PUNTIUS FILAMENTOSUS (Valenciennes) FILAMENTED BARB

Leuciscus filamentosus Cuvier and Valenciennes, Hist. Nat. Poiss., xvii, 1844, p. 96. *Barbus filamentosus, Leuciscus mahecola. Barbus mahicolor* Day, Fish. India, 1875-78, Pl. cxl, fig. 5 (plate 8).

D. IV, 8. A. III, 5. P. I, 15. L. lat. complete, 22-23. Tr. 4½+2½. Predorsals 7. Preventrals 12-14. A pair of rudimentary maxillary barbels. Oblong compressed body with equally convex dorsal and ventral profiles. Tips of dorsal rays filamentous in adult males. Olive dorsally with emerald green above lateral line. Sides silvery or yellow. Caudal fin orange with white tips. Ventrals green, other fins orange. Large blotch above posterior part of anal. Attains 4½ inches. Inhabits rivers and tanks.
Dankolla petiya, Goma kadeya, Goma kolla, Ipilli kadaya (S).

117. PUNTIUS CHOLA (Hamilton-Buchanan) GREEN BARB

Cyprinus chola Hamilton-Buchanan, Fish. Ganges, 1822, p. 312. *Leuciscus* or *Barbus thermalis. Barbus chola* Day, Fish. India, 1875-78, Pl. cxlii, fig. 4 (plate 8).

D. IV, 8. A. III, 5. P. I, 15. L. lat. complete, 24-27. Tr. 5½+3½. Predorsals 9-10. Preventrals 12. A short pair of maxillary barbels. Body rounded in outline. Mouth terminal. Pale olive dorsally with green scale pockets, opercles golden, belly and sides silvery. Dorsal fin pale orange with 2 rows dark blotches. Black spot equal to eye over 22-24 L. lat. scale. Caudal orange, pectoral and ventrals hyaline to orange. Attains 3½ inches. Inhabits streams and tanks in low-country, including hot-springs.
Kotu ipilla, Kotu petiya (S).

118. PUNTIUS AMPHIBIUS (Valenciennes) SCARLET-BANDED BARB

Capoëta amphibia Cuvier and Valenciennes, Hist. Nat. Poiss., xvi, 1842, p. 478. (?) *Barbus layardi. Barbus amphibius* Day, Fish. India, 1875-78, Pl. cxlii, fig. 8 (plate 8).

D. IV, 8. A. III, 5. P. I, 15. L. lat. complete, 23-24. Tr. 4½+2½. Predorsals 7-8. Preventrals 11-13. A pair of maxillary barbels, shorter than eye. Elongate body, more convex dorsally than ventrally. Mouth small, subterminal. Dorsally pale olive to blue-green. Sides silvery with yellow reflections. Dark blotch at end of lateral line. Sexually mature fish with scarlet lateral band from eye to tail. Dorsal fin orange with oblique band. Attains 3 inches. Inhabits streams and ponds in low-country.
Ipilli kadaya, Madda ipilla (S).

119. PUNTIUS BIMACULATUS (Bleeker) TWO-SPOT BARB

Gnathopogon bimaculatus Bleeker, Verh. Nat. Holl. Maatsch. Haarlem (2), xx, 1864, p. 17. *Barbus bimaculatus. Gnathopogon bimaculatus* Bleeker, *loc. cit.*, Pl. iv, fig. 1 (plate 8).

D. IV, 7-8. A. III, 5. P. I, 15. L. lat. complete or interrupted after 16-17 scales, 24-25. Tr. 3½+(2½-3½). Predorsals 8-10. Preventrals 12-14. A pair of maxillary barbels, less than eye. Elongate body more convex dorsally than ventrally. Mouth small, subterminal with thick lower lip. Olive green dorsally with silvery sides. Dorsal with black spot at base of 3rd to 8th rays. Black spot on lateral line on 23-24 scales. Sometimes a black spot also on

anal. A crimson lateral band on spawning individuals. Attains 2¼ inches. Inhabits ponds and streams from sea level to mountain torrents.
Ipilli kadaya (S).

120. PUNTIUS NIGROFASCIATUS (Günther)
THREE-BANDED CRIMSON BARB; BLACK RUBY

Barbus nigrofasciatus Günther, Cat. Fish., vii, 1868, p. 155. *Rasbora nigrifasciata. Puntius nigrofasciatus* Deraniyagala, Spolia Zeyl., xvi, 1930, Pl. iii, fig. 1 (plate 8).

D. III, 8. A. III, 5. P. I, 12. Lateral line complete, 20-22. Tr. 4½+3½. Predorsals 8. Preventrals 11. Barbels absent. Ovate fish with dorsal and ventral convexities increasing with age. Mouth subterminal, oblique. Head orange red. Body drab olive green with golden spots on some of scale rows and 3 black vertical cross-bands. Dorsal and anal fins blackish. When stimulated males assume a glowing dark cherry red on forward part of body and fins. Attains 2¼ inches. Inhabits shady parts of slow flowing streams.
Bulath sapeya, Manamalaya (S).

121. PUNTIUS PLEUROTAENIA Bleeker SIDE-STRIPED BARB

Puntius (Barbodes) pleurotaenia Bleeker, Verh. Nat. Holl. Maatsch. Haarlem (2), xx, p. 13. *Barbus pleurotaenia. Puntius pleurotaenia* Deraniyagala, Spolia Zeyl., xvi, 1930, Pl. v, fig. 2 (plate 8).

D. IV, 8. A. III-IV, 4-5. P. I, 15. L. lat. complete, 28-30. Tr. (4½-5½)+2½. Predorsals 9-10. Preventrals 16-17. Rostral barbels as long as eye, maxillary barbels longer. Elongate body, more convex dorsally than ventrally. Back green-brown, rest of body silvery. Dark band from eye to tail along lateral line; below this sometimes a row of black dots and a yellowish longitudinal streak above. Dorsal pale orange, other fins pale yellow. Attains 6 inches. Inhabits mountain streams.
Hitha massa (S).

122. PUNTIUS SARANA (Hamilton-Buchanan) OLIVE BARB

Cyprinus sarana Hamilton-Buchanan, Fish. Ganges, 1822, p. 307. *Cyclocheilichthys* or *Barbus pinnauratus*, *Puntius (Barbodes)* or *Barbus chrysopoma. Barbus sarana* Day, Fish. India, 1875-78, Pl. cxxxvi, fig. 2 (plate 8).

D. IV, 8. A. III, 5. P. I, 12-16. Lateral line complete, 26-29. Tr. (4½-5½)+3½. Predorsals 8-10. Preventrals 15-17. Maxillary and rostral barbels longer than eye. Oblong deep body. Mouth terminal. Dorsally olive, sides with silver and gold reflections. Barbels red-brown. Cheeks golden. Fins dusky-brown to orange. Dark caudal blotch behind anal. Young with 2 or 3 additional spots. Attains 12 inches. Inhabits streams, rivers, ponds, lakes and tanks.
Mas petiya, Petiya, Vellan kola petiya (S).

123. PUNTIUS VITTATUS Day STRIPED BARB

Puntius vittatus Day, Proc. Zool. Soc., 1865, p. 303. *Barbus vittatus* Day, Fish. India, 1875-78, Pl. cxliv, fig. 2 (plate 8).

D. II, 8. A. II, 5. P. I, 11. L. lat. incomplete, disappearing after 3-6 scales, 20-22. Tr. $3\frac{1}{2}+3\frac{1}{2}$. Predorsals 7-8. Preventrals 11-12. Barbels only on maxillaries, often absent. Oblong body. Mouth terminal. Dorsally pale olive, silvery or metallic blue sides. Yellowish green median band from vertex to caudal and a similar lateral band. Dorsal with black band. Fins orange or yellow. Black spots on base of tail and near vent. Attains 2 inches. Inhabits ponds, streams and tanks in low-country. Withstands muddy water during drought.
Bandi titteya (S).

124. PUNTIUS TITTEYA Deraniyagala CHERRY BARB; CRIMSON CARPLET

Puntius titteya Deraniyagala, Spolia Zeylanica, xv, 1929, p. 73, Pl. xxi (plate 8).

D. III, 7. A. III, 5. P. I, 10. L. lat. 19-20, incomplete, disappearing after 3-4 scales. Tr. $3\frac{1}{2}+3\frac{1}{2}$. Predorsals 7-8. Preventrals 11. Barbels on maxillary only. Body elongate, dorsal convexity greatest anteriorly. Mouth small subterminal. Colour variable. Dorsally purplish brown, sides silvery with sometimes a blush of deep red in males. A brown lateral stripe from eye to caudal fork, above which is another of yellow. Fins yellowish hyaline to deep red. Attains $1\frac{7}{8}$ inches. Inhabits rills in foot-hills.
Kondhaya, Dola titteya, Lai titteya, Vairan titteya (S).

125. PUNTIUS CUMINGI (Günther) CUMING'S TWO-BANDED BARB

Barbus cumingii Günther, Cat. Fish., vii, 1868, p. 155. *Puntius phytunio* Bleeker. *Puntius cumingi* Deraniyagala, Spolia Zeyl., xvi, 1930, Pl. ii, fig. 2 (plate 8).

D. II, 8. A. III, 5. P. I, 11. L. lat. 19-21, incomplete, disappearing after 5-8 scales. Tr. $3\frac{1}{2}+3\frac{1}{2}$. Predorsals 8. Preventrals 11. Body ovate with convex dorsal and ventral profiles. Barbels absent. Mouth small, subterminal. Body silvery with yellow reflections on shoulders. Two rhomboid black patches on each side, one above pectoral, other above end of anal. Dorsal and ventral orange. Dorsal with 2 horizontal rows of 5-7 black spots on dorsal. Other fins yellow. Attains $1\frac{1}{2}$-2 inches. Inhabits hill streams.
Pothaya (S).

126. PUNTIUS TICTO (Hamilton-Buchanan) FIRE-FIN BARB

Cyprinus ticto Hamilton-Buchanan, Fish. Ganges, 1822, p. 314. *Barbus ticto* Day, Fish. India, 1875-78, Pl. cxliv, fig. 7 (plate 8).

D. II, 8. A. III, 5. P. I, 11. L. lat. 20-22, incomplete, disappears after 7-15 scales. Tr. $4\frac{1}{2}+(2\frac{1}{2}-3\frac{1}{2})$. Predorsals 8-9. Preventrals 12. Barbels absent. Body oblong. Mouth small, terminal. Pale olive dorsally with silver sides. Black spots under 3rd-4th lateral line scale and on 16th-17th lateral line scales. Two rows of indistinct spots on dorsal. Male with arched band of red on dorsal fin, becoming more vivid in breeding season. Attains $1\frac{1}{2}$ inches or more. Inhabits ponds and tanks chiefly in dry zone.

127. LABEO DUSSUMIERI (Valenciennes) COMMON LABEO

Rohita dussumieri Cuvier and Valenciennes, Hist. Nat. Poiss., xvi, 1842, p. 258. *Labeo (Morulius) dussumieri*. *Labeo dussumieri* Deraniyagala, Col. Atlas Vert. Ceylon, i, 1952, Pl. ix (plate 7).

D. III, 12-13. A. III, 5. P. I, 15. L. lat. 48-53, complete. Tr. $(8\frac{1}{2}-9\frac{1}{2})+(6\frac{1}{2}-7\frac{1}{2})$. Predorsals 22-24. Preventrals 32-35. Rostral and maxillary barbels present, rostrals smallest. Oblong body, 2-3 rows large tubercles on snout and preorbital. Dorsally golden-brown, paler beneath and pinkish in thoracic region. Fins golden brown. A large dark spot on caudal peduncle. Attains 14 inches. Very common in larger tanks.
Hiri kanaya, Gan kanaya (S).

128. LABEO FISHERI Jordan and Starks

GREEN LABEO, MOUNTAIN LABEO

Labeo fisheri Jordan and Starks, Ann. Carnegie Mus., xi, 1917, p. 436. *Labeo (Morulius) gadeya* Deraniyagala, Spolia Zeyl., xv, 1929, Pl. xxii (plate 7).

D. III, 10-12. A. III, 5. P. I, 15. Paired fins inserted low down. Scales rudimentary on chest. L. lat. 40-42, complete. Tr. $7\frac{1}{2}+5\frac{1}{2}$. Predorsals 17-18. Preventrals 30-35. Maxillary and rudimentary mental barbels, hidden in post-labial grooves. Thick set body. Well-developed tubercles on snout and cheeks. Olive dorsally, sides golden with red reflections. A greenish lateral stripe along upper half of body. Dark brown spot on caudal peduncle. Dorsal and paired fins olive. Caudal and anal from olive to bright orange. Attains 15 inches. Inhabits mountain streams.
Gadaya, Kalu gadaya, Vali gadaya (S).

129. LABEO PORCELLUS LANKAE Deraniyagala — ORANGE-FIN LABEO

Labeo porcellus lankae Deraniyagala, Coloured Atlas Vertb. Ceylon, i, 1952, p. 41. *Labeo* or *Morulius porcellus*, *Labeo calabasu*. *Labeo porcellus lankae* Deraniyagala, *loc. cit.*, Pl. x (plate 7).

D. III, 11-12. A. III, 5. P. I, 16. L. lat. 40-42, complete. Tr. $(7\frac{1}{2}-8\frac{1}{2})+(5\frac{1}{2}-6\frac{1}{2})$. Predorsals 14-17. Preventrals 23-25. Maxillary barbels present, rostral sometimes absent. Oblong body. Small tubercles on snout and preorbital. Olive dorsally, white ventrally and pinkish on thorax. Fins vermilion, caudal edged with olive. Dark blotch on caudal peduncle. Attains 15 inches. Inhabits tanks in dry zone.
Tambala vanna, Tambalaya (S).

130. GARRA CEYLONENSIS Bleeker — STONE-SUCKER

Garra ceylonensis Bleeker, Versl. Med. Akad. Amsterdam, xv, 1863, p. 8. *Discognathus lamta*, (?) *Crossochilus reba*. *Garra ceylonensis* Deraniyagala, Spolia Zeyl., xvi, 1930, Pl. v, fig. 1 (plate 7).

D. III, 8-9. A. II, 5. P. I, 9-12. Chest scales rudimentary or absent. L. lat. 33-35, complete. Tr. $4\frac{1}{2}+(2\frac{1}{2}-3\frac{1}{2})$. Predorsals 9-12. Preventrals 15-20. Intermaxillary and maxillary barbels present. Body elongate. Lower lip unites with submental disc to form an adhering organ. Olive-greenish dorsally, silvery yellow underneath. A dark shoulder spot and a purplish longitudinal lateral streak. Fins yellowish with rays orange. Attains 6 inches. Inhabits flowing streams, more commonly in hills than plains. There are two subspecies differing in minor details—*Garra ceylonensis ceylonensis* Deraniyagala and *Garra ceylonensis phillipsi* Deraniyagala.
Gal panderuwa, Gal pandiya, Pandiya (S). Kal koravai (T).

LOACHES

Family COBITIIDAE

Air bladder enclosed in bone. Barbels 6-8. Usually suborbital or preorbital spine. Inferior mouth. Pharyngeal teeth uniserial. Scales present or absent. Adipose dorsal fold on tail.

1. Erectile suborbital spine present .... *Lepidocephalus thermalis*
   No spine on cheeks 2
2. Body scaleless .... *Noemacheilus notostigma*
   Body scaly 3
3. Pectorals reaching ventrals; dorsal base longer than pectorals .... *Noemacheilus botia botia*
   Pectorals not reaching ventrals; dorsal base shorter than pectorals.. *Noemacheilus botia aureus*

131. LEPIDOCEPHALUS THERMALIS (Valenciennes) LESSER LOACH

Cobitis thermalis Cuvier and Valenciennes, Hist. Nat. Poiss., xviii, 1846, p. 78. *Lepidocephalichthys thermalis. Leptocephalus thermalis* Deraniyagala, Col. Atlas Vertb. Ceylon, i, 1952, Pl. xii, fig. 1 (plate 13).

D. II, 6. A. II, 5. P. I, 6. L. lat. absent. Scales small or inconspicuous. Rostral, intermaxillary, maxillary and mental pairs of barbels. Yellow with iridocytes on head. Brown spots and reticulations on head, back and sides. 11-12 irregular brown blotches along lateral line. Caudal with black spot and 6 V-shaped bands. Attains 2 inches. Inhabits quiet pools up to 1600 feet, including hot springs.

Ahirava, Ahiraya, Pulli ahirava (S). Aiarai (T).

132. NOEMACHEILUS NOTOSTIGMA Bleeker SPOTTED LOACH

Nemacheilus notostigma Bleeker, Versl. Akad. Amsterdam, xv, 1863, p. 254. *Noemacheilus notostigma* Deraniyagala, Col. Atlas Vertb. Ceylon, i, 1952, Pl. xii, fig. 3 (plate 13).

D. III, 6-8. A. III, 5. P. I, 7-9. Scales rudimentary, L. lat. incomplete. Rostral, maxillary and intermaxillary pairs of barbels. Olive green dorsally, sides yellow. 13-14 broad vertical dark bands on sides. Black vertical streak at base of caudal which has 4 indistinct bars. Fins yellowish with 2 rows of black spots on dorsal. Attains $2\frac{1}{4}$ inches. Inhabits mountain streams above 600 feet.

Gomera ahirava, Kandhu ahirava, Pol ahirava (S).

133. NOEMACHEILUS BOTIA (Hamilton-Buchanan) STRIPED LOACH

Cobitis botia Hamilton-Buchanan, Fish. Ganges, 1822, p. 350. *Noemacheilus botia* Deraniyagala, Col. Atlas Vertb. Ceylon, i, 1952, Pl. xii, fig. 2 (plate 13).

D. III-IV, 10-11. A. III, 5. P. I, 10-12. L. lat. complete or incomplete. Tr. (16-18) + (11-15). Predorsals 52. Rostral, maxillary and intermaxillary barbels. Brownish yellow with about 17 broad black vertical bars. A black white-edged ocellus on upper part of caudal peduncle. Dorsal with 5 rows of black spots, caudal with 7 V-shaped black bars, yellowish. Other fins yellow to orange. Two subspecies differing in minor details—*Noemacheilus botia botia* Deraniyagala, $1\frac{1}{2}$ inches inhabiting small streams in low-country up to

1000 feet, and *Noemacheilus botia aureus* Day, 2½ inches also inhabiting small streams in low-country.
Pol ahirava, Vairan ahirava, Vairan pol ahirava (S).

FRESH WATER CAT-FISH

Family SILURIDAE

Elongate compressed body without scales. Head conical or with depressed snout, covered with skin. Head with maxillary and mandibular barbels. No adipose fin. Anal long with numerous rays. Pectorals with spine, dorsal without. Caudal fin forked. Inhabits fresh waters.

Mouth subterminal, ends behind eye .... *Wallago attu*
Mouth superior, ends in front of eye .... *Ompok bimaculatus*

134. WALLAGO ATTU (Bloch and Schneider) FRESH WATER SHARK

Silurus attu Bloch and Schneider, Syst. Ichth., 1801, p. 378. *Wallagonia attu. Wallago russellii* Bleeker, Atlas Ichth. Ind. Neerl., ii, 1862, Pl. lxxxvi, fig. 1 (plate 10).

D. I, 4. A. 79-81. P. I, 13-14. Maxillary barbels reach ventral tip in adult or anal in young. Mental barbels 2·5-3·5 times eye. Dorsally greyish-brown, head purplish, sides leaden or silvery white with broad purplish reflection under lateral line of abdomen. Fins pale yellow. Attains 5 feet. Inhabits deep water of rivers and tanks in low-country and hills. Predatory.
Maha vallaya, Vallaya (S).

135. OMPOK BIMACULATUS (Bloch) BUTTER CAT-FISH

Silurus bimaculatus Bloch, Nat. Ausl. Fische, viii, 1794, p. 24. *Callichrous ceylonensis. Callichrous bimaculatus* Bleeker, Atlas Ichth. Ind. Neerl., ii, 1862, Pl. lxxxvii, fig. 3 (plate 10).

D. I, 3-4. A. 61-66. P. I, 12-15. Maxillary barbels reach pectoral tip or 6th anal ray. Mental barbels 0·5-2 times eye. Dorsally pale bluish olive, sides dull silvery white. A yellow shoulder patch which becomes brown after death. Opercles shot with gold, chest pink, fins yellow, lower caudal lobe silvery, base of anal purplish. Attains 16 inches. Inhabits fresh water.
Kokussa, Pena valapotta, Valapotta (S).

FRESH WATER CAT-FISH

Family CLARIIDAE

Elongate body. Head depressed with top and sides covered with osseus plates, forming chamber containing a dendritic organ for aerial breathing. Four pairs of barbels. Pectorals with spine, dorsal without. No adipose fin. Caudal fin rounded.

136. CLARIAS TEYSMANNI BRACHYSOMA (Günther)
TEYSMANN'S SPOTTED CAT-FISH

Clarias brachysoma Günther, Cat. Fish., v, 1864, p. 19. *Clarias teysmanni, teijsmanni* or *magur. Clarias batrachus* (closely related form) Bleeker, Atlas Ichth. Ind. Neerl., ii, 1862, Pl. xcviii, fig. 2 (plate 10).

D. 70. A. 53-60. P. I, 8. Nasal barbels shorter than head, maxillaries equal to head. Mandibulars reach maxilla, ventrals not so far. Supra-branchial chamber with two dendritic respiratory organs arising from 2nd and 4th gill-arches. Ventrals behind origin of dorsal. Olive green, brown or yellow with small irregular dark spots as large as eye. Young dark brown or copper red. Attains 12 inches but some as large as 8 lb. Common in muddy streams and ponds of low-country and hills. As an air breather comes on land during rain or at night.
Kaha magura, Magura, Vel magura (S).

FRESH WATER CAT-FISH

Family HETEROPNEUSTIDAE

Elongate, compressed body. Head depressed with top and sides covered with osseus plates expanded to form supra-branchial chambers into which open elongate lung-like sacs which extend under dorsal muscles almost to tail and used for aerial breathing. Four pairs of barbels. Pectorals with spine, dorsals without. No adipose fin. Caudal fin rounded.

137. HETEROPNEUSTES FOSSILIS (Bloch) STINGING CAT-FISH

Silurus fossilis Bloch, Ichthyologie, 1797, pl. 370, fig. 2. *Saccobranchus microcephalus. Saccobranchus* or *Heteropneustes microps. Saccobranchus fossilis* Day, Fish. India, 1875-78, Pl. cxiv, fig. 1 (plate 10).

D. 7-8. A. 68-78. P. I, 7. Nasal and mental barbels reach tip of pectoral spine, maxillaries reach beyond pectorals but not as far as ventral fins, mandibulars slightly shorter. Copper red in young. Dark brown in adult with two lateral yellow bands. Ventrally lighter. Attains 10 inches. Inhabits ponds and tanks in low-country, sometimes entering brackish water. Pectoral spines inflict bad wounds.
Hunga, Kaha hunga, Lai hunga, Vel hunga (S). Shunken (T).

CAT-FISH EELS

Family PLOTOSIDAE

Elongate, eel-like body with tail tapering to a point. No scales. Head depressed. Dorsal and pectorals with serrated spines. No adipose fin. 4 pairs of barbels. An aborescent organ in a pit behind vent, movable, of unknown function.

Dorsal with 130-140 rays; nasal barbel reaches beyond eye .... *Plotosus canius*
Dorsal with 80-100 rays; nasal barbel does not reach hindborder of eye .... *Plotosus anguillaris*

138. PLOTOSUS CANIUS Hamilton-Buchanan CANINE CAT-FISH EEL

Plotosus canius Hamilton-Buchanan, Fish. Ganges, 1822, p. 142; Bleeker, Atlas Ichth. Ind. Neerl., ii, 1862, Pl. xcvi, fig. 6 (plate 10).

D. I, 4-5; 130-140. A. 106-118. P. I, 11-12. Lips and chin fleshy, papillate. 4 pairs of barbels. G.R. 6+17. Back smoke grey to drab grey, below soiled creamy-buff. Barbels grey. Fins grey, first dorsal and pectorals darker. Attains occasionally 3 feet. Marine, entering rivers and brackish waters.

Kalapu magura, Kana magura (S). Keduthal (T).

139. PLOTOSUS ANGUILLARIS (Bloch) STRIPED CAT-FISH EEL

Platystacus anguillaris Bloch, Nat. Ausl. Fische, viii, 1794, p. 61. *Plotosus lineatus. Plotosus arab* Bleeker, Atlas Ichth. Ind. Neerl., ii, 1862, Pl. xcv, fig. 2 (plate 10).

D. I, 4-5; 80-100. A. 70-77. P. I, 10-11. Lips thick, papillose. 4 pairs of barbels. G.R. 8+(23-24). Uniform brownish, vertical fins with blackish edges. Young with longitudinal line from snout along each side to caudal; a second below eye, above pectorals, sometimes reaching tail. Attains 30 inches. Marine.

Mudhu hunga (S). Irung keleru, Kadal schunguan (T).

ESTUARINE CAT-FISH, SALMON CAT-FISH

Family TACHYSURIDAE*

Elongate body without scales. Head depressed or conical, usually covered with conspicuous bony plates. Adipose fin present, opposite anal. Anal with 14-26 rays. Dorsal and pectorals with serrated pungent spines. Caudal fin forked. Head generally with maxillary, mandibular and mental barbels. Mostly estuarine, inhabiting mouths and tidal parts of rivers and adjacent coastline. Fin spines can inflict dangerous wounds.

1. Maxillary and mandibulary barbels present 3
   Only a pair of stiff, osseous, maxillary barbels present 2
   Only a pair of small mandibulary barbels present .... *Batrachocephalus mino*
2. Head $4\frac{1}{4}$-$4\frac{1}{2}$ in total length; greatest width of head equals its length behind angle of mouth .... *Osteogeniosus militaris*
   Head $4\frac{3}{4}$ in total length; greatest width of head equals half its length .... *Osteogeniosus sthenocephalus*
3. Teeth on palate villiform, conic or pointed 4
   Teeth on palate granular 9
4. Teeth in 1 group on each side of palate 5
   Teeth in 2 groups on each side in a transverse row on palatines .... *Hexanematichthys sona*
   Teeth in 3 groups on each side of palate, forming a large triangular patch 8
5. Head shields smooth; pectoral spine as long as dorsal spine .... *Tachysurus venosus*
   Head shields granular 6
6. Palatine teeth in 2 widely separated patches 7
   Palatine teeth in 2 large patches .... *Tachysurus maculatus*
7. Pectoral spine shorter than dorsal .... *Tachysurus caelatus*
   Pectoral spine longer than dorsal .... *Tachysurus subrostratus*

* *Arius tenuispinis* Day has been doubtfully listed from Ceylon.

8. Head 4½-5 in length; maxillary barbels reach base of pectoral fin .... *Netuma thalassinus*
   Head 5¼ in length; maxillary barbels reach middle of pectoral fin .... *Netuma serratus*
9. Palatine teeth in 4 groups; 14-16 anal rays .... *Ariodes dussumieri*
   Palatine teeth in 2 groups; 17-19 anal rays 10
10. Maxillary barbels longer than head .... *Pseudarius platystomus*
    Maxillary barbels shorter than or subequal to head 11
11. Eye 4½-5 in head; maxillary barbels shorter than head; palatine teeth reduced to small oval posterior patches or absent .... *Pseudarius jatius*
    Eye 6 in head; maxillary barbels subequal to head; palatine teeth in convex, oblong-ovate patches, slightly convergent behind .... *Pseudarius jella*

140. BATRACHOCEPHALUS MINO (Hamilton-Buchanan)

FROG-HEADED CAT-FISH

Ageneiosus mino Hamilton-Buchanan, Fish. Ganges, 1822, p. 159. *Batrachocephalus mino* Day, Fish. India, 1875-78, Pl. cviii, fig. 1 (plate 9).

D. I, 7. A. 19-21. P. I, 9. Head shields forming a cross. A pair of very small mandibulary barbels. Dorsal and pectoral fin spines serrated. Silvery, dark brown above. Fins yellowish, powdered with dark. 10 inches. Sea, estuaries and tidal rivers.

141. OSTEOGENEIOSUS MILITARIS (Linnaeus) SOLDIER CAT-FISH

Silurus militaris Linnaeus, Syst. Nat., ed. 10, i, 1758, p. 305. *Osteogeneiosus militaris* Day, Fish. India, 1875-78, Pl. cviii, fig. 4 (plate 9).

D. I, 7. A. 19-21. P. I, 9-10. Head covered with smooth skin. A pair of stiff bony maxillary barbels, rather longer than head. Dorsal and pectoral fin spines serrated. Back bluish, sides silvery with obscure grey dots. Fins whitish, tips of dorsal and adipose fins dark blue. 12 inches. Sea, estuaries and tidal rivers.
Gal anguluva, Goraka anguluva, Manaval anguluva (S).

142. OSTEOGENEIOSUS STHENOCEPHALUS Day

STRONG-HEADED CAT-FISH

Osteogeneiosus sthenocephalus Day, Fish. India, 1875-78, p. 469, Pl. cviii, fig. 3 (plate 9).

D. I, 7. A. 20. P. I, 9. Head covered with smooth skin. A pair of stiff bony maxillary barbels, as long as head. Differs from preceding species in having a shorter and more slender head and longer anal rays. Silvery, darker above. Sea, estuaries and tidal rivers.

143. HEXANEMATICHTHYS SONA (Hamilton-Buchanan) DUSKY CAT-FISH

Pimelodus sona Hamilton-Buchanan, Fish. Ganges, 1822, p. 172. *Arius gagora* or *gagorides*. *Arius sona* Day, Fish. India, 1875-78, Pl. cv, fig. 2 (plate 9).

D. I, 7. A. 17. P. I, 12. Head with bony striate plates. Maxillary barbels reach to end of head, mandibular shorter. Palatine teeth villiform in 2 small and 2 large patches. Leaden grey, little paler below. Fins dusky to black terminally. Sides sometimes bluish with gold. 3 feet. Estuaries and tidal rivers.

144. TACHYSURUS VENOSUS (Valenciennes) VEINED CAT-FISH

Arius venosus Cuvier and Valenciennes, Hist. Nat. Poiss., xv, 1840, p. 69. *Arius sumatranus* or *malabaricus*. *Arius venosus* Bleeker, Atlas Ichth. Ind. Neerl., ii, 1862, Pl. liv, fig. 1 (plate 9).

D. I, 7. A. 22. P. I, 10. Bony head plates mostly smooth, in part rugosely striate or granular. Maxillary barbels extend to base of pectoral fin, mandibular shorter. Villiform teeth on palate in 2 triangular patches as large as eye. Back and upper surfaces drab brown to bluish, soiled white below. Fins whitish with grey tips. 13 inches. Estuaries and tidal rivers.

145. TACHYSURUS MACULATUS (Thunberg) SPOTTED CAT-FISH

Silurus maculatus Thunberg, Kon. Akad. Handl. Stockholm, xiii, 1792, p. 31. *Tachysurus falcarius*. *Arius falcarius* Day, Fish. India, 1875-78, Pl. cvi, fig. 5 (plate 9).

D. I, 7. A. 18-20. P. I, 10. Head shields rugose striate, but smooth to touch. Maxillary barbels black, as long as head, others shorter. Villiform teeth in broad bands across jaws, palatines with large equilateral triangular patches. Back dusky to slate black, sides grey, white underneath. Maxillary barbel black, others mostly whitish. Fins whitish with black tip to dorsal. Pectoral dusky above, white beneath. 14 inches. Estuaries and tidal rivers.
Gal anguluva (S). Kalla keleru (T).

146. TACHYSURUS CAELATUS (Valenciennes) ENGRAVED CAT-FISH

Arius caelatus Cuvier and Valenciennes, Hist. Nat. Poiss., xv, 1840, p. 66; Bleeker, Atlas Ichth. Ind. Neerl., ii, 1862, Pl. liii (plate 9).

D. I, 7. A. 19-21. P. I, 9. Head shields strongly rugose striate and strongly granulated. Maxillary barbel reaches middle of pectoral fin, external mandibular pair one fifth shorter. Villiform teeth in bands of 5-6 rows in jaws and in ovate areas on palatine. Back dusky brown or bluish, lower surfaces white or soiled brownish. Fins pale, blackish terminally with age. Barbels dusky. Reaches large size. Sea, estuaries and tidal rivers.
Anguluva (S). Keleru (T).

147. TACHYSURUS SUBROSTRATUS (Valenciennes)
SHORT-NOSED CAT-FISH

Arius subrostratus Cuvier and Valenciennes, Hist. Nat. Poiss., xv, 1840, p. 62; Day, Fish. India, 1875-78, Pl. cvi, fig. 6 (plate 9).

D. I, 7. A. 17-20. P. I, 9. Head shields roughly granulated. Barbels short, not reaching eye. Palatine teeth villiform, in 2 small patches wide apart. Pectoral spine as long as or longer than dorsal spine. Silvery leaden on back with about 15 minutely spotted vertical bands. Fins greyish. 12 inches. Sea, estuaries and tidal rivers.
Ura anguluva (S).

148. NETUMA THALASSINUS (Rüppell) GIANT CAT-FISH

Bagrus thalassinus Rüppell, Neue Wirbelth., Fische, 1835, p. 75. *Arius* or *Tachysurus thalassinus*. *Netuma nasuta* Bleeker, Atlas Ichth. Ind. Neerl., ii, 1862, Pl. lxi (plate 9).

D. I, 7. A. 15-17. P. I, 12. Head shield striate or granulated. Maxillary and inner mental barbels reach beyond eye. Teeth fine, in rather broad bands in jaws. On palate in 2 large triangular patches composed of 3 groups. Silvery, darkest above. Adipose fin with large blackish brown blotch. Sometimes back brown with gold on head. Attains large size. Seas, estuaries and tidal rivers.

Thora anguluva, Mudhu anguluva (S). Ven keleru (T).

149. NETUMA SERRATUS (Day) SAW-EDGED CAT-FISH

Arius serratus Day, Fish. India, 1875-78, p. 462, Pl. cv, fig. 3 (plate 9).

D. I, 7. A. 16. P. I, 10. Head shield granulated and a serrated ridge on occipital process. Maxillary barbels reach to middle of pectoral fin, others to gill-opening. Villiform teeth in 3 distinct patches on each side of palate. Upper surface and head coppery, sides silvery. Fins yellowish, adipose with a dark spot. Dark band edged with white on anal. Caudal lobes darker at tips. Perhaps the young of the preceding species.

150. ARIODES DUSSUMIERI (Valenciennes) DUSSUMIER'S CAT-FISH

Arius dussumieri Cuvier and Valenciennes, Hist. Nat. Poiss., xv, 1840, p. 84. *Hemipimelodus dussumieri. Arius dussumieri* Day, Fish. India, 1875-78, Pl. cvii, fig. 7 (Head only) (plate 9).

D. I, 7. A. 14-16. P. I, 12. Few scattered granulations on posterior part of head. Maxillary barbels reach just beyond base of pectorals, others shorter. Palatine teeth obtusely conical, in 4 widely separated patches. Bluish along back, lighter on sides and below. Fins blackish externally. Brackish water, entering estuaries.

151. PSEUDARIUS PLATYSTOMUS (Day) FLAT-MOUTHED CAT-FISH

Arius platystomus Day, Fish. India, 1875-78, p. 464, Pl. cvii, fig. 3 (plate 9).

D. I, 7. A. 19. P. I, 10. Very few scattered granulations over posterior part of head. Maxillary barbels reach just beyond base of pectorals, others shorter. Villiform teeth in premaxillaries 6 times longer than wide. Palatine teeth obtusely conic in 2 pyriform patches. Bluish on back, lighter on sides and below. Fins with slight tinge of grey. No black mark on adipose. Sea and estuaries.

152. PSEUDARIUS JATIUS (Hamilton-Buchanan) RIVER CAT-FISH

Pimelodus jatius Hamilton-Buchanan, Fish. Ganges, 1822, p. 171. *Arius jatius* Day, Fish. India, 1875-78, Pl. cvi, fig. 4 (plate 9).

D. I, 7. A. 18. P. I, 10. Posterior part of head sparsely granulated but occipital processes more thickly so. Maxillary barbels shorter than head, outer mandibulars reach gill-opening. Villiform teeth in bands on premaxillaries, 6 times long as wide. Palatine teeth globular, in small oval patch. Back dark bluish, lighter on sides and below. Fins yellowish. Dorsal and caudal edged with black. Anal and adipose with a dark spot. Estuaries and rivers, ascending above tidal influence. 12 inches.

153. PSEUDARIUS JELLA (Day) SMALL-EYE CAT-FISH

Arius jella Day, Fish. India, 1875-78, p. 467, Pl. cvi, fig. 3 (plate 9).

D. I, 7. A. 17-18. P. I, 10. Most of occiput closely granulated. Maxillary barbels almost as long as head, others shorter. Teeth on palate globular and in 2 convex, oblong-ovate patches. Back silver grey, white on sides and below. Fins grey with white edge to anal. Adipose dorsal with dark blotch. In sea and estuaries.

FRESH WATER CAT-FISH

Family BAGRIDAE

Body subcylindrical. Head subconical or depressed, covered with skin or naked. 4 pairs of barbels, nasal, mental, maxillary and mandibular. Dorsal and pectorals with spines. Adipose fin present. Caudal fin forked.

1. Two distinct cephalic fontanels 2
   Only one cephalic fontanel .... *Macrones gulio*
2. Fontanels equal .... *Macrones keletius*
   Posterior fontanel small .... *Macrones vittatus*

154. MACRONES GULIO (Hamilton-Buchanan)

LONG-WHISKERS CAT-FISH

Pimelodus gulio Hamilton-Buchanan, Fishes Ganges, 1822, p. 201. *Arius*, *Aoria* or *Mystus gulio*, *Bagrus albilabris*. *Aspidobagrus gulio* Bleeker, Atlas Ichth. Ind. Neerl., ii, 1862, Pl. lxxiv, fig. 2 (plate 9).

D. II, 6-7. A. III-IV, 10-12. P. I, 7-9. Nasal barbels reach hind edge of preopercle; maxillaries reach middle of ventrals or anal; mandibulars reach tip of pectoral; mentals reach insertion of pectoral. G.R. 29-33. Adipose fin small with base shorter than dorsal. Brown shot with bronze dorsally, yellow shading to white ventrally. Fins dull orange with dark tips. 10 inches. Estuarine, moving into fresh water rivers.
Mada anguluva, Mana ankutta, Vel anguluva (S).

155. MACRONES KELETIUS (Valenciennes) DWARF CAT-FISH

Bagrus keletius Cuvier and Valenciennes, Hist. Nat. Poiss., xiv, 1839, p. 411. *Aoria* or *Mystus keletius*, (?) *Macrones cavasius*. *Macrones keletius* Day, Fish. India, 1875-78, Pl. xcviii, fig. 5 (plate 10).

D. II, 7-8. A. II, 8. P. I, 8. Nasal barbels reach base of occipital process; maxillaries reach anal or caudal; mandibulars reach tip of pectoral; mentals reach insertion of pectoral. G.R. 17-18. Adipose fin stretches from behind dorsal to above base of anal. Dorsally olive brown, sides yellow, belly white. Fins yellow. 5 inches. Common in fresh water rivers and streams.
Pat ankutta (S).

156. MACRONES VITTATUS (Bloch) STRIPED DWARF CAT-FISH

Silurus vittatus Bloch, Nat. Ausländ. Fische, 1794, Pl. 371. *Aoria* or *Mystus vittatus*, (?) *Macrones tengara*. *Macrones vittatus* Day, Fish. India, 1875-78, Pl. xcix, fig. 4 (plate 10).

D. II, 7-8. A. II-III, 7-8. P. I, 8. Nasal barbels reach hind edge of preopercle; maxillaries reach middle of anal; mandibulars reach tip of pectoral; mentals a little shorter. G.R. 25-29. Adipose fin remote from dorsal and ends above anal. Dorsally olive brown, with dark shoulder patch and 3 dark lateral bands. Sometimes blotch on caudal peduncle. Sides yellow. Belly white. Fins pale yellow. 4 inches. Fresh water ponds and streams up to 3000 feet.

Hiri ankutta, Iri ankutta (S).

EELS, MORAYS, CONGERS

Order ANGUILLIFORMES

Elongate powerful snake-like body. Usually naked or with minute embedded scales. No pelvic fins. Dorsal and anal fins usually long and low, rays often concealed by skin, confluent with caudal, sometimes absent. No fin spines. Gill-openings usually much reduced. Inhabit marine and fresh waters. Many species live in holes in coral and amongst weeds, are aggressive and vicious and can inflict bad wounds.

1. Distance of anus from gill-openings much greater than head length; jaws not exceedingly slender and not much produced .... (*Suborder* ANGUILLOIDEI) 2
   Distance of anus from gill-openings less than head length; jaws exceedingly slender and much produced .... (*Suborder* NEMICHTHYOIDEI) *Family* NEMICHTHYIDAE, p. 69
2. Tongue and pectorals absent; opercles vestigial; fin in thick skin .... *Family* MURAENIDAE, p. 59
   Tongue present; opercular and branchial bones developed 3
3. Scales and pectoral fins present .... *Family* ANGUILLIDAE, p. 57
   Scales absent 4
4. Posterior nostrils are holes in head 5
   Posterior nostrils are slits in upper lip 7
5. Dorsal and caudal fins developed 6
   Dorsal on tail only and partly suppressed .... *Family* RATABOURIDAE, p. 63
6. Tongue free; teeth close set on jaws .... *Family* CONGRIDAE, p. 64
   Tongue adnate; anterior vomerine teeth caninoid .... *Family* MURAENESOCIDAE, p. 64
7. Caudal fin present .... *Family* ECHELIDAE, p. 65
   Caudal fin absent .... *Family* OPHICHTHYIDAE, p. 65

FRESH WATER EELS

Family ANGUILLIDAE

Jaws undershot. Teeth small and in cardiform bands. Tongue free. Gill-openings vertical slits below insertion of pectoral fins. Dorsal caudal and anal fins well developed, confluent. Tail compressed, body cylindrical. Anus in anterior half of length. Scales small. Lateral line distinct. Catadromous fishes.

Dorsal origin approximately above anus; no edentulous grooves in teeth bands ...... *Anguilla bicolor bicolor*
Dorsal overlaps anal fin by more than half a head length; edentulous groove in teeth bands .... *Anguilla nebulosa nebulosa*

157. ANGUILLA BICOLOR BICOLOR McClelland — LEVEL-FINNED EEL

Anguilla bicolor McClelland, Calcutta Journ. Nat. Hist., v, 1845, p. 178. *Anguilla australis* and *spengeli*. *Anguilla bicolor* Day, Fish. India, 1875-78, Pl. clxvii, fig. 3 (plate 12).

Head 6-8 in length. Eye 8-13 in head length. Gape extends beyond eye. Depth 14-17 in length. Light blue dorsally, brown on sides, shading from pale cream ventrally in young to bright yellow in adults. Pectorals transparent. In spawning phase eye increases to 6-7 in head, snout becomes blunter, gape shortens, caudal becomes rounded and colour changes

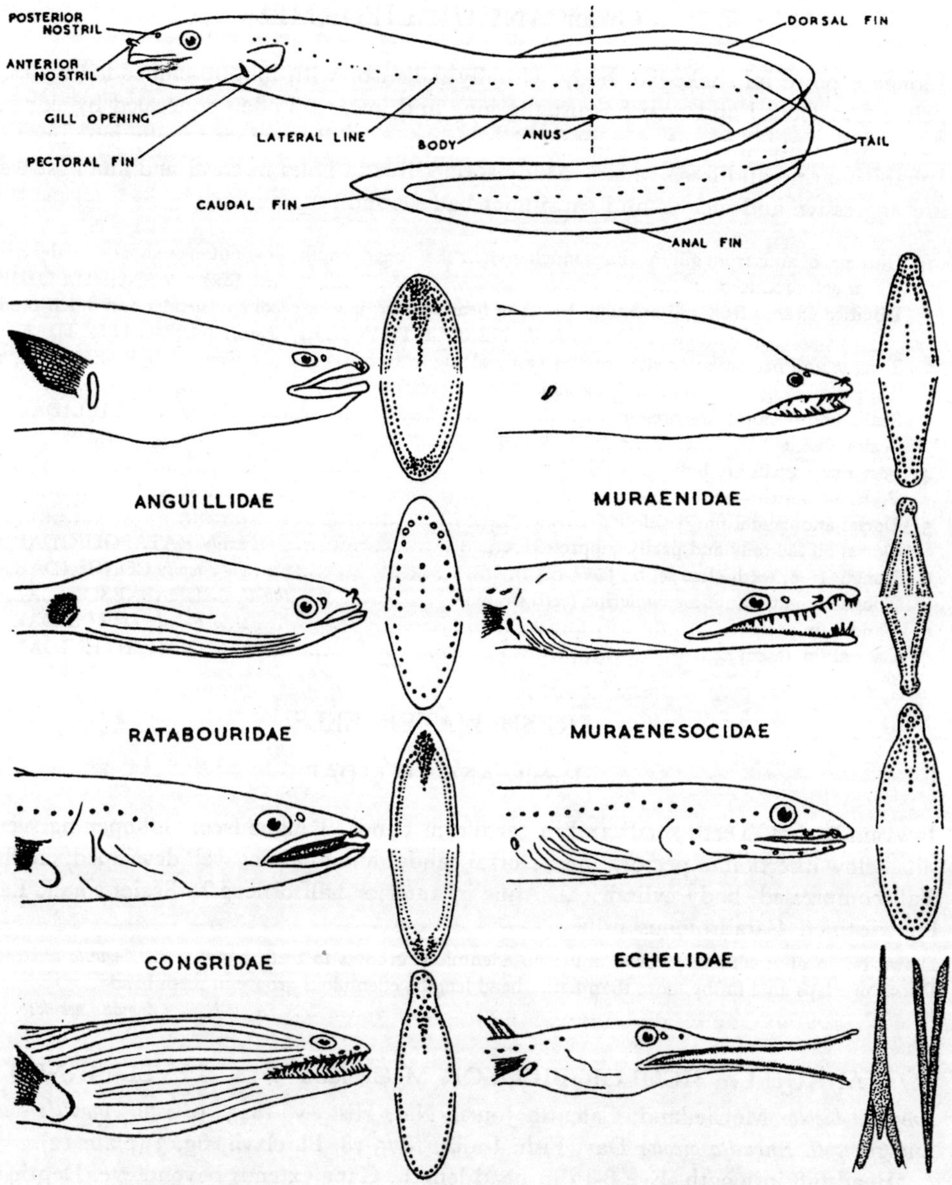

Fig. 8.

Explanation of the characters of the Order Anguilliformes (Eels), and principal differences between the eight component families of the order. To the right of each head is a diagrammatic representation of arrangement of teeth in upper and lower jaws of a typical genus of each family.

to bronze and later to silver. Attains 2 feet. Catadromous, namely inhabiting fresh waters and spawning at sea. Commoner on coastal plain.
Kakkuta arndha, Kalu arndha, Mada arndha (S).

158. ANGUILLA NEBULOSA NEBULOSA McClelland — LONG-FINNED EEL

Anguilla nebulosa McClelland, Calcutta Journ. Nat. Hist., v, 1845, p. 177. *Anguilla elphinstonei, bengalensis* or *mauritiana. Muraena maculata* Bleeker, Atlas Ichth. Ind. Neerl., iv, 1864, Pl. cxlv, fig. 2 (plate 11).

Head 6-8·25 in length. Eye 9·5-14 in head length. Gape extends beyond eye. Depth 9-17 in length. Body brownish dorsally mottled with darker brown. Ventrally yellowish. Fins yellowish mottled with brown. 3½ feet. Catadromous, descending to sea to spawn in rainy season. Commoner in mountain pools.
Gal arndha, Ganga arndha, Kabara arndha, Pulli arndha,
Kaha arndha, Pol mal arndha, Polon arndha, Vali arndha (S). Pulli vilangu (T).

MORAY EELS

Family MURAENIDAE

Eels with large mouth and powerful teeth, sometimes fang-like and depressible. Gill-openings small. Posterior nostril above eye. Pectorals absent. Body robust, naked with thick tough skin. Mostly brilliantly marked. Marine shallow water eels living amongst coral reefs.

1.	Bony subdermal scutes on tail	*Arndha zebra*
	No bony subdermal scutes on tail	2
2.	Some teeth blunt	3
	Teeth fang-like	4
3.	White with 2-3 rows of stellate blotches	*Echidna nebulosa*
	Brown with numerous round or oval patches	*Echidna xanthospilus*
	Pale yellow with fine brown mottling	*Echidna delicatula*
4.	Lateral line present	*Thyrsoidea macrura*
	Lateral line absent	5
5.	Median (dorsal and anal) fins absent	6
	Median fins present	7
6.	Uniform brown	*Uropterygius concolor*
	Yellow or brownish, mottled	*Uropterygius marmoratus*
7.	Depth more than 35 in length	*Pseudechidna brummeri*
	Depth less than 30 in length	8
8.	Mid-front teeth short, fixed, blunt	9
	Depressible fangs in mid-front of upper jaw	10
9.	Maxillary teeth in 1 series	*Sideria picta*
	Maxillary teeth in 2 series	*Sideria thyrsoidea*
10.	Body uniform brown	*Gymnothorax boschi*
	Brown with 15-20 encircling rings	*Gymnothorax rüppelli*
	White with irregular black blotches	*Gymnothorax polyuranodon*
	Blotches or streaks longer than 2 orbits	11
	Spots about as large as orbit	12

11. Brown with fine white or yellow lines .... *Gymnothorax pseudothyrsoidea*
Yellow with irregular russet blotches .... *Gymnothorax undulatus*
White with polygonal dark chocolate blotches .... *Gymnothorax favagineus*
12. Light brown with irregular black spots .... *Gymnothorax fimbriatus*
Purplish with regular white spots .... *Gymnothorax punctatus*

159. ARNDHA ZEBRA (Shaw) ZEBRA MORAY

Gymnomuraena zebra Shaw, Nat. Miscell., ix, 1797, pl. 322. *Echidna zebra* Bleeker, Atlas Ichth. Ind. Neerl., iv, 1864, Pl. clxxi, fig. 1 (plate 11).

Depth 17-21, head 8½-10 in length. Granular teeth in 2 short series in upper and 3-4 series in lower jaw. Pear-shaped patch on vomer. Two sets of inter-fitting subdermal bony scutes laterally in tail region. Chocolate with about 112 white or pale yellow encircling rings. 4 feet. Reefs near shore.
Vairan gal-gulla (S).

160. ECHIDNA NEBULOSA (Ahl) STARRY MORAY

Muraena nebulosa Ahl, Dissert. Muraena & Ophichth., 1789, p. 5. *Echidna nebulosa* Jordan & Evermann, Bull. U.S. Fish. Comm., xxiii (1), 1905, Pl. i (plate 11).

Depth 16-21, head 9-10 in length. 1-2 series of blunt conical teeth in both jaws. 2 series on vomer. Pale yellowish brown with 2-3 rows of stellate blotches having one or more ocelli in centre. Interspaces mottled with irregular lines and spots. 30 inches. Coral reefs near shore.
Pulli gal-gulla (S).

161. ECHIDNA XANTHOSPILUS (Bleeker) YELLOW-SPOTTED MORAY

Muraena xanthospilus Bleeker, Nat. Tijds. Ned. Ind., xix, 1859, p. 348. *Echidna xanthospilus* Bleeker, Atlas Ichth. Ind. Neerl., iv, 1864, Pl. clxvii, fig. 1 (plate 11).

Depth 13½-14½, head 7-7½ in length. Teeth granular in 1 series in upper and 2-3 series in lower jaw. Vomer with 3 series anteriorly and 2 series posteriorly. Brownish with numerous yellowish round or oval patches and irregular streaks. 26 inches. Coral reefs.

162. ECHIDNA DELICATULA (Kaup) MOTTLED MORAY

Poecilophis delicatulus Kaup, Cat. Apodal Fish, 1856, p. 102. *Echidna delicatula* Bleeker, Atlas Ichth. Ind. Neerl., iv, 1864, Pl. clxvii, fig. 3 (plate 11).

Depth 14-17, head 7½-9 in length. Teeth conical, in 2 series in upper and 1 series in lower jaw. Intermaxillary with conical marginal teeth and 2 stout mesial teeth. Vomer with 2 rows. Greyish with fine dark mottling of irregular blotches and marbling. 18 inches. Reefs near shore.

163. THYRSOIDEA MACRURA (Bleeker) GIANT MORAY

Muraena macrurus Bleeker, Nat. Tijds. Ned. Ind., vii, 1854, p. 324. *Thyrsoidea macrurus* Bleeker, Atlas Ichth. Ind. Neerl., iv, 1864, Pl. clxvi, fig. 2 (plate 12).

Depth 38-47, head 10-14 in length. Teeth in 2 series in upper and lower jaws, inner series strongest. Intermaxillary with peripheral row of 9 teeth and 4 mesial depressible fangs. Dark brown, lighter below. Largest in the family, attaining 10 feet. Shallow water along coast, entering lagoons.

164. UROPTERYGIUS CONCOLOR Rüppell BROWN MORAY

Uropterygius concolor Rüppell, Neue Wirbelth., Fische, 1835, p. 83. *Gymnomuraena concolor*. *Uropterygius concolor* Deraniyagala, Spolia Zeyl., xv, 1929, Pl. xi, fig. 1 (plate 11).

Depth 20-33, head 7-8½ in length. No visible dorsal or anal fins. Teeth in upper and lower jaws fine, in single row with larger teeth posteriorly. Upper jaw with single strong mesial canine. Uniform reddish brown. 20 inches. Reefs and shallow waters of pearl banks.

165. UROPTERYGIUS MARMORATUS (Lacépède) MARBLED MORAY

Gymnomuraena marmorata Lacépède, Hist. Nat. Poiss., v, 1803, p. 648. *Gymnomuraena pantherina* Bleeker, Atlas Ichth. Ind. Neerl., iv, 1864, Pl. clxxv, fig. 3 (plate 11).

Depth 19-26, head 9-11 in length. Teeth in both jaws in 3 series, outer with numerous small close-set teeth, inner 2 series with fewer, longer depressible teeth. Yellowish or brownish, lighter below and finely or coarsely marbled with darker. 36 inches. Reefs and shallow waters of pearl banks.

166. PSEUDECHIDNA BRUMMERI (Bleeker) BRUMMER'S MORAY

Muraena brummeri Bleeker, Nat. Tijds. Ned. Ind., xvii, 1858, p. 137. *Strophidon brummeri*. *Pseudechidna brummeri* Bleeker, Atlas Ichth. Ind. Neerl., iv, 1864, Pl. clxii, fig. 1 (plate 12).

Depth 40-51, head 12-17 in length. Extremely elongate. Dorsal and anal prominent, dorsal twice as high as anal. Teeth in both jaws in single row, conical. Intermaxillary with 3 mesial depressible fangs. Vomer with 7 large conical teeth in single row. Pale dirty white with hyaline fins. Anterior part of head with black dots. 30 inches. Inhabits coastal lagoons.

167. SIDERIA PICTA (Ahl) PAINTED MORAY

Muraena picta Ahl, Dissert. Muraena & Ophichth., 1789, p. 6. *Gymnothorax pictus* Jordan and Evermann, Bull. U.S. Fish. Comm., xxiii (1), 1905, Pl. xix (plate 11).

Depth 15-20, head 6-8 in length. Dorsal 3 times as high as anal. Teeth in single series in both jaws. Intermaxillary with 3 mesial depressible fangs. Vomer with 3 teeth. Yellow with 3 longitudinal rows of brown stellate blotches, somewhat larger than eye. Single row of blotches on dorsal. Interspaces scattered with numerous small spots. 30 inches. Coral reefs of pearl banks.

168. SIDERIA THYRSOIDEA (Richardson) SLENDER MORAY

Muraena thyrsoidea Richardson, Voy. Sulphur, Ichth., 1844, p. 111. *Gymnothorax prosopeion* Bleeker, Atlas Ichth. Ind. Neerl., iv, 1864, Pl. clxxxiii, fig. 3 (plate 12).

Depth 17-21, head 9-10 in length. Teeth in 2 series in both jaws, outer series more numerous, inner series stouter. Intermaxillary with 1-2 mesial conic canines. 2 series on vomer. Brownish, dotted and speckled with darker brown and white. 2 feet. Coral reefs near shore.

169. GYMNOTHORAX BOSCHI (Bleeker) BOSCH'S MORAY

Muraena boschi Bleeker, Verh. Bat. Gen., xxv, 1853, p. 52. *Gymnothorax afra. Gymnothorax monochrous* Bleeker, Atlas Ichth. Ind. Neerl., iv, 1864, Pl. cxci, fig. 2 (plate 11).

Depth 12-20, head 8 in length. Dorsal 3 times as high as anal. Teeth in single series in both jaws. Intermaxillary with 3 mesial teeth and mandible with 2 inner symphysial canines. No vomerines but infrapharyngeals with 2 parallel double rows. Face yellow. Body light brown with darker dorsal and anal. Black around eyes, mouth and gill-cleft. 2 feet. Pedro Bank.

170. GYMNOTHORAX RÜPPELLI (McClelland) RÜPPELL'S MORAY

Dalophis rupelliae McClelland, Calcutta Journ. Nat. Hist., v, 1845, p. 213. *Muraena ruppellii. Muraena reticularis* Bleeker, Atlas Ichth. Ind. Neerl., iv, 1864, Pl. clxxxiii, fig. 2 (plate 11).

Depth 16-25, head 8-9 in length. Teeth in single series in both jaws but sometimes with rudimentary inner series. Intermaxillaries with 2 mesial depressible compressed fangs. Vomer with 6 or more conical teeth. Yellow or reddish brown with 15-20 encircling rings of black, narrower than interspaces. 22 inches. Coral reefs.

171. GYMNOTHORAX POLYURANODON (Bleeker) FRESH WATER MORAY

Muraena polyuranodon Bleeker, Nat. Tijds. Ned. Ind., v, 1853, p. 248. *Gymnothorax polyuranodon* Bleeker, Atlas Ichth. Ind. Neerl., iv, 1864, Pl. clxxiv, fig. 2 (plate 11).

Depth 19-28, head 9-11 in length. Teeth conical, in 2 series in both jaws. Intermaxillary with 6 mesial depressible fangs. White with irregular black blotches which are closer together and more confluent dorsally. Coalesce on head to form 6 longitudinal bands. 3 feet. Catadromous. Only species inhabiting fresh water rivers.

172. GYMNOTHORAX PSEUDOTHYRSOIDEA (Bleeker)
YELLOW-LINED MORAY

Muraena pseudothyrsoidea Bleeker, Nat. Tijds. Ned. Ind., iii, 1852, p. 778. *Gymnothorax pseudothyrsoidea* Bleeker, Atlas Ichth. Ind. Neerl., iv, 1864, Pl. cxc, fig. 2 (plate 11).

Depth 14-20, head 7-8 in length. Teeth in single series in both jaws, sometimes short inner row of larger teeth. Intermaxillary with 3 mesial canines. Vomer with 3-9 small teeth in single row. Brownish with very fine white or yellow lines forming a network. 2 feet.

173. GYMNOTHORAX UNDULATUS (Lacépède) LEOPARD MORAY

Muraenophis undulata Lacépède, Hist. Nat. Poiss., v, 1803, p. 629. *Muraena undulata. Gymnothorax undulatus* Jordan & Evermann, Bull. U.S. Fish. Comm., xxiii (1), 1905, Pl. xvi (plate 11).

Depth 14-22, head 6½-8 in length. Skin thick. Savage with powerful jaws. Teeth in single series in both jaws, symphysial ones fang-like. Intermaxillary with 3 long mesial depressible fangs. 2 rows of infrapharyngeal teeth forming a V. Yellow or orange with irregular russet brown blotches in more or less regular rows. 3 feet. Reefs and paars in deeper water.

Divi gal-gulla, Kabara gal-gulla (S). Aajalai (T).

174. GYMNOTHORAX FIMBRIATUS (Bennett) DARK-SPOTTED MORAY

Muraena fimbriata Bennett, Proc. Comm. Zool. Soc., i, 1831, p. 168. *Gymnothorax undulatus* var. *fimbriatus*. *Muraena fimbriata* Day, Fish. India, 1875-78, Pl. clxxii, fig. 1 (plate 12).

Generally considered a colour variety of *G. undulatus* from which it differs by being greenish with numerous brown spots as large as eye, or light reddish brown with irregular lines of black spots. Wadge Bank.

175. GYMNOTHORAX FAVAGINEUS Bloch and Schneider
BLACK-LINED MORAY

Gymnothorax favaginea Bloch and Schneider, Syst. Ichth., 1801, p. 525. *Muraena favaginea*. *Muraena tessellata* Day, Fish. India, 1875-78, Pl. clxxi, fig. 4 (plate 11).

Depth $15\frac{1}{2}$-$21\frac{1}{2}$, head $7\frac{1}{2}$-9 in length. Both fins under thick skin. Teeth in single row in both jaws, canines directed backwards. 5 larger canines at mandibular symphysis and 3 elongate mesial fangs on intermaxillary. Vomerines small in single row. Polygonal dark chocolate spots bordered by narrow white or yellowish reticulation. 3 feet. Shallow water and coral reefs.
Kalu iriya (S).

176. GYMNOTHORAX PUNCTATUS (Bloch and Schneider)
WHITE-SPOTTED MORAY

Muraena punctatus Bloch and Schneider, Syst. Ichth., 1801, p. 526. *Muraena tile*. *Muraena punctata* Day, Fish. India, 1875-78, Pl. clxxiii, fig. 1 (plate 11).

Depth 16-23, head $7\frac{1}{2}$-8 in length. Dorsal fin 4 times as high as anal fin. Teeth in single row in upper and lower jaws. Intermaxillary with 3 mesial fangs. Vomer with 4-9 teeth in single row. 2 double rows in V-shape on infrapharyngeals. Purplish brown with white dots smaller than eye and increasing in size posteriorly. 3 feet. Shallow water and reefs of pearl banks. Flesh considered poisonous.

THREAD EELS

Family RATABOURIDAE

Differ from other eels by having anus in posterior third or fourth of total length. Elongate cylindrical body. Dorsal and anal fins feeble. Pectorals feeble or absent. Small conical recurved teeth in one series in both jaws and on vomer.

177. RATABOURA BICOLOR (Kaup) BLACK-TAILED THREAD EEL

Moringua bicolor Kaup, Cat. Apodal Fish, 1856, p. 107. *Moringua macrochir*. *Moringua bicolor* Bleeker, Atlas Ichth. Ind. Neerl., iv, 1864, Pl. cxlvii, fig. 1 (plate 12).

Body fine and cylindrical. Lateral line present. Dorsal and anal fins confined to tail region. Depth 60-70, head $12\frac{1}{2}$-16 in length. Greenish brown above, yellow or purplish ventrally, the two colours sharply contrasted. Fins yellowish, caudal black. 2 feet. Marine pelagic from pearl banks.

FALSE CONGER EELS

Family MURAENESOCIDAE

Body very long wi h only tail compressed. Snout elongate, mouth large, extending far behind eye. Teeth large and powerful. Pectorals present. Large eels inhabiting estuaries and entering fresh water. Excellent eating.

178. MURAENESOX CINEREUS (Forskål) SILVER CONGER EEL

Muraena cinerea Forskål, Descript. Animal., 1775, p. 22. *Muraenesox bagio* Bleeker, Atlas Ichth. Ind. Neerl., iv, 1864, Pl. clxx, fig. 2 (plate 12).

Depth 20, head 6-7 in length. Vent midway in body. Pectoral about 3 in head. Constriction behind intermaxillary plate giving the jaw a resemblance to that of a crocodile. Teeth in 1-2 series in upper and 3 series in lower jaw. 3 rows of tricuspid teeth on vomer. Anterior teeth in both jaws are canines. Silvery with grey or yellow sheen. 5-7 feet. Brackish and salt waters of lagoons and shallow coastal inlets.
Mudu luhulla, Mudu teliya, Puliviya, Vel luhulla (S).

CONGER EELS

Family CONGRIDAE

Robust body without scales. Wide gill-openings. Dorsal and anal confluent with caudal and covered with thick skin. Mouth not extending beyond eye. Teeth strong but no canines in jaw. Powerful and vigorous eels.

1. Front nostril tubular 2
   Front nostril not tubular .... *Uroconger lepturus*
2. Outer teeth close together forming a cutting edge .... *Conger cinereus*
   Teeth not forming a cutting edge .... *Ariosoma anago*

179. CONGER CINEREUS Rüppell INDIAN CONGER EEL

Conger cinereus Rüppell, Atlas Reise Nord. Afrika, Fische Roth. Meers, 1828, p. 115. *Leptocephalus cinereus*. *Conger Noordzieki* Bleeker, Atlas Ichth. Ind. Neerl., iv, 1864, Pl. clxvii, fig. 2 (plate 11).

Depth 20, head 9 in length. Head sharp and pointed. Nostrils are tubes at tip of snout. Outer teeth in both jaws close-set and forming a cutting edge. Young golden. Adults dark brown, paler underneath with black edges to fins. 3-5 feet. Reefs close to shore.
Mudu arndha (S).

180. ARIOSOMA ANAGO (Schlegel) SEA CONGER

Conger anago Schlegel, Fauna Japonica, Pisces, 1846, p. 259. *Congrellus anago*. *Congromuraena anago* Day, Fish. India, 1875-78, Pl. clxix, fig. 2 (plate 12).

Depth 16-18, head 6 in length. Anterior nostril tubular. Teeth small and pointed, not forming a cutting edge, in several series in both jaws. Dorsal starts above pectoral base. Brown or greyish, fins yellowish hyaline. 2 feet. Shallow waters of pearl banks.

181. UROCONGER LEPTURUS (Richardson) SLENDER CONGER EEL

Conger lepturus Richardson, Voy. Sulphur, Zool., 1845, p. 106. *Uroconger lepturus* Day, Fish. India, 1875-78, Pl. clxxi, fig. 1 (plate 12).

Depth 20-24, head 7-9 in length. Cleft of mouth extending beyond middle of eye. Snout 4 in head. Nostril not tubular. Teeth moderate, unequal in size; caniniform, in 1-2 series on vomer. Blackish or brownish, fins darker. 17 inches. Gulf of Manaar.

WORM EELS

Family ECHELIDAE

Elongate slender cylindrical naked body. Gill-openings small. No pectorals. Dorsal originates far back. Hind nostril below eye. Teeth conical, not fang-like. Small, nearly colourless eels.

Dorsal much in advance of vent .... *Muraenichthys gymnopterus*
Dorsal behind vent .... *Muraenichthys vermiformis*

182. MURAENICHTHYS GYMNOPTERUS (Bleeker)
LONG-FINNED WORM EEL

Muraena gymnopterus Bleeker, Verh. Bat. Gen., xxv, 1853, p. 52. *Muraenichthys macropterus. Muraenichthys gymnopterus* Bleeker, Atlas Ichth. Ind. Neerl., iv, 1864, Pl. cl, fig. 1 (plate 12).

Depth 30, head 7-8 in length. Origin of dorsal ahead of anus and nearer to anus than gill-openings. Teeth in 1-2 rows in both jaws. Single strong mesial tooth in premaxillaries. Vomerines in 2 series anteriorly, strong. Grey brown. 7 inches. Marine shallow waters.

183. MURAENICHTHYS VERMIFORMIS (Peters)
SHORT-FINNED WORM EEL

Chilorhinus (*Muraenichthys*) *vermiformis* Peters, Monatsb. Akad. Wiss. Berlin, 1866, p. 524.

Origin of dorsal behind anus. Angle of mouth a little behind eye. Teeth in jaw and on vomer in a single series. An imperfectly known species.

SNAKE EELS

Family OPHICHTHYIDAE

Gill-openings are moderate slits. Tongue present. No scales. Dorsal and anal poorly developed. Tail pointed, free, without fin at end. Mainly live in sand or near reefs. Burrow in sand or mud, tail first. Marine.

1. Origin of dorsal before gill-opening 2
   Origin of dorsal above or behind gill-opening 4
2. Protuberance on upper lip between nostrils; intermaxillary teeth absent .... *Callechelys longipinnis*
   No protuberance on upper lip; teeth on intermaxillary 3
3. White with black spots and marbled with brown .... *Callechelys marmoratus*
   Brownish olive, yellowish below .... *Callechelys kirki*

4. Teeth uniserial in jaws; upper lip with fringe of papillae .... *Brachysomophis cirrhocheilus*
   Teeth pluriserial on jaws; no fringe on lip 5
5. No teeth on vomer .... *Leiuranus semicinctus*
   Teeth present on vomer 6
6. Pectoral fins present 7
   Pectorals absent 10
7. Teeth granular .... *Pisoodonophis cancrivorus*
   Teeth conical 8
8. Maxillary teeth in 2 series .... *Ophichthys rhytidodermatoides*
   Maxillary teeth in 1 series 9
9. Origin of dorsal behind root of pectorals .... *Ophichthys apicalis*
   Origin of dorsal above gill-openings .... *Ophichthys altipinnis*
10. Gill-openings without duplicature .... *Caecula lumbricoides*
    Gill-openings with duplicature of membrane 11
11. Distance of dorsal from gill-openings 2·3-4 in head .... *Caecula polyophthalmus*
    Distance of dorsal from gill-openings 7 in head .... *Caecula orientalis*

184. CALLECHELYS LONGIPINNIS (Kner and Steindachner)

LONG-FIN SNAKE EEL

Sphagebranchus longipinnis Kner and Steindachner, Sitzb. Akad. Wiss. Wien, liv, 1867, p. 390, Pl. v, fig. 14 (plate 12).

Depth 55, head 21 in length. Dorsal originates midway between front of eye and gill-opening. Pectorals normally absent. Blunt protuberance between anterior tubular nostril and posterior nostril which is slit in upper lip. Teeth uniserial on jaws and palate. No intermaxillary teeth. Olive brown dorsally, yellow ventrally. 3 feet. Estuarine, inhabiting river mouths and coastal lagoons.

185. CALLECHELYS MARMORATUS (Bleeker) MARBLED SNAKE EEL

Dalophis marmoratus Bleeker, Verh. Bat. Gen., xxv, 1852, p. 37. *Callechelys marmoratus* Bleeker, Atlas Ichth. Ind. Neerl., iv, 1864, Pl. clv, fig. 2 (plate 12).

Depth 38, head 15-16 in length. Dorsal originates above angle of mouth. Pectorals absent. Teeth pointed, recurved and in single series but in 2 series on vomer and intermaxillary. White with crowded black spots or irregularly marbled with brown and yellow. 3 feet. Marine.

186. CALLECHELYS KIRKI (Günther) KIRK'S SNAKE EEL

Ophichthys kirkii Günther, Cat. Fish., viii, 1870, p. 89.

Depth 64, head 20 in length. Dorsal originates midway between angle of mouth and gill-opening. Uniserial teeth in jaws and palate and cluster of 3 on intermaxillary between tubular anterior nostrils. Olive-brown dorsally, pale yellow on sides and belly. 29 inches. Pedro Bank. Marine.

187. BRACHYSOMOPHIS CIRRHOCHEILUS (Bleeker)

YELLOW-FIN SNAKE EEL

Ophisurus cirrhocheilus Bleeker, Act. Soc. Sci. Ind. Neerl., ii, 1859, p. 89. *Brachysomophis cirrhocheilus* Bleeker, Atlas Ichth. Ind. Neerl., iv, 1864, Pl. cliii (plate 11).

Depth 26, head 9-10 in length. A series of fleshy, bifid papillae fringe both lips. Gill-openings large with membranous posterior half. Dorsal originates $\frac{2}{3}$ head length behind gill-openings. Series of well-defined grooves on body. Teeth in 1 series in lower and 2 series in upper jaw. Largest caniniform teeth on vomer. Brown with large irregular dark transverse spots and bands. Fins yellow edged with brown. 5 feet. Marine.
Mudu polonga, Ran teliya (S).

188. LEIURANUS SEMICINCTUS (Lay and Bennett)
HALF-BANDED SNAKE EEL

Ophisurus semicinctus Lay and Bennett, Beechey's Voy., Fishes, 1839, p. 66. *Leiuranus columbrinus* Bleeker, Atlas Ichth. Ind. Neerl., iv, 1864, Pl. clxiii, fig. 1 (plate 12).

Depth 46-55, head 12-15 in length. Tip of tail finless. Upper jaw prominently overshot. Dorsal fin originates behind gill-opening. Teeth in single row in upper jaw, rudimentary in lower jaw and absent on vomer. Pale cream with 22-27 black cross-bands which are broader than interspaces and fade ventrally. Bands continued on dorsal fin. 20 inches. Marine, from pearl banks.

189. PISOODONOPHIS CANCRIVORUS (Richardson)
BURROWING SNAKE EEL

Ophisurus cancrivorus Richardson, Voy. Erebus and Terror, Fish., 1844, p. 97, Pl. xl, fig. 6 (plate 11).

Depth 22-33, head 8-10 in length. Dorsal originates behind gill-opening. 6 pores along mandible. Gape extends beyond eye. Anterior nostril tubular followed by short papilla. Teeth granular in several bands in both jaws. 8 intermaxillary teeth in cluster. 3 rows on vomer. Brown to yellow dorsally, yellow ventrally. 30 inches. Burrowing form inhabiting estuaries and coastal lagoons.
Gulliviya (S).

190. OPHICHTHYS RHYTIDODERMATOIDES (Bleeker)
WRINKLED-SKIN SNAKE EEL

Ophisurus rutidodermatoides Bleeker, Verh. Bat. Gen., xxv, 1852, p. 31. *Ophichthys rutidodermatoides* Bleeker, Atlas Ichth. Ind. Neerl., iv,' 1864, Pl. clx, fig. 1 (plate 12).

Depth 46, head 15 in length. Posterior nostrils are solid papillae. Gill-openings lateral. Tip of tail finless. Teeth pointed and biserial except on front of maxilla and hind part of vomer which are in a single series. Olive brown above and yellowish below. 27 inches. Burrowing form inhabiting estuaries and coastal lagoons.
Mada panuwa, Pannu malu (S).

191. OPHICHTHYS APICALIS (Bennett) POINTED-TAIL SNAKE EEL

Ophisurus apicalis Bennett, Mem. Life of Raffles, 1830, p. 692. *Ophichthys Diepenhorsti* Bleeker, Atlas Ichth. Ind. Neerl., iv, 1864, Pl. clix, fig. 4 (plate 12).

Depth 27-35, head 8½-9 in length. Dorsal originates above middle of pectoral. Dorsal and anal highest nearest tail tip which is finless. Posterior nostrils slits in lip between papillae. Teeth conical in 2 rows in maxilla and vomer and 1 row in lower jaw. 3-5 on intermaxillary. Brownish yellow, darkest dorsally. 17 inches. Burrowing form from coastal waters, ascending to surface at night.

192. OPHICHTHYS ALTIPINNIS (Kaup) HIGH-FINNED SNAKE EEL

Microdontophis altipinnis Kaup, Cat. Apodal Fish, 1856, p. 6. *Ophichthys altipinnis* Bleeker, Atlas Ichth. Ind. Neerl., iv, 1864, Pl. clvii, fig. 1 (plate 12).

Depth 33, head 9½ in length. Posterior nostril in slit between papillae. Dorsal originates slightly in advance of gill-opening. Dorsal and anal as high as half body depth. Teeth in jaws and vomer uniserial but in 2 series anteriorly on mandibles and intermaxillary. Brownish above, yellowish below. Fins with blue-black margins. 3 feet. Pearl banks.

193. CAECULA LUMBRICOIDES (Bleeker) WORM EEL

Sphagebranchus lumbricoides Bleeker, Ned. Tijds. Dierk., ii, 1864, p. 46. *Ophichthys timorensis. Ophichthys lumbricoides* Bleeker, Atlas Ichth. Ind. Neerl., iv, 1864, Pl. clviii, fig. 3 (plate 12).

Depth 60, head 17 in length. Dorsal and anal very low, almost absent. Mouth cleft reaching far behind eye. Teeth pointed, uniserial, those on intermaxillary and anterior part of vomer biserial. 9 inches. An imperfectly known probably burrowing form from pearl banks.

194. CAECULA POLYOPHTHALMUS (Bleeker) OCELLATED WORM EEL

Dalophis polyophthalmus Bleeker, Nat. Tijds. Ned. Ind., iv, 1853, p. 299. *Sphagebranchus polyophthalmus* Bleeker, Atlas Ichth. Ind. Neerl., iv, 1864, Pl. cliv, fig. 1 (plate 12).

Depth 28-35, head 8-9 in length. Dorsal originates ¼ head length behind gill-openings. Posterior nostril on blunt fleshy papilla on upper lip. Teeth pointed, uniserial in jaws and vomer with V-shaped group on intermaxillary. Yellowish dusted with brown with double row of 10 ocelli from snout to origin of dorsal fin. Vertex with yellow transverse streak. 14 inches. Estuarine form burrowing tail first in sand between tide marks.
Panu malu (S).

195. CAECULA ORIENTALIS (McClelland) ORIENTAL WORM EEL

Dalophis orientalis McClelland, Calcutta Journ. Nat. Hist., v, 1845, p. 213. *Sphagebranchus orientalis. Ophichthys orientalis* Day, Fish. India, 1875-78, Pl. clxxi, fig. 1 (plate 12).

Depth 34, head 8 in length. Dorsal and anal fins very low. Dorsal originates above hind end of gill-opening. Gill-openings with accessory membranes. Teeth sharply pointed, uniserial. Brownish above, otherwise yellow. 14 inches. Burrows in sand along shores of inlets and bays.
Oteliya (S).

SNIPE EELS

Family NEMICHTHYIDAE

Body long, slender and compressed with fine tapering tail and elongated needle-like jaws. Dorsal and anal confluent with rayed caudal at tip of tail. Pectorals present. Vent far forward. Rare, fragile deep-water eels.

196. GAVIALICEPS MICROPS Alcock SNIPE EEL

Gavialiceps microps Alcock, Ann. Mag. Nat. Hist. (6), iv, 1889, p. 461.

Body cylindrical with long pointed tail. Anus situated a snout length behind the gill-opening. Snout in the form of a long, rigid, needle-like beak. Upper jaw serrated, lower with slightly recurved teeth. No scales. Vertical fins confluent. No pectorals. Grey-brown, belly yellowish. Branchiostegal region and base of beak black. 10½ inches. Deep water to 900 fathoms.

DEEP-SEA LIZARD-FISH

Order HALOSAURIFORMES

Family HALOSAURIDAE

Elongate compressed fishes with long tapering tail without caudal fin. Head conical with flattened snout and inferior mouth. No barbels. Teeth in villiform bands in jaws. Lateral line along lower profile, sometimes with luminous organs in scale pouches. One dorsal fin, short and behind ventrals. Anal long and occupying entire length of tail.

197. HALOSAUROPSIS AFFINIS (Günther) RAT-TAILED LIZARD-FISH

Halosaurus affinis Günther, Ann. Mag. Nat. Hist., xx, 1877, p. 444. *Halosaurus anguilliformis*. *Halosaurus affinis* Günther, Challenger Repts., Zool., xxii, 1887, Pl. lix, fig. B (plate 14).

D. 10-12. A. ca. 175. V. I, 8. P. 10-13. L. lat. 30 between head and vent. Tr. 13-14. Depth 17-23. Scales deciduous, those of lateral line enlarged and provided each with a luminous organ. Light brown, end of tail somewhat darker as also upper side of head and snout. Head and ventral surface of body black. 21 inches. 600-675 fathoms.

GARFISH, NEEDLE-FISH, HALF-BEAKS, FLYING FISH

Order BELONIFORMES

Body elongate, nearly quadrangular in section. Dorsal and anal fins opposite and inserted towards the posterior end of the body. Pectorals inserted high up. Ventrals abdominal with 6 rays. Lateral line near lower profile. Jaws sometimes produced and beak-like. Scales cycloid and deciduous.

1. Both jaws produced and teeth enlarged and needle-like .... (*Suborder* SCOMBERESCOIDEI) *Family* BELONIDAE, p. 71
   Both jaws not produced and with minute teeth .... (*Suborder* EXOCOETOIDEI) 2
2. Lower jaw produced into a beak .... *Family* HEMIRHAMPHIDAE, p. 73
   Lower jaw normal .... *Family* EXOCOETIDAE, p. 75

GARFISH, NEEDLE-FISH, LONG-TOMS

Family BELONIDAE

Elongate slender body. Jaws much elongated with small and large needle-like teeth. No detached finlets. No gill-rakers. Pelagic species mostly entering estuaries.

1. Operculum scaly, or if not the tail is forked; origin of dorsal behind that of anal 2
   Operculum not scaly; caudal truncate; origin of dorsal opposite that of anal .... *Xenentodon cancila*
2. Body strongly compressed; height twice breadth of body; intermaxillaries swollen at base ... *Ablennes hians*
   Body at most moderately compressed, usually cylindrical; intermaxillaries not swollen at base 3
3. Caudal forked; lateral line forming keel on caudal peduncle .... *Tylosurus crocodilus*
   Caudal rounded or truncate; lateral line not forming keel on caudal peduncle 4
4. Anal 2, 13-15; Dorsal 2, 10-13. L.lat. 170 .... *Tylosurus strongylurus*
   Anal 2, 21-23; Dorsal 2, 16-18. L.lat. 185-200 .... *Tylosurus leiurus*

198. XENENTODON CANCILA (Hamilton-Buchanan)

FRESH WATER GARFISH

Esox cancila Hamilton-Buchanan, Fish. Ganges, 1822, p. 213. *Tylosurus cancila. Belone cancila* Day, Fish. India, 1875-78, Pl. cxviii, fig. 5 (plate 13).

D. 15-18. A. 16-18. Depth 8-12, head $2\frac{3}{4}$ in body. Deep longitudinal groove along upper surface of head. Caudal truncate. Greenish grey above, whitish beneath. Silvery lateral streak with dark margin and fine black dots over upper two thirds of body. 12 inches. Inhabits coastal fresh waters.

Yona (S).

199. ABLENNES HIANS (Valenciennes) BARRED LONG-TOM

Belone hians Cuvier and Valenciennes, Hist. Nat. Poiss., xviii, 1846, p. 432. *Athlennes caeruleofasciatus* Ogilby, Mem. Qld. Mus., v, 1916, Pl. xiv (plate 13).

D. 23-26. A. 25-27. Depth 12-18, head $3\frac{1}{2}$-4 in length. Caudal peduncle compressed. Caudal fin forked. Operculum smooth. Back greenish, sides and belly silvery with 6-8 dark cross-bands behind ventrals. Dorsal blackish. Pectorals with black tips, ventral yellow. 40 inches. Marine.
Moralla (S). Mural (T).

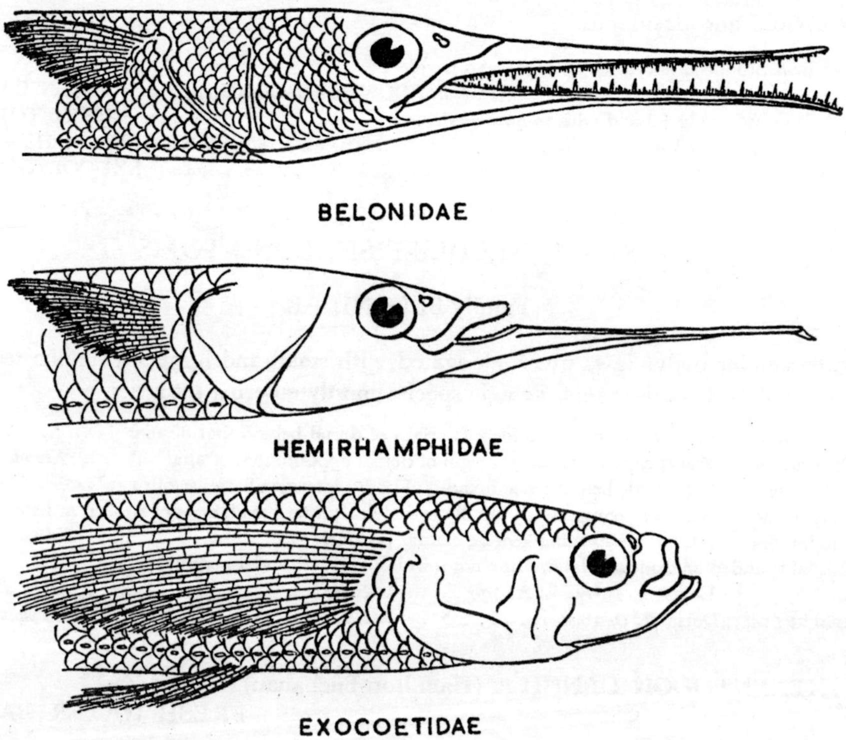

FIG. 9.
Principal structural differences between the three component families of the Order Beloniformes, showing relative development of upper and lower jaws.

200. TYLOSURUS CROCODILUS (Le Sueur) FORK-TAIL ALLIGATOR GAR

Belone crocodila Le Sueur, Journ. Acad. Nat. Sci. Philad., ii, 1821, p. 129. *Strongylura crocodila. Tylosurus crocodilus* Fowler, Journ. Acad. Nat. Sci. Phila. (2), xii (4), 1904, Pl. ix, fig. 1 (plate 13).

D. 22-23. A. 19-21. Depth 15, head 3 in length. Operculum not scaly. Teeth strong. Caudal forked. Blue-green above, silvery below. Fins mostly yellowish. 40 inches. Marine.
Dhandu theliya, Kallingu moralla, Theliya (S). Koliya mural, Periya mural (T).

201. TYLOSURUS STRONGYLURUS (Van Hasselt)

ROUND-TAIL ALLIGATOR GAR

Belone strongylura Van Hasselt, Alg. Konst- en Letterbode, i, 1823, p. 130. *Strongylura strongylura. Belone strongylurus* Day, Fish. India, 1875-78, Pl. cxviii, fig. 6 (plate 13).

D. 11-15. A. 15-17. L. lat. 170. Depth 16-17, head 3 in length. Operculum scaly. Teeth small, directed backwards. Caudal fin rounded. Greenish-brown above, silvery below. Silvery lateral band, more distinct posteriorly where it is bordered with a dark stripe. Blue-black spot at base of caudal fin. 18 inches. Marine, entering estuaries.

Dhiya moralla (S). Pambu mural (T).

202. TYLOSURUS LEIURUS (Bleeker) SQUARE-TAIL ALLIGATOR GAR

Belone leiurus Bleeker, Nat. Tijds. Ned. Ind., i, 1850, p. 94. *Strongylura leiurus. Tylosurus leiurus* Smith, Fishes Sth. Africa, 1950, Pl. vii, fig. 228 (plate 13).

D. 18-20. A. 23-25. L. lat. 185-200. Depth 18-29, head 3-3½ in length. Operculum scaly. Teeth slender, curved backwards. Caudal sub-truncate. Dorsal originates above 7th anal ray. Back bluish, sides and belly silvery. Silvery lateral streak bordered above with dark line. Fins yellowish. Caudal dusky. 48 inches. Marine.

Habareliya (S). Valai mural (T).

HALF-BEAKS, GARFISH

Family HEMIRHAMPHIDAE

Lower jaw developed into a long bony beak bearing a sensitive fringe. Upper jaw a triangular expansion. Teeth minute and cardiform. Body cylindrical or slightly compressed.

1. Caudal truncate or rounded; anal rays modified in male .... *Zenarchopterus dispar*
   Caudal forked; anal rays normal in male 2
2. Origin of anal opposite origin of dorsal; ventrals arising halfway between caudal base and eye 3
   Origin of anal behind origin of dorsal; ventrals arising much nearer caudal base than pectorals 5
3. Beak red-tipped; 7 rows of scales between dorsal base and lateral line .... *Hyporhamphus xanthopterus*
   Beak uniformly coloured; 5 rows of scales between dorsal base and lateral line 4
4. No scales on anal or dorsal .... *Hyporhamphus gaimardi*
   Scales present on anal and dorsal .... *Hyporhamphus unifasciatus*
5. Triangular part of upper jaw longer than broad .... *Hemirhamphus georgii*
   Triangular part of upper jaw broader than long .... *Hemirhamphus marginatus*

203. ZENARCHOPTERUS DISPAR (Valenciennes)

VIVIPAROUS HALF-BEAK

Hemirhamphus dispar Cuvier and Valenciennes, Hist. Nat. Poiss., xix, 1846, p. 58; Day, Fish. India, 1875-78, Pl. cxix, fig. 5 (plate 13).

D. 10-11. A. 11-12. L. lat. ca. 40. Caudal obtusely rounded. Anal originates below 3rd dorsal ray. Anal normal in females but 6th and 7th rays enormously enlarged and thickened in males. Back brownish with silver lateral band bordered above by black line. Fins dusky. 6½ inches. Rivers and estuaries.

204. HYPORHAMPHUS XANTHOPTERUS (Valenciennes)
RED-TIPPED HALF-BEAK

Hemirhamphus xanthopterus Cuvier and Valenciennes, Hist. Nat. Poiss., xix, 1846, p. 47. *Hemirhamphus erythrorhynchus. Hemirhamphus xanthopterus* Day, Fish. India, 1875-78, Pl. cxix, fig. 2 (plate 13).

D. 14-15. A. 16-17. L. lat. 56-58. Caudal forked, lower lobe longer. Anal originates opposite dorsal. No scales on dorsal and anal fins. Upper jaw longer than wide. Brilliant lateral band. End of beak coral-red. 6 inches. Coastal waters.

205. HYPORHAMPHUS GAIMARDI (Valenciennes) GAIMARD'S HALF-BEAK

Hemirhamphus gaimardi Cuvier and Valenciennes, Hist. Nat. Poiss., xix, 1846, p. 36. *Hemirhamphus limbatus* Day, Fish. India, 1875-78, Pl. cxix, fig. 3 (plate 13).

D. 14-15. A. 14-16. L. lat. 50. Caudal deeply forked. Anal originates opposite dorsal. No scales on dorsal and anal fins. Upper jaw broader than long. Yellowish green above with narrow silvery lateral band, bordered above by black line. Front edge of dorsal and borders of caudal black. 13 inches. Sea and estuaries.
Marandha, Moralla (S).

205a. HYPORHAMPHUS UNIFASCIATUS (Ranzani)
SILVER-LINED HALF-BEAK

Hemirhamphus unifasciatus Ranzani, Nov. Comm. Acad. Sc. Inst. Bonon., v, 1842, p. 326. *Hemirhamphus neglectus.*

D. 14-15. A. 14-16. L. lat. 50-55. Caudal forked, lower lobe longer. Anal originates opposite dorsal. Scales present on both dorsal and anal fins. Upper jaw wider than long. Bluish with a rather narrow silvery lateral band. Dorsal and caudal with black borders. Coastal waters. 8 inches.

206. HEMIRHAMPHUS GEORGII (Valenciennes) LONG-BILLED HALF-BEAK

Hemirhamphus georgii Cuvier and Valenciennes, Hist. Nat. Poiss., xix, 1846, p. 37; Day, Fish. India, 1875-78, Pl. cxx, fig. 2 (plate 13).

D. 13-17. A. 14-15. L. lat. 57-60. Caudal forked. Anal originates below 2nd or 3rd divided ray of dorsal. Ventrals originate twice nearer caudal base than corner of mouth. Yellowish green dorsally with broad silvery lateral band, bordered above by narrower black band. Fringe on lower jaw black. Dorsal and caudal bordered with black. 12 inches. Coastal waters.

207. HEMIRHAMPHUS MARGINATUS (Forskål) BARRED HALF-BEAK

Esox marginatus Forskål, Descript. Animal., 1775, p. 67. *Hemirhamphus marginatus* Bleeker, Atlas Ichth. Ind. Neerl., vi, 1872, Pl. ccliv, fig. 4 (plate 13).

D. 13-14. A. 10-13. L. lat. 53-57. Caudal forked. Anal originates below 5th dorsal ray. Ventrals originate twice nearer caudal base than tip of upper jaw. Blue-green above, silvery below with lateral silver band, bordered above by narrower black band. Dorsal

and caudal bordered with black; ventrals and anal sometimes tipped with black. Young with vertical bars on body and darker fins. 18 inches. Sea and estuaries.
Marandha, Thani hote moralla (S).

FLYING FISH

Family EXOCOETIDAE

Small surface fishes resembling the above family but without the lower jaw produced into a beak. Pectorals and sometimes the ventrals greatly enlarged to form wings used for gliding over the surface of the sea. Caudal fin deeply forked and with lower lobe longer. At least three species in seas surrounding Ceylon, probably considerably more.

1. Ventrals much shorter than head, much nearer to snout than caudal base .... *Exocoetus volitans*
   Ventrals longer than head, nearer caudal base than to snout 2
2. Only first ray of pectorals unbranched .... *Cypselurus comatus*
   At least first two rays of pectorals unbranched .... *Prognichthys gibbifrons*

208. EXOCOETUS VOLITANS Linnaeus TWO-WINGED FLYING FISH

Exocoetus volitans Linnaeus, Syst. Nat., ed. 10, 1758, p. 316. *Exocoetus evolans* Day, Fish. India, 1875-78, Pl. cxx, fig. 5 (plate 13).

D. 13-15. A. 13-14. L. lat. 40-43. 17-20 scales before dorsal. G.R. 24. Origins of dorsal and anal opposite. Pectorals reaching caudal base. Ventrals small, originating nearer tip of snout than caudal base. Bluish above, silvery below. Pectorals blackish with white border. 10 inches. Open sea.
Maha piha massa (S).

209. CYPSELURUS COMATUS (Mitchill) FOUR-WINGED FLYING FISH

Exocoetus comatus Mitchill, Trans. Lit. Philos. Soc. N.Y., i, 1815, p. 448. *Exocoetus bahiensis* Day, Fish. India, 1875-78, Pl. cxxi, fig. 10 (plate 13).

D. 12-14. A. 9-11. L. lat. 37-40. Predorsal scales 30-35. Origin of anal opposite 5th or 6th dorsal ray. Pectorals reaching little beyond dorsal origin. Ventrals longer than head, originating nearer caudal than eye. Bluish above, silvery below. Pectorals black with narrow border of white. Dorsal with dark blotch. 18 inches. Open sea.

209a. PROGNICHTHYS GIBBIFRONS (Valenciennes)
BLUNT-NOSED FLYING FISH

Exocoetus gibbifrons Cuvier and Valenciennes, Hist. Nat. Poiss., xix, 1847, p. 118. *Cypselurus hallbergi.*

D. 12-13. A. 9-10. Predorsal scales 20-24. Origin of anal opposite 4th dorsal ray. Pectorals reaching beyond end of dorsal. Ventrals much longer than head, originating about midway between caudal and eye. The first two and sometimes first four pectoral rays simple. Brown above, white below. Dorsal and caudal brown. Pectorals grey with light middle region. Ventral with darker rays in middle. 6 inches. Open sea.

DEEP-SEA COD

Order GADIFORMES

Family GADIDAE

Elongate body with small scales. Free caudal fin. Dorsal and anal fins long, the dorsal sometimes divided into 2 or 3 parts. No true fin spines. Deep water fishes.

210. PHYSICULUS ARGYROPASTUS Alcock — MANAAR COD

Physiculus argyropastus Alcock, Journ. Asiatic Soc. Bengal., lxii (2), 1893, p. 489; Illust. Zool. Investigator, v, 1898, Pl. xxii, fig. 1 (plate 14).

D. 8-9, 55. A. 57. V. 6. Dorsal divided into two parts. Large head with inconspicuous barbel on chin. All body and head with thin deciduous cycloid scales. Second ray of ventrals approximately as long as head. Light pinkish brown with silvery sheen. Fins scarlet. Belly, throat and gill-membranes black. 9 inches. 128-217 fathoms.

RAT-TAILS, WHIP-TAILS

Order MACRURIFORMES

Family MACRURIDAE

Elongate body with small scales. Tail tapering to a sharp point without free caudal fin. Short elevated dorsal in front, followed by a long fin which joins a similar anal fin around end of tail. Often barbel below chin. No true fin spines, except 2nd dorsal ray which can be spiniform and serrated. Deep water fishes.

1. Two conspicuous lens-shaped bodies in mid-ventral line of abdomen; more than 15 gill-rakers on lower limb of second arch .... *Hymenocephalus cavernosus*
   No lens-shaped bodies on abdomen; less than 15 gill-rakers on lower limb of second arch 2
2. Second dorsal fin better developed than anal; second ray of first dorsal not spine-like .... *Bathygadus furvescens*
   Anal fin better developed than second dorsal; second ray of first dorsal modified into a spine 3
3. First dorsal fin with long serrate spine .... *Macrurus wood-masoni*
   First dorsal fin spine smooth .... *Coelorhynchus parallelus*

211. HYMENOCEPHALUS CAVERNOSUS (Goode and Bean)

SILVER RAT-TAIL

Bathygadus cavernosus Goode and Bean, Proc. U.S. Nat. Mus., 1885, p. 598. *Macrura (Mystaconurus) heterolepis* Alcock, Illust. Zool. Investigator, 1892, Pl. iii, fig. 3 (plate 14).

D. II, 10; 133. A. 27 in a space equal to length of head. V. 11. P. 11-13. Depth 7, head 6 in total length. Teeth in villiform bands. Gill-rakers very short, about 18 below angle. Two lens-shaped probably luminous organs in mid-ventral region connected by streak of dark tissue. Light silvery. 12 inches. 180-217 fathoms.

212. BATHYGADUS FURVESCENS Alcock

DUSKY RAT-TAIL

Bathygadus furvescens Alcock, Journ. Asiatic Soc. Bengal, xliii, 1894, p. 128; Alcock, Illust. Zool. Investigator, 1895, Pl. xvi, fig. 1 (plate 14).

D. II, 8-9. P. 15. V. 8. Head 5 in total length. Operculum with 2 weak prominent spines. Teeth in maxillary and intermaxillary in villiform bands. Gill-rakers setiform, 5+(17-19) on first arch. No mandibulary barbel. Uniform dark brown. Fins, branchial and buccal cavities black. 21 inches. 140-719 fathoms.

213. MACRURUS WOOD-MASONI Alcock

WOOD-MASON'S RAT-TAIL

Macrurus wood-masoni Alcock, Ann. Mag. Nat. Hist., vi, 1890, p. 301; Illust. Zool. Investigator, 1894, Pl. xiii, fig. 1 (plate 14).

D. II, 9; ca. 100. A. ca. 105. P. 21. V. 8. L. lat. ca. 130. Tr. 4+ca. 22. Depth $\frac{3}{4}$ head, which is $4\frac{1}{2}$ in total length. Snout with strong tubercles. Small conical teeth in broad bands in both jaws. Barbel reduced to small papilla. Chocolate with black fins and black lining to mouth and pharynx. 19 inches. 738-902 fathoms.

214. COELORHYNCHUS PARALLELUS (Günther) SPINY RAT-TAIL

Macrurus parallelus Günther, Ann. Mag. Nat. Hist. (2), xx, 1877, p. 439; Challenger Repts., xxii, 1887, Pl. xxix, fig. A (plate 14).

D. II, 9. A. 80-90. P. 16-17. Snout pointed in front. Head and body with scales which are strongly spinose. Teeth in villiform bands. Second spine of first dorsal weak, filamentous in distal part. Uniform brown. 18 inches. 597 fathoms.

PIPE FISH, SEA HORSES, FLUTE MOUTHS, RAZOR FISH

Order SYNGNATHIFORMES

Curious specialized fishes characterized externally by a small mouth at the end of a tube-like snout. Body with either scales or a protective bony sheath, or encased in stout bony rings. Mostly small sluggish fishes concealing themselves amongst weeds.

1. Mouth toothed; body covered with scales, minute spines or naked; continuous lateral line .... (*Suborder* AULOSTOMOIDEI) 2
   Mouth toothless; body covered in dermal plates; no lateral line ... (*Suborder* SYNGNATHOIDEI) 3
2. Body compressed, scaly; dorsal fin divided into spinous and long soft parts; anus far behind ventrals .... *Family* AULOSTOMIDAE, p. 80
   Body depressed, without scales; short soft dorsal fin only; anus close to ventrals .... *Family* FISTULARIDAE, p. 80
3. Two dorsal fins; trunk encased in cuirass of bony plates .... *Family* CENTRISCIDAE, p. 81
   Two dorsal fins; skin with series of large stellate bony plates ...... *Family* SOLENOSTOMIDAE, p. 81
   One or two dorsal fins; ventrals absent; body completely encased in armour of bony rings .... *Family* SYNGNATHIDAE, p. 81

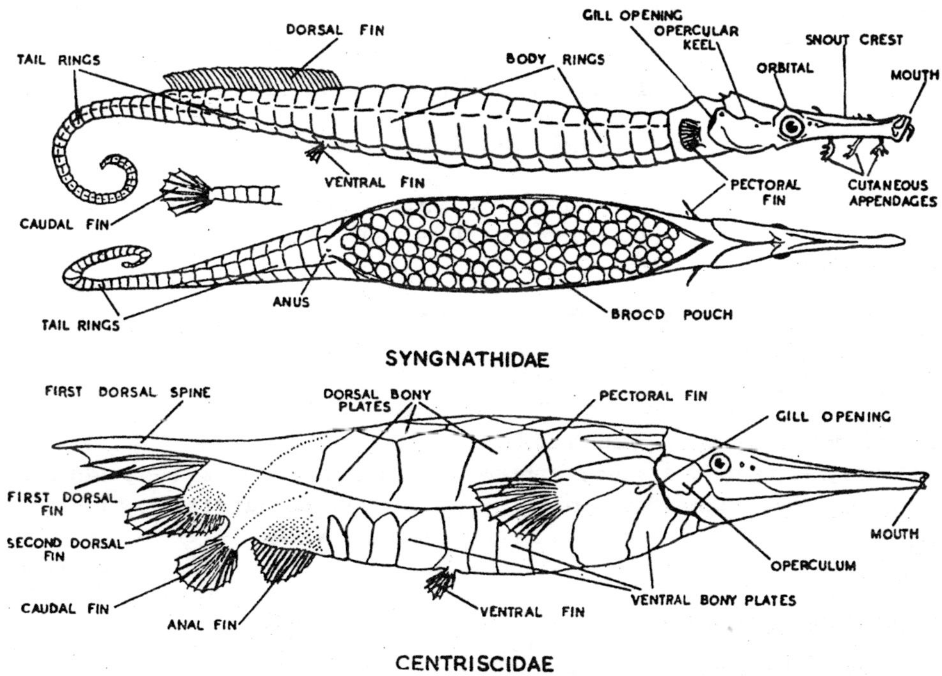

FIG. 10.

Explanation of characters of two families of the Order Syngnathiformes. Above.—Lateral view of Pipe fish showing general external anatomy, and ventral view of male showing detail of eggs implanted in the brood-pouch. Below.—External anatomy of a Razor fish.

PAINTED FLUTE MOUTHS, CORNET FISH

Family AULOSTOMIDAE

Pronounced tube-like snout with small mouth at end. Body elongate, compressed and with minute scales. Spinous dorsal fin present but reduced to a row of spines. A barbel below chin.

215. AULOSTOMA CHINENSIS Lacépède — PAINTED FLUTE MOUTH

Aulostomus chinensis Lacépède, Hist. Nat. Poiss., v, 1803, p. 357. *Aulostoma valentini. Aulostomus chinensis* Günther, Jour. Mus. Godeffroy, iv (15), 1881, Pl. cxxiii, fig. B (plate 14).

D. VIII-XII, 24-27. A. 4, 22-25. P. 17. V. 6. No gill-rakers. Second dorsal and anal at extreme posterior. Ventrals beneath spinous dorsal and midway between eye and caudal base. Colour variable, usually rather colourful. Generally brownish or yellowish with cross-bars or narrow longitudinal white patches. Sometimes black spot on caudal. 20 inches. Shallow water marine.

Maha nalava (S). Kovinchi (T).

HAIR-TAILED FLUTE MOUTHS

Family FISTULARIDAE

Elongate fleshy depressed body, wider than deep, without scales, but with bony plates beneath the skin in parts. Mid-caudal rays filamentous.

Naked; no scales midway on back .... *Fistularia petimba*

Body rough; narrow scales down back before and behind dorsal .... *Fistularia villosa*

216. FISTULARIA PETIMBA Lacépède — SMOOTH FLUTE MOUTH

Fistularia petimba Lacépède, Hist. Nat. Poiss., v, 1803, p. 349.

D. 14-17. A. 14-16. Body much depressed. Interorbital flat. Skin smooth. Brown above, lighter below. Two narrow blue stripes laterally along upper part of sides. 5 feet. Shallow water marine, swimming head down amongst mud and weeds.

Malava (S). Kovinchi (T).

217. FISTULARIA VILLOSA Klunzinger — ROUGH FLUTE MOUTH

Fistularia villosa Klunzinger, Abh. Zool. Bot. Ges. Wien, xxi, 1871, p. 516. *Fistularia serrata* Day, Fish. India, 1875-78, Pl. lxxvi, fig. 3 (plate 14).

D. 13-15. A. 13-15. Body not greatly depressed. Interorbital concave, ridges on head strongly serrate. Brownish above, white below with traces of broad cross-bars. Tips of fins dusky. 5 feet. Shallow water marine.

RAZOR FISH

Family CENTRISCIDAE

Elongate blade-like body covered with bony plates which are expansions of vertebral column. Sharp edge below. First dorsal spine is at end of body and most of dorsal and caudal fins are on lower surface.

218. CENTRISCUS SCUTATUS Linnaeus — RAZOR FISH

Centriscus scutatus Linnaeus, Syst. Nat. ed. 10, 1753, p. 336. *Amphisile scutata. Centriscus scutatus* Weber and De Beaufort, Fish. Indo-Aust. Archipel., iv, 1922, fig. 11 (plate 14).

D. III, 10-12. A. 11-12. Body form unmistakable. Silvery with dark lateral band. Ventral plates golden with 7-8 conspicuous cross-bars. 6 inches. Marine shores to 80 metres including pearl banks. Swim in shoals, head up.
Ihi maluva (S).

FALSE PIPE FISHES

Family SOLENOSTOMIDAE

Compressed rather knobbly body with large fins, the ventrals being below the 1st dorsal. Females have brood pouch on belly formed by the ventrals.

219. SOLENOSTOMUS PARADOXUS (Pallas) — SHORT-BODIED PIPE FISH

Fistularia paradoxa Pallas, Spicil. Zool., viii, 1870, p. 32. (?) *Solenostoma lacinatum. Solenostomus paradoxus* Weber, Siboga Exped., lvii, Fische, 1913, fig. 33b (plate 14).

D. V, 18-22. A. 18-23. P. 24-26. V. 7. Characteristic appearance with short body with dermal ossifications and large fins. Light brown with irregular orange spots. Black blotches on spinous dorsal. Caudal dusky. 6½ inches. Pearl banks.

PIPE FISHES, SEA HORSES

Family SYNGNATHIDAE

Moderate to very elongate body encased in bony rings. Fins small. Caudal absent sometimes. Ventrals absent. Degenerate slow swimming fishes with tube-like mouths and living amongst weeds. Eggs are attached to belly of male, sometimes enclosed in a pouch.

1. Caudal fin present 2
   Caudal fin absent 11
2. Dorsal edges of trunk and tail continuous .... *Ichthyocampus carce*
   Dorsal edges of trunk and tail discontinuous 3
3. Opercular keel rectilinear; base of dorsal not elevated 4
   Opercular keel convex, bent upwards to gill-opening, with radial lines; base of dorsal elevated 9
4. Egg-pouch abdominal 5
   Egg-pouch sub-caudal 7

5. Anus in posterior half of body; trunk and tail rings equal in number; snout longer than rest of head .... *Microphis brachyurus*
   Anus in anterior half of body; more tail rings than trunk rings; snout shorter than rest of head 6
6. Dorsal 37-40; ocelli along sides of trunk .... *Dorichthys ocellatus*
   Dorsal 50-53; diffuse spots along upper edges of sides of trunk .... *Dorichthys cunculus*
7. Dorsal fin far behind vent .... *Urocampus southwelli*
   Dorsal fin opposite or near vent 8
8. Operculum crossed by bony ridge horizontally; orbital region prominent .... *Corythoichthys fasciatus*
   Operculum without horizontal ridge except anteriorly; orbital region not prominent .... *Syngnathus spicifer djarong*
9. Edges of shields and head ridges spinous .... *Halicampus koilomatodon*
   Edges of shields and head ridges smooth 10
10. Snout less than half head length, with serrated crest .... *Trachyrhamphus serratus*
   Snout more than half head length, with low rough median ridge .... *Trachyrhamphus longirostris*
11. Body depressed; head in line with trunk .... *Syngnathoides biaculeatus*
   Body compressed; head and trunk at right angles .... *Hippocampus kuda*

220. ICHTHYOCAMPUS CARCE (Hamilton-Buchanan)

FRESH WATER PIPE FISH

Syngnathus carce Hamilton-Buchanan, Fish. Ganges, 1822, p. 13. *Ichthyocampus carce* Day, Fish. India, 1875-78, Pl. clxxv, fig. 2 (plate 14).

D. 23-27. A. 2-3. P. 13-17. Rings (14-15)+(37-40). Dorsal on (2-3)+(7-9) tail rings. Tail more than twice trunk. No cutaneous appendages. Brown with white spot or black-edged ocellus on each trunk ring. Ventral surface of tail with brown and yellow dots. Caudal black. 4½ inches. Sea, estuaries and enters fresh water.

221. MICROPHIS BRACHYURUS (Bleeker) SHORT-TAILED PIPE FISH

Syngnathus brachyurus Bleeker, Verh. Bat. Genootsch., xxv, 1853, p. 16. *Microphis brachyurus* Weber and De Beaufort, Fish. Indo-Aust. Arch., iv, 1922, fig. 21 (plate 14).

D. 36-48. A. 3-4. P. 18-23. Rings (19-22)+(20-24). Dorsal on (1-3)+(6-9). Tail without caudal shorter than trunk. Median caudal ray enlarged. Brownish, lighter below. Operculum and anterior part of trunk with black spots. Dark diffuse lateral band. Estuaries and fresh water rivers. 9 inches.
Loku ata theliya, Vatakeya moralla (S).

222. DORICHTHYS OCELLATUS Duncker OCELLATED PIPE FISH

Dorichthys ocellatus Duncker, Spolia Zeyl., vii, 1910, p. 28, Pl., fig. A (plate 14).

D. 37-40. A. 4. P. 17-19. Rings (15-16)+(31-32). Dorsal on (1-2)+(7-8). Lower lateral edges of trunk and tail discontinuous. Operculum with straight keel and often 2 weaker ones. Anal fin in front of middle of total length. Rings smooth in adult, spiny in young. Trunk rings each with a black white-bordered ocellus. Back light reddish brown, sides yellow grey. Dark lateral stripe on head through eye. Caudal with yellow margins. 4 inches. Estuarine, entering fresh water.
Punchi ata theliya (S).

223. DORICHTHYS CUNCULUS (Hamilton-Buchanan)
CROCODILE-TOOTH PIPE FISH

Syngnathus cunculus Hamilton-Buchanan, Fish. Ganges, 1822, p. 12. *Dorichthys cunculus* Day, Fish. India, 1875-78, Pl. clxxiv, fig. 4 (plate 14).

D. 50-53. A. 2-3. P. 18-20. Rings (17-18)+(25-28). Dorsal on (2-3)+(7-9). Green above, dirty white below. Network of dusky lines on back. Longitudinal spotted stripe along sides. Opercle silver. Caudal reddish. 7 inches. Estuaries and fresh water rivers.

224. UROCAMPUS SOUTHWELLI Duncker SOUTHWELL'S PIPE FISH

Urocampus southwelli Duncker, Spolia Zeyl., vii, 1910, p. 30, Pl., fig. C (plate 14).

D. 14. A.o. P. 8-10. Rings 8+(45-50). Dorsal on o+(7-12). Opercle keeled for anterior $\frac{2}{3}$. No cutaneous appendages. Rostrum longer than postorbital region of head. Uniformly yellow brown. $1\frac{3}{4}$ inches. Pelagic in harbours.

225. CORYTHOICHTHYS FASCIATUS (Gray) BANDED PIPE FISH

Syngnathus fasciatus Gray, Illustr. Indian Zool., i, 1830, p. 89. *Syngnathus* or *Corythoichthys conspicillatus*. *Corythroichthys waitei* Jordan and Seale, Bull. U.S. Bur. Fish., xxv, 1906, fig. 17 (plate 14).

D. 25-32. A. 3-4. P. 14-18. Rings (15-18)+(33-37). Dorsal on (0-1)+(5-6). Shields transversely striated and corrugated. Crest on each side of head above eyes. Tail about twice length of trunk. Light grey with numerous broad dark cross-bands. Operculum with dark longitudinal bands. 7 inches. Around coral reefs and near shore.

226. SYNGNATHUS SPICIFER DJARONG (Bleeker) SLENDER PIPE FISH

Syngnathus djarong Bleeker, Verh. Bat. Genootsch., xxv, 1853, p. 22; Duncker and Mohr, Mitteil. Zool. Mus. Hamburg, xli, 1925, Pl. ii, fig. 5 (plate 14).

D. 23-29. A. 2-3. P. 13-16. Rings (14-16)+(39-43). Dorsal on 2nd-6th or 3rd-8th tail rings. Shields transversely striated with smooth edge. Tail about twice as long as trunk. Trunk reticulated with brown, tail spotted. 3 black bands radiating from eye. $5\frac{1}{2}$ inches. Estuaries, entering fresh water.
Ata theliya (S).

227. HALICAMPUS KOILOMATODON (Bleeker) SPINY PIPE FISH

Syngnathus koilomatodon Bleeker, Act. Soc. Sci. Indo-Neerl., v, 1858, p. 10. *Halicampus koilomatodon* Weber and De Beaufort, Fish. Indo-Aust. Arch., iv, 1922, fig. 43 (plate 14).

D. 19-22. A. 3-5. P. 16-19. Rings (17-18)+(33-36). Dorsal on (2-3)+(1-3). Shields transversely striated, edges prominent and serrated. Cutaneous appendages, best developed in male. Brownish with white marblings. Operculum with or without white stripes. 6 inches. Eggs in complete brood pouch. In sea, deeper water near shore.

228. TRACHYRHAMPHUS SERRATUS (Schlegel) CRESTED PIPE FISH

Syngnathus serratus Schlegel, Fauna Japonica, Poiss., 1847, p. 272; Day, Fish. India, 1875-78, Pl. clxxiii, fig. 4 (plate 14).

D. 25-29. A. 3-4. P. 14-19. Rings (21-23)+(44-50). Dorsal on (2-4)+(2-3). Eye large. Spiny serrated crest on head. Cutaneous appendages. Brown with 9-12 broad dark diffuse cross-bands. Also white spots. 12 inches. Marine, along coastal shores.

229. TRACHYRHAMPHUS LONGIROSTRIS Kaup
LONG-SNOUTED PIPE FISH

Trachyrhamphus longirostris Kaup, Cat. Lophobr. Fish., 1856, p. 24. *Syngnathus longirostris* Day, Fish. India, 1875-78, Pl. clxxiii, fig. 5 (plate 14).

D. 27. A. 4. P. 18. Rings 23+49. Length of snout more than half length of head. No serrated crest on head. Egg-pouch not half as long as tail. Greyish brown with darker rings. Under surface of snout light with dark spots. Pearl banks.

230. SYNGNATHOIDES BIACULEATUS (Bloch) TWO-BARBEL PIPE FISH

Syngnathus biaculeatus Bloch, Nat. Ausl. Fische, i, 1785, p. 10. *Gastrotokeus biaculeatus. Syngnathoides biaculeatus* Smith, Fish. Sth. Africa, 1950, fig. 357 (plate 14).

D. 37-50. A. 4-6. P. 20-23. Rings (15-17)+(40-54). Dorsal on (1-2)+(8-10). Skinny flaps on body and head, sometimes a pair of branched barbels on chin. Dark brown or green, spotted. 10 inches. Shallow water marine.
Moralla (S).

231. HIPPOCAMPUS KUDA Bleeker SEA HORSE

Hippocampus kuda Bleeker, Nat. Tijds. Ned. Ind., iii, 1852, p. 82. *Hippocampus guttulatus* and (?) *villosus. Hippocampus fisheri* Jordan and Evermann, Bull. U.S. Fish. Comm., xxiii (1), 1905, fig. 36 (plate 14).

D. 15-18. A. 4. P. 17-18. Rings 11+(33-37). Dorsal on 2+2 rings. Body very deep in thoracic region. Head at right angles to trunk. Dark brown or blackish with darker transverse bands or elongate spots. Sometimes with white dots. 12 inches, usually less. Marine littoral.
Mudhu asvaya (S). Kodai kutherai (T).

TOP-MINNOWS, KILLIFISH

Order CYPRINODONTIFORMES

Family CYPRINODONTIDAE

Small fish with no lateral line. Head depressed. Jaws dentigerous. Surface feeding fishes. Fresh water and estuarine. Useful mosquito larvivores.

1. Anal with 13-17 radials 2
   Anal with 20-24 radials .... *Panchax melastigma*
2. L.lat. 32-33; ventral rays produced in male .... *Panchax lineatus dayi*
   L.lat. 27-29; ventral rays not produced .... *Panchax panchax blochii*

232. PANCHAX MELASTIGMA McClelland ESTUARINE TOP-MINNOW

Aplocheilus melastigmus McClelland, As. Res., xix, 1839, p. 301, 427. *Oryzias melastigma, Haplochilus melanostigma. Haplochilus melastigma* Day, Fish. India, 1875-78, Pl. cxxi, fig. 4 (plate 15).

D. 6-7. A. 20-24. L. lat. 27. L. tr. 9-11. Ventrals without prolonged rays. Anal rays sometimes prolonged. Caudal rounded. Back dull green, white on abdomen and outer portion of anal rays white edged. Narrow dark line along middle of side terminating with spot on base of caudal. 1½ inches. Estuarine.
Diya pita handheya, Handhe titteya (S).

233. PANCHAX LINEATUS DAYI (Steindachner) STRIPED TOP-MINNOW

Haplochilus dayi Steindachner, Denschr. Akad. Wien, lix, 1892, p. 357. *Aplochilus* or *Panchax lineatus*, (?) *Aplocheilus rubrostigma. Haplochilus lineatum* Day, Fish. India, 1875-78, Pl. cxxi, fig. 6 (plate 15).

D. 3-4. A. 13-16. L. lat. 32-33. L. tr. 9. Predorsals 25. Ventral rays in male prolonged, reaching half way to base of anal fin. Anal elongated to reach middle of caudal fin. Dorsally olive with bright green reflections. Sides yellow with red spots. 1-4 dark cross-bars on body. Fins yellow with red or orange rays, caudal and anal with red margins. Females with more (6-8) cross-bars and hyaline fins. 2¼ inches. Fresh waters of coastal plains, principally wet zone.
Handhe naleya, Iri nale handheya, Iri udda (S).

234. PANCHAX PANCHAX BLOCHII (Arnold) LESSER TOP-MINNOW

Haplochilus panchax var. *blochii* Arnold, Wochenschr. Aquarien und Terrarienkunde, viii, 1911, p. 672. *Panchax panchax, Panchax* or *Aplocheilus parvus. Haplochilus panchax* Day, Fish. India, 1875-78, Pl. cxxi, fig. 3 (plate 15).

D. 3-6. A. 14. L. lat. 27-29. L. tr. 9. Predorsals 22-23. Ventral fins not elongated. Light olive with black spot extending from 2nd to 4th dorsal fin rays. 1½ inches. Fresh and brackish water of coastal plains.
Udda (S).

RED-FISH, SLIME HEADS, KNIGHT FISH, SOLDIER FISH

Order BERYCIFORMES

Fairly deep body with normal scales or bony scutes, often spiny. Often bony ridges with an elaborate muciferous canal system on head. Some deep water species but mostly shallow water marine fishes of coral reefs.

1. Scales forming bony case .... *Family* MONOCENTRIDAE, p. 86
   Scales normal but with strong ctenoid border 2
2. Dorsal spines 5-8 .... *Family* TRACHICHTHYIDAE, p. 86
   Dorsal spines 11-12 .... *Family* HOLOCENTRIDAE, p. 87

KNIGHT FISH

Family MONOCENTRIDAE

Small plump fishes with scales enlarged to form a solid rough armour.

235. MONOCENTRUS JAPONICUS (Houttuyn) KNIGHT FISH

Gasterosteus japonicus Houttuyn, Verh. Holl. Maatsch. Wet. Haarlem, xx, 1782, p. 329. *Monocentrus japonicus* Bleeker, Atlas Ichth. Ind. Neerl., ix, 1877, Pl. ccclvi, fig. 4 (plate 15).

D. V-VI, 11-12. A. I-II, 8-9. L. lat. 15. Dorsal spines stick out sideways when erect. Phosphorescent organ each side of lower jaw. Yellowish-brown. 5 inches. Marine, 20-100 fathoms.

RED-FISH, SLIME HEADS

Family TRACHICHTHYIDAE

Body fairly deep, moderately compressed. Head large, with thin surface bones bearing bony ridges bordering mucous canals. Fins with few spines but body rough. Deep water marine fishes.

Dorsal spines 6 .... *Hoplostethus mediterraneus*
Dorsal spines 8 .... *Gephyroberyx darwini*

236. HOPLOSTETHUS MEDITERRANEUS Valenciennes
MEDITERRANEAN RED-FISH

Hoplostethus mediterraneus Cuvier and Valenciennes, Hist. Nat. Poiss., iv, 1829, p. 469. Goode and Bean, Oceanic Ichth., 1895, Pl. lvi, fig. 208 (plate 15).

D. VI, 12-15. A. III, 9-11. L. lat. 28-30. Belly with 11-13 enlarged scutes. G.R. 15. Rosy to reddish. 12 inches. 296-320 fathoms off Ceylon coast.

237. GEPHYROBERYX DARWINI (Johnson) DARWIN'S RED-FISH

Trachichthys darwinii Johnson, Proc. Zool. Soc., 1866, p. 311, Pl. xxxii (plate 15).

D. VIII, 14. A. III, 11-12. L. lat. 27-30. Belly with 10-12 scutes. 18 inches. Reddish. 296-320 fathoms off Ceylon coast.

SOLDIER FISH, SQUIRREL FISH

Family HOLOCENTRIDAE*

Brilliantly coloured shallow water fishes, with spiny fins, serrations and spines on operculum and scales with rough ctenoid border. Anal spines 4.

1. Long spine at angle of preoperculum 2
   No spine at preoperculum angle 6
2. L.lat. 46-48 .... *Holocentrus diadema*
   L.lat. 33-44 3
3. Nasalia rounded in front 4
   Nasalia ending anteriorly in 2 divergent spines 5
4. L.lat. 39-40; dark brown spot on anterior part of spinous dorsal .... *Holocentrus ittodai*
   L.lat. 41-44; red margin to spinous dorsal .... *Holocentrus lacteo-guttatus*
5. Snout blunt, much shorter than eye; no silvery patch on caudal peduncle .... *Holocentrus rubrum*
   Snout sharp, sub-equal to eye; silvery patch on caudal peduncle .... *Holocentrus caudimaculatum*
6. L.lat. 36-40 .... *Myripristis pralinius*
   L.lat. 28-32 .... *Myripristis murdjan*

238. HOLOCENTRUS DIADEMA Lacépède CROWNED SOLDIER FISH

Holocentrus diadema Lacépède, Hist. Nat. Poiss., v, 1803, p. 372. *Holocentrum diadema* Bleeker, Atlas Ichth. Ind. Neerl., ix, 1877, Pl. ccclvi, fig. 1 (plate 15).

D. XI-XII, 12-13. A. IV, 8-9. L. lat. 46-48. Tr. 3+8. Depth 3¼ in length. 8 series of scales on preoperculum. Reddish with light bands along body. Membrane of spinous dorsal black with white band and white margin. Membrane between 3rd and 4th anal spines black. 9 inches. Marine, littoral.

239. HOLOCENTRUS ITTODAI Jordan and Fowler

JAPANESE SOLDIER FISH

Holocentrus ittodai Jordan and Fowler, Proc. U.S. Nat. Mus., xxvi, 1903, p. 16; Vasiliu, Jen. Zeitsch., lxvi, 1931, Pl. vii, fig. 1 (plate 15).

D. XI-XII, 13-14. A. IV, 9. L. lat. 38-40. Tr. 3+6. Depth 4 in length. Reddish with lighter bands along body. Membrane of spinous dorsal white with dark spot between 1st and 3rd spines. 4-5 inches. Marine, littoral.

240. HOLOCENTRUS LACTEO-GUTTATUS Cuvier

WHITE-SPOTTED SOLDIER FISH

Holocentrum lacteo-guttatum Cuvier and Valenciennes, Hist. Nat. Poiss., iii, 1829, p. 214. *Holocentrus argenteus. Holocentrum punctatissimum* Bleeker, Atlas Ichth. Ind. Neerl., ix, 1877, Pl. ccclix, fig. 2 (plate 15).

D. XI-XII, 12-13. A. IV, 8-10. L. lat. 40-44. Tr. 3+8. Depth 3·3 to 3·7 in length. Pinkish silvery, lighter stripes being wider than ground colour. Dorsal whitish with reddish marginal band. 5 inches. Marine, littoral.
Pookooranah (S).

* The identity of *Myripristis blochii* of Malpas (1926), p. 34 from Pedro Bank, is not clear.

241. HOLOCENTRUS RUBRUM (Forskål) RED SOLDIER FISH

Sciaena rubra Forskål, Descript. Anim., 1775, p. 48. *Holocentrum rubrum* Bleeker, Atlas Ichth. Ind. Neerl., ix, 1877, Pl. ccclvii, fig. 4 (plate 15).

D. XI-XII, 12-14. A. IV, 8-10. L. lat. 33-36. Tr. 3+6. Depth 3-3½ in length. Dark reddish brown with silvery bands along scale rows. Upper part of spinous dorsal membrane blackish. Membrane between 3rd and 4th anal spines black. 8 inches. Marine, littoral.
Ratoo pahayah, Siripunakkaliya (S). Chenguli, Kithy panakanni (T).

242. HOLOCENTRUS CAUDIMACULATUS Rüppell
SPOT-TAILED SOLDIER FISH

Holocentrum caudimaculatum Rüppell, Atlas Reise Nordl. Afrika, Fische, 1826-1831, p. 97. (?) *Holocentrum spiniferum. Holocentrum caudimaculatum* Bleeker, Atlas Ichth. Ind. Neerl., ix, 1877, Pl. ccclvi, fig. 3 (plate 15).

D. XI, 14. A. IV, 9. L. lat. 41. Tr. 3+8. Depth 3-3¼ in length. Reddish or yellowish with conspicuous silvery patch on upper part of caudal peduncle, behind dorsal. Reddish-brown band along base of dorsal and anal. 8 inches. Marine, littoral.

243. MYRIPRISTIS PRALINIUS Cuvier
SMOOTH-CHEEKED SOLDIER FISH

Myripristis pralinius Cuvier and Valenciennes, Hist. Nat. Poiss., iii, 1829, p. 170. *Myripristis bleekeri* Bleeker, Atlas Ichth. Ind. Neerl., ix, 1877, Pl. ccclvi, fig. 5 (plate 15).

D. X, 15-17. A. IV, 12-15. L. lat. 36-40. Tr. 3+7. Depth 3-3½ in length. No spine at angle of preoperculum. Reddish-brown with gold reflections and broad silvery longitudinal bands along scale rows. 7 inches. Marine, littoral.
Kalu varal punakkalļiya (S). Karuppu val putchakanni (T).

244. MYRIPRISTIS MURDJAN (Forskål) BLACK-TIPPED SOLDIER FISH

Sciaena murdjan Forskål, Descript. Animal., 1775, p. 48. *Myripristis murdjan* Bleeker, Atlas Ichth. Ind. Neerl., ix, 1877, Pl. ccclx, fig. 3 (plate 15).

D. XI, 13-15. A. IV, 12-14. L. lat. 28-30. Tr. 3+7. Depth 2¾ in length. No spine in angle of preoperculum. Dark red with dusky longitudinal bands above lateral line. Black border to operculum and black tips to soft dorsal, anal and caudal fin lobes. Fins otherwise dark red. Black markings sometimes absent. 12 inches. Marine, littoral.
Balal maruva, Lota, Vatta kyava (S). Parullam, Punakanni (T).

BOAR FISH

Order ZEIFORMES

Family CAPROIDAE

Compressed deep body with small ctenoid scales. Mouth small with bands of fine teeth in jaws and sometimes palate. Deep water oceanic fishes.

245. ANTIGONIA RUBESCENS (Günther) BOAR FISH

Hypsinotus rubescens Günther, Cat. Fish., ii, 1860, p. 63. *Antigonia capros. Antigonia rubescens* Smith, Fish. Sth. Africa, 1950, Pl. xxix, fig. 579 (plate 15).

D. IX, 26-29. A. III, 25-27. L. lat. 55-60. Tr. 14+10. G.R. 14. Brownish orange, silvery below. 7 inches. Deep water marine, off coast in 296-320 fathoms.

PIKE, BARRACUDA, GREY MULLET, HARDYHEADS

Order MUGILIFORMES

Elongate body with small to large scales. Two dorsal fins, first of a few spines. Caudal forked to emarginate. Coastal and estuarine species, some entering fresh water.

1. Lateral line well developed; jaws with large fang-like teeth; gill-rakers obsolete .... *Family* SPHYRAENIDAE, p. 90
   Lateral line absent or rudimentary; teeth very small; gill-rakers long and slender 2
2. Dorsal with 4 spines; anal with 3 spines; eyes more easily visible from below ... *Family* MUGILIDAE, p. 91
   Dorsal with 5-6 spines; anal with 1 spine; eyes lateral .... *Family* ATHERINIDAE, p. 94

SEA PIKE, BARRACUDA

Family SPHYRAENIDAE

Long cylindrical body with small scales. First dorsal fin with 5 spines. Large mouth, with strong canines in jaws and palate. Lower jaw projecting. Large, voracious predatory fishes.

1. L.lat. 80-90 2
   L.lat. 110-130 4
2. Angle of preoperculum rounded .... *Sphyraena picuda*
   Angle of preoperculum rectangular 3
3. Depth 6; spinous dorsal originates before tip of pectorals .... *Sphyraena obtusata*
   Depth 7-8; spinous dorsal originates above or behind tip of pectorals .... *Sphyraena langsar*
4. L. tr. 18+22; operculum with 2 points .... *Sphyraena jello*
   L. tr. 14+18; operculum with 1 point .... *Sphyraena acutipinnis*

246. SPHYRAENA PICUDA Bloch — COMMERSON'S SEA PIKE

Sphyraena sphyraena var. *picuda* Bloch and Schneider, Syst. Ichthyol., 1801, p. 110. *Sphyraena commersoni.* Sketch of a New Guinea specimen by A. M. Rapson (plate 16).

D. V; II, 8. A. II, 7-8. L. lat. 80-90. Tr. 13+14. Angle of preoperculum rounded. Operculum with 2 spines. Dorsal originates above tip of pectorals and behind ventrals origin. Silvery, bluish above. Dorsal, caudal and anal black with white tips. 6 feet. Coastal.

247. SPHYRAENA OBTUSATA Cuvier — BLUNT-JAWED SEA PIKE

Sphyraena obtusata Cuvier and Valenciennes, Hist. Nat. Poiss., iii, 1829, p. 350; Günther, Journ. Mus. Godeffroy, iv (13), 1877, Pl. cxix, fig. B (plate 16).

D. V; I, 9. A. II-III, 8-9. L. lat. 82-85. Tr. 8+10. Angle of preoperculum rectangular. Operculum without spines. Dorsal originates before tip of pectorals and behind origin of ventrals. Fleshy knob at tip of lower jaw. Green-grey above with indistinct darker bars, white below. Pectorals grey, other fins yellow. 16 inches. Coastal, entering estuaries. Theliya, Ulava (S).

248. SPHYRAENA LANGSAR Bleeker SHORT-JAWED SEA PIKE

Sphyraena langsar Bleeker, Nat. Tijds. Ned. Ind., vii, 1854, p. 367. *Sphyraena brachygnathus. Sphyraena langsar* Whitley, Aust. Mus. Mag., ix (10), 1949, fig. on p. 343 (plate 16).

D. V; I, 8-9. A. II-III, 8-9. L. lat. 82-87. Tr. 6+7. Angle of preoperculum rectangular. Operculum without spines. Dorsal originates slightly behind tip of pectorals and behind origin of ventrals. Greyish above, silvery below. Fins yellowish. 12 inches. Inshore marine.

249. SPHYRAENA JELLO Cuvier GIANT SEA PIKE

Sphyraena jello Cuvier and Valenciennes, Hist. Nat. Poiss., iii, 1829, p. 349; Weber and De Beaufort, Fish. Indo-Aust. Archip., iv, 1922, fig. 66 (plate 16).

D. V; II, 8. A. II, 8. L. lat. 123-130. Tr. (17-18)+(20-23). Angle of preoperculum rounded. Operculum with 2 spines. Dorsal originates above tip of pectorals and behind origin of ventrals. Grey above, white below with short vertical cross-bars on upper part. Ventrals whitish and other fins dusky. 5 feet. Coastal waters and pearl banks.
Silava (S). Jeela (T).

250. SPHYRAENA ACUTIPINNIS Day POINTED-FINNED SEA PIKE

Sphyraena acutipinnis Day, Fish. India, 1875-78, p. 342, Pl. lxxix, fig. 1 (plate 16).

D. V; I, 9. A. I, 8. L. lat. 120. Tr. 14+18. Operculum ending in single fleshy point. Dorsal originates above tip of pectoral and anterior to ventral origin. Leaden grey above, white below. Fins grey. 20 inches. Coastal waters and pearl and trawling banks.

GREY MULLET

Family MUGILIDAE

Fishes without lateral line. Elongate body. Head usually broad and flattened. Scales with a long perforation. Teeth minute or absent, gill-rakers long and slender. Sometimes eye with adipose lid of variable development. 4 spines in dorsal fin.

1. Maxillary hidden 2
   Maxillary exposed 5
2. Adipose eyelid well-developed; anal before 2nd dorsal 3
   Adipose eyelid absent; anal level with 2nd dorsal 4
3. L.lat. 37-42; Tr. 14-15 .... *Mugil cephalus*
   L.lat. 33; Tr. 10 .... *Mugil kelaartii*
4. Dorsal and anal falcate; L.lat. 33-36 .... *Valamugil buchanani*
   Dorsal and anal not falcate; L.lat. 39-41 .... *Valamugil seheli*
5. Adipose eyelid well-developed 6
   Adipose eyelid very small or absent 9
6. L.lat. 28-31; anterior half of anal before second dorsal .... *Liza dussumieri*
   L.lat. 33-43; anterior third or less of anal before second dorsal 7
7. A long axillary scale at pectoral .... *Liza strongylocephalus*
   No long scale in pectoral axilla 8
8. Eye 1/7 in head; dark line along scale rows .... *Liza tade*
   Eye 2/7 in head; golden spot on operculum .... *Liza parsia*

9. L.Tr. 16-18 .... *Liza cascasia*
L.Tr. 9-12 10
10. Anal with 8 soft rays; L.tr. 9-10 .... *Liza waigiensis*
Anal with 9 soft rays; L.tr. 10-12 11
11. L.lat. 26-28 .... *Liza oligolepis*
L.lat. 30-34 12
12. Pectorals with an axillary scale .... *Liza ceramensis*
Pectorals without axillary scale .... *Liza macrolepis*

251. MUGIL CEPHALUS Linnaeus GREY MULLET

Mugil cephalus Linnaeus, Syst. Nat. ed. 10, 1758, p. 316. *Mugil dobula* Günther, Jour. Mus. Godeffroy, iv (13), 1877, Pl. cxx, fig. A (plate 16).

D. IV; I, 6-8. A. III, 8. L. lat. 39-42. Tr. 14-15. Adipose eyelids well-developed. Head broad and flat. Lower lip with thin edge. Pectorals folded forwards scarcely reach eye. Scale process at base of spinous dorsal 5-6½ in predorsal distance. Greyish above, silvery below. 18 inches. Estuarine, moving into the sea and entering fresh water.

Is barri godeya, Kitheya, Thel godeya (S). Manalei (T).

252. MUGIL KELAARTII Günther KELAART'S MULLET

Mugil kelaartii Günther, Cat. Fish., iii, 1861, p. 472; Day, Fish. India, 1875-78, Pl. lxxv, fig. 1 (plate 16).

D. IV; I, 8. A. III, 9. L. lat. 33. Tr. 10. Adipose eyelids moderately developed. Dorsal profile of head convex. Pectoral with long axillary scale. Dorsals originate over 10th and 20th scales. Silvery above, lighter below. Dark spot in pectoral axilla. Dorsal and caudal tips dark. Marine, coastal.

253. VALAMUGIL BUCHANANI (Bleeker)
BUCHANAN'S BLUE-TAIL MULLET

Mugil buchanani Bleeker, Verh. Bat. Genootsch., xxv, 1853, p. 99. *Mugil ceylonensis*. *Valamugil buchanani* Smith, Fish. Sth. Africa, 1950, Pl. lxviii, fig. 888 (plate 16).

D. IV; I, 8. A. III, 9. L. lat. 33-36. Tr. 13. No adipose eyelids. Head markedly broad and blunt. Pectorals folded forwards reach beyond eye. Soft dorsal and anal, high in front, falcate and longer than ventrals. Teeth absent. Upper lip thin. Short pointed axillary scale. Greenish above, silvery below. Dark spot at base of pectorals. Caudal blue. Paired fins yellowish. Iris golden. 20 inches. Marine, entering rivers.

254. VALAMUGIL SEHELI (Forskål) BLUE-SPOT MULLET

Mugil seheli Forskål, Descript. Anim., 1775, p. 73. *Valamugil seheli* Smith, Fish. Sth. Africa, 1950, fig. 889 (plate 16).

D. IV; I, 8. A. III, 9. L. lat. 39-41. Tr. 14. No adipose eyelids. Pectorals folded forwards reach front of eyes. Soft dorsal not longer than ventrals. Scale at base of first dorsal 4¼-5 in predorsal distance. No teeth. Greenish blue above, silvery below. Pectorals yellowish with black spot in axilla. 12 inches, often less. Marine, coastal.

255. LIZA STRONGYLOCEPHALUS (Richardson) ROUND-HEAD MULLET

Mugil strongylocephalus Richardson, Rept. 15th. Meet. Brit. Assn. Adv. Sci., 1846, p. 249. *Mugil longimanus* or *cunnesius. Mugil strongylocephalus* Smith, Fish. Sth. Africa, 1950, fig. 879 (plate 16).

D. IV; I, 8. A. III, 9. L. lat. 33-35. Tr. 11-12. Adipose eyelids well-developed. Head moderately broad. Pectorals folded forwards reach beyond eye. Scale process at base of first dorsal 4-4½ in predorsal distance. Lower lip with thin edge. Greenish above, silvery below. Diffuse dark mark on pectorals. 10 inches. Coastal waters and lagoons.
Pandhura godeya (S).

256. LIZA DUSSUMIERI (Valenciennes) DUSSUMIER'S MULLET

Mugil dussumieri Cuvier and Valenciennes, Hist. Nat. Poiss., xi, 1836, p. 147; Day, Fish. India, 1875-78, Pl. lxxiv, fig. 4 (plate 16).

D. IV; I, 8-9. A. III, 9. L. lat. 28-30. Tr. 10-11. Adipose eyelid broad posteriorly, narrow anteriorly. Upper lip ciliated. Dorsals originating above 9-10th and 18-20th scales. Unpaired fins thickly scaled. No large axillary scale on pectoral. Greenish above, silvery below. Often 3-6 longitudinal dark stripes. Outer edge of caudal dark. 12 inches. Marine, entering fresh water.

257. LIZA TADE (Forskål) GREEN-BACK MULLET

Mugil tade Forskål, Descript. Anim., 1775, p. 74. *Mugil planiceps. Liza planiceps* Tosh, Rept. Marine Dept. Qld., 1903, Pl. i, fig. 3 (plate 16).

D. IV; I, 8-9. A. III, 9. L. lat. 33-35. Tr. 11. Adipose eyelid broad posteriorly, narrow anteriorly. Snout somewhat pointed. Dorsals originating above 10-11th and 21-23rd scales. Small scales on unpaired fins. No enlarged pectoral axillary. Green-brown above, silvery below. 5-7 indistinct longitudinal dark streaks. 18 inches. Coastal, entering estuaries and tidal rivers.

258. LIZA PARSIA (Hamilton-Buchanan) GOLD-SPOT MULLET

Mugil parsia Hamilton-Buchanan, Fish. Ganges, 1822, p. 215; Day, Fish. India, 1875-78, Pl. lxxv, fig. 2 (plate 16).

D. IV; I, 8. A. III, 8-9. L. lat. 34-35. Tr. 11-12. Adipose eyelid broad posteriorly. Snout depressed, upper lip thin. Dorsals originate above 11th and 22nd scales. No scale in pectoral axilla. Fine scales over unpaired fins. Golden spot on upper portion of opercle. Narrow dark edge on both dorsal fins. Caudal base yellow, dark at margin. 5-6 inches. Coastal, entering estuaries.

259. LIZA CASCASIA (Hamilton-Buchanan) YELLOW-TAIL MULLET

Mugil cascasia Hamilton-Buchanan, Fish. Ganges, 1822, p. 217; Day, Fish. India, 1875-78, Pl. lxxv, fig. 6 (plate 16).

D. IV; I, 8. A. III, 8-9. L. lat. 36-39. Tr. 16-18. Adipose eyelids absent. Head flat above. Operculum with strong spine. Dorsal fins originate above 7th and 24th scales. Second dorsal originates opposite anal. No enlarged pectoral axillary scale. Head yellow-olive. Back dusky grey, silvery below. Bases of pectoral and caudal deep yellow. Fins hyaline. 4 ins. Enters rivers.

260. LIZA WAIGIENSIS (Quoy and Gaimard) DIAMOND-SCALE MULLET

Mugil waigiensis Quoy and Gaimard, Voy. Uranie Physic., 1824, p. 337. *Ellochelon vaigiensis* Smith, Fish. Sth. Africa, 1950, fig. 883 (plate 16).

D. IV; I, 8. A. III, 8. L. lat. 26-28. Tr. 9-10. No adipose eyelids. Scale process at base of first dorsal 6-8 in predorsal distance. Dorsals originate above 7-8th and 17-18th scales. Caudal nearly truncate. Brownish above, silvery below with dark longitudinal streaks along scale rows. Fins, particularly pectoral, almost black. 15 inches. Coastal waters, entering estuaries and rivers.

Kora godeya (S). Punnakanni manalei, Thiruven manalei (T).

261. LIZA OLIGOLEPIS (Bleeker) LARGE-SCALED MULLET

Mugil oligolepis Bleeker, Nat. Tijds. Ned. Ind., xix, 1859, p. 437. *Mugil olivaceus*. *Liza oligolepis* Smith, Fish. Sth. Africa, 1950, fig. 885 (plate 16).

D. IV; I, 8. A. III, 9. L. lat. 26-28. Tr. 10-11. Adipose eyelids rudimentary. Scale process at base of spinous dorsal 4½ in predorsal distance. Pectorals folded forwards reach beyond eye. No pectoral axillary scale. Greenish above, silvery below. 12 inches. Coastal waters, entering estuaries.

Godeya (S). Manalei (T).

262. LIZA CERAMENSIS (Bleeker) CERAM MULLET

Mugil ceramensis Bleeker, Nat. Tijds. Ned. Ind., iii, 1852, p. 699.

D. IV; I, 8-9. A. III, 9. L. lat. 30-31. Tr. 10-11. Adipose eyelid forming narrow ring around eye. Upper lip thick and ciliated. Dorsal fins originate above 10th and 18-19th scales. Pectorals shorter than head and with axillary scale. Caudal not deeply concave. Olive above, silvery below. Dorsal and caudal dusky. 7 inches. Coastal waters.

263. LIZA MACROLEPIS (Smith) BORNEO MULLET

Mugil macrolepis Smith, Illust. S. Afric. Zool., Pisces, 1849, p. 28. *Mugil borneensis* or *troscheli*. *Liza macrolepis* Smith, Fish. Sth. Africa, 1950, fig. 886 (plate 16).

D. IV; I, 8. A. III, 9. L. lat. 31-35. Tr. 10-12. No adipose eyelids. Scale process at base of spinous dorsal 5-6 in predorsal distance. Pectorals folded forward do not reach front of eye. Dorsals originate above 11-12th and 19-20th scales. No pectoral axillary scale. Silvery, above green. Fins blackish along margins. 14 inches. Coastal waters and estuaries.

Kana rath godeya, Kiri godeya, Panu godeya, Vali godeya (S).

HARDYHEADS, SILVERSIDES

Family ATHERINIDAE

Small semi-transparent fishes with two dorsal fins, the first with a few slender spines. Resemble the Mugilidae but distinguished by a broad silvery stripe along the body. Usually congregate in dense shoals in shallow water.

L.lat. 42-45; anus opposite 12-14th scale .... *Allanetta forskåli*
L.lat. 35-40; anus opposite 9-10th scale .... *Pranesus duodecimalis*

264. ALLANETTA FORSKÅLI (Rüppell) FORSKÅL'S HARDYHEAD

Atherina Forskålii Rüppell, Neue Wirbelth., Fische, 1835, p. 132; Jordan and Hubbs, Ann. Carneg. Mus., xi, 1917, Pl. xlvi (plate 16).

D. V; II, 8. A. II, 11-13. L. lat. 42-45. Tr. 6. Dorsals originate above 15-16th and 24-26th scales, separated by 6-8 interdorsal scales. Anus situated between tips of ventrals or immediately behind their end. Upper surface pale green, remainder silvery with prominent metallic silvery lateral stripe. Fins and parts of head somewhat dusky. 5 inches. Coastal waters, harbours and estuaries.

Kili, Korala babba (S). Therakham (T).

265. PRANESUS DUODECIMALIS (Valenciennes) SMALL-MOUTH HARDYHEAD

Atherina duodecimalis Cuvier and Valenciennes, Hist. Nat. Poiss., x, 1835, p. 458.

D. V-VI; I-II, 8-9. A. II, 9-12. L. lat. 35-40. Tr. 6. Dorsals originate above 13-14th and 21-23rd scales, separated by 5 interdorsal scales. Anus situated near middle of length of ventrals. Silvery, darker above. A metallic silvery band along 3rd row of scales and a green band above it. Fins dusky, especially caudal and pectoral. 3½ inches. Coastal waters, entering harbours and estuaries.

THREAD-FINS, TASSEL FISH

Order POLYNEMIFORMES

Family POLYNEMIDAE

Body moderately deep with two dorsal fins. Mouth moderately large but below snout, with small teeth. Adipose eyelids present. Pectoral fin in two parts, the lower rays filamentous and used as feelers, being longer in young and shorter in adults. Carnivorous. Enter estuaries.

1. Lower lip restricted to corner of mouth; teeth extend on exterior of jaws; 3-4 pectoral filaments .... *Eleutheronema tetradactylum*
   Lower lip fully developed; no external teeth; 5 or more pectoral filaments 2
2. With 7 pectoral filaments .... *Polynemus heptadactylus*
   With 6 pectoral filaments 3
   With 5 pectoral filaments 5
3. Body silvery; swim-bladder absent .... *Polynemus xanthonemus*
   Body golden; swim-bladder present 4
4. Black blotch on lateral line; pectoral filaments reach middle of ventrals .... *Polynemus sextarius*
   No black blotch on lateral line; pectoral filaments extend beyond tips of ventrals ... *Polynemus sexfilis*
5. L.lat. 60-65; pectoral rays simple .... *Polynemus plebeius*
   L.lat. 70-75; most pectoral rays divided .... *Polynemus indicus*

266. ELEUTHERONEMA TETRADACTYLUM (Shaw)

FOUR-THREAD TASSEL FISH

Polynemus tetradactylus Shaw, Gen. Zool., v, 1804, p. 155. *Eleutheronema tetradactylum* Weber and De Beaufort, Fish. Indo-Aust. Arch., iv, 1922, fig. 64 (plate 17).

D. VIII; I-II, 13-15. A. II, 15-17. P. (16-17)+4. L. lat. 78-80. Tr. (9-10)+(13-14). Upper lip absent, lower developed in angle of mouth. Anal originates opposite second dorsal. Upper pectoral filament reaches just beyond ventral origin. Silvery green above, creamy below. Dorsal and caudal grey, dusky at edges. Ventral and anal orange. Dark mark on operculum. 6 feet, usually less. In coastal and brackish waters.
Kalawa (S). Kalemeen (T).

267. POLYNEMUS HEPTADACTYLUS Cuvier

SEVEN-THREAD TASSEL FISH

Polynemus heptadactylus Cuvier and Valenciennes, Hist. Nat. Poiss., iii, 1829, p. 300. *Trichidion heptadactylum. Polynemus heptadactylus* Day, Fish. India, 1875-78, Pl. xlii, fig. 5 (plate 17).

D. VIII; I, 11-12. A. III, 11-12. P. (13-15)+7. L. lat. 48-50. Tr. 5+10. Origin of second dorsal in advance of that of anal. All pectoral rays simple. 3rd, 4th and 5th pectoral filaments longest, extend to tip of ventrals or beyond. Swim-bladder absent. Brownish above, golden beneath. Pectoral black. Margins of other fins black. 6 inches. Coastal and brackish waters.

268. POLYNEMUS XANTHONEMUS Valenciennes

SILVER SIX-THREAD TASSEL FISH

Polynemus xanthonemus Cuvier and Valenciennes, Hist. Nat. Poiss., vii, 1831, p. 517.

D. VIII; I, 11. A. III, 11-12. Pectoral filaments extend beyond ventral tips. Swim-bladder absent. Greenish above, silvery beneath. Fins yellow with a black border. 6 inches. Coastal waters.

Kalava (S).

269. POLYNEMUS SEXTARIUS Bloch

BLACK-SPOT SIX-THREAD TASSEL FISH

Polynemus sextarius Bloch and Schneider, Syst. Ichth., 1801, p. 18. *Trichidion sextarium. Polynemus sextarius* Day, Fish. India, 1875-78, Pl. xlii, fig. 6 (plate 17).

D. VIII; I, 12-13. A. II-III, 12-13. P. (13-14)+6. L. lat. 48-50. Tr. 5+9. Second dorsal originates above anal. Longest pectoral filament reaches tip of ventrals. 12-13 rays of pectorals divided. Swim-bladder small. Golden-olive above, silvery beneath. Fins dusky or with dark margins. Dark spot at beginning of lateral line. 12 inches. Coastal waters, and trawling grounds.

270. POLYNEMUS SEXFILIS Valenciennes

GOLDEN SIX-THREAD TASSEL FISH

Polynemus sexfilis Cuvier and Valenciennes, Hist. Nat. Poiss., vii, 1831, p. 515. *Polydactylus sexfilis. Polynemus sexfilus* Day, Fish. India, 1875-78, Pl. xliii, fig. 1 (plate 17).

D. VIII; I, 12-13. A. III, 11-12. L. lat. 46. Tr. 5+10. P. 15+6. Pectoral rays all simple. Pectoral filaments reach beyond tips of ventrals. Large swim-bladder. Golden. Pectoral black. Anal with black margin. Ventrals dark in middle. Coastal waters, pearl and trawling banks.

Gatha (S). Kala (T).

271. POLYNEMUS PLEBEIUS (Broussonet) COMMON TASSEL FISH

Polynemus plebeius Broussonet, Ichth. Decas., i, 1782, Pl. viii. *Trichidion plebium* or *Polydactylus plebejus*. *Polynemus plebeius* Günther, Jour. Mus. Godeffroy, ii (9), 1875, Pl. lxxvii, fig. A (plate 17).

D. VIII; I, 13. A. II, 11. P. (17-18)+5. L. lat. 60+65. Tr. (6-7)+12. All pectoral rays simple. Pectoral filaments reach little beyond ventral tips. Soft dorsal originates little in advance of anal. Dusky golden-olive above with few narrow dusky stripes. Fins greyish. 18 inches. Coastal waters, entering estuaries.

Bandi kalava (S). Barmeen (T).

272. POLYNEMUS INDICUS Shaw INDIAN TASSEL FISH

Polynemus indicus Shaw, Gen. Zool., v, 1804, p. 155; *Polynemus sele* Hora, Mem. Indian Mus., ix (4), 1929, Pl. xvii, fig. 4 (plate 17).

D. VIII; I, 13-14. A. II-III, 11-12. P. (2-3)+(12)+5. L. lat. 70-75. Tr. 7+(12-13). First 2-3 pectoral rays simple, remainder branched. Pectoral filaments reach almost to anal origin. Second dorsal originates in front of anal. Swim-bladder thick and very long. Body golden, much darker above, with faint lines along scale rows. Fins yellowish. 45 inches. Coastal waters and rivers.

SNAKE-HEADS

Order OPHIOCEPHALIFORMES

Family OPHIOCEPHALIDAE

Body scaly, elongate, sub-cylindrical anteriorly, compressed posteriorly. Head depressed, with large shield-like scales above. Mouth large and protractile. Dependent equally on aerial and branchial respiration. Two accessary pharyngeal air cavities. Fins spineless; dorsal and anal long. Can migrate across land from one fresh water pool to another. Nest builders.

1. **Ventrals absent** .... *Channa orientalis*
   Ventrals present 2
2. Cephalic pits multiple; prenasals present 3
   Cephalic pits simple; prenasals absent 4
3. Six transverse rows of dorsicephalic scales before level of opercles .... *Ophiocephalus marulius ara*
   Ten transverse rows of dorsicephalic scales before level of opercles .... *Ophiocephalus striatus*
4. Dorsal rays 29-30 .... *Ophiocephalus punctatus*
   Dorsal rays 30-33 .... *Ophiocephalus gachua kelaarti*

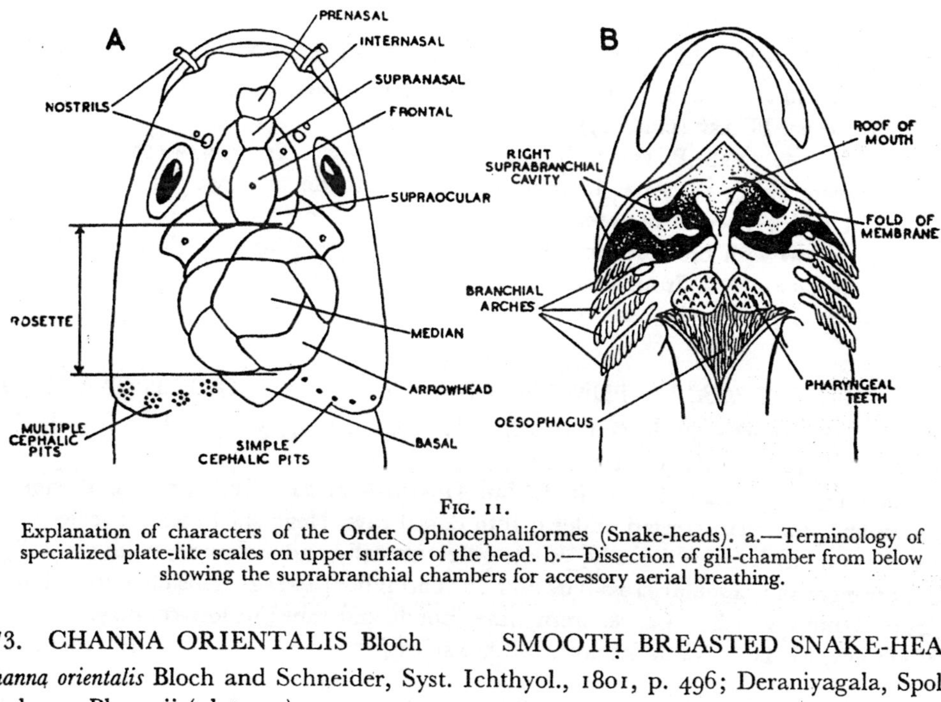

FIG. 11.

Explanation of characters of the Order Ophiocephaliformes (Snake-heads). a.—Terminology of specialized plate-like scales on upper surface of the head. b.—Dissection of gill-chamber from below showing the suprabranchial chambers for accessory aerial breathing.

273. CHANNA ORIENTALIS Bloch SMOOTH BREASTED SNAKE-HEAD

Channa orientalis Bloch and Schneider, Syst. Ichthyol., 1801, p. 496; Deraniyagala, Spolia Zeyl., xv, Pl. xxvii (plate 17).

D. 31-34. A. 20-22. P. 13-15. L. lat. (9-12)+(27-30). Tr. (5-6)+(6-7). Predorsals 6-7. Prenasals 17-20. Anal originates below 10-11th dorsal rays. Body yellow brown, greenish

dorsally, and with bluish cross-bands. Dorsal orange with greenish longitudinal band. Anal greenish. Pectorals and caudal yellow with blue cross-bands. 4 inches. Clean fresh water pools close to streams.
Gas kannaya, Kola kannaya (S).

274. OPHIOCEPHALUS MARULIUS ARA Deraniyagala
GIANT SNAKE-HEAD

Ophiocephalus marulius ara Deraniyagala, Spolia Zeyl., xxiv, 1945, p. 93. *Ophiocephalus marulius* Deraniyagala, Spolia Zeyl., xv, 1929, Pl. xxiv (plate 17).

D. 45-49. A. 23-31. P. 16-18. L. lat. (16-21)+(40-44). Tr. (6-7)+(10-12). Predorsals 7-8. Preventrals 19-20. Dorsal longer than anal by half length of anal base. Head shields with 1, 2 or 3 internasals and 2-3 scales between rosette and basal. Pale olive to yellow with diffuse violet lateral band crossed by 5-6 incomplete cross-bars. Fins dark olive to dusky violet with white spots. Ventrals white. 32 inches. Inland fresh waters up to 1500 feet.
Ara, Gangara, Kalumaha (S). Iru viral (T).

275. OPHIOCEPHALUS STRIATUS Bloch STRIPED SNAKE-HEAD; MURREL

Ophiocephalus striatus Bloch, Nat. Ausländ. Fische, vii, 1793, p. 141; Deraniyagala, Spolia Zeyl., xv, 1929, Pl. xxv (plate 17).

D. 42-46. A. 25-28. P. 15-17. L. lat. (15-18)+(39-42). Tr. (5-8)+(10-14). Predorsals 8-10. Preventrals 15-18. Anal inserted below 16-19th dorsal rays. Head shield with 1 internasal and 4-5 scales between rosette and basal. Olive to dark brown dorsally and white to yellow with brownish mottling below. About 15 W-shaped dark cross-bars. Fins mostly olive. 27 inches. Fresh water ponds, tanks and streams.
Hal path maha, Lulla (S). Viral (T).

276. OPHIOCEPHALUS PUNCTATUS Bloch GREEN SNAKE-HEAD

Ophiocephalus punctatus Bloch, Nat. Ausländ. Fische, vii, 1794, p. 358; Deraniyagala, Spolia Zeyl., xv, 1929, Pl. xxvi (plate 17).

D. 29-30. A. 20-22. P. 15-18. L. lat. (13-15)+(21-24). Tr. (5-6)+8. Predorsals 5. Preventrals 10. Anal inserted under 9-10th dorsal rays. Head shield with 2 scales between rosette and basal. Olive dorsally, yellow-green laterally and yellow to white ventrally. 6-7 dark transverse triangular cross-bars and 10 ventral dark streaks. Fins olive to yellow, with darker bands. 8 inches. Larger fresh water ponds and tanks in low-country.
Mada ara, Madaya, Mada kanaya, Mada kariya (S). Korruvai (T).

277. OPHIOCEPHALUS GACHUA KELAARTI Günther
BROWN SNAKE-HEAD

Ophiocephalus kelaarti Günther, Cat. Fish., iii, 1861, p. 472. *Ophiocephalus gachua* Bleeker, Atlas Ichth. Ind. Neerl., ix, 1877, Pl. cccxcvii fig. 4 (plate 17).

D. 30-33. A. 20-23. P. 14-15. L. lat. (10-12)+(28-32). Tr. (5-7)+(7-8). Predorsals 5-6. Preventrals 9-12. Anal inserted below 9-12th dorsal ray. Head shield with rosette touching basal. Dorsally brown with W-shaped cross-bars. Margin of dorsal orange. Anal green-blue. Often 1-2 ocelli on dorsal. 6 inches. Fresh water streams and stagnant pools of coastal plains to 3600 feet. Hot springs.

Parandal kanaya (S). Para korruvai (T).

SHORE EELS

Order SYMBRANCHIFORMES

Family SYMBRANCHIDAE

Small eel-like fishes with a single gill-opening which is on the ventral surface.

278. SYMBRANCHUS BENGALENSIS (McClelland) PIGMY EEL

Ophisternon bengalensis McClelland, Calcutta Journ. Nat. Hist., v, 1845, p. 197. *Symbranchus bengalensis* Day, Fish. India, 1875-78, Pl. clxvii, fig. 2 (plate 12).

Head 9-12 in distance from snout to anus. Lips fleshy. Gape extends behind orbit. Teeth in upper jaw fine and pointed, not continuous at symphysis, but terminating in triangular patch. Teeth larger in lower jaw. Palatines in a band. Conspicuous lateral line. Brownish red in brackish water, dark greenish dorsally in fresh water. 17 inches. Brackish and tidal fresh waters and paddy fields.

Potta arndha (S).

PERCH-LIKE FISHES

Order PERCIFORMES

All the more typical or less specialized bony fishes. Most have two dorsal fins which may or may not be united at least basally. The ventrals have not more than 6 rays and are usually, although not always, thoracic. There is no Weberian apparatus. In Ceylon the order embraces 82 families which are grouped in 13 suborders. The principal characters of the suborders and families refer to the detailed structure of the skeleton, and therefore are of limited practical use for rapid field identification.

The characters of each suborder together with a provisional key to the contained families are provided under separate headings below. The general key on pages 18-20 subdivides the Perciformes into its component suborders.

Suborder PERCOIDEI

Fins with spines. Ventrals thoracic or jugular, with 1 spine and not more than 5 rays. Caudal rays do not overlap hypural. Second infraorbital not united with preopercular.

1. A single nostril on each side 2
   Two nostrils on each side 3
2. Anal fin with 2 spines .... *Family* POMACENTRIDAE, p. 177
   Anal fin with 3 or more spines .... *Family* CICHLIDAE, p. 176
3. Gills $3\frac{1}{2}$, with slit behind, last small or wanting; lower pharyngeals united into a characteristic structure which normally bears molar teeth 4
   Gills 4, with a long slit behind last; lower pharyngeals not united and not bearing molar teeth 5
4. Teeth in jaws quite separate and normal .... *Family* LABRIDAE, p. 183
   Teeth in jaws partly or wholly coalesced .... *Family* SCARIDAE, p. 194
5. Not more than one weak anal spine 6
   Always 2 or 3 strong anal spines 9
6. Body extremely elongate; no clear distinction between spines and rays of dorsal fin, anterior ones filamentous .... *Family* TRICHONOTIDAE, p. 199
   Body moderate in length; spines and rays distinguishable in dorsal fin, anterior spines not filamentous ... 7
7. Robust body with heavy bony head; eyes entirely dorsal and mouth vertical .... *Family* URANOSCOPIDAE, p. 200
   Body slender with normally proportioned head; eyes lateral and mouth horizontal 8
8. Head cylindrical; jaws subequal; first dorsal with 5 spines, all lower than rays .... *Family* PARAPERCIDAE, p. 198
   Head depressed; lower jaw projecting; first dorsal with 6 spines, anterior ones as high as rays .... *Family* BEMBROPIDAE, p. 199
9. Lower pectoral rays simple and thickened .... *Family* CIRRHITIDAE, p. 197
   Lower pectoral rays branched, not specially thickened 10
10. Body orbicular or deeper than long 11
    Body oval or elongate, always longer than deep 15
11. Dorsal and anal spines absent in adults; body apparently naked .... *Family* MENIDAE, p. 132
    Dorsal and anal spines always present; body obviously scaly 12
12. Ventral fins minute or vestigial .... *Family* MONODACTYLIDAE, p. 165
    Ventral fins moderately large or greatly elongated 13
13. Dorsal spines confluent with rays; first spine not directed forwards ...... *Family* PLATACIDAE, p. 168
    Spinous dorsal differentiated from soft dorsal; first spine directed forwards 14

14. Mouth protractile; pectoral falcate and elongate .... *Family* DREPANIDAE, p. 169
Mouth not protractile; pectoral rounded and short .... *Family* EPHIPPIDAE, p. 167
15. Dorsal spines absent .... *Family* CORYPHAENIDAE, p. 132
Dorsal spines present 16
16. Two separated dorsal fins 17
One continuous dorsal fin, spinous and dorsal parts at least connected at base 22
17. Two spines detached from anal fin .... *Family* CARANGIDAE, p. 123
No spines or simple rays detached from anal fin 18
18. Two long barbels behind symphysis of lower jaw .... *Family* MULLIDAE, p. 162
No barbels behind symphysis of lower jaw 19
19. Dorsal with 8 to 10 rays .... *Family* APOGONIDAE, p. 118
Dorsal with 16 or more rays 20
20. Dorsal spines short, strong, disconnected; scales small .... *Family* RACHYCENTRIDAE, p. 131
Dorsal spines long and connected by wide membrane; scales moderately large 21
21. Body oval, compressed; teeth on palate .... *Family* LACTARIIDAE, p. 122
Body elongate, subcylindrical; teeth on vomer .... *Family* SILLAGINIDAE, p. 121
22. Two spines detached from anal fin .... *Family* CARANGIDAE, p. 123
No spines detached from anal fin 23
23. Anal with 1 or 2 spines 24
Anal with 3 to 12 spines 26
24. Dorsal more or less deeply notched 25
Dorsal without notch .... *Family* MALACANTHIDAE, p. 122
25. Body elongate; long soft dorsal with more than 20 rays; opercle with 2 soft flat spines .... *Family* SCIAENIDAE, p. 152
Body deep oval; short soft dorsal with 12-15 rays; opercle with 3 pungent spines .... *Family* DIPLOPRIONIDAE, p. 115
26. A scaly process in axil of ventrals, or ventrals minute or vestigial 27
No scaly process in axil of ventrals 45
27. Teeth on palate well developed 28
No teeth on palate, or a few small ones on vomer only 32
28. Dorsal fin deeply notched .... *Family* LATIDAE, p. 105
Dorsal fin not notched 29
29. Anal with 40 or more rays .... *Family* PEMPHERIDAE, p. 166
Anal with 7 to 18 rays 30
30. Origin of dorsal behind middle of body .... *Family* TOXOTIDAE, p. 166
Origin of dorsal before middle of body 31
31. Outer row of teeth narrow lanceolate incisors; inner rows in a villiform band .... *Family* KYPHOSIDAE, p. 167
Several rows of conical pointed teeth, outer row enlarged; canines present or absent .... *Family* LUTIANIDAE, p. 133
32. Body elongate and subcylindrical; dorsal with a long deep notch .... *Family* EMMELICHTHYIDAE, p. 133
Body moderately elongate, oblong or very deep, compressed; dorsal without a notch 33
33. Mouth protractile 34
Mouth not protractile 36
34. Teeth in brush-like bands in jaws .... *Family* CHAETODONTIDAE, p. 172
Minute setiform teeth in one (seldom more) series, often with small symphysial canines 35
35. Top of head with bony ridges and a nuchal spine; scales small; gill-membranes attached to isthmus .... *Family* LEIOGNATHIDAE, p. 144
Top of head smooth and without nuchal spine; scales large and deciduous; gill-membranes free of isthmus .... *Family* GERRIDAE, p. 142

36. Teeth small, in brush-like or villiform bands in jaws 37
    Teeth larger, conical or molariform, in several series; canines or incisors sometimes present 38
37. Preopercle serrated; anal with 3 spines .... *Family* LOBOTIDAE, p. 141
    Preopercle smooth; anal with 4 spines .... *Family* SCATOPHAGIDAE, p. 169
38. Preoperculum without scales .... *Family* LETHRINIDAE, p. 157
    Preoperculum with several rows of scales 39
39. A group of enlarged teeth in front of both jaws, canines or incisors; inner rows of rounded molars .... *Family* SPARIDAE, p. 160
    Small pointed teeth in one or more rows in jaws, external row sometimes enlarged and canines sometimes present anteriorly 40
40. Preorbital naked; preopercle entire; dorsal and anal spines weak 41
    Preorbital scaly; preopercle serrated; dorsal and anal spines robust 44
41. Distinct bony process on premaxilla sliding beneath maxilla .... *Family* CAESIODIDAE, p. 140
    No such process on premaxilla 42
42. Suborbital with a distinct spine below eye .... *Family* SCOLOPSIDAE, p. 149
    Suborbital without a distinct spine 43
43. Eye large; gill-rakers knob-like .... *Family* PENTAPODIDAE, p. 159
    Eye small; gill-rakers not knob-like .... *Family* NEMIPTERIDAE, p. 140
44. A pit behind the chin and 2 pores; head straight or convex; scales small .... *Family* POMADASYIDAE, p. 147
    No pit behind chin but 6 pores; head parabolic; scales larger ... *Family* PLECTORHYNCHIDAE, p. 150
45. Dorsal deeply notched; body glass-like .... *Family* AMBASSIDAE, p. 106
    Dorsal not notched; body not glass-like 46
46. Lateral line divided into an upper and lower portion, or incomplete 47
    Lateral line complete 48
47. Dorsal with 11-12 spines .... *Family* PLESIOPIDAE, p. 116
    Dorsal with 2-7 spines .... *Family* PSEUDOCHROMIDAE, p. 114
48. Teeth setiform in brush-like bands, tricuspid or sub-tricuspid; strong spine in preopercular angle .... *Family* POMACANTHIDAE, p. 169
    Teeth not in brush-like bands nor tricuspid; no strong spine in preopercular angle 49
49. Dorsal with 12-14 spines .... *Family* THERAPONIDAE, p. 115
    Dorsal with 6-11 spines 50
50. Inner series of teeth in jaws enlarged and depressible .... *Family* SERRANIDAE, p. 107
    No depressible teeth 51
51. Maxillary scaly; eyes very large .... *Family* PRIACANTHIDAE, p. 117
    Maxillary naked; eyes moderate 52
52. Dorsal with 6 or 7 spines .... *Family* SERRANIDAE, p. 107
    Dorsal with 10 spines .... *Family* KUHLIIDAE, p. 117

SEA PERCH, COCK-UP

Family LATIDAE

Perch-like fishes with 2 almost separated dorsal fins and a preopercular edge partly serrated or spiniferous. Opercles ending in a single spine. Scales small to moderate, 40-70 in longitudinal series. 8-9 anal rays. Pectorals and ventrals subequal.

Lower edge of preopercle with spines or serrae; tongue smooth; lower gill-rakers 16-17 .... *Lates calcarifer*

Lower edge of preopercle entire; tongue with patch of fine short teeth; lower gill-rakers 11-13 .... *Psammoperca waigiensis*

279. LATES CALCARIFER (Bloch) COCK-UP, GIANT PERCH

Holocentrus calcarifer Bloch, Naturg. Ausländ. Fische, iv, 1790, p. 100. *Lates calcarifer* Day, Fish. India, 1875-78, Pl. i, fig. 1 (plate 20).

D. VII-VIII; I, 10-11. A. III, 7-8. P. 17. L. lat. 52-61. Tr. 6+13. Predorsals 27-28. Compressed, elongate body. Large protractile mouth with exposed maxillary. Opercular edge with spine. Preopercle with spine in angle and serrated edge behind. Caudal rounded. Dorsal and anal with scaly sheaths. Golden brown above, silvery below. Fins dark brown, paired ones paler. 5 feet. Coastal waters, estuaries and lagoons, entering brackish water. Modha, Modhe koleya (S). Koduva (T).

280. PSAMMOPERCA WAIGIENSIS (Cuvier) WAIGEU SEA PERCH

Labrax waigiensis Cuvier and Valenciennes, Hist. Nat. Poiss., ii, 1828, p. 83. *Psammoperca waigiensis* Bleeker, Atlas Ichth. Ind. Neerl., vii, 1876, Pl. cccvi, fig. 2 (plate 20).

D. VII; I, 12. A. III, 8. P. 16-17. L. lat. 45-50. Tr. 6+11. Predorsals 23-24. Cheek scales 9-10. Compressed elongate body. Large protractile mouth with exposed maxillary. Opercle with spine. Preopercle with strong spine in angle and serrated hind edge and smooth lower edge. Caudal rounded. Dorsal and anal with scaly sheaths. Reddish brown above, silvery below. Fins reddish brown, ventrals darkish. 14 inches. Coastal waters and estuaries.

GLASSY PERCHLETS

Family AMBASSIDAE

Small perch-like fishes with compressed body with easily shed cycloid scales. Oblique mouth with fine teeth in bands on jaws and palate. Spinous dorsal strong, elevated in front, with 7 spines and joined at base to second dorsal. Small spine pointing forwards before spinous dorsal. Scaly sheath at bases of dorsal and anal. Translucent silvery yellow often with silvery band along body. Estuarine, entering fresh water.

1. Greenish with silvery lateral band .... *Ambassis thermalis*
   Silvery with burnished silver lateral band 2
2. Interopercle serrated .... *Ambassis dayi*
   Interopercle not serrated 3
3. Lateral line continuous, smooth .... *Ambassis commersoni*
   Lateral line interrupted or bent 4
4. Predorsal scales 7-9 .... *Ambassis urotaenia*
   Predorsal scales 14-16 .... *Ambassis gymnocephalus*

281. AMBASSIS THERMALIS Cuvier HOT-SPRING GLASSY PERCHLET

Ambassis thermalis Cuvier and Valenciennes, Hist. Nat. Poiss., iii, 1829, p. 493.

D. VII; I, 9. A. III, 9. Depth not quite one third length excluding caudal. Large scales. Greenish with silvery lateral band. A dark longitudinal streak along each caudal lobe. An imperfectly known species from the thermal springs of Cannia but not since noticed.

282. AMBASSIS DAYI Bleeker DAY'S GLASSY PERCHLET

Ambassis dayi Bleeker, Nat. Verh. Holl. Maat. Wet., ii, 1874, p. 95; Day, Fish. India, 1875-78, Pl. xv, fig. 7 (plate 17).

D. VII; I, 10-11. A. III, 10. L. lat. 30. Depth 3 in total length. Preopercle and interopercle serrated. A spine at upper posterior angle of orbit. Lateral line continuous. Silvery glossed with purple and a broad burnished lateral band. Membrane between second and third dorsal spines dark. Soft dorsal, caudal and anal with dark edges. 7 inches. Fresh waters.

283. AMBASSIS COMMERSONI Cuvier COMMERSON'S GLASSY PERCHLET

Ambassis commersonii Cuvier and Valenciennes, Hist. Nat. Poiss., ii, 1828, p. 176. *Ambassis safgha. Ambassis commersonii* Smith, Fishes Sth. Africa, 1949, fig. 635 (plate 17).

D. VII; I, 10. A. III, 10. L. lat. 28-30. Tr. 4+(8-9). Predorsals 12-18. Gill-rakers 17-20. Depth 3¼-3½ in total length. Preopercle serrated, interopercle smooth. Lateral line continuous. Silvery with purplish reflections and bright silver lateral band. Membrane between 2nd and 3rd dorsal spines dusky. 6 inches. Estuarine, ascending rivers.

284. AMBASSIS UROTAENIA Bleeker BANDED-TAIL GLASSY PERCHLET

Ambassis urotaenia Bleeker, Nat. Tijds. Ned. Ind., iii, 1852, p. 257; Day, Fish. India, 1875-78, Pl. xv, fig. 8 (plate 17).

D. VII; I, 9-10. A. III, 9-10. L. lat. 26-28, sharply bent or interrupted at about 13th scale. Tr. (2-3)+9. Predorsals 7-9. Gill-rakers 20-22. Depth 3¼ in total length. Preopercle serrated, interopercle smooth. Preorbital with 7 teeth. Silvery with burnished lateral band. Membrane between 2nd and 3rd dorsal spines black. Caudal lobes with dark longitudinal bands. 4 inches. Estuarine, entering rivers.

285. AMBASSIS GYMNOCEPHALUS (Lacépède)
NAKED-HEAD GLASSY PERCHLET

Lutjanus gymnocephalus Lacépède, Hist. Nat. Poiss., iii, 1802, p. 23; iv, p. 216. *Ambassis gymnocephalus* Day, Fish. India, 1875-78, Pl. xv, fig. 6 (plate 17).

D. VII; I, 9-10. A. III, 9-10. L. lat. 28-30, interrupted at about 14th scale. Tr. 2+9. Predorsals 14-16. Gill-rakers 21. Preorbital with 6-7 teeth. Preopercle serrated, interopercle smooth. Small spine on shoulder. Silvery with bright lateral band. Blackish on membrane between 2nd and 3rd dorsal spines. Black edge to caudal. 4 inches. Estuarine, entering fresh water.
Gon katilla (S).

ROCK-CODS, GROUPERS

Family SERRANIDAE

Perch-like fish with robust, elongate body, covered with small adherent ctenoid scales. Mouth large with small teeth on bands in jaws and palate and enlarged caniniform teeth in front. 6-8 dorsal spines. 3 anal spines. Caudal rounded, truncate or weakly emarginate. Margin of preopercle normally serrated. Mostly inhabit coral reefs.

1. Dorsal spines 6-8 2
   Dorsal spines 9 4
   Dorsal spines 11 11
2. Anal spines absent; caudal rounded .... *Grammistes sexlineatus*
   Three anal spines; caudal weakly emarginate 3
3. L.lat. 83-86; Tr. (17-18)+(40-43); 7 dorsal spines .... *Plectropomus maculatus*
   L.lat. 73-81; Tr. (20-21)+(34-44); 8 dorsal spines .... *Plectropomus leopardus*
4. Mandible with 1-2 lateral curved canines; caudal emarginate .... *Variola louti*
   Mandible without lateral canines; caudal rounded or truncate 5
5. Caudal truncate; body dusky, orange inside mouth .... *Aethaloperca rogaa*
   Caudal rounded; body with blue and white spots 6
6. Body with 6-8 dark transverse bars 7
   Body without transverse bars 9
7. Scales above lateral line 18-22 .... *Cephalopholis boenack*
   Scales above lateral line 12-17 8
8. Cheek scales 34-36; more than 90 scales in longitudinal series to caudal base; without longitudinal blue lines .... *Cephalopholis argus*
   Cheek scales 22; less than 90 scales in longitudinal series to caudal base; with longitudinal blue lines .... *Cephalopholis pachycentron*
9. L.lat. 81-105; Tr. 8-10 above .... *Enneacentrus miniatus*
   L.lat. 100-112; Tr. 11-15 above 10
10. Meshwork of dark brown on head; caudal with dark border .... *Enneacentrus sonnerati*
    Head dotted with yellow; pectorals and caudal edged with white .... *Enneacentrus nigripinnis*
11. L.lat. scale tubules with 4-6 radiating ridges; dorsal spines low .... *Promicrops lanceolatus*
    L.lat. tubules simple, without radiating ridges; dorsal spines moderately high 12
12. Mandibular teeth in 2 series 13
    Mandibular teeth in 3 series; caudal rounded 21
13. Caudal rounded; 40-50 cheek scales; 70-80 predorsal scales 14
    Caudal truncate or emarginate; 20-36 cheek scales; 57-70 predorsal scales 15
14. Yellow-brown with black-brown spots on body and fins .... *Epinephelus fario*
    Red-brown with 6-7 darker transverse bands and margin of spinous dorsal black .... *Epinephelus fasciatus*
15. Dorsal rays 18-19; head with spots confluent into 6-12 longitudinal bluish lines on body .... *Epinephelus undulosus*
    Dorsal rays 14-17; body without fine longitudinal streaks 16
16. Height 2·4-2·7; body purple with minute dots; fins and lips yellow .... *Epinephelus flavocaeruleus*
    Height 2·3-3·0; body with large spots, sub-vertical bars or a network of light lines 17
17. Dorsal rays 14-15; predorsal scales 78; body with sub-vertical broad bands and horizontal streaks .... *Epinephelus morrhua*
    Dorsal rays 15-17; predorsal scales 56-70; body with small or large spots or network of light lines ... 18
18. Body with dark transverse bands; cheek scales 21-25 .... *Epinephelus diacanthus*
    Body spotted or with network of lighter lines; cheek scales 30-36 19
19. Tr. above L.lat. 12-15; caudal emarginate .... *Epinephelus areolatus*
    Tr. above L.lat. 15-16; caudal truncate 20
20. Body and fins with hexagonal spots or light network; caudal margin light; preopercle rectangular; 53 scale rows below L.lat. .... *Epinephelus chlorostigma*
    Body and fins with rows of large spots; lower half of caudal dark; preopercle obtuse; 40 scale rows below L.lat. .... *Epinephelus bleekeri*
21. Less than 60 predorsal scales; cheek scales 25-27; body with large dark polygonal spots separated by a network of lighter lines .... *Epinephelus merra*
    More than 60 predorsal scales; cheek scales 31-40; body without honeycomb of lighter lines 22

22. Dark blotches on body and caudal peduncle; body marbled and speckled with small round dots or ocelli 23
Body with 4-5 transverse vertical bands on a ground of small dots or blotches 24
23. Gill-rakers 10+17; cheek scales 31-32; predorsal scales 85-100; maxilla reaches posterior border of eye .... *Epinephelus caeruleopunctatus*
Gill-rakers 13+17; cheek scales 35-37; predorsal scales 75-78; maxilla reaches beyond posterior border of eye .... *Epinephelus fuscoguttatus*
24. Dorsal rays 15-16; gill-rakers 10+15; opercular spines equidistant .... *Epinephelus tauvina*
Dorsal rays 14-15; gill-rakers 7+14; middle opercular spine nearer lower than upper .... *Epinephelus malabaricus*

286. GRAMMISTES SEXLINEATUS (Thunberg) SIX-LINED PERCH

Perca sexlineata Thunberg, Kon. Vet. Akad. Handl. Stockholm, xiii, 1792, p. 142. *Grammistes orientalis*. *Grammistes sexlineatus* Jordan and Seale, Bull. U.S. Bur. Fish., xxv, 1906, Pl. xxxviii, fig. 2 (plate 18).

D. VI-VII; 13-15. A. 8-11. L. lat. 70. G.R. 7-9 plus rudiments. Chin with rudimentary dermal appendage. Tr. 14+27. Predorsals 20. Cheek scales 15-18. Skin slimy in life. Markings variable. Ground colour brown with 3-9 yellowish-white longitudinal bands. Fins pale brown. 10 inches. Coastal waters.

287. PLECTROPOMUS MACULATUS (Bloch) SPOTTED CORAL-TROUT

Bodianus maculatus Bloch, Nat. Ausländ. Fische, iv, 1790, p. 48. *Acanthistius maculatus* Bleeker, Atlas Ichth. Ind. Neerl., vii, 1876, Pl. ccxci, fig. 3 (plate 18).

D. VII-VIII; 10-12. A. III, 7-8. L. lat. 125, in lateral line 83-86. Tr. (17-18)+(40-43). Predorsals 42-44. Cheek scales 26-28. G.R. 7+13. Colour variable, usually brilliant orange with numerous blue spots on body, head and fins except ventrals and part of pectorals. 42 inches. Coastal waters around coral reefs.

288. PLECTROPOMUS LEOPARDUS (Lacépède) LEOPARD CORAL-TROUT

Holocentrus leopardus Lacépède, Hist. Nat. Poiss., iv, 1802, p. 332. *Acanthistius leopardinus* Bleeker, Atlas Ichth. Ind. Neerl., vii, 1876, Pl. ccxcvi, fig. 3 (plate 18).

D. VII-VIII; 10-12. A. III, 7-8. L. lat. 120-130, lateral line with 73-81. Tr. (20-21)+(34-44). Predorsals 40-52. Cheek scales 23-25. G.R. 6+(10-11). Brownish with numerous fine, slightly darker spots. Pectoral usually unspotted, sometimes with small blue spots. Ventrals darker brown. 26 inches or more. Coastal waters around coral reefs.

289. VARIOLA LOUTI (Forskål) PAINTED CORAL-TROUT

Perca louti Forskål, Descript. Anim., 1775, p. 40. *Serranus louti* or *punctulatus*. *Variola louti* Bleeker, Atlas Ichth. Ind. Neerl., viii, 1876-77, Pl. cccxlviii, fig. 3 (plate 18).

D. IX; 13-14. A. III, 8. L. lat. 83-120, lateral line with 64-65. Tr. (15-20)+(30-50). Predorsals minute. Cheek scales 20-28. G.R. reduced to knobs, (7-10)+(14). Markings and shape vary with age. Older fish with deeply emarginate tail. Orange marked with numerous small rosy purple spots. Paler below. Fins spotted, caudal with yellow margin. 30 inches. Coastal waters around coral reefs.

290. AETHALOPERCA ROGAA (Forskål) DUSKY ROCK-COD

Perca rogaa Forskål, Descript. Animal., 1775, p. 38. *Epinephelus rogaa. Cephalopholis rogaa* Smith, Fish. Sth. Africa, 1950, fig. 420 (plate 18).

D. IX; 17-18. A. III, 9. L. lat. 114-120, lateral line with 52-57. Tr. (22-23)+(29-44). Predorsals 80. Cheek scales 27-29. G.R. 11+15. Depth about $2\frac{1}{4}$ in length. Uniform dark brown to blackish. Fins dark. Pectoral axilla livid orange. 24 inches. Coastal waters.

291. CEPHALOPHOLIS BOENACK (Bloch) BLUE-LINED CORAL-COD

Bodianus boenak Bloch, Naturg. Ausländ. Fische, iv, 1790, p. 43. *Serranus boelang, Serranus* or *Epinephelus boenack. Epinephelus formosus* Bleeker, Atlas Ichth. Ind. Neerl., viii, 1877, Pl. cccxl, fig. 3 (plate 18).

D. IX; 16. A. III, 8. L. lat. 98-108, lateral line with 52-68. Tr. (18-22)+(25-27). Predorsals 48-56. Cheek scales 30. G.R. 8+16. Reddish brown with or without darker vertical bars. Head and body with numerous undulose narrow blue lines, often extending on to fins. Inside gill-opening and mouth orange. 13 inches. Coastal waters.
Kangan kossa (S). Verri kaleva (T).

292. CEPHALOPHOLIS ARGUS Schneider PEACOCK CORAL-COD

Cephalopholis argus Bloch and Schneider, Syst. Ichth., 1801, p. 311. *Epinephelus argus. Serranus guttatus* Day, Fish. India, 1875-78, Pl. vi, fig. 3 (plate 18).

D. IX; 15-17. A. III, 8-9. L. lat. 103-115, lateral line with 44-56. Tr. (12-16)+(22-32). Cheek scales 34-36. G.R. (9-11)+17. Colour variable, brownish with blue spots to uniform blackish. Fins also spotted. Edges of unpaired fins often bluish or yellow. 18 inches. Coastal waters.

293. CEPHALOPHOLIS PACHYCENTRON (Valenciennes)
BROWN CORAL-COD

Serranus pachycentron Cuvier and Valenciennes, Hist. Nat. Poiss., ii, 1828, p. 295. *Epinephelus pachycentron. Epinephelus microprion* Bleeker, Atlas Ichth. Ind. Neerl., vii, 1876, Pl. cclxxx, fig. 1 (plate 18).

D. IX; 15-17. A. III, 8. L. lat. 95-115, lateral line with 46-53. Tr. (15-17)+(29-30). Predorsals 45. Cheek scales 22. G.R. 7+14. Dark brownish with about 8 poorly defined darker vertical cross-bars. Anterior part of body and head sometimes with small blue dots. Fins dark brown. 10 inches. Coastal waters.

294. ENNEACENTRUS MINIATUS (Forskål) VERMILION CORAL-COD

Perca miniata Forskål, Descript. Animal., 1775, p. 41. *Epinephelus* or *Cephalopholis miniatus, Serranus amboinensis. Serranus miniatus* Day, Fish. India, 1875-78, Pl. vi, fig. 2 (plate 18).

D. IX; 14-16. A. III, 9. L. lat. 105-121, lateral line with 51-56. Tr. (14-17)+(27-34). Predorsal 54-56. Cheek scales 24-28. G.R. (8-9)+(14-17). Orange-red to brownish. Body and head except paired fins with small dark brown ringed or ocellated blue spots. Paired fins deep orange red. 18 inches. Coastal waters and pearl banks.

295. ENNEACENTRUS SONNERATI (Valenciennes) RED CORAL-COD

Serranus sonnerati Cuvier and Valenciennes, Hist. Nat. Poiss., ii, 1828, p. 299. *Epinephelus* or *Cephalopholis sonnerati*. *Serranus sonnerati* Day, Fish. India, 1875-78, Pl. vii, fig. 1 (plate 18).

D. IX; 14-15. A. III, 9. L. lat. 114-130, lateral line with 72-82. Tr. (22-23)+(35-38). Predorsals 62-68. Cheek scales 33. G.R. 9+13. Yellow or reddish brown. Head and anterior part of body spotted paler or with fine network of pale blue lines. Fins with blackish margins. 24 inches. Coastal waters.

Ran thambuva (S). Segepu kaleva (T).

296. ENNEACENTRUS NIGRIPINNIS (Valenciennes) DARK-FINNED CORAL-COD

Serranus nigripinnis Cuvier and Valenciennes, Hist. Nat. Poiss., ii, 1828, p. 339. *Cephalopholis nigripinnis* Bleeker, Atlas Ichth. Ind. Neerl., vii, 1876, Pl. cclxxxiv, fig. 2 (plate 18).

D. IX; 14-16. A. III, 9. L. lat. 107-112. Tr. 15+(42-50). Dark reddish brown. Head and anterior part of body dotted with yellowish or blue. Unpaired fins with dark brown broad submarginal band. Paired fins dark brown. Pectorals and caudals broadly edged with white. 17 inches. Coastal waters.

297. PROMICROPS LANCEOLATUS (Bloch) GROUPER

Holocentrus lanceolatus Bloch, Naturg. Ausländ. Fische, iv, 1790, p. 92. *Epinephelus lanceolatus* Bleeker, Atlas Ichth. Ind. Neerl., viii, 1877, Pl. cccxxxii, fig. 3 (plate 18).

D. XI; 14-15. A. III, 8. L. lat. 94-106, lateral line with 60-70. Tr. (21-22)+(28-31). Predorsals 50. Cheek scales 30. G.R. 10+(15-16). Colours change with growth. Large adults uniform blackish. Younger fish greyish variegated with obscure darker blotchings, spots and bars. Unpaired fins brownish yellow with black markings. Paired fins lemon yellow blotched with black. Largest of group attaining 12 feet and said to attack man. Coastal waters around coral reefs.

Hali valan kossa, Sellan kossa (S). Komari kaleva (T).

298. EPINEPHELUS FARIO (Thunberg) TROUT REEF-COD

Perca fario Thunberg, Kon. Vet. Akad. Nya. Handl. Stockholm, xiv, 1793, p. 296. *Serranus fario*, *Epinephelus maculatus*. *Epinephelus fario* Tanaka, Fishes Japan, xxxviii, 1927, Pl. clxii, fig. 451 (plate 18).

D. XI; 15-17. A. III, 8. L. lat. 115-125, lateral line with 53-55. Tr. (20-24)+(33-36). Predorsals 76-80. Cheek scales 40-50. G.R. 7+16. Colour variable. Brownish yellow, usually with brown or blackish spots, or sometimes bluish blotches. Fins similar. 16 inches. Coastal waters and pearl banks.

Pulli kossa (S). Pulli kaleva (T).

299. EPINEPHELUS FASCIATUS (Forskål) BANDED REEF-COD

Perca fasciata Forskål, Descript. Animal., 1775, p. 11. *Epinephelus marginalis*. *Epinephelus fasciatus* Bleeker, Atlas Ichth. Ind. Neerl., viii, 1877, Pl. cccxxvi, fig. 3 (plate 19).

D. XI; 16. A. III, 8. L. lat. 105-127, lateral line with 47-83. Tr. (13-15)+(26-28). Predorsals 70-74. Cheek scales 43-45. G.R. 6+14. Dark reddish with darker 6-7 transverse bands, which fade with age. Fins also reddish with black margin to spinous dorsal. 14 inches. Coastal waters.

300. EPINEPHELUS UNDULOSUS (Quoy and Gaimard) BROWN-LINED REEF-COD

Bodianus undulosus Quoy and Gaimard, Voy. Uranie Zool., 1824, p. 310. *Serranus amboinensis* or *undulosus*, *Epinephelus lineolatus*. *Epinephelus undulosus* Bleeker, Atlas Ichth. Ind. Neerl., vii, 1876, Pl. cclxxxviii, fig. 3 (plate 19).

D. X-XI; 18-20. A. III, 8. L. lat. 108-119, lateral line with 65-73. Tr. (20-21)+(34-36). Predorsals 57-67. Cheek scales 30-32. G.R. 14+20. Brownish with 12 dark brown longitudinal lines, little inclined upward posteriorly with paler lines in interspaces. Spinous dorsal with black margin. Pectorals light yellow. 15 inches. Coastal waters, trawling and pearl banks.
Thambeleya, Thambuwa (S). Panchi kaleva (T).

301. EPINEPHELUS FLAVOCAERULEUS (Lacépède) BLUE AND YELLOW REEF-COD

Holocentrus flavo-caeruleus Lacépède, Hist. Nat. Poiss., iv, 1802, p. 331. *Perca flavopurpurea*. *Serranus flavocaeruleus* Day, Fish. India, 1875-78, Pl. iii, fig. 1 (plate 19).

D. XI; 15-17. A. III, 8. L. lat. 133-140, lateral line with 75-77. Tr. (22-25)+(42-43). Predorsals 60-66. Cheek scales 30-33. G.R. 10+18. Body deep blue or purplish with numerous fine dots. Fins and lips yellow. Extent of colours varies. 18 inches. Coastal waters and trawling banks.
Kaha laveya (S). Manjel kaleva (T).

302. EPINEPHELUS MORRHUA (Valenciennes) BANDED-CHEEK REEF-COD

Serranus morrhua Cuvier and Valenciennes, Hist. Nat. Poiss., ix, 1833, p. 434. *Serranus angularis*. *Epinephelus döderleini* Tanaka, Fish. Japan, xxxiv, 1925, Pl. cliii, fig. 423 (plate 19).

D. XI; 14-15. A. III, 7-8. L. lat. 118, lateral line with 62. Tr. 22+33. Predorsals 78. Cheek scales 33. G.R. 7+14. Younger fish reddish with few indistinct subvertical darker bands and yellowish fins. Older fish with brownish grey ground colour, and yellowish fins. Several dark thin streaks horizontally across cheeks, and along body also in young. 24 inches. Coastal waters.
Raja laveya (S). Kallu kaleva (T).

303. EPINEPHELUS DIACANTHUS (Valenciennes) SIX-BARRED REEF-COD

Serranus diacanthus Cuvier and Valenciennes, Hist. Nat. Poiss., ii, 1828, p. 319; Day, Fish. India, 1875-78, Pl. iii, fig. 4 (plate 19).

D. XI; 15-17. A. III, 7-8. L. lat. 93-108, lateral line with 57-60. Tr. (17-18)+(33-36). Predorsals 56-60. Cheek scales 21-25. G.R. 8+16. Two marked spines at preopercle angle. Brownish sometimes with obscure darker patches. Six dark brown broad vertical bars. Fins sometimes clouded or spotted with darker colour. 22 inches. Trawling banks.

304. EPINEPHELUS AREOLATUS (Forskål) AREOLATED REEF-COD

Perca summana areolata Forskål, Descript. Animal., 1775, p. 42. *Serranus areolatus. Epinephelus angularis. Serranus angularis* Day, Fish. India, 1875-78, Pl. v, fig. 2 (plate 19).

D. XI; 15-17. A. III, 8. L. lat. 105-111, lateral line with 48-54. Tr. (19-20)+(32-33). Predorsals 60-70. Cheek scales 30-31. G.R. 9+16. Caudal truncate. Brown with hexagonal darker spots, smaller on caudal. Interspaces small, or equal to diameter of spots. Older fish with pale edge to caudal. 16 inches. Coastal waters and trawling banks.

305. EPINEPHELUS CHLOROSTIGMA (Valenciennes)
BROWN-SPOTTED REEF-COD

Serranus chlorostigma Cuvier and Valenciennes, Hist. Nat. Poiss., ii, 1828, p. 352. *Serranus wandersi. Epinephelus chlorostigma* Chevey, Mem. Instit. Oceanogr. Indochine, iv, 1932, Pl. xxvi (plate 19).

D. XI; 16-17. A. III, 8. L. lat. 114, lateral line with 56. Tr. 20+33. Predorsals 68. Cheek scales 36. G.R. 10+21. Caudal truncate. Generally brownish with numerous close-set dark hexagonal or rounded spots, all greater than pale interspaces. Hind edge of caudal light. 25 inches. Pearl banks.

306. EPINEPHELUS BLEEKERI (Vaillant and Bocourt)
BLEEKER'S REEF-COD

Serranus bleekeri Vaillant and Bocourt, Mission Scient. Mexique, Poiss., 1877, p. 69. *Serranus coromandelicus. Serranus waandersi* Day, Fish. India, 1875-78, Pl. viii, fig. 1 (plate 19).

D. XI; 16-17. A. III, 8. L. lat. 87-132, lateral line with 60-67. Tr. (17-22)+(30-31). Predorsals 50-63. Cheek scales 24-28. G.R. 10+16. Caudal truncate. Reddish brown, lighter below. Head unspotted or with pale spots. Back and sides with obscure spots smaller than eye. Dark stripe above maxillary. Lower half of caudal brown contrasting with upper half. Young with six transverse bands. 26 inches. Pearl banks and trawling grounds.

307. EPINEPHELUS MERRA Bloch WIRE-NETTING REEF-COD

Epinephelus merra Bloch, Naturg. Ausländ. Fische, vii, 1793, p. 17. *Serranus merra* or *faveatus, Epinephelus hexagonatus. Serranus hexagonatus* Day, Fish. India, 1875-78, Pl. ii, fig. 3 (plate 19).

D. XI; 15-17. A. III, 8. L. lat. 102-106, lateral line with 50-55. Tr. (15-17)+(27-31). Predorsals 56-62. Cheek scales 25-27. G.R. 7+15. Light brown with large dark hexagonal brown spots with narrow light interspaces forming a honey-comb. Some spots fused to form cross-bars. Pattern very variable. 18 inches. Coastal waters and trawling grounds.
Pulli kossa (S). Pulli kaleva (T).

308. EPINEPHELUS CAERULEOPUNCTATUS (Bloch)
WHITE-SPOTTED REEF-COD

Holocentrus caeruleopunctatus Bloch, Naturg. Ausländ. Fische, iv, 1790, p. 94. *Epinephelus Hoevenii* Bleeker, Atlas Ichth. Ind. Neerl., vii, 1876, Pl. cclxxxii, fig. 1 (plate 19).

D. XI; 15-16. A. III, 8. L. lat. 110-117, lateral line with 53-68. Tr. (17-18)+(22). Predorsals 85-100. Cheek scales 31-32. G.R. 10+17. Colour variable, usually purple brown with white or blue spots of various types, usually large and chalky white in adults and ocelli in young. 30 inches. Coastal waters.

309. EPINEPHELUS FUSCOGUTTATUS (Forskål)
BROWN-MARBLED REEF-COD

Perca summana fusco-guttata Forskål, Descript. Animal., 1775, p. 42. *Serranus fuscoguttatus. Epinephelus fuscoguttatus* Bleeker, Atlas Ichth. Ind. Neerl., vii, 1876, Pl. cccvii, fig. 3 (plate 19).

D. XI; 13-15. A. III, 8. L. lat. 85-113, lateral line with 68-74. Tr. (15-22)+ (27-32). Predorsals 75-78. Cheek scales 35-37. G.R. 13+17. Brownish with small darker spots on head, body, tail and fins. Irregular large dark blotches or bands on base of dorsal and others along sides. Fins yellowish. Markings fade with growth, adults often uniform brown. 40 inches. Trawling grounds and Gulf of Manaar.

310. EPINEPHELUS TAUVINA (Forskål) GREASY REEF-COD

Perca tauvina Forskål, Descript. Animal., 1775, p. 39. *Epinephelus polypodonphilus* Bleeker, Atlas Ichth. Ind. Neerl., vii, 1876, Pl. cclxxxiii, fig. 1 (plate 19).

D. XI; 15-16. A. III, 8. L. lat. 95-123, lateral line with 61-65. Tr. (17-19)+(28-29). Predorsals 66-70. Cheek scales 36-38. G.R. 10+15. Old fish uniform dark brown. Younger fish brown with cross-bars and large spots on body and fins. Sometimes also black spots, which increase with age. 7 feet 6 inches. Coastal waters.
Gal bola, Gal kossa (S).

311. EPINEPHELUS MALABARICUS (Schneider) MALABAR REEF-COD

Holocentrus malabaricus Bloch and Schneider, System. Ichth., 1801, p. 319. *Serranus bontoo* or *malabaricus, Epinephelus pantherinus* or *salmoides. Serranus gilberti* Bleeker, Atlas Ichth. Ind. Neerl., viii, 1877, Pl. cccxxxi, fig. 3 (plate 19).

D. XI; 14-15. A. III, 8. L. lat. 112-120, lateral line with 60-64. Tr. (16-18)+(30-33). Predorsals 60-80. Cheek scales 33-40. G.R. 7+14. Brownish, paler below. Everywhere with obscure ill-defined dark blotches. Fins sometimes darker than body and abundantly spotted. May be a variety of *E. tauvina*. 18 inches. Coastal waters, pearl banks and trawling grounds.
Gal bola, Gal kossa (S).

DOTTYBACKS

Family PSEUDOCHROMIDAE

Perch-like fish with elongate compressed body. Mouth moderate with fine teeth in band in front of each jaw. Anal and dorsal each with 2-3 spines. Lateral line in two separate parts.

312. PSEUDOCHROMIS FUSCUS Müller and Troschel

BROWN DOTTYBACK

Pseudochromis fuscus Müller and Troschel, Horae Ichthyol., iii, 1849, p. 23. *Pseudochromis nebulosus* McCulloch, Mem. Qld. Mus., iii, 1915, Pl. xvi, fig. 2 (plate 20).

D. III; 25-28. A. III, 14-15. P. 18-19. L. lat. 40, with 30 in upper and 8-10 in lower lateral line. Lateral lines separated by 4 scale rows. Dorsal with 14-18 anterior rays simple, remainder branched. Unpaired fins in scaly sheath. Brownish above, lighter below. Scales with dark spot forming longitudinal streaks along sides. Darker bands on caudal and pectorals. 4 inches. Coastal waters.

SEA PERCH

Family DIPLOPRIONIDAE

Perch-like fish with oblong-oval body, much compressed. Scales small, ctenoid, covering preoperculum and operculum. One lateral line which is strongly arched. Lower edges of preoperculum, suboperculum and operculum serrated. Operculum with three strong spines. Two dorsals connected only at base, with 8 strong spines.

313. DIPLOPRION BIFASCIATUM Cuvier TWO-BAND SEA PERCH

Diploprion bifasciatum Cuvier and Valenciennes, Hist. Nat. Poiss., ii, 1828, p. 137; Bleeker, Atlas Ichth. Ind. Neerl., viii, 1876-77, Pl. cccxlvi, fig. 3 (plate 20).

D. VIII; 12-15. A. II, 12-13. P. 15-16. L. lat. 102-106, lateral line with 69-75. Tr. (17-18)+(43-46). Predorsals 18-19. Cheek scales 11. Teeth in bands in jaws, vomer and palatines are villiform. G.R. 9+21. Yellowish with two broad blackish cross-bars, through eye and 5th dorsal spines. 9½ inches. Coastal waters.
China thambuva, Hini gandeya (S).

THERAPON PERCH, GRUNTERS

Family THERAPONIDAE

Perch-like fishes with oblong to oblong-ovate body, compressed. Scales fairly small. Mouth moderate with fine teeth in bands in jaws, outer series larger, sometimes on palate also. Preopercle serrate, sometimes with spines. Dorsal spines XII-XIV, strong.

1. Teeth rather flattened, in 2 series in lower jaw .... *Pelates quadrilineatus*
   Teeth conical, in bands 2
2. Scales above L.lat. 12-16 3
   Scales above L.lat. 9-10 .... *Eutherapon theraps*
3. Depth less than head; 4 straight lateral bands .... *Autisthes puta*
   Depth greater than head; 3 bent lateral bands .... *Therapon jarbua*

314. PELATES QUADRILINEATUS (Bloch) TRUMPETER PERCH

Holocentrus quadrilineatus Bloch, Naturg. Ausländ. Fische, iv, 1790, p. 82. *Helotes polytaenia*. *Therapon quadrilineatus* Day, Fish. India, 1875-78, Pl. xviii, fig. 5 (plate 20).

D. XII; 10. A. III, 9-10. P. 15. L. lat. 75-80. Tr. (12-14)+(23). Predorsals 15-17. Cheek scales 5. G.R. 18+22. Caudal emarginate. Brownish olive, silver beneath. 4-6 horizontal brown bands, first along spinous dorsal base. Pattern variable. Dark blotch on spinous dorsal and sometimes another on shoulder. 8 inches. Estuaries.

315. EUTHERAPON THERAPS (Cuvier)

LARGE-SCALED BANDED GRUNTER

Therapon theraps Cuvier and Valenciennes, Hist. Nat. Poiss., iii, 1829, p. 129; Chevey, Mem. Instit. Oceanogr. Indochine, iv, p. 932, Pl. xxx (plate 20).

D. XI-XII; 9-10. A. III, 8-9. L. lat. 49-57. Tr. (9-10)+(14-15). Predorsals 12-15. Cheek scales 5-6. G.R. 8+16. Pale brown, silvery below. 3-4 dark stripes along body, and a black blotch on 3rd-7th dorsal spines. 2 oblique bars across each caudal lobe. Anal with dark blotch. 7 inches. Estuarine.

316. AUTISTHES PUTA (Cuvier) SMALL-SCALED BANDED GRUNTER

Therapon puta Cuvier and Valenciennes, Hist. Nat. Poiss., iii, 1829, p. 131. *Therapon trivittatus*, Bleeker, Atlas Ichth. Ind. Neerl., viii, 1877, Pl. cccxl, fig. 2 (plate 20).

D. XII; 10. A. III, 8. L. lat. 88. Tr. 12+24. Predorsals 19-20. Cheek with 6 scale rows. G.R. 9+20. Brownish olive, lighter below. 4 nearly straight longitudinal dark stripes along sides of body, first (sometimes absent) along base of spinous dorsal. Caudal with 2 oblique bars across lobes. Usually dark blotch between 3rd and 7th dorsal spines. 6½ inches. Estuaries and pearl banks.

Gon kili, Vairan kili (S). Kove kitchan (T).

317. THERAPON JARBUA (Forskål) CRESCENT PERCH

Sciaena jarbua Forskål, Descript. Animal., 1775, p. 50. *Therapon servus*. *Therapon jarbua* Chevey, Mem. Instit. Oceanogr. Indochine, iv, 1932, Pl. xxxi (plate 20).

D. XI-XII; 10-11. A. III, 7-9. L. lat. 82-84. Tr. (15-16)+(23-25). Predorsals 20-25. Cheek scales 6-7. G.R. 6+13. Silvery, brownish above. 3 dark longitudinal cross-bands, curved downwards, first from origin of dorsal. Bands continued across caudal and upper lobes tipped with black. Large black blotch between 4th and 8th dorsal spines. 10 inches. Coastal waters and estuaries.

Gonga, Iri bateya, Pol bateya, Kili (S). Palin kitchan (T).

ROUND-HEADS

Family PLESIOPIDAE

Perch-like fishes with heavy robust body with blunt head. Fairly large strong scales. Lateral line in 2 parts, upper ending below soft dorsal. Fairly large mouth with fine teeth in bands, outer enlarged. Teeth on palate. Pelvics with 1 spine and 4 rays, front usually thickened.

318. PLESIOPS NIGRICANS (Rüppell) BLACK ROUND-HEAD

Pharopteryx nigricans Rüppell, Atlas Reise Nordl. Afrika, Fische, 1828, p. 15. *Plesiops nigricans* Bleeker, Atlas Ichth. Ind. Neerl., ix, 1877, Pl. cccxc, fig. 3 (plate 21).

D. XI-XII; 6-7. A. III, 6-8. L. lat. 25-28; lower tubules 13-18, total series 28-30. Tr. 2+12. G.R. 11 lower. Dark brown with minute blue spots. 10 inches. Coastal waters.

FLAGTAILS

Family KUHLIDAE

Elongate oval body, fairly compressed. Moderate ctenoid scales. Mouth moderate, with fine teeth in bands in jaws and palate. Fin spines 10, strong. Dorsals united with deep notch before soft dorsal.

Caudal with 5 dark stripes; L. lat. 55-56 .... *Kuhlia taeniurus*
Caudal spotted with black margin; L. lat. 42-45 .... *Kuhlia marginatus*

319. KUHLIA TAENIURUS (Cuvier) BARRED FLAGTAIL

Dules taeniurus Cuvier and Valenciennes, Hist. Nat. Poiss., iii, 1829, p. 114. *Perca argentea*, *Dules bennetti*. *Moronopsis taeniurus* Bleeker, Atlas Ichth. Ind. Neerl., viii, 1877, Pl. cccxlv, fig. 5 (plate 20).

D. X; 10-11. A. III, 11. P. 14. L. lat. 55-56. Tr. (5-6)+(12-14). Predorsals 12. Cheek scales 5. G.R. 12+14. Caudal deeply forked. Brownish above, silvery beneath. Fins yellowish. Caudal with 5 black cross stripes. Young with black margins to dorsals. 8 inches. Fresh water, estuaries and coral reefs.

This nagula (S).

320. KUHLIA MARGINATUS (Cuvier) SPOTTED FLAGTAIL

Dules marginatus Cuvier and Valenciennes, Hist. Nat. Poiss., iii, 1829, p. 116. *Kuhlia marginatus* Weber and De Beaufort, Fish. Indo-Aust. Archipel., v, 1929, fig. 72 (plate 20).

D. X; 10-12. A. III, 11-13. P. 13-14. L. lat. 42-45. Tr. (5-6)+(9-10). Predorsals 11-12. Cheek scales 5. G.R. 9+18. Caudal moderately emarginate. Silvery, brownish above. Back and sides with brown spots extending to vertical fins and caudal. Soft dorsal, anal and caudal dark edged. 8½ inches. Fresh waters.

Gal pulunna, Katu massa (S).

BULLSEYES

Family PRIACANTHIDAE

Ovoid compressed body with small rough scales. Large mouth, oblique, with fine teeth in bands in jaws and palate. Dorsals united, spines well developed. Caudal truncate. Preopercular margin serrate, with spines at angle. Small carnivorous fishes, usually red in colour.

1. Ventrals enlarged, long as or longer than head .... *Cookeolus boops*
   Ventrals shorter than head 2
2. Lower gill-rakers 17-18; vertical fins not dusky .... *Priacanthus tayenus*
   Lower gill-rakers 20-24; fins dark or dusky .... *Priacanthus hamrur*

321. COOKEOLUS BOOPS (Schneider) LONG-FINNED BULLSEYE

Anthias boops Bloch and Schneider, Syst. Ichth., 1801, p. 308. *Priacanthus boops* or *fax. Priacanthus velabundus* McCulloch, Endeavour Rept., iii (3), 1915, Pl. xix (plate 20).

D. X; 12-13. A. III, 12-13. L. lat. 85-86, with 62-65 in lateral line. Tr. (13-14)+(30-32). Predorsals 50-53. Cheek scales 12-15. G.R. 7+17. Depth about 2. Pelvics usually longer than head. Red, lighter below, fins dusky marginally, pelvics blackish. 20 inches. Coastal waters in 32 fathoms.

322. PRIACANTHUS TAYENUS Richardson SPOT-FINNED BULLSEYE

Priacanthus tayenus Richardson, Rept. 15th meeting Brit. Assn., 1846, p. 237. *Priacanthus holocentrum. Priacanthus tayenus* Chevey, Mem. Institut. Oceanogr. Indochine, iv, 1932, Pl. xxvii (plate 20).

D. X; 12. A. III, 13. L. lat. 73-78, with 60-63 in lateral line. Tr. (10-11)+(30-33). Predorsals 42-43. Cheek scales 14-15. G.R. 4+(17-18). Depth about 3. Pelvics shorter than head. Strong serrated spine at angle of preopercle. Caudal crescentic with lobes produced. Reddish above, silvery below. Ventrals whitish finely dotted with black. Other fins pale with dark edge. 10 inches. Trawling banks.

323. PRIACANTHUS HAMRUR (Forskål) DUSKY-FINNED BULLSEYE

Sciaena hamrur Forskål, Descript. Animal., 1775, p. 45. *Priacanthus hamrur* Bleeker, Atlas Ichth. Ind. Neerl., viii, 1876-77, Pl. cccliii, fig. 3 (plate 20).

D. X; 14-15. A. III, 14-16. L. lat. 120-129, with 84-88 in lateral line. Tr. (14-15)+(41-43). Predorsals 60-66. Cheek scales 15-17. G.R. 8+24. Depth about $2\frac{3}{4}$. Pelvics shorter than head. Caudal emarginate in young, becoming crescentic with age. Short blunt spine at preopercle angle. Red to purplish red, lighter below, fins usually with dusky margins, pelvics dark basally. 16 inches. Coastal waters.

CARDINAL FISHES

Family APOGONIDAE

Body oblong to elongate, sometimes fairly compressed. Scales rather large. Mouth large, teeth fine or caniniform, sometimes also on vomer and palatines. Two dorsal fins, spinous with 5-9 spines. Anal with 2 spines. Mostly small and vividly coloured. Carnivorous.

1. Jaws with distinct canine teeth .... *Cheilodipterus quinquelineatus*
   Jaws without canines 2
2. Anal with 13-16 rays .... *Archamia lineolatus*
   Anal with 8 rays 3

3. Caudal rounded or subtruncate .... *Apogonichthys ellioti*
   Caudal emarginate or forked 4
4. First dorsal with six spines .... *Apogon thermalis*
   First dorsal with seven spines 5
5. Body without longitudinal bands 6
   Body with longitudinal bands 7
6. Round black, white-edged spot above pectoral .... *Apogon enneastigma*
   Broad dark brown vertical band, 4 scales wide, at caudal base .... *Apogon aureus*
7. Maxillary reaches to below middle of eye .... *Apogon septemstriatus*
   Maxillary reaches to below posterior half of eye 8
8. Two bands on upper part of body only; orbital rim serrated .... *Apogon quadrifasciatus*
   Bands on lower as well as upper part of body; orbital rim without serrations 9
9. Snout pointed, $\frac{3}{4}$ of eye; preorbital height $\frac{1}{3}$ eye .... *Apogon novemfasciatus*
   Snout rounded, $\frac{2}{3}$ of eye; preorbital $\frac{1}{2}$ eye .... *Apogon endekataenia*

324. CHEILODIPTERUS QUINQUELINEATUS Cuvier
TOOTHED CARDINAL FISH

Cheilodipterus quinquelineatus Cuvier and Valenciennes, Hist. Nat. Poiss., ii, 1828, p. 167. *Paramia quinquelineata* Bleeker, Atlas Ichth. Ind. Neerl., viii, 1876-77, Pl. cccxxvi, fig. 2 (plate 21).

D. VI; I, 9. A. II, 8. L. lat. 25-26. Tr. 2+7. Depth 4-4$\frac{1}{2}$. Preoperculum serrated. Caudal emarginate. Jaws with narrow bands of villiform teeth and canines present anteriorly in both jaws. G.R. 1+9. Yellowish with 5 lateral broad or narrow dark stripes, median one through eye with 2 above and 2 below. A sixth median stripe along ventral profile to anus. Median lateral band ending in a yellow-edged black spot on caudal peduncle. Anterior border of dorsal blackish, other unpaired fins dusky and paired fins yellowish. 4 inches. Coastal waters.

325. ARCHAMIA LINEOLATUS (Cuvier) CROSS-BANDED CARDINAL FISH

Apogon lineolatus Cuvier and Valenciennes, Hist. Nat. Poiss., ii, 1828, p. 160. *Apogon zeylonicus* or *Ceylonicus. Archamia lineolatus* Jordan and Seale, Bull. U.S. Bur. Fish., xxv, 1906, Pl. xxxviii, fig. 1 (plate 21).

D. VI; I, 9. A. II, 13-17. P. 2, 12. L. lat. 23-27. Tr. 2+8. Preoperculum serrated. Caudal slightly emarginate. Gill-rakers 17. Depth 2$\frac{1}{2}$-3 in total length. Colour variable, usually yellowish, pinkish-brown above and with narrow brownish vertical cross bands. Black spot on caudal base. 3$\frac{1}{2}$ inches. Shallow water marine.

326. APOGONICHTHYS ELLIOTI (Day) ELLIOT'S CARDINAL FISH

Apogon ellioti Day, Fish. India, 1875-78, p. 63, Pl. xvii, fig. 1 (plate 21).

D. VII; I, 9. A. II, 8. P. 15. L. lat. 25-26. Tr. 2+7. Depth 3$\frac{1}{2}$-4 in total length. Vertical border of preoperculum smooth, lower border finely serrated. Caudal rounded or subtruncate. Silvery-grey with traces of 4-5 faint cross-bars. Upper part of first dorsal black. Soft dorsal and anal fins with dark longitudinal bands. Indistinct blotch on tail in young. 4 inches. Coastal waters.

327. APOGON THERMALIS Cuvier TRANSLUCENT CARDINAL FISH

Apogon thermalis Cuvier and Valenciennes, Hist. Nat. Poiss., iii, 1829, p. 492. (?) *Apogon hyalosoma* Day, Fish. India, 1875-78, Pl. xvii, fig. 5 (plate 21).

D. VI; I, 9. A. II, 8. P. 2, 12. L. lat. 24-26. Tr. (2-3)+(7-8). Depth 3-3½ in total length. Preopercular borders feebly serrated. Caudal slightly emarginate. Yellowish-brown, lighter below. Large round black spot near caudal base. Similar spots at ends of soft dorsal and anal fins. Fin membrane between 2nd and 3rd spines of first dorsal black. 6½ inches. Marine and fresh waters. (*A. thermalis* recorded from the thermal springs of Cannia is insufficiently characterized and accepted here as a synonym of *A. hyalosoma* Bleeker, to which this description and figure pertains.)

328. APOGON ENNEASTIGMA Rüppell NINE-SPOT CARDINAL FISH

Apogon enneastigma Rüppell, Neue Wirbelth., Fische, 1835, p. 87, Pl. xxii, fig. 3 (plate 21).

D. VII; I, 9-10. A. II, 8-9. Reddish with lighter fins. Spinous dorsal with black front border. A round black, white-edged spot above pectoral. Five black specks along base of dorsal fins. Another black spot on each side of tail. Pearl banks. Marine.

329. APOGON AUREUS (Lacépède) GOLDEN CARDINAL FISH

Centropomus aureus Lacépède, Hist. Nat. Poiss., iv, 1802, p. 253. *Apogon roseipinnis* or *annularis* var. *roseipinnis*. *Amia fleurieu*. *Amia aurea* Bleeker, Atlas Ichth. Ind. Neerl., viii, 1877, Pl. cccxxxvii, fig. 1 (plate 21).

D. VII; I, 9. A. II, 8. P. 2, 11. L. lat. 26-27. Tr. 2+7. Depth 2·6-2·8 in total lergth. Preopercular margins finely serrated. Caudal feebly incised. Gill-rakers 16. Yellowish-brown, lighter below. A dark brown stripe through eye from snout to operculum. A broad dark brown band, about 4 scales wide on caudal peduncle. Anterior part of spinous dorsal dusky. Fins golden with dusky margins. 5 inches. Coastal waters.

Gal punakkaliya, Gorakeya (S). Puna (T).

330. APOGON SEPTEMSTRIATUS Günther SEVEN-BAND CARDINAL FISH

Apogon septemstriatus Günther, Challenger Rept., Zool., i, 1880, p. 38, Pl. xvi, fig. A (plate 21).

D. VII; I, 8. A. II, 8. L. lat. 28. Preoperculum serrated, its ridge smooth. Light coloured with 3 well-defined black streaks on each side. Lowest passing through eye to middle of caudal fin. A seventh, median streak on mid-line of head and nape. No round black spot at caudal base. Narrow blackish line along base of soft dorsal and anal fins. 3½ inches. Coastal waters.

331. APOGON QUADRIFASCIATUS Cuvier FOUR-BAND CARDINAL FISH

Apogon quadrifasciatus Cuvier and Valenciennes, Hist. Nat. Poiss., ii, 1828, p. 153. *Apogon monogramma* Günther, Challenger Rept., Zool., i, 1880, Pl. xvi, fig. B (plate 21).

D. VII; I, 9. A. II, 8. P. 2, 13-14. L. lat. 25-27. Tr. 2+7. Depth 3·1-3·6 in total length. Preoperculum serrated and with 2 rows of scales. Gill-rakers 14. Caudal feebly incised with rounded lobes. Pale yellowish-brown, silvery below. Two dark longitudinal bands on each side in upper half of body, lower passing through eye, and bordered by silvery edges. Fins brownish, sometimes with darker markings. 4 inches. Coral reefs.

332. APOGON NOVEMFASCIATUS Cuvier NINE-BAND CARDINAL FISH

Apogon novemfasciatus Cuvier and Valenciennes, Hist. Nat. Poiss., ii, 1828, p. 154. *Amia novemfasciatus* Jordan and Seale, Bull. U.S. Bur. Fish., xxv, 1906, fig. 36 (plate 21).

D. VII; I, 9. A. II, 8. P. 2, 11-12. L. lat. 26-27. Tr. 2+7. Depth 3·3-3·6 in total length. Preoperculum serrated with 2 feeble rows of scales. Gill-rakers 14. Caudal incised with rounded lobes. Colour variable, usually light reddish-brown. Five distinct longitudinal dark bands along sides, one passing through eye. Usually dark spot at caudal base or more elongate streak extending on to caudal fin. Fins pale. Second dorsal and anal with dark basal bands. 3 inches. Tidal pools extending into deeper coastal waters.

333. APOGON ENDEKATAENIA Bleeker ELEVEN-BAND CARDINAL FISH

Apogon endekataenia Bleeker, Nat. Tijds. Ned. Ind., iii, 1852, p. 449. *Amia endeketaenia. Amia fasciata stevensi* McCulloch, Endeavour Rept., iii (3), 1915, Pl. xvi, fig. 2 (plate 21).

D. VII; 9-10. A. II, 8. P. 2, 12. L. lat. 26. Tr. 2+7. Depth 3½ in total length. Preoperculum finely serrated with 2 indistinct vertical rows of scales. Orbital rim rough. Light reddish-brown with five distinct longitudinal bands along sides, the two principal ones always present. Always dark blotch at caudal base. Fins pale. Superficially similar to the above species. Coastal waters.

WHITING

Family SILLAGINIDAE

Rather robust elongate tapering body. Fairly small scales. Small mouth with fine teeth in bands in jaws and vomer. None on palatine. Two dorsals, first with 10-11 spines. Anal with 2 spines.

Uniform pearly .... *Sillago sihama*
With dark blotches .... *Sillago maculata*

334. SILLAGO SIHAMA (Forskål) SILVER WHITING

Atherina sihama Forskål, Descript. Animal., 1775, p. 70. *Sillago acuta. Sillago sihama* Bleeker, Atlas Ichth. Ind. Neerl., ix, 1877, Pl. ccclxxxix, fig. 4 (plate 21).

D. X-XI; I, 20-23. A. II, 21-24. L. lat. 70-73. Tr. (5-6)+(10-12). G.R. 7-9 plus few rudimentary ones on lower limb of first arch. Scales on cheek cycloid. Olivaceous or grey above, silvery below with a diffuse silver band along body. 12 inches. Coastal waters, estuaries and trawling banks.

Kalanda (S). Kilaken (T).

335. SILLAGO MACULATA Quoy and Gaimard TRUMPETER WHITING

Sillago maculata Quoy and Gaimard, Voy. Uranie et Physic., 1824, p. 261; Bleeker, Atlas Ichth. Ind. Neerl., ix, 1877, Pl. ccclxxxix, fig. 5 (plate 21).

D. X-XI; I, 18-21. A. II, 20-23. L. lat. 70-74. Tr. (5-6)+(10-12). G.R. 9-10 on lower limb of first arch. Scales on cheek cycloid. Silvery grey with indefinite dark blotches along side. Fins with dusky to greyish markings. Silver lateral band on body. 12 inches. Coastal waters and estuaries.

FALSE WHITING

Family MALACANTHIDAE

Body elongate, almost cylindrical. Scales small. Mouth moderate with fine teeth in bands in both jaws. Some small canines. Palate without teeth. Dorsal and anal fins long, spines feeble.

336. MALACANTHUS LATOVITTATUS (Lacépède) BLUE WHITING

Labrus latovittatus Lacépède, Hist. Nat. Poiss., iii, 1802, p. 455. *Oceanops latovittata* Jordan and Seale, Bull. U.S. Bur. Fish., xxv, 1906, Pl. xxxix (plate 21).

D. 47-51, with first 4-6 rays spiniform. A. 39-41 with first ray spiniform. L. lat. ca. 130. Tr. 10+45. A few granular teeth behind the others in each jaw. Head and back pale blue. A black lateral band from operculum to caudal base. Under surface, anal and ventral fins white. Dorsal fin pale grey with yellowish marginal band. Upper lobe of caudal similar to dorsal, lower white largely obscured by black. Pectoral blue. 15 inches. Coastal waters.

WHITEFISH

Family LACTARIIDAE

Body oblong, compressed. Scales moderate, cycloid, deciduous. Mouth large with small pointed teeth in jaws, vomer, palatines and tongue. Small canines anteriorly in jaws. Two dorsal fins, first with 7-8 weak spines.

337. LACTARIUS LACTARIUS (Schneider) WHITEFISH

Scomber lactarius Bloch and Schneider, Syst. Ichth., 1801, p. 31. *Lactarius delicatulus* Day, Fish. India, 1875-78, Pl. liii, fig. 2 (plate 23).

D. VII-VIII; I, 20-22. A. III, 25-28. L. lat. 74-80. Maxillary about 2 in head. 3rd dorsal spine longest. Lateral line parallel to dorsal profile and without shields or scutes. Silvery, greyish above. Black spot on operculum. Dorsal and caudal with dusky border, other fins yellowish. 11 inches. Coastal waters and trawling grounds.
Pulunna (S).

TREVALLY, HORSE-MACKEREL, QUEENFISH, DARTS

Family CARANGIDAE

More or less compressed body, sometimes naked, usually with small thin cycloid scales, those on lateral line sometimes enlarged and spiny. Mouth moderate, teeth usually feeble, sometimes absent. Front 2 anal spines separate from rest of fin. Detached finlets behind dorsal and anal fins sometimes present. Caudal powerful, forked, with slender peduncle. Swift carnivorous fishes, some reaching large size.

1. Large scutes on at least part of lateral line 2
   No enlarged scutes on lateral line 26
2. Dorsal and anal fins with posterior rays separated as finlets 3
   Dorsal and anal without true finlets 4
3. Finlets 6-9; L.lat. scutes 53-58, reaching below spinous dorsal fin .... *Megalaspis cordyla*
   Finlets 1; L.lat. scutes 40, not reaching origin of soft dorsal fin .... *Decapterus russelli*
4. Less than 7 dorsal spines, rudimentary, unconnected by a membrane; scales obscured; L.lat. scutes feeble 5
   Dorsal spines 7-8, connected by membrane; scales present; L.lat. scutes prominent 6
5. Preorbital less than eye; gill-rakers long and slender .... *Alectis ciliaris*
   Preorbital nearly twice eye; gill-rakers short and stout .... *Alectis indica*
6. Abdomen with deep median groove containing vent and detached anal spines and receiving ventrals; some dorsal and anal rays filamentous .... *Atropus atropus*
   Abdomen without deep groove; no dorsal or anal rays filamentous 7
7. Upper jaw without teeth 8
   Teeth in upper jaw, vomer and palatines 9
8. Minute teeth in single series in lower jaw and some rudimentary teeth on tongue ... *Selaroides leptolepis*
   Teeth entirely absent .... *Gnathanodon speciosus*
9. Breast completely scaled 10
   Breast naked ventrally and sometimes laterally 13
10. Under operculum a groove in shoulder girdle; teeth equal and small 11
    No groove in shoulder girdle; teeth in outer series enlarged, often caniniform anteriorly 22
11. Last dorsal and anal rays finlet-like and a little separated .... *Selar mate*
    Last dorsal and anal rays not separated 12
12. Lateral line with 48-56 scutes, the broadest 9-10 in body height .... *Selar malam*
    Lateral line with 40-46 scutes, the broadest 6-7 in body height .... *Selar kalla*
13. Breast naked only in median line; anterior dorsal and anal rays not elevated to form a falciform lobe ... 14
    Breast completely naked ventrally and mostly so laterally; anterior dorsal and anal rays prolonged to form a falciform lobe 15
14. Scales conspicuous; lateral line with 25-30 scutes .... *Carangoides praeustus*
    Scales minute, inconspicuous; lateral line with 36-40 small, unarmed scutes ...... *Carangoides plagiotaenia*
15. At least anteriorly, teeth in several rows in lower jaw 16
    Teeth in single series in lower jaw 24
16. More than 30 scutes in lateral line .... *Carangoides oblongus*
    Less than 30 scutes in lateral line 17
17. Curved part of lateral line more than 1·5 times length of straight part 18
    Curved part of lateral line less than 1·5 times length of straight part 19
18. Gill-rakers 16-17 .... *Carangoides chrysophrys*
    Gill-rakers 23-25 .... *Carangoides malabaricus*
19. Anterior dorsal rays much longer than head 20
    Anterior dorsal rays equal to or less than head 21

20. Lateral line with 20 feeble scutes; dentition complete .... *Carangoides armatus*
Lateral line with 25 scutes; dentition reduced or absent .... *Carangoides dinema*
21. No opercular spot .... *Carangoides gymnostethoides*
Opercular spot distinct .... *Carangoides ferdau*
22. Anal rays 15-17 .... *Caranx sexfasciatus*
Anal rays 18-20 23
23. Lateral line with 36-38 scutes; no spots .... *Caranx melampygus*
Lateral line with 30-35 scutes; spotted .... *Caranx stellatus*
24. Lateral line with 28-30 scutes .... *Caranx ignobilis*
Lateral line with 33-37 scutes 25
25. Eye 4·0-4·75 in head; cleft of mouth commences opposite edge of eye; breast scaly laterally .... *Caranx sansun*
Eye 3·5-4·0 in head; cleft of mouth commences opposite lower third of eye; few breast scales near pectoral .... *Caranx carangus*
26. Dorsal and anal fins with several partly detached finlets 27
Dorsal and anal fins without finlets 31
27. Snout blunt; maxilla extending behind eye .... *Chorinemus lysan*
Snout pointed; maxilla not extending behind eye 28
28. Dark blotches intersected by lateral line 29
Dark blotches not intersected by lateral line 30
29. Elongate, height equal to head, 4·5 in length .... *Chorinemus tol*
Ovate, height greater than head, 3·5 in length .... *Chorinemus tala*
30. Height greater than head length; scales rhombic .... *Chorinemus tolooparah*
Height equal to head length; scales lanceolate .... *Chorinemus sancti-petri*
31. Anal and dorsal about equal 32
Anal shorter than dorsal 34
32. Height 2 or less in length; immaculate .... *Trachinotus blochi*
Height more than 2 in length; 2-6 spots 33
33. Snout blunt; 2-5 small spots on lateral line .... *Trachinotus bailloni*
Snout pointed; 3-6 large spots above lateral line .... *Trachinotus russelli*
34. Dorsal spines separate .... *Naucrates ductor*
Dorsal spines connected .... *Zonichthys nigrofasciata*

338. MEGALASPIS CORDYLA (Linnaeus) TORPEDO TREVALLY

Scomber cordyla Linnaeus, Syst. Nat., ed. 10, 1758, p. 298. *Caranx rottleri. Megalaspis cordyla* Weber and De Beaufort, Fish. Indo-Aust. Archipel., vi, 1931, fig. 40 (plate 23).

D. I, VIII; I, 10-11; 7-9. A. II, I, 8-10; 6-8. G.R. about 20. L. lat. with 50-58 spiny scutes. Easily recognized by its torpedo-shaped, tuna-like body and numerous detached finlets. Bluish green above, silvery below. Fins yellowish, mostly fringed with black. 20 inches. Coastal waters.

Pothu giralava, Giralava (S). Vangadi (T).

339. DECAPTERUS RUSSELLI (Rüppell) RUSSELL'S SCAD

Caranx Russellii Rüppell, Atlas Reise Nordl. Afrika, 1828, p. 99. *Decapterus russelli* Ogilby, Mem. Qld. Mus., iii, 1915, Pl. xix (plate 23).

D. I, VIII; I, 30-32; 1. A. II, I, 24-27; 1. G.R. 30. Lateral line with 40-45 scutes. Depth about 5. Teeth in one row in each jaw, some on vomer and palatines. Pectoral shorter than head. Silvery blue-green. Fins yellowish or reddish. A dark opercular spot present. 18 inches. Coastal waters.
Am paratiya, Koralava (S).

340. ALECTIS CILIARIS (Bloch) THREAD-FINNED TREVALLY

Zeus ciliaris Bloch, Nat. Ausl. Fische, vi, 1788, p. 29. *Caranx ciliaris. Alectis ciliaris* Ogilby, Mem. Qld. Mus., iii, 1915, Pl. xxvii (plate 22).

D. VI; I, 18-22. A. II, I, 15-17. Depth 1-1¼. G.R. 14. L. lat. scutes 10-15, feeble. Front anal spines vanish with growth. Body much compressed, angular and rhomboidal. Anterior rays of soft dorsal, anal and ventral much prolonged and filamentous. Bluish above with sides silvery. Immature specimens with 5-7 dark transverse bands. 10 inches. Coastal waters, and trawling grounds.
Kannadi parava, Kendhi parava, Minikkuva (S). Perum pareh (T).

341. ALECTIS INDICA (Rüppell) INDIAN THREAD-FINNED TREVALLY

Scyris indica Rüppell, Atlas Reise Nordl. Afrika, 1828, p. 128. *Caranx* or *Alectis gallus, Selene vomer. Alectis indica* Ogilby, Mem. Qld. Mus., iii, 1915, Pl. xxvi (plate 22).

D. VI; I, 19. A.II, I, 16. Depth 1½. G.R. 20. L. lat. with 8-12 feeble scutes. Front dorsal and anal spines vanish with growth. Body strongly compressed. Angular and rhombic when young, becoming more elongate with age. Anterior rays of dorsal, anal and ventrals produced and filamentous particularly in young. Silvery with golden hue. 5 feet. Coastal waters and pearl banks.
Handhe panna (S).

342. ATROPUS ATROPUS (Bloch) KUWEH

Brama atropus Bloch and Schneider, Syst. Ichth., 1801, p. 98. *Caranx* or *Citula atropus. Atropus atropus* Weber and De Beaufort, Fish. Indo-Aust. Archipel., vi, 1931, fig. 42 (plate 22).

D. I, VIII; I, 22-23. A. II, I, 18. L. lat. with 30-35 scutes. Depth 2-2½. Middle rays of soft dorsal and anal fins, especially in old males, are prolonged into filaments. Ventrals as long as head. Blue-green above, silvery below. Young with vertical bands. Older fish with an opercular spot. Ventrals black, other fins hyaline. 15 inches. Coastal waters.

343. SELAROIDES LEPTOLEPIS (Cuvier) SLENDER-SCALED SCAD

Caranx leptolepis Cuvier and Valenciennes, Hist. Nat. Poiss., ix, 1833, p. 63; McCulloch, Endeavour Repts., iii (3), 1915, Pl. xxi (plate 23).

D. VIII; I, 25. A. II, I, 20. G.R. 26. L. lat. with 25-30 small scutes. Depth 3-3½. Body oblong with small but conspicuous scales. L. lat. not prominently arched. Ventrals half length of head. Greenish above, silvery below and with broad golden lateral band. Distinct black opercular spot. Fins hyaline. 6½ inches. Coastal waters, pearl banks and trawling grounds.

344. GNATHANODON SPECIOSUS (Forskål)

GOLDEN TOOTHLESS TREVALLY

Scomber speciosus Forskål, Descr. Animal., 1775, p. 54. *Caranx speciosus* Ogilby, Mem. Qld. Mus., iii, 1915, Pl. xxi (plate 23).

D. I, VIII; I, 18-21. A. II, I, 15-17. G.R. 20. Depth 2½. L. lat. with 10-16 very feeble scutes. No teeth. Scales small. Breast scaly. Bright golden yellow on body and fins. 8-10 transverse black bands across body, usually wide ones alternating with narrower ones. 40 inches. Pearl banks.
Kabara parava (S).

345. SELAR MATE (Cuvier) ONE-FINLET SCAD

Caranx mate Cuvier and Valenciennes, Hist. Nat. Poiss., ix, 1833, p. 54. *Caranx* or *Selar affinis, Caranx xanthurus. Caranx mate* Fowler, Journ. Acad. Nat. Sc. Phila. (2), xii, 1904, Pl. xiii (plate 23).

D. I, VIII; I, 23-24. A. II, I, 19-21. P. 21-23. L. lat. with 40-45 scutes which increase in size posteriorly. Body oblong. Depth 3½. Breast totally scaled. Grey above, silvery white below. Sometimes indications of transverse bars. Large dark blotch on shoulder girdle. Recognized by detached finlet terminally on soft dorsal and anal fins. 10½ inches. Coastal waters and trawling grounds.

346. SELAR MALAM Bleeker BANDED SCAD

Selar malam Bleeker, Nat. Tijds. Ned. Ind., i, 1851, p. 362. *Caranx nigripinnis. Caranx malam* McCulloch, Endeavour Repts., iii (3), 1915, Pl. xxii (plate 23).

D. I, VIII; I, 23-24. A. II, I, 20-21. P. 20-22. L. lat. with 50-56 scutes, posterior ones largest, being 9-10 in body depth. Body oblong. Depth 3. Breast totally scaled. A few scales on cheeks and opercle. Dark green or brownish above, lighter below. 6-9 fairly conspicuous cross-bars above lateral line. A distinct black shoulder spot. Dorsal and lobes of caudal black. 8 inches. Coastal waters and trawling grounds.

347. SELAR KALLA (Cuvier) GOLDEN SCAD

Caranx kalla Cuvier and Valenciennes, Hist. Nat. Poiss., ix, 1833, p. 49. *Selar* or *Caranx calla. Alepes kalla* Ogilby, Mem. Qld. Mus., iii, 1915, Pl. xx (plate 23).

D. I, VIII; I, 23-24. A. II, I, 18-20. L. lat. with 40-46 scutes, highest being 6-7 in body depth. G.R. 27. Curved part of lateral line half length of straight part. Depth 2½-3¼. Greenish silver with yellow reflections. A black spot on opercle. Caudal yellow. Traces of cross-bars above lateral line. 10 inches. Coastal waters and trawling grounds.

348. CARANGOIDES PRAEUSTUS (Bennett) BROWN-BACKED TREVALLY

Caranx praeustus Bennett, Cat. Fish. Sumatra, Life of Raffles, 1830, p. 689. *Selar ire. Caranx ire* Day, Fish. India, 1875-78, Pl. xlix, fig. 6 (plate 23).

D. I, VIII; I, 23-24. A. II, I, 19-20. L. lat. with 25-30 scutes, largest in mid-region, 8-9 in body height. Straight part of lateral line shorter than curved portion. G.R. 23. Brownish above, silvery below. No opercular spot. Fins yellowish. Soft dorsal with black patch anteriorly. 8 inches. Coastal waters.

349. CARANGOIDES PLAGIOTAENIA Bleeker OBLIQUE-BANDED TREVALLY

Carangoides plagiotaenia Bleeker, Acta. Soc. Sc. Indo-Neerl., ii, 1857, p. 59.

D. I, VIII; I, 22-24. A. II, I, 19-20. L. lat. with 36-40 small scutes, broadest on caudal peduncle. Curved portion of lateral line one third longer than straight part. Depth 2·2. G.R. 25. Brownish above, silvery below. 6 broad brownish somewhat oblique cross-bands, which fade in larger specimens. No opercular spot. Fins hyaline. 4½ inches. Coastal waters.

350. CARANGOIDES OBLONGUS (Cuvier) OBLONG TREVALLY

Caranx oblongus Cuvier and Valenciennes, Hist. Nat. Poiss., ix, 1833, p. 128. *Caranx oblongatus. Citula gracilis* Ogilby, Mem. Qld. Mus., iii, 1915, Pl. xxiii (plate 23).

D. I, VIII; I, 20-22. A. II, I, 18-20. G.R. 18. L. lat. with 35-40 scutes, 9-10 in body depth. Curved part of lateral line as long as straight part. Depth 2½. Breast naked. Fine teeth in bands in both jaws. Pectoral much longer than head. Anterior rays of dorsal and anal fins produced to form falciform lobes. Second ray of dorsal often filamentous. Silvery blue above. Young often with cross-bars. No opercular spot. 15 inches. Trawling grounds and pearl banks.

351. CARANGOIDES CHRYSOPHRYS (Cuvier) DUSKY TREVALLY

Caranx chrysophrys Cuvier and Valenciennes, Hist. Nat. Poiss., ix, 1833, p. 77. *Caranx nigrescens. Citula chrysophrys* Ogilby, Mem. Qld. Mus., iii, 1915, Pl. xxiv (plate 22).

D. I, VIII; I, 18-20. A. II, I, 16-17. G.R. 16-17. L. lat. with 25 feeble scutes. Curved part of lateral line much longer than the straight part. Breast naked. Pectoral much longer than head. Soft dorsal and anal slightly falcate, but low. Fine teeth in bands in both jaws. Slaty green above, with diffuse blotch on opercle. 14 inches. Coastal waters and trawling grounds and pearl banks.

352. CARANGOIDES MALABARICUS (Bloch) MALABAR TREVALLY

Scomber malabaricus Bloch and Schneider, Syst. Ichth., 1801, p. 31. *Caranx talamparoides. Caranx malabaricus* Day, Fish. India, 1875-78, Pl. l, fig. 2 (plate 22).

D. I, VIII; I, 21-22. A. II, I, 18-19. G.R. 24. L. lat. with 25 feeble scutes. Curved part of lateral line much longer than the straight part. Breast naked. Pectoral longer than head. Fine teeth in bands in both jaws. Distinguished from preceding species by larger eye. Silvery blue-green. 15 inches. Coastal waters, pearl banks and trawling banks.
Goliya parava, Labu parava (S).

353. CARANGOIDES ARMATUS (Forskål) ARMED TREVALLY

Sciaena armata Forskål, Descript. Animal., 1775, p. 53. *Caranx armatus, Citula armata. Caranx plumbeus* Jordan & Seale, Bull. U.S. Bur. Fish., xxv, 1906, fig. 28 (plate 22).

D. I, VIII; I, 20-21. A. II, I, 16-17. G.R. 21. L. lat. with 20 feeble scutes. Curved part of lateral line about 1¼ times straight part. Depth 2. Breast naked. Dorsal and anal rays filamentous, about half body length in adults. Pectoral much longer than head. Fine teeth in narrow bands in each jaw. Blue-green above. Opercular spot obscure. Caudal yellow. Young have six cross-bars. Recognizable by filamentous rays of dorsal and anal. 24 inches. Coastal waters, pearl banks and trawling grounds.

354. CARANGOIDES DINEMA Bleeker — TWO-THREAD TREVALLY

Carangoides dinema Bleeker, Nat. Tijds. Ned. Ind., i, 1851, p. 365.

D. I, VIII; I, 18-21. A. II, I, 17-18. G.R. 18. L. lat. with 25 scutes. Curved part of lateral line 1⅕ times as long as straight part. Depth 2⅓. Breast naked. Pectoral longer than head in adults. Fine teeth in bands in both jaws. Greenish above. Opercle with obscure blotch. Young with cross-bars. 30 inches. Coastal waters.

355. CARANGOIDES GYMNOSTETHOIDES Bleeker — NAKED-BREAST TREVALLY

Carangoides gymnostethoides Bleeker, Nat. Tijds. Ned. Ind., i, 1851, p. 364. *Caranx gymnostethoides* Day, Fish. India, 1875-78, Pl. xlviii, fig. 6 (plate 23).

D. I, VIII; I, 29-31. A. II, I, 24-26. Body rather elongate. Depth 3-3½. L. lat. with 24 small scutes. Curved and straight parts of lateral line about equal. Breast naked. Fine teeth in bands in both jaws. Pectoral longer than head. Snout about twice eye diameter. Greenish above. No opercular spot. 30 inches. Coastal waters.
Vattiya (S). Komalien pareh (T).

356. CARANGOIDES FERDAU (Forskål) — FERDAU'S TREVALLY

Scomber ferdau Forskål, Descript. Animal., 1775, p. 55. *Caranx ferdau. Caranx laticaudis* McCulloch, Mem. Qld. Mus., viii, 1924, Pl. xii (plate 22).

D. I, VII; I, 26-30. A. II, I, 22-26. G.R. 18. L. lat. scutes variable usually 18-28, diminishing with age. Depth 3¼. Fine teeth in bands in both jaws. Breast naked. Greenish-blue above. Fins yellow. Young with 6-7 cross-bars. 30 inches. Trawling grounds.

357. CARANX SEXFASCIATUS Quoy and Gaimard — SIX-BANDED TREVALLY

Caranx sexfasciatus Quoy and Gaimard, Voy. Uranie et Physic., 1824, p. 358. *Caranx hippos* or *heberi*, *Tricropterus forsteri*. *Caranx paraspistes* Richardson, Voy. Erebus and Terror, 1848, Pl. lviii, fig. 6 (plate 23).

D. I, VII-VIII; I, 20-22. A. II, I, 16-18. G.R. 15. L. lat. with 33-38 scutes. Depth 2½. Breast naked, but a scaly patch before pelvics. One row of teeth in lower jaw. Upper jaw with band with outer row enlarged. Silvery blue-green above. Young with cross-bars. Sometimes a small opercular spot. 30 inches. Coastal waters.
Inguru parava (S).

358. CARANX MELAMPYGUS Cuvier BLACK-TIPPED TREVALLY

Caranx melampygus Cuvier and Valenciennes, Hist. Nat. Poiss., ix, 1833, p. 116; McCulloch, Mem. Qld. Mus., viii (1), 1924, Pl. xi, fig. 2 (plate 23).

D. I, VIII; I, 22-25. A. II, I, 18-20. G.R. 18. Lateral line with 34-38 scutes, the curved part being ⅕ shorter than straight part. Breast scaly. Pectoral longer than head. Blue-green above, silvery below. Pectorals yellow. Other fins pale, often with black tips. Young with cross-bars. 12 inches. Pearl banks.

359. CARANX STELLATUS Eydoux and Souleyet STARRY TREVALLY

Caranx stellatus Eydoux and Souleyet, Voy. "La Bonite", Zool., ii, 1841, p. 167; Weber and De Beaufort, Fish. Indo-Aust. Archipel., vi, 1931, fig. 48 (plate 23).

D. I, VII; I, 24-25. A. II, I, 18-20. G.R. 17-18. L. lat. with 30-35 scutes. Breast scaly. Lower jaw with one row of teeth; upper jaw in a band with outer row enlarged. Silvery with irregular dark spots which enlarge with age and become more numerous. Ventrals dusky. Lobes of other fins dusky with age. 27 inches. Coastal waters.

360. CARANX IGNOBILIS (Forskål) YELLOW-FIN TREVALLY

Scomber ignobilis Forskål, Descript. Animal., 1775, p. 55. *Caranx ignobilis* Jordan and Evermann, Bull. U.S. Fish Comm., xxiii, 1905, fig. 72 (plate 22).

D. I, VIII; I, 19-21. A. II, I, 16-17. G.R. 15. L. lat. scutes about 27-30. Depth 2¼ or more, slender. Breast naked, a few scales before pelvics. A single row of teeth in both jaws. Olive green above. No opercular spot. Fins yellow. 30 inches. Coastal waters.

361. CARANX SANSUN (Forskål) GREEN-BACK TREVALLY

Scomber sansun Forskål, Descript. Animal., 1775, p. 55. *Tricropterus jarra. Caranx sansun* Day, Fish. India, 1875-78, Pl. l, fig. 5 (plate 22).

D. I, VIII; I, 22-23. A. II, I, 18. L. lat. with 36 scutes, broadest 8-9 in body depth. Curved part of lateral line shorter than straight part. Depth 2·2-2·8. Breast scaly laterally but naked ventrally. Teeth in lower jaw in single row; upper jaw in band with outer row enlarged. Greenish above, lighter below. Margin of soft dorsal and upper lobe of caudal dusky. 12 inches. Coastal waters and pearl banks.

Atanagul parava (S). Pareh (T).

362. CARANX CARANGUS (Bloch) BLACK-TAILED TREVALLY

Scomber carangus Bloch, Nat. Ausl. Fische, vii, 1793, p. 69. *Caranx carangus* Day, Fish. India, 1875-78, Pl. l, fig. 4 (plate 23).

D. I, VIII; I, 20-22. A. II, I, 16-18. L. lat. with 33-37 scutes, broadest 7-9 in body depth, curved part much shorter than straight part. Depth 2½. Ventral part of breast naked, a few scales laterally. Teeth in lower jaw in single row; in upper jaw forming a band with outer row enlarged. Golden above, silvery below. Young with 4-5 dark cross-bars. Small opercular spot. Entire margin of soft dorsal and upper lobe of caudal blackish. 16 inches. Coastal waters and pearl banks.

363. CHORINEMUS LYSAN (Forskål) TALANG, LEATHERSKIN

Scomber lysan Forskål, Descript. Animal., 1775, p. 12. *Chorinemus lysan* Blegvad and Løppenthin, Fishes Iranian Gulf, 1944, fig. 56 (plate 24).

D. I, VI-VII; I, 19-20. A. II, I, 18. G.R. 11-12. Depth 3. Maxilla extends well behind eye. Scales elongate. Mainly silver, darker above. 6-8 dark spots above lateral line. Fins yellowish. 48 inches. Coastal waters and pearl banks.

Nil kattava, Gona kattava, Katu bollu kattava (S). Katta (T).

364. CHORINEMUS TOL Cuvier SLENDER QUEENFISH

Chorinemus tol Cuvier and Valenciennes, Hist. Nat. Poiss., viii, 1831, p. 385. *Chorinemus moadetta* Day, Fish. India, 1875-78, Pl. li,B, fig. 1 (plate 24).

D. I, VII; I, 19-21. A. II, I, 17-20. G.R. 19-20. Depth 4½. A few sharp teeth. Scales needle-like. Mainly silvery. Eye, pectoral and caudal yellowish. 24 inches. Pearl banks.

365. CHORINEMUS TALA Cuvier DEEP QUEENFISH

Chorinemus tala Cuvier and Valenciennes, Hist. Nat. Poiss., viii, 1851, p. 377. *Scomberoides toloo. Chorinemus toloo* Day, Fish. India, 1875-78, Pl. li,A, fig. 3 (plate 24).

D. I, VII; I, 20. A. II, I, 18. Depth 3 in length. Maxilla extending beyond hind border of eye. Scales small, lanceolate. Mainly silvery, darker above. 4-8 vertically oblong dark blotches touching lateral line above or intersected by it. Vertical fins dark. 18 inches. Coastal waters and pearl banks.

Han kattava, Pothu kattava (S). Thol pareh, Katta (T).

366. CHORINEMUS TOLOOPARAH (Rüppell) BLACK-TIPPED LEATHERSKIN

Lichia toloparah Rüppell, Atlas Reise Nordl. Afrika, 1828, p. 91. *Scomberoides toloo-parah. Chorinemus tolooparah* Weber and De Beaufort, Fish. Indo-Aust. Archipel., vi, 1931, fig. 54 (plate 24).

D. I, VI; 19-20. A. II, I, 18-19. Depth 4¼-4½. Oblong-elongate. Maxilla extends at least to posterior third of eye. Scales conspicuous, rhombic and pointed. Bluish above, silvery below. A double series of dusky round blotches above and below lateral line, which may be transversely connected. Tips of dorsal and anal black. 12 inches. Coastal waters.

367. CHORINEMUS SANCTI-PETRI Cuvier SAINT PETER'S LEATHERSKIN

Chorinemus sancti petri Cuvier and Valenciennes, Hist. Nat. Poiss., viii, 1831, p. 379, Pl. ccxxxvi (plate 24).

D. I, VI-VII; I, 20-21. A. II, I, 18-19. G.R. 16-17. Maxilla extends below hind margin of eye. Scales elongate but not needle-like. Depth 4. Mainly silver. 6-8 spots above lateral line and 3-5 fainter ones below. Fins yellowish. Black blotch on anterior part of soft dorsal, 30 inches. Coastal waters.

368. TRACHINOTUS BLOCHI (Lacépède) BLOCH'S DART

Caesiomorus blochii Lacépède, Hist. Nat. Poiss., iii, 1802, p. 95. *Trachinotus ovatus* Ogilby, Mem. Qld. Mus., v, 1916, Pl. xix (plate 22).

D. I, V-VI; I, 18-23. A. II, I, 16-20. G.R. 9-12. Depth 2. Deep ovate body. Golden tinge on silvery body, darker above. Fins golden yellow with dusky tips. Pectorals darker. 30 inches. Coastal waters.
Kota dharuva, Kukulamaha (S). Kutili (T).

369. TRACHINOTUS BAILLONI (Lacépède) BAILLON'S DART

Caesiomorus bailloni Lacépède, Hist. Nat. Poiss., iii, 1802, p. 93. *Trachinotus bailloni* Ogilby, Mem. Qld. Mus., v, 1916, Pl. xviii (plate 22).

D. I, V-VI; I, 22-24. A. II, I, 22-24. G.R. 13. Depth 2½. Silvery, greenish above. 2-5 rather small black spots in a series along the lateral line. Tips of unpaired fins black. 24 inches. Coastal waters.

370. TRACHINOTUS RUSSELLI Cuvier RUSSELL'S DART

Trachinotus russellii Cuvier and Valenciennes, Hist. Nat. Poiss., viii, 1831, p. 436. *Trachinotus oblongus*. *Trachinotus botla* Ogilby, Mem. Qld. Mus., iii, 1915, Pl. xxviii (plate 22).

D. I, VI; I, 22-23. A. II, I, 20-21. G.R. 12. Depth 2⅔. Silvery, greenish above. 3-6 more or less indistinct dull round patches, mostly above lateral line. Tips of unpaired fins dusky. 24 inches. Coastal waters.
Kade mariya, Koliyava (S). Mukali (T).

371. NAUCRATES DUCTOR (Linnaeus) PILOT FISH

Gasterosteus ductor Linnaeus, Syst. Nat., ed. 10, 1758, p. 295. *Naucrates indicus* Valenciennes, Illust. Poiss. Regne Animal, 1837, Pl. liv, fig. 1 (plate 24).

D. III-V; I, 26-28. A. II, I, 16-18. G.R. 16. Depth 4¼. Body elongate. Young very spiny in opercular region. Greyish above, lighter below. 6-7 dark cross-bars, less conspicuous with age. Caudal blackish with white tips. Inside pectoral and anal dusky. 24 inches. Coastal waters. Young shelter under jellyfish, adults often accompany sharks.

372. ZONICHTHYS NIGROFASCIATA (Rüppell)
BLACK-BANDED KINGFISH

Nomeus nigrofasciatus Rüppell, Atlas Reise Nordl. Afrika, 1831, p. 28. *Seriola nigrofasciata*. *Zonichthys nigrofasciata* Smith, Fish. Sth. Africa, 1949, Pl. xxiv, fig. 536 (plate 22).

D. I, V-VI; I, 31-35. A. II, I, 15-17. G.R. 8. Ventrals enlarged in young. Bluish above, pale below. Young with 5-7 dark cross-bars. Spinous dorsal fin black. Ventral and anal fins dusky as is also part of soft dorsal, Caudal yellow. 24 inches. Coastal waters down to 60 fathoms.

BLACK KINGFISH

Family RACHYCENTRIDAE

Elongate cylindrical fish with broad depressed head. Mouth large with fine teeth in bands in jaws, palate and tongue. Dorsal spines low, separate. Anal spines 2-3.

373. RACHYCENTRON CANADUS (Linnaeus) BLACK KINGFISH

Gasterosteus canadus Linnaeus, Syst. Nat., ed. 12, 1766, p. 491. *Rachycentron lineatus* or *nigra. Elecate nigra* Day, Fish. India, 1875-78, Pl. lv, fig. 2 (plate 24).

D. VII-IX; I-III, 26-33. A. II-III, 22-28. Depth variable, about 6½. Caudal rounded in young, becoming truncate with age, and lunate in adults with upper lobe longest. Dark to brownish above, silvery white below. Young specimens with one or two silvery lateral bands along lower part of brown upper colouration. Fins brownish. 5 feet. A pelagic species from coastal waters and trawling grounds.

Mudhu luhula, Mudhilla (S). Kedal viral (T).

MOONFISH

Family MENIDAE

Contains a single curious species with a highly compressed disc-like body and sharp breast. Scales present but not visible to naked eye. Mouth protrudes upwards and contains minute teeth.

374. MENE MACULATA (Bloch) MOONFISH

Zeus maculatus Bloch and Schneider, Syst. Ichth., 1801, p. 95. *Mene maculata* Day, Fish. India, 1875-78, Pl. liii, fig. 5 (plate 24).

D. III-IV; 40-43. A. 30-33; V. I, 5 with small spine and long rays. Depth 1⅓. L. lat. parallel with dorsal profile. Maxilla nearly vertical. Bluish above, sides and belly silver. 2-4 irregular series of lead coloured round spots above and below lateral line. Dorsal and caudal dusky as are also ventral filaments. Other fins hyaline. 8 inches. Coastal waters, entering estuaries and deeper water reefs of trawling grounds.

Ambatta kattiya, Panna, Handhe panna, Panara (S).

Amuttan pareh, Kannadi karel, Ambattankathi (T).

DOLPHIN FISH

Family CORYPHAENIDAE

Elongate compressed. Small scales present. Mouth large, oblique, with fine teeth in bands in jaws and on palate. Dorsal fin long, originating on head. Forehead prominent in large males. Caudal deeply forked.

375. CORYPHAENA HIPPURUS Linnaeus DOLPHIN

Coryphaena hippurus Linnaeus, Syst. Nat., ed. 10, 1758, p. 261. *Coryphaena hippurus hippurus* and *hippurus equisetis. Coryphaena hippurus* Walford, Marine Game Fish. Pacif. Coast, 1937, fig. 43 (plate 24).

D. 55-67. A. 25-30. Body proportions change with age from elongate in juveniles to fairly deep and compressed in older fish and forehead becomes more prominent. Colour variable. Young with a series of vertical cross-bars extending on to dorsal fin. Older fish

greenish or bluish above, golden below with small dark spots on anterior part. Colours fade on death. Dorsal usually greenish, other fins usually yellowish to orange. 5 feet. Pelagic, taken mostly by trolling. Coastal and deeper waters.
Galvannava, Dhiya vannava, Vannava, Radheva (S).
Avlis, Kal vanna, Ailai, Badahlen, Vanna (T).

REDBAIT

Family EMMELICHTHYIDAE

Elongate, slender, little compressed fish. Small scales covering body. Dorsal fin continuous or in two parts with spines between. Oblique extensible mouth. Maxilla expanded. Teeth feeble or absent. Inhabit fairly deep water.

376. DIPTERYGONOTUS LEUCOGRAMMICUS Bleeker REDBAIT

Dipterygonotus leucogrammicus Bleeker, Jour. Indian Archipel., iii, 1849, p. 71. *Erythrocles leucogrammicus, Smaris balteatus. Erythrichthys leucogrammicus* Johnstone, Rept. Ceylon Pearl Oyster, Suppl. xv, 1904, Pl. i, fig. 6 (plate 21).

D. XIV; I, 9. A. III, 10. P. 15-17. L. lat. 80. Tr. 10+16. Depth 4½-5. Eye with narrow adipose lid. Maxilla rounded, naked. Greyish above, silvery below. Back with 3 pale longitudinal bands. 4 inches. Marine and brackish water to about 7 fathoms, preferring sandy bays. Pearl banks and trawling grounds.

SNAPPERS, SEA PERCH, HUSSARS, BASS

Family LUTIANIDAE

More or less typical perch-like fishes with body moderately compressed to robust with moderate or small ctenoid scales. Dorsal with 10-12 spines. Mouth fairly large with canine teeth in jaws and fine teeth on palate. Preopercle sometimes serrate. Caudal truncate to forked. Active predacious fishes, often brightly coloured and living in environments varying from tidal estuaries to coral reefs.

1. Vomer and palatines with teeth 2
   Palate toothless 22
2. Interorbital flat 3
   Interorbital more or less convex 4
3. Pectoral about half head length, not falcate .... *Aprion virescens*
   Pectoral about as long as head, falcate .... *Pristipomoides typus*
4. Soft dorsal and anal scaleless; last ray of each extended, filamentous .... *Apsilus fuscus*
   Soft dorsal and anal at least partly scaly; last rays not extended 5
5. Scales above lateral line in series parallel to its course; those below lateral line in horizontal series .... *Lutianus johni*
   Scales above lateral line in oblique rows, at least for some part posteriorly; those below lateral line in horizontal series 6
   Scales above and below lateral line in oblique rows crossing its course 19

6. Predorsal scales not extending forward in interorbital space 7
Predorsal scales extend forward over interorbital or to front of eye 15
7. Anterior scales above lateral line parallel with its course, only posterior rows oblique .... *Lutianus argentimaculatus*
Scales above lateral line in oblique rows throughout 8
8. Five dusky longitudinal bands, crossed by dark transverse bands .... *Lutianus decussatus*
Body without dark transverse bands 9
9. Body with two dark longitudinal bands .... *Lutianus lemniscatus*
Body without dark transverse bands 10
10. Light coloured blotch on back below spinous dorsal and another below soft dorsal .... *Lutianus bohar*
No light coloured blotch on back below each dorsal fin 11
11. Head with numerous longitudinal blue lines; pearly blotch with black edge on lateral line .... *Lutianus rivulatus*
No blue lines on head nor pearly blotch on lateral line 12
12. Black lateral blotch on body 13
No black lateral blotch on body 14
13. Greater part of blotch below lateral line; 14 scale rows below lateral line .... *Lutianus fulviflamma*
Greater part of blotch above lateral line; 17 scale rows below lateral line .... *Lutianus russelli*
14. Dark crescent-shaped cross-bar on caudal .... *Lutianus lunulatus*
Body with golden longitudinal stripes .... *Lutianus waigiensis*
15. Preopercle with distinct notch; interopercle armed with strong spur .... *Lutianus kasmira*
Preopercle at most with a slight notch; interopercle spine feeble or absent 16
16. Deep brown or blackish median lateral bands 17
No black lateral bands 18
17. Single black lateral band; no pearly blotches on back .... *Lutianus vitta*
Usually two black lateral bands; pearly blotches below each dorsal fin .... *Lutianus biguttatus*
18. Preorbital and infraorbital very narrow .... *Lutianus lineolatus*
Preorbital and infraorbital much wider than maxillary .... *Lutianus lutianus*
19. Soft dorsal longer than high 20
Soft dorsal higher than long, angular behind 21
20. Preopercle entire; broad dark band across head .... *Lutianus sanguineus*
Preopercle deeply gashed; no dark band on head .... *Lutianus gibbus*
21. Body with longitudinal golden streaks and caudal peduncle with dark blotch .... *Lutianus malabaricus*
Three broad black obliquely transverse bands across body .... *Lutianus sebae*
22. Gill-rakers 10+(16-18); L.lat. 68-70 .... *Aphareus furcatus*
Gill-rakers (16-19)+(30-34); L.lat. 70-73 .... *Aphareus rutilans*

377. APRION VIRESCENS Valenciennes — GREEN JOBFISH

Aprion virescens Cuvier and Valenciennes, Hist. Nat. Poiss., vi, 1830, p. 544. *Chaetopterus microchir* Bleeker, Atlas Ichth. Ind. Neerl., vii, 1873-76, Pl. ccxciii, fig. 3 (plate 24).

D. X; 11. A. III, 8. L. lat. 45-49. Tr. 7+14. G.R. 13-14. Depth about 4. Pectorals short and rounded, equal to snout. Bluish or greenish, darker above. Vertical fins brownish with series of violet blotches basally along dorsal fin membrane. 40 inches. A pelagic species of shallow coastal waters and trawling grounds.

Dhiulava (S). Dhiula (T).

378. PRISTIPOMOIDES TYPUS Bleeker — SHARP-TOOTHED BASS

Pristipomoides typus Bleeker, Nat. Tijds. Ned. Ind., iii, 1852, p. 575. *Aprion pristipoma* Bleeker, Atlas Ichth. Ind. Neerl., viii, 1876-77, Pl. cccxxxvi, fig. 3 (plate 24).

D. X; 11. A. III, 8. L. lat. 51. Tr. 6+13. G.R. (7-8)+(15-16). Depth about 4½. Pectorals long and pointed, approximately equal to head. Narrow villiform bands of small curved teeth in jaws with outer series of distant canines. Caudal deeply forked, upper lobe produced. Uniform rosy, paler below. About 6 longitudinal yellow bands along body. Fins reddish. Attains large size. Coastal waters. One of commonest species taken on trawling grounds.

Kalamiya, Lagu muva, Lomassa, Paremuva Ramuva, Lomaha (S). Lomia (T).

379. APSILUS FUSCUS Valenciennes BROWN SEA-PERCH

Apsilus fuscus Cuvier and Valenciennes, Hist. Nat. Poiss., vi, 1830, p. 549, Pl. clxviii (plate 24).

D. X; 10. A. III, 8. L. lat. 65. Tr. 7+16. Depth 3. Teeth on vomer and palatines, absent from tongue. Pectoral equal to head. Caudal lobes attenuate, longer than head. Interorbital convex, not scaly. Soft dorsal and anal not scaly. Brownish with darker mottlings. 30 inches. Coastal waters.

380. LUTIANUS JOHNI (Bloch) MOSES PERCH

Anthias johnii Bloch, Naturg. Ausländ. Fische, vi, 1792, p. 113. *Mesoprion johnii. Lutjanus johni* Bleeker, Atlas Ichth. Ind. Neerl., viii, 1876-77, Pl. cccxxxviii, fig. 3 (plate 25).

D. X; 13-14. A. III, 8. L. lat. 48-50. Tr. 6+(15-16). G.R. 9+12. Depth 3·1-3·5. Teeth on vomer and palatines, and outer row of canines in jaws. All rows of scales above lateral line are parallel to it. Back olive brown. Belly silvery. Some brown or gold lines following the horizontal scale rows. A large oval black spot on lateral line below end of spinous dorsal. Unpaired fins yellowish. Cheeks, pectoral and pelvics mauve or pinkish. 15 inches. Coastal waters.

381. LUTIANUS ARGENTIMACULATUS (Forskål) RED SNAPPER

Sciaena argentimaculatus Forskål, Descript. Animal., 1775, p. 47. *Mesoprion* or *Lutjanus rangus, Lutjanus roseus, sillaoo, argentimaceatus* or *jahngarah. Lutianus argentimaculatus* Day, Fish. India, 1875-78, Pl. xi, fig. 5 (plate 25).

D. X; 13-15. A. III, 8-9. L. lat. 40-47. Tr. 7+14. G.R. 9-12, some greatly reduced. Depth 2¾. Feeble notch in preopercle. Adults blood red or sometimes silvery pink darkening to deep red after death. Fins similar. Young with whitish cross-bars. 36 inches. Coastal waters, entering estuaries. Also from trawling grounds.

Thambalaya, Guru thambuwa, Dhala (S). Adallu (T).

382. LUTIANUS DECUSSATUS (Cuvier) CROSS-BANDED SNAPPER

Mesoprion decussatus Cuvier and Valenciennes, Hist. Nat. Poiss., ii, 1828, p. 487. *Lutianus decussatus* Day, Fish. India, 1875-78, Pl. xiv, fig. 4 (plate 25).

D. X; 13. A. III, 8. L. lat. 49-52. Tr. 7+17. G.R. 6+10. Depth 3½. Preopercular notch shallow. Whitish with 6 longitudinal black bands along body crossed by 6 indistinct short vertical bars on upper part of body. Large blackish blotch at root of caudal. Fins yellowish, dorsal and caudal rather dusky. 12 inches. Pearl banks.

383. LUTIANUS LEMNISCATUS (Valenciennes) RIBBONED SNAPPER

Serranus lemniscatus Cuvier and Valenciennes, Hist. Nat. Poiss., ii, 1828, p. 240. *Lutjanus melanotaenia* Bleeker, Atlas Ichth. Ind. Neerl., vii, 1873-76, Pl. cclxxxv, fig. 2 (plate 26).

D. X; 13. A. III, 8. L. lat. 52-53. Tr. 7+19. Depth 3·3. Hind and lower border of preoperculum denticulate; notch very shallow. Dark brownish red above, paler or silvery below. A broad black band through eye to caudal. Band separated from dark back by a white band. Sometimes a second black band below. 5 inches. Possibly the young of *L. janthinuropterus*. Coastal waters.

384. LUTIANUS BOHAR (Forskål) TWO-SPOT SNAPPER

Sciaena bohar Forskål, Descript. Animal., 1775, p. 46. *Lutjanus bohar* Bleeker, Atlas Ichth. Ind. Neerl., viii, 1876-77, Pl. cccxlviii, fig. 4 (plate 25).

D. X; 14-15. A. III, 8-9. L. lat. 48-52. Tr. 8+17. G.R. 14. Depth 2½. Feeble notch in preopercle. Bronzy above, lighter below. Two large pearly blotches on back, one below 8th dorsal spine, other below last soft dorsal rays. Base of pectoral and pelvics rose, otherwise fins dusky. 30 inches. Coastal waters and Gulf of Manaar.

385. LUTIANUS RIVULATUS (Cuvier) BLUE-LINED SNAPPER

Diacope rivulata Cuvier and Valenciennes, Hist. Nat. Poiss., ii, 1828, p. 414. *Genyoroge caeruleopunctatus*. *Lutianus rivulatus* Fowler, Bull. U.S. Nat. Mus., 100, xi, fig. 11 (plate 25).

D. X; 15-16. A. III, 8-9. L. lat. 40-45. Tr. 8+17. G.R. 13. Fairly deep notch in preopercle. Depth 2¼. Olive brown on back, paler to silvery below. Head with numerous wavy white or bluish bands across cheeks. Pearly blotch with black edge on lateral line below anterior part of soft dorsal. Body scales each with white or bluish dot. Fins largely yellowish with shading of dusky brown. Sometimes a dark brown cross-bar at level of pectoral base. 24 inches. Coastal waters, trawling grounds and pearl banks.
Badava, Rumasse, Kuruviliya (S). Kuruvilla, Baddau (T).

386. LUTIANUS FULVIFLAMMA (Forskål) ONE-SPOT GOLDEN SNAPPER

Sciaena fulviflamma Forskål, Descript. Animal., 1775, p. 45. *Mesoprion auro-lineatus*. *Lutjanus fulviflamma* Bleeker, Atlas Ichth. Ind. Neerl., viii, 1876-77, Pl. cccxliv, fig. 3 (plate 25).

D. X; 13-14. A. III, 8-9. L. lat. 46-50. Tr. 7+14. G.R. 7-9. Preopercle feebly notched. Depth 2¾. Yellow or rosy along back with golden lines along each scale row from head to caudal. Large black spot on lateral line, mostly below it and opposite commencement of soft dorsal fin. Fins yellowish to olive, dorsals and caudals often edged with pink. 12 inches. Coastal waters entering estuaries and fresh water. Also pearl banks and trawling grounds.

387. LUTIANUS RUSSELLI (Bleeker) RUSSELL'S ONE-SPOT SNAPPER

Mesoprion russelli Bleeker, Verh. Bat. Gen., xxii, 1849, p. 41. *Lutjanus russelli* Bleeker, Atlas Ichth. Ind. Neerl., vii, 1873-76, Pl. ccc, fig. 2 (plate 25).

D. X; 14-15. A. III, 8. L. lat. 50. Tr. 7+17. G.R. 8. Depth 2⅗. Feeble notch in preopercle. Golden olive, darker above, silvery below. A dark oval blotch on lateral line, mostly above it and opposite anterior part of soft dorsal. Fins yellowish, borders of dorsal and caudal sometimes dusky. Young specimens with 3 dark longitudinal bands, which are replaced by about 6 golden longitudinal bands in adults. 15 inches. Trawling grounds.

388. LUTIANUS LUNULATUS (Mungo-Park) CRESCENT SNAPPER

Perca lunulata Mungo-Park, Trans. Linn. Soc. Lond., iii, 1797, p. 37. *Lutianus lunulatus* Day, Fish. India, 1875-78, Pl. xii, fig. 4 (plate 25).

D. X; 13-14. A. III, 8-9. L. lat. 47-51. Tr. 7+(18-21). G.R. 7. Depth 3·3. Preopercular notch very shallow. Reddish brown above, silvery below with indications of longitudinal stripes. Fins light reddish. Caudal with broad crescent-shaped dark vertical band. 10 inches. Coastal waters.

389. LUTIANUS WAIGIENSIS (Quoy and Gaimard) WAIGEU SNAPPER

Diacope vaigiensis Quoy and Gaimard, Voy. Uranie Physic., 1824, p. 307. *Genyoroge marginata, Diacope xanthopus. Lutjanus marginatus* Bleeker, Atlas Ichth. Ind. Neerl., viii, 1876-77, Pl. cccxlix, fig. 2 (plate 25).

D. X; 13-14. A. III, 7-9. L. lat. 46-48. Tr. 8+14. G.R. 10-11. Depth 2½. Preopercle deeply notched. Colouration variable. Back olive to brownish, sides yellowish and belly silvery, each scale with lighter spot. Side of head with longitudinal grey or bluish lines, sometimes broken into spots. Sometimes about 7 silvery to golden orange longitudinal lines. Dorsals with dark edges; caudal dusky; ventrals and anal yellowish to brown. 24 inches. Coastal waters and trawling grounds.
Padalla, Pen dhalla (S).

390. LUTIANUS KASMIRA (Forskål) BLUE AND YELLOW SNAPPER

Sciaena kasmira Forskål, Descript. Animal., 1775, p. 46. *Lutianus bengalensis* or *spilurus, Genyoroge bengalensis* or *spilura*; *Diacope decemlineata* or *spilura. Lutjanus quinquelinearis* Bleeker, Atlas Ichth. Ind. Neerl., viii, 1876-77, Pl. cccxliii, fig. 4 (plate 25).

D. X; 13-16. A. III, 8. L. lat. 50. Tr. 10+17. G.R. 15. Depth 2¾. Deep notch in preopercle. Generally canary yellow, darker above. 4 or 5 narrow blue lines, often bordered with brown, cross cheek and body longitudinally, some bifurcating on body. Fins yellow, dorsal with a blue median band, others brownish basally. 15 inches. Coastal waters and trawling grounds.
Irri ranna, Vali ranna (S).

391. LUTIANUS VITTA (Quoy and Gaimard) OLIVE STRIPED SNAPPER

Serranus vitta Quoy and Gaimard, Voy. Uranie, Physic., 1824, p. 315. *Lutianus vitta* Day, Fish. India, 1875-78, Pl. xiv, fig. 2 (plate 26).

D. X; 13. A. III, 7-8. L. lat. 50-51. Tr. (6-7)+18. Depth 3½. Preopercular notch shallow. G.R. 6+12. Yellowish red on back, rosy below lateral line. Stripes along scale rows olive above lateral line and yellow below. Sometimes broad black band from eye to caudal. Fins orange; dorsal, anal and caudal tips with white margins. 15 inches. Coastal waters and trawling grounds.

392. LUTIANUS BIGUTTATUS (Valenciennes)

TWO-SPOT BANDED SNAPPER

Serranus biguttatus Cuvier and Valenciennes, Hist. Nat. Poiss., vi, 1830, p. 507. *Lutjanus biguttatus* Bleeker, Atlas Ichth. Ind. Neerl., viii, 1876-77, Pl. cccxxxviii, fig. 5 (plate 26).

D. XI; 12. A. III, 8. L. lat. 51-53. Tr. 5+14. Depth 3½. G.R. 8+18. Preopercle notch very slightly developed. Yellowish grey above, lighter below. Broad black bands from eye to middle of caudal and from below jaws to end of anal base. Pearly white spots under middle of spinous dorsal and under front of soft dorsal. Fins yellowish. 11 inches. Coastal waters.

393. LUTIANUS LINEOLATUS (Rüppell) GOLDEN STRIPED SNAPPER

Diacope lineolata Rüppell, Atlas Reise Nordl. Afrika, Fische, 1828, p. 76. *Lutianus erythropterus. Lutianus lineolatus* Day, Fish. India, 1875-78, Pl. xi, fig. 1 (plate 26).

D. XI; 12-13. A. III, 7-9. L. lat. 45-50. Tr. 7+13. G.R. 16-18. Depth 3. No obvious notch in preopercle. Purplish red along back, golden yellow below. Upper third of body with oblique and often sinuous golden lines along the scale rows. Below lateral line, the longitudinal lines are pink and horizontal. Fins yellow with light band along centre of dorsal. Lines darken with age. 15 inches. Mostly from trawling grounds.
Hunu ranna (S).

394. LUTIANUS LUTIANUS Bloch ROSY SNAPPER

Lutjanus lutjanus Bloch, Nat. Ausl. Fische, iv, 1790, p. 107. *Lutianus madras* Day, Fish. India, 1875-78, Pl. xiv, fig. 3 (plate 26).

D. X; 13-14. A. III, 8. L. lat. 48-52. Tr. 6+(14-15). G.R. 6+15. Depth 3½. Preopercular notch slightly developed. Generally pinkish with olive on back and yellow to silvery below. Darker lines following scale rows, oblique above lateral line and horizontal below; the lower may be rosy. Fins pale yellowish. 12 inches. Coastal waters and trawling grounds.

395. LUTIANUS SANGUINEUS (Cuvier) BLOOD-RED SNAPPER

Diacope sanguinea Cuvier and Valenciennes, Hist. Nat. Poiss., ii, 1828, p. 437. *Mesoprion angularis, Loxolutjanus erythropterus. Lutianus erythropterus* Day, Fish. India, 1875-78, Pl. x, fig. 1 (plate 25).

D. XI; 13-15. A. III, 9-10. L. lat. 55-65. Tr. 12+20. G.R. 12-13. Depth 2¼ in young to 2⅔ in adults. Feeble notch in preopercle. Blood-red with orange reflections and broad blackish band from eye to dorsal origin. Older fish more or less uniform red on body and fins. Young with red head and shoulder band and with body yellowish marked with longitudinal reddish streaks. A black saddle on caudal peduncle. Ventrals and vertical fins dark reddish brown. Pectorals orange. 36 inches, larger fish developing a bump on the occipital region. Trawling grounds and pearl banks.

396. LUTIANUS GIBBUS (Forskål) HUNCHED SNAPPER

Sciaena gibba Forskål, Descript. Animal., 1775, p. 46. *Genyoroge gibba. Mesoprion gibbus* Günther, Jour. Mus. Godeffroy, ii (3), 1873, Pl. xii (plate 26).

D. X; 13-15. A. III, 8-9. L. lat. 47-52. Tr. 8+17. G.R. 18. Depth 2½. Deep notch in preopercle. All scale rows oblique. Orange-red uniformly on body and fins. Sometimes with darker stripes on body following scale rows. Fins darker to brown in young specimens. Pectorals paler, concealing a greyish spot. Young more gibbous than adults. 24 inches. Reported as having poisonous flesh. Coastal reefs, trawling grounds and pearl banks.

397. LUTIANUS MALABARICUS (Schneider) MALABAR SNAPPER

Sparus malabaricus Bloch and Schneider, Syst. Ichth., 1801, p. 278. *Lutianus dodecacanthus*. *Lutjanus malabaricus* Bleeker, Atlas Ichth. Ind. Neerl., vii, 1873-76, Pl. ccxciii, fig. 1 (plate 25).

D. XI; 14-15. A. III, 8-9. L. lat. 57-62. Tr. (7-8)+(18-22). G.R. 5+14. Depth 3¼. Preopercular notch shallow receiving a well-developed interopercular knob. Reddish, lighter below. Narrow yellow streaks along scale rows, oblique above lateral line and horizontal below. A dark blotch on caudal peduncle, preceded by a pale pearly spot. Fins light red; dorsal and caudal with narrow black margin. Ventral light. 19 inches. Pearl banks and trawling grounds.

Gola, Para galla, Rathu gal malu (S). Konde sevalai (T).

398. LUTIANUS SEBAE (Cuvier) EMPEROR; GOVERNMENT BREAM

Diacope sebae Cuvier and Valenciennes, Hist. Nat. Poiss., ii, 1828, p. 411. *Genyoroge sebae*. *Lutjanus sebae* Bleeker, Atlas Ichth. Ind. Neerl., viii, 1876-77, Pl. cccl, fig. 2 (plate 25).

D. XI-XII; 15-16. A. III, 9-11. L. lat. 45-50. Tr. 10+20. G.R. 12. Deep notch in preopercle. Depth 2½. Reddish with three oblique dusky cross-bars; one through eye to snout, one subvertically from ventrals to spinous dorsal, and third from soft dorsal to lower part of caudal, forming an arrow-head. Fins reddish with black on tips of soft dorsal, anal and upper caudal lobe. Ventrals blackish. All bands may be indistinct. 40 inches. Coastal waters and trawling grounds.

399. APHAREUS FURCATUS (Lacépède) OLIVE SMALL-TOOTHED JOBFISH

Labrus furcatus Lacépède, Hist. Nat. Poiss., iii, 1802, p. 429. *Aphareus flavivultus* Jenkins, Bull. U.S. Bur. Fish., xix, 1901, fig. 4 (plate 26).

D. X; 11. A. III, 8. L. lat. 68-70. Tr. 8+20. G.R. 10+(16-18). Depth 4½. Caudal deeply forked. Last rays of soft dorsal and anal extended. Uniformly olive on back, paler below. Fins brownish to dusky. Dorsal with faint light yellow spots. 14 inches. Coastal waters.

400. APHAREUS RUTILANS Cuvier RUSTY SMALL-TOOTHED JOBFISH

Aphareus rutilans Cuvier and Valenciennes, Hist. Nat. Poiss., vi, 1830, p. 490; Bleeker, Atlas Ichth. Ind. Neerl., vii, 1873-76, Pl. ccxcix, fig. 2 (plate 24).

D. X; 11. A. III, 8. L. lat. 70-73. Tr. (7-8)+(19-21). G.R. (16-19)+(30-34). Depth 4·3-4·9. Caudal deeply forked. Last rays of soft dorsal and anal extended. Light brown above, paler to silvery below. Head bronze with silvery reflections. Fins uniformly pale brown. 15 inches. Coastal waters.

FUSILIERS

Family CAESIODIDAE

Body elongate with moderate scales. Mouth small, with fine teeth, none on palate. Dorsals continuous, spines slender. A distinct bony process on premaxilla sliding beneath the maxilla, clearly visible inside mouth.

Caudal with black tips to each lobe .... *Caesio chrysozonus*
Caudal with black longitudinal band on each lobe .... *Caesio caerulaureus*

401. CAESIO CHRYSOZONUS Cuvier GOLDEN-BANDED FUSILIER

Caesio chrysozona Cuvier and Valenciennes, Hist. Nat. Poiss., vi, 1831, p. 440; Bleeker, Atlas Ichth. Ind. Neerl., vii, 1873-76, Pl. cccvii, fig. 2 (plate 24).

D. X; 15. A. III, 12. L. lat. 67-77. Tr. 8+(16-17). G.R. 10+25. Depth $4\frac{1}{4}$-$4\frac{3}{4}$. Caudal deeply forked with lobes pointed. Bluish above with golden band along lateral line and silvery below. A black tip to each caudal lobe. Black spot in axilla. Fins pale yellowish or reddish. 9 inches. Trawling grounds.
Hemala, Ilitta (S).

402. CAESIO CAERULAUREUS Lacépède BLUE AND GOLD FUSILIER

Caesio caerulaureus Lacépède, Hist. Nat. Poiss., iii, 1802, p. 85; Bleeker, Atlas Ichth. Ind. Neerl., viii, 1876-77, Pl. cccxlvii, fig. 4 (plate 24).

D. X; 14-15. A. III, 11-13. L. lat. 65-70. Tr. (7-8)+(15-16). G.R. 8+25. Depth 4-$4\frac{1}{2}$. Caudal deeply forked with pointed lobes. Bluish above, silvery below with wide golden-yellow band along lateral line. Fins yellowish. Each caudal lobe with broad dark longitudinal band. Dorsal with indistinct dark margin. Pectoral axilla black-blue as is also its outer basal part. 11 inches. Coastal waters.

THREADFIN-BREAM

Family NEMIPTERIDAE

Elongate body with fairly large scales which are easily shed. Mouth moderate, with rows of small conical teeth and a few canines in front and none on palate. Pelvics and upper caudal lobe sometimes with filamentous rays. Preopercular flange naked. Caudal forked. Small brightly coloured species.

1. Membranes of spinous dorsal notched marginally .... *Odontoglyphis tolu*
   Spinous dorsal edge entire, membranes not notched 2
2. Caudal deeply forked, upper lobe with filamentous prolongation .... *Nemipterus japonicus*
   Caudal moderately forked, lobes equal 3
3. Preorbital subequal with eye; a brilliant spot anteriorly below lateral line .... *Nemipterus furcosus*
   Preorbital $1\frac{3}{8}$ in eye; a bluish spot on opercle .... *Nemipterus bleekeri*

403. ODONTOGLYPHIS TOLU (Valenciennes)

NOTCHED THREADFIN-BREAM

Dentex tolu Cuvier and Valenciennes, Hist. Nat. Poiss., vi, 1830, p. 248. *Synagris tolu. Dentex tolu* Bleeker, Atlas Ichth. Ind. Neerl., vii, 1873-76, Pl. cccxv, fig. 1 (plate 26).

D. X; 9. A. III, 7. L. lat. 48-49. Tr. 4+11. G.R. (3-4)+6. Depth 4-4½. Head 4½. Dorsal interspinous membrane deeply notched. Caudal forked, upper lobe very slightly longer. Rosy with 4-5 yellowish longitudinal bands between scale rows from immediately above base of pectoral. Fins pink. Tips of dorsal spines orange. 8 inches. Pearl banks and trawling grounds.

Kunda (S). Kundan, Kundil (T).

404. NEMIPTERUS JAPONICUS (Bloch) JAPANESE THREADFIN-BREAM

Sparus japonicus Bloch, Naturg. Ausländ. Fische, v, 1791, p. 110. *Synagris japonicus* Day, Fish. India, 1875-78, Pl. xxiv, fig. 2 (plate 26).

D. X; 9. A. III, 6-8. L. lat. 48. Tr. (3-4)+(10-11). G.R. (5-6)+(7-10). Depth 3½-4 excluding filamentous prolongation of upper lobe. Head 3½-4. Upper lobe of caudal very much prolonged. Yellowish red with longitudinal yellowish or red lines along each scale row, usually numbering 8. Dorsal and anal fins with yellow streak along centres. 10 inches. Coastal waters and trawling grounds.

405. NEMIPTERUS FURCOSUS (Valenciennes)

RED-SPOT THREADFIN-BREAM

Dentex furcosus Cuvier and Valenciennes, Hist. Nat. Poiss., vi, 1830, p. 224. *Synagris furcosus. Synagris notatus* Day, Fish. India, 1875-78, Pl. xxiv, fig. 3 (plate 26).

D. X; 9. A. III, 7. L. lat. 47-50. Tr. 4+(10-11). G.R. 5+8. Depth 4 or up to 5·4. Head 3¾-4. Lobes of caudal subequal, deeply forked. Rosy with brilliant red and yellow spot on first five scales below lateral line. 5-6 longitudinal yellow bands and 3 silvery bands below lateral line. Fins pink with yellow band along base of dorsal and anal. 11 inches. Coastal waters.

406. NEMIPTERUS BLEEKERI (Day) BLEEKER'S THREADFIN-BREAM

Synagris bleekeri Day, Fishes of India, 1875-78, p. 92, Pl. xxiv, fig. 1 (plate 26).

D. X; 9. A. III, 7. L. lat. 48. Tr. 4+11. Depth 4⅓. Head 4-4½. Caudal lobes subequal. Dorsal fin not notched. Outer ventral ray enlarged nearly reaching anal. Reddish above, silvery with yellow bands on sides and beneath. Blue spot on opercle. Fins reddish. Dorsal edged with orange and having a golden band along its base. 8 inches. Coastal waters and trawling grounds.

TRIPLE-TAILS

Family LOBOTIDAE

Body oblong, elevated and laterally compressed with large scales, which are smaller on head and bases of fins. Mouth wide with narrow bands of villiform teeth in jaws, with an outer series of enlarged conical ones; absent on vomer, palatines and tongue. Preopercle serrate. Caudal rounded to slightly emarginate.

Eye in anterior third of head; preoperculum coarsely dentate; third anal spine longest ... *Lobotes surinamensis*
Eye in anterior half of head; preoperculum finely serrate; second anal spine longest ... *Datnioides quadrifasciatus*

407. LOBOTES SURINAMENSIS (Bloch) BROWN TRIPLE-TAIL

Holocentrus surinamensis Bloch, Nat. Ausländ. Fische, iv, 1790, p. 98. *Lobotes auctorum. Lobotes erate* Bleeker, Atlas Ichth. Ind. Neerl., vii, 1873-76, Pl. cccxi, fig. 4 (plate 26).

D. XII; 15-16. A. III, 11. L. lat. 42-44. Tr. (9-10)+(16-18). Depth more than 2, head $2\frac{1}{2}$-3. G.R. 13-15. Preoperculum coarsely dentate. Olive to brown, darker above and silvery below. Sometimes dusky lines and faint spots on body. Pectorals and caudal margin pale yellow; other fins olivaceous. 40 inches. Coastal waters entering estuaries and rivers.

408. DATNIOIDES QUADRIFASCIATUS (Sevastianov) FOUR-BAND TRIPLE-TAIL

Chaetodon quadrifasciatus Sevastianov, Mem. Acad. Sci. St. Petersbourg, i, 1809, p. 448. *Datnoides polota* Day, Fish. India, 1875-78, Pl. xxiv, fig. 6 (plate 26).

D. XII; 13-14. A. III, 8-9. L. lat. 55. Tr. (10-12)+(18-19). Depth 2 or slightly more, head 2·2-2·8. Preoperculum is finely serrated. Caudal slightly emarginate. Olivaceous, darker above, lighter below with 8-10 dark brown transverse bands on sides. Bands become broader and are reduced to 4 in older fish. Fins olive to blackish. 12 inches. Coastal species preferring brackish water and ascending rivers to above tidal influence. Its occurrence in Ceylon needs further confirmation.

SILVER-BIDDIES

Family GERRIDAE

Body compressed, oval. Head and body with moderate deciduous scales. Gill-membranes free from isthmus. Mouth strongly protractile, downwards, and with fine teeth in jaws and none on palate or tongue. Small silvery fishes of shallow waters close inshore and in estuaries.

1. Anal fin short, with 2-4 spines and 7-10 rays ... 2
   Anal fin long, with 5-6 spines and 12-13 rays ... *Pentaprion longimanus*
2. Dorsal spines 10 ... *Gerreomorpha setifer*
   Dorsal spines 9 ... 3
3. Second dorsal spine greatly elongated, usually greater than body depth ... *Pertica filamentosa*
   Second dorsal spine normal, usually greatly less than body depth ... 4
4. Body slender; depth $3\frac{1}{4}$-$3\frac{3}{8}$ ... *Gerres oblongus*
   Body deep; depth $1\frac{7}{8}$-$2\frac{7}{8}$ ... 5
5. Second dorsal spine less than 1·5 in head ... *Gerres abbreviatus*
   Second dorsal spine more than 1·5 in head ... *Gerres oyena*

409. PENTAPRION LONGIMANUS (Cantor) LONG-FINNED SILVER-BIDDY

Equula longimana Cantor, Jour. Asiatic Soc. Bengal, xviii, 1850, p. 1134. *Pentaprion longimanus* Day, Fish. India, 1875-78, Pl. lii, fig. 6 (plate 27).

D. IX-X; 14-15. A. V-VI, 12-13. L. lat. 44-48. Tr. 6+(11-12). G.R. (4-5)+(13-14). Depth ca. 3. Readily distinguished by the large number of radials in the long anal fin. Bright silvery with silvery stripe from eye to caudal. 5 inches. Trawling grounds.

410. GERREOMORPHA SETIFER (Hamilton-Buchanan) BLACK-TIPPED SILVER-BIDDY

Chanda(?) *setifer* Hamilton-Buchanan, Fish. Ganges, 1822, p. 105. *Gerres lucidus* or (?) *limbatus*, *Xystaema lucidum*. *Gerres setifer* Day, Fish. India, 1875-78, Pl. xxv, fig. 1 (plate 28).

D. X; 10. A. III, 7. L. lat. 38. Tr. 5+10. Depth 3. Head 4. Lower preopercle edge serrated on posterior half. Cheek with 3 rows of scales. Silvery with indistinct vertical dark band over nape followed by three across body under dorsal fins. Fins canary yellow. Black margin to dorsal fin membrane and a black area between 2nd and 5th dorsal spines. 5 inches. Estuarine.

Oleya, Katu oleya, Udassa, Udegamuva (S). Kari moral, Uddan, Udakhan, Uduvan (T).

411. PERTICA FILAMENTOSA (Cuvier) LONG-RAYED SILVER-BIDDY

Gerres filamentosus Cuvier, Regne Animal, ed. 2, ii, 1829, p. 188. *Gerres punctatus* or *Xystaema punctatum*. *Gerres filamentosus* Day, Fish. India, 1875-78, Pl. xxv, fig. 3 (plate 28).

D. IX; 10-11. A. III, 7-8. L. lat. 45-48. Tr. (5-6)+(12-13). G.R. (4-6)+7. Depth 2-2½. Head pointed, slightly concave before eyes. Second dorsal spine not elongated in young, but increases with age and may become almost as long as body without caudal. Blue-grey above, silvery below. Several series of bluish blotches on sides. These are elongate and may almost unite to form bars. Fins grey to yellow. 10 inches. Coastal waters, trawling grounds and pearl banks.

412. GERRES OBLONGUS Cuvier OBLONG SILVER-BIDDY

Gerres oblongus Cuvier and Valenciennes, Hist. Nat. Poiss., vi, 1830, p. 479. *Gerres oblongatus*. *Gerres oblongus* Day, Fish. India, 1875-78, Pl. xxv, fig. 2 (plate 28).

D. IX; 10. A. III, 7. L. lat. 44-48. Tr. (6-7)+10. G.R. 4+8. Depth 3¼-3½. Dorsal spines normal. Purplish olive above, otherwise silvery. Dorsal with faint dark margin. Young with very indistinct cross-bars. 12 inches. Pearl banks.

413. GERRES ABBREVIATUS Bleeker BLUE-BACKED SILVER-BIDDY

Gerres abbreviatus Bleeker, Verh. Batav. Gen., xxiii, 1850, p. 11; Day, Fish. India, 1875-78, Pl. xxv, fig. 6 (plate 28).

D. IX; 10. A. III, 7. L. lat. 33-36. Tr. (4-6)+10. G.R. 5+7. Depth 2-2½. Body oblong, elevated anteriorly. Dorsal spines not elongated. Silvery, greenish or bluish above. Indistinct darker lines longitudinally along scale rows. Dorsal with black margin. 9 inches. Coastal waters and Gulf of Manaar.

414. GERRES OYENA (Forskål) LINED SILVER-BIDDY

Labrus oyena Forskål, Descript. Animal., 1775, p. 35. *Diapterus oyena* Bleeker, Atlas Ichth. Ind. Neerl., ix, 1877, Pl. ccclxi, fig. 5 (plate 28).

D. IX; 10-11. A. III, 7. L. lat. 35-38. Tr. (4-5)+(10-11). G.R. 6+7. Depth $2\frac{1}{2}$-3. Silvery to olive. Spinous dorsal with dark margin. 10 inches. Shallow waters of Gulf of Manaar.

PONY FISH

Family LEIOGNATHIDAE

Body compressed and slimy with minute scales which are not easily visible. Gill-membranes united with isthmus. Mouth extremely protractile. Teeth usually small in jaws, sometimes with larger canines; absent from palate. Dorsal and anal with basal sheath. Extrude large amounts of mucus after death. Small shallow water species shoaling in estuaries and entering fresh water.

1. Teeth minute 2
   Teeth include distinct canines 11
2. Mouth very oblique 3
   Mouth horizontal 4
3. Depth $1\frac{1}{2}$-$1\frac{3}{4}$ .... *Secutor ruconius*
   Depth 2-$2\frac{1}{4}$ .... *Secutor insidiator*
4. Snout truncate; dorsal profile more convex than ventral; gape of mouth commences below lower border of eye; notch above eye followed by sharp ascent to dorsal fin 5
   Snout not truncate; dorsal profile subequal in convexity to ventral; gape of mouth commences opposite lower third of eye; gradual ascent from snout to dorsal fin without prominent notch 8
5. Breast covered with conspicuous scales; mandible slightly concave 6
   Breast apparently naked; mandible strongly concave 7
6. Snout black; spinous dorsal with black margin; maxillary reaches below middle of eye .... *Leiognathus splendens*
   Snout and spinous dorsal without black; maxillary reaching front border of eye ... *Leiognathus dussumieri*
7. Second dorsal spine elongate; subequal or longer than body depth .... *Leiognathus fasciatus*
   Second dorsal spine more than twice in body depth .... *Leiognathus equulus*
8. Lateral line continued to base of caudal .... *Leiognathus brevirostris*
   Lateral line ending near end of dorsal 9
9. Breast naked .... *Leiognathus daura*
   Breast scaly 10
10. Depth 2 in length without caudal; spinous dorsal with orange spot, inferiorly edged with black .... *Leiognathus bindus*
    Depth $2\frac{1}{2}$-3 in length without caudal; back with dark irregular spots .... *Leiognathus lineolatus*
11. Body with fine scales except in a triangular area between pectoral and anal base .... *Gazza minuta*
    Body completely naked ahead of a line from origin of soft dorsal to origin of anal .... *Gazza achlamys*

415. SECUTOR RUCONIUS (Hamilton-Buchanan) DEEP-BODIED PONY FISH

Chanda ruconius Hamilton-Buchanan, Fish. Ganges, 1822, p. 106. *Leiognathus ruconius* or *niconius*. *Equula ruconius* Day, Fish. India, 1875-78, Pl. li, C, fig. 4 (plate 27).

D. VIII; 16. A. III, 14. Body ovate, depth $1\frac{1}{2}$-$1\frac{3}{4}$. Lateral line ceases below middle of soft dorsal. Back bluish silver, more silvery below with pink and blue iridescence. Upper third of body with black or bronze transverse bars or series of spots. A dark line from angle of mouth to chin and dark opercular spot. Fins blue. Dorsal with black tip. 5 inches. Coastal waters.

Kum panna, Valipanna (S). Karel (T).

416. SECUTOR INSIDIATOR (Bloch) SLENDER BARRED PONY FISH

Zeus insidiator Bloch, Nat. Ausländ. Fische, iii, 1787, p. 41. *Leiognathus insidiatrix. Equula insidiatrix* Day, Fish. India, 1875-78, Pl. li, C, fig. 5 (plate 27).

D. VIII-IX; 15-17. A. III, 14. G.R. 17-18. Depth 2-2¼. Scales barely visible. Lateral line finishes short of caudal base. Mouth protrudes upwards. Back bluish-silver, more silvery below with bronze reflections. 3-4 horizontal lines of black spots, with bronze reflections, form 8-10 vertical bands on upper half of body. Black streak from eye to throat. Dark mark in axilla. Dorsal tipped with black. Caudal and pectoral yellowish. 4 inches. Trawling grounds.

417. LEIOGNATHUS SPLENDENS (Cuvier) SPLENDID PONY FISH

Equula splendens Cuvier, Regne Animal, ii, ed. 2, 1829, p. 212; Day, Fish. India, 1875-78, Pl. lii, fig. 3 (plate 27).

D. VII-VIII; 16-17. A. III, 14. Depth 1¾. L. lat. about 55, terminating at base of caudal. Breast markedly scaled. Ventrals with large axillary scale. Silvery. End of snout dotted with black. Pectoral axilla black. Spinous dorsal membrane between 2nd to 6th spines black. 5 inches. Coastal waters and trawling grounds.

Katu karalla (S). Karel (T).

418. LEIOGNATHUS DUSSUMIERI (Valenciennes) DUSSUMIER'S PONY FISH

Equula dussumieri Cuvier and Valenciennes, Hist. Nat. Poiss., x, 1835, p. 77; Day, Fish. India, 1875-78, Pl. lii, fig. 2 (plate 27).

D. VIII; 16. A. III, 14. Strongly elevated and curved behind occiput. Depth 2½. Two rather strong spines above front border of eye. Scales conspicuously developed on breast. L. lat. about 60, less distinct posteriorly. Silvery. No black on dorsal fin or snout. Dark narrow vertical bands on back crossing lateral line. 8 inches. Reported from fresh water.

419. LEIOGNATHUS FASCIATUS (Lacépède) BANDED PONY FISH

Clupea fasciata Lacépède, Hist. Nat. Poiss., v, 1803, p. 460. *Equula fasciata* Day, Fish. India, 1875-78, Pl. li, C, fig. 2 (plate 27).

D. VII-VIII; 16. A. III, 14. Depth about 2. Mouth horizontal forming tube which points down when protracted. Back strongly elevated behind occiput. Breast appears naked. Lateral line with 65 tubes, terminating at caudal base. Second dorsal spine elongated, filiform and equal to body depth. Second anal spine similarly elongate. Silvery with diffuse irregular vertical streaks on upper half of body. Pectoral axilla dark. Fins hyaline. 6 inches. Coastal waters, harbours and trawling grounds.

420. LEIOGNATHUS EQUULUS (Forskål) GREATER PONY FISH

Scomber equula Forskål, Descript. Animal., 1775, p. 75. *Leiognathus edentula. Equula edentula* Day, Fish. India, 1875-78, Pl. lii, fig. 1 (plate 27).

D. VIII; 15-16. A. III, 14. G.R. 14-16. Breast with very thin scales thus appearing naked. Depth 2-2½. Lateral line 60, terminating at caudal base. Mandible strongly concave. Mouth when protracted points downwards. Teeth very fine, in single series. Silvery with iridescence. Faint vertical narrow cross-bands on upper part of body. Fins hyaline, soft dorsal and pectoral base greyish. 12 inches. Coastal waters, ascending rivers, and trawling grounds.

Hotu panna, Mas karalla (S).

421. LEIOGNATHUS BREVIROSTRIS (Valenciennes) SHORT-NOSED PONY FISH

Equula brevirostris Cuvier and Valenciennes, Hist. Nat. Poiss., x, 1835, p. 83. *Equula blochii* Day, Fish. India, 1875-78, Pl. lii, fig. 5 (plate 27).

D. VIII; 16. A. III, 14. Depth 2⅔. Body oval with dorsal and ventral profiles equal. Mandible very concave below. Naked on breast. Lateral line 58-64, reaching base of caudal. Oval black blotch on upper half of spinous dorsal between 3rd and 7th spine. A dark transverse blotch across nape. Pectoral black at base. Narrow yellow band from above eye to caudal base. 4½ inches. Trawling grounds.

422. LEIOGNATHUS DAURA (Cuvier) GOLDEN-STRIPED PONY FISH

Equula daura Cuvier, Regne Animal, ii, 1829, p. 212; Day, Fish. India, 1875-78, Pl. lii, fig. 4 (plate 27).

D. VIII; 16. A. III, 14-15. Depth 2·8. Dorsal profile slightly more convex than ventral. Mandibles slightly concave. Scales small, absent from breast. Lateral line 60, not extending beyond end of caudal. G.R. 15-16. Silvery with golden stripe along side. A dark line along base of dorsal. A dark triangular spot between occiput and dorsal fin. Upper half of spinous dorsal black from 2nd to 5th spines. 5 inches. Coastal waters and pearl banks.

423. LEIOGNATHUS BINDUS (Valenciennes) ORANGE-FINNED PONY FISH

Equula bindus Cuvier and Valenciennes, Hist. Nat. Poiss., x, 1835, p. 78. *Leiognathus virgatus* Fowler, Jour. Acad. Nat. Sci. Phila., xii(4), 1904, Pl. xv, fig. 4 (plate 27).

D. VIII; 16-17. A. III, 14. Depth 2½. Head somewhat pointed. Abdominal profile more convex than dorsal, which is somewhat concave above eyes. Ventrals short, scarcely reaching half way to anal. Lateral line tubules disappearing behind dorsal. Silvery, olive on nape. Fins hyaline. Upper third of spinous dorsal orange, edged inferiorly with black. 3½ inches. Gulf of Manaar and trawling grounds.

424. LEIOGNATHUS LINEOLATUS (Valenciennes) LINED PONY FISH

Equula lineolata Cuvier and Valenciennes, Hist. Nat. Poiss., x, 1853, p. 86; Day, Fish. India, 1875-78, Pl. li, C, fig. 3 (plate 27).

D. VIII; 16. A. III, 14. Depth 3-3½. Dorsal and ventral profiles equally convex. A slight concavity over occiput. Breast scaly. Lateral line continues to caudal base with about 60 tubules. Silvery with vertical zigzag lines passing down the back. Base of pectoral black behind. Upper edge of dorsal darkish. 4 inches. Trawling grounds.

425. GAZZA MINUTA (Bloch) TOOTHED PONY FISH

Scomber minutus Bloch, Ichth., 1797, Pl. 429, fig. 2. *Gazza equulaeformis. Gazza minuta* Day, Fish. India, 1875-78, Pl. liii, fig. 1 (plate 27).

D. VIII; 16. A. III, 14. G.R. 14-16. Depth 2⅓-2½. Eye 3 in head. Scales very fine, barely visible but present on body except a triangular space between pectorals and anal base. Head naked. Dusky above, silvery on sides and below. Sometimes irregular blotches or wavy lines above lateral line. Axil black, confluent with black blotch on shoulder girdle. Spinous dorsal membrane dusky at tip. Other fins hyaline. 7 inches. Pearl banks and trawling grounds.

Mas panna, Mas karalla, Pulunu karalla (S).

426. GAZZA ACHLAMYS Jordan and Starks NAKED TOOTHED PONY FISH

Gazza achlamys Jordan and Starks, Ann. Carnegie Mus., xi, 1917, p. 446, Pl. xlv (plate 27).

D. VIII; 16. A. III, 13-14. G.R. 14. Depth 1⅘. Eye 2-2½ in head. Head naked. Body naked anterior to the line which joins the origins of soft dorsal and anal fins. Dusky above. Sides silvery and never with wavy lines. Axil dusky. Also sometimes a dusky blotch on upper part of opercle. Spinous dorsal tip very slightly dusky. Other fins hyaline. 5½ inches. Coastal waters.

GRUNTERS, JAVELIN FISH

Family POMADASYIDAE

Small to moderate sized fish of typical perch-like appearance. Mouth small. Teeth feeble. A pit beneath the chin. Preopercular margin serrate. Dorsal and anal spines usually strong. Caudal not deeply forked. Silvery or olive to dusky. Enter estuaries.

1. At most one faint stripe along the body 2
   Distinct stripes present along body .... *Rhonciscus furcatus*
2. Second anal spine nearly as long as third; L.lat. 55; nine scales between lateral line and dorsal .... *Pomadasys olivaceus*
   Second anal spine conspicuously longer than third; L.lat. 44-53; 4-7 scales between lateral line and dorsal 3
3. Dorsal with black blotch on spinous part; back often with large transverse bands ... *Pomadasys maculatus*
   Dorsal spotted; body with longitudinal rows of spots or transverse bands .... *Pomadasys hasta*
   Dorsal without spots; body uniform silvery .... *Pomadasys argyreus*

427. RHONCISCUS FURCATUS (Schneider) BANDED GRUNTER

Grammistes furcatus Bloch and Schneider, Syst. Ichth., 1801, p. 187. *Pomadasys furcatum* or *dussumieri. Pristipoma furcatum* Bleeker, Atlas Ichth. Ind. Neerl., viii, 1876-77, Pl. ccclii, fig. 2 (plate 29).

D. XII-XIII; 15-16. A. III, 8-9. L. lat. 55-60. Tr. 9+(15-16). G.R. 7+10. Depth about 3. Second anal spine conspicuously longer and stronger than third. Silvery with 6 horizontal blackish bands, 3 of which are above lateral line. Spinous part of dorsal fin with 3 rows of brown spots and a dark edge. Upper half of soft dorsal dark and a dark band along anal. 15 inches. Coastal waters, Gulf of Manaar and trawling grounds.

428. POMADASYS OLIVACEUS (Day) — OLIVE GRUNTER

Pristipoma olivaceum Day, Fish. India, 1875-78, p. 73. *Plectorhinchus olivaceum. Pristipoma olivaceum* Day, *loc. cit.*, Pl. xix, fig. 1 (plate 29).

D. XII; 15-17. A. III, 11-12. L. lat. 55. Tr. (8-9)+15. G.R. 13-15. Depth 2$\frac{3}{4}$. Second anal spine strongest and nearly as long as third. Olive grey with purplish head. A large black blotch bordered anteriorly with yellow on upper angle of operculum. Fins yellowish but to some extent clouded with black. 12 inches. Trawling grounds.

429. POMADASYS MACULATUS (Bloch) — SPOTTED GRUNTER

Anthias maculatus Bloch, Nat. Ausländ. Fische, vii, 1797, p. 9. *Pristipoma maculatum* Day, Fish. India, 1875-78, Pl. xix, fig. 5 (plate 29).

D. XII-XIII; 12-14. A. III, 7-8. L. lat. 49-53. Tr. (5-6)+(12-15). G.R. 13-14. Depth 3$\frac{1}{2}$-3$\frac{3}{4}$. Second anal spine much longer and stronger than third. Silvery grey with purplish head. Nape and back with series of incomplete cross-bars, often broken into blotches, that on the nape being most prominent. Spinous dorsal with large black blotch. Dorsals and caudal edged with black, otherwise fins yellowish. 18 inches. Coastal waters, Gulf of Manaar and trawling grounds.

430. POMADASYS HASTA (Bloch) — LINED SILVER GRUNTER

Lutjanus hasta Bloch, Nat. Ausländ. Fische, iv, 1790, p. 109. *Pristipoma hasta. Pomadasys hasta* Bleeker, Atlas Ichth. Ind. Neerl., viii, 1876-77, Pl. cccxxv, fig. 3 (plate 29).

D. XII-XIII; 13-15. A. III, 6-8. L. lat. 45-52. Tr. (4-5)+(12-14). G.R. 12-13. Depth 3-3$\frac{1}{2}$. Second anal spine much stronger and longer than third. Silver grey. 4-5 interrupted dark grey lines along sides, 3-4 of which are above lateral line. Dorsal with 2-3 rows of brown spots which may reduce to 1 row at base of fin in adults. 18 inches. Harbours and Gulf of Manaar.

Iri bateya (S). — Kithalu valayen (T).

431. POMADASYS ARGYREUS (Valenciennes) — SILVER GRUNTER

Pristipoma argyreum Cuvier and Valenciennes, Hist. Nat. Poiss., ix, 1833, p. 485. *Pomadasys guoraca. Pristipoma guoraca* Bleeker, Atlas Ichth. Ind. Neerl., viii, 1876-77, Pl. cccxxiii, fig. 1 (plate 29).

D. XII; 13-14. A. III, 7-8. L. lat. 44-50. Tr. 5+(11-13). G.R. 7+14. Depth 2$\frac{3}{4}$-3$\frac{1}{4}$. Second anal spine longer but not always stronger than third. Silvery, slightly darker on back. Fins yellowish brown. Young with large dusky blotch on operculum. 8 inches. Coastal waters and trawling grounds.

MONOCLE-BREAM

Family SCOLOPSIDAE

Perch-like fishes with rather large scales. Principal character is the backwardly directed spine on the suborbital bone, which is also serrated below. Dorsal spines 10. Small brilliantly coloured fishes resembling the Lutianidae.

1. Suborbital spine feeble .... *Parascolopsis eriomma*
   Suborbital spine distinct 2
2. White transverse band from occiput over opercle .... *Scolopsis vosmeri*
   No white band across opercle 3
3. Dark median horizontal band, often broken into 2 large lateral blotches .... *Scolopsis bimaculatus*
   Narrow light band close along base of dorsal fin and a wide blue band through eye to pectoral axilla .... *Scolopsis phaeops*

432. PARASCOLOPSIS ERIOMMA (Jordan and Richardson) ROSY MONOCLE-BREAM

Scolopsis eriomma Jordan and Richardson, Mem. Carnegie Mus., iv (4), 1909, p. 188, Pl. lxx (plate 28).

D. X; 9. A. III, 7. L. lat. 35. Tr. 3+11. G.R. 11, moderately long, spiny and close-set. Depth 2¾. Rosy pink, lighter below with broad yellowish lateral stripe from operculum to caudal base. Dorsals, pectorals, caudal and ventrals rosy. Soft dorsal and caudal with yellowish tips. Spinous dorsal with red tips. Anal yellow. 12 inches. Coastal waters in shallows.

433. SCOLOPSIS VOSMERI (Bloch) WHITE-CHEEKED MONOCLE-BREAM

Anthias vosmeri Bloch, Nat. Ausländ. Fische, vi, 1792, p. 120. *Scolopsis japonicus. Scolopsis vosmaeri* Bleeker, Atlas Ichth. Ind. Neerl., viii, 1876-77, Pl. cccxxxix, fig. 5 (plate 28).

D. X; 9-10. A. III, 7-8. L. lat. 38-41. Tr. 4+(11-14). G.R. 6, knob-like. Depth 2¼. Suborbital spine short but strong. Pale brownish red with a prominent white transverse band across opercles. A blood-red spot behind opercular spine. Fins yellowish to orange; dorsal and anal spine bright red. Inside of mouth red. 10 inches. Coastal waters and pearl banks.

Andiya, Kiri vavoula (S). Andiyan (T).

434. SCOLOPSIS BIMACULATUS Rüppell TWO-SPOT MONOCLE-BREAM

Scolopsis bimaculatus Rüppell, Atlas Reise Nordl. Afrika, 1828, p. 8. (?) *Scolopsis monogramma. Scolopsis bimaculatus* Day, Fish. India, 1875-78, Pl. xxii, fig. 1 (plate 28).

D. X; 9. A. III, 6-7. L. lat. 45. Tr. 5+12. G.R. 5, small. Depth 2¾. Dull olive brown, paler below. A brown band across snout. One or two dark brown blotches on lateral line, the first large or the two may be joined. Fins pinkish or orange becoming reddish towards the margins. 12 inches. Coastal waters, Gulf of Manaar and trawling grounds.

Pol ranna (S).

435. SCOLOPSIS PHAEOPS (Bennett) BLUE-CHEEKED MONOCLE-BREAM

Scolopsides phaeops Bennett, Proc. Zool. Soc., 1831, p. 165. *Scolopsis phaeops* Day, Fish. India, 1875-78, Pl. xxii, fig. 2 (plate 28).

D. X; 9. A. III, 7. L. lat. 46-48. Tr. 5+(16-17). G.R. 5, low tubercles. Depth 3-3¾. Suborbital spine strong, flat and pointed. Greenish-olive above, yellowish below. A narrow light band along back close to base of dorsal fin. A wide bright blue band from eye to upper jaw, and from posterior edge of eye to the axilla, where it ends in a blue spot. 9½ inches. Gulf of Manaar.

SWEETLIPS

Family PLECTORHYNCHIDAE

Perch-like fishes with compressed body and moderate to small scales. Dorsals united with 9-14 spines and 15-25 rays. Anal short. Mouth small and lips usually thick. Feeble teeth in bands in jaws, absent on palate. Distinct pores on chin. Preopercle margin serrated. Scale rows tend to be irregular. Mostly moderate sized fishes inhabiting shallow waters.

1. Dorsal spines 9-10 .... *Spilotichthys pictus*
   Dorsal spines 12-14 2
2. Spines strong .... *Pseudopristipoma nigra*
   Spines normal 3
3. Body uniform or with golden spots .... *Gaterin schotaf*
   Body with dark longitudinal stripes, persisting with age 4
4. Fins without dark spots or bands .... *Gaterin cinctus*
   Fins with dark spots or blotches 5
5. Body with 3-4 longitudinal bands .... *Gaterin orientalis*
   Body with 6-7 black longitudinal bands 6
6. Pectoral with large black median blotch .... *Gaterin lineatus*
   Pectoral without median blotch .... *Gaterin diagrammus*

436. SPILOTICHTHYS PICTUS (Thunberg) PAINTED SWEETLIP

Perca picta Thunberg, Kon. Vet. Akad. Stockholm, xiii, 1792, p. 143. *Diagramma blochi, poecilopterum, cinerascens,* or *pictus. Plectorhynchus pictus* Tanaka, Fish. Japan, xxi, 1915, Pl. cv, fig. 325 (plate 29, B, adult). *Diagramma punctatum* Day, Fish. India, 1875-78, Pl. xxi, fig. 4 (plate 29, A, young).

D. IX-X; 21-25. A. III, 6-7. L. lat. 90-110. Tr. (17-19)+(30-32). G.R. 11-14. Depth 2½-3. Young with 2-3 broad bands along sides, which diminish in size and break into spots. Young are yellow orange with brown or black longitudinal markings. Fins yellow streaked and blotched with black. Adults are greyish with darker spots and grey fins. 18 inches. Harbours, pearl banks and trawling grounds.

437. PSEUDOPRISTIPOMA NIGRA (Cuvier) BLACK SWEETLIP

Pristipoma nigrum Cuvier and Valenciennes, Hist. Nat. Poiss., v, 1830, p. 258. *Euelatichthys* or *Plectorhynchus crassispina. Diagramma crassispinum* Day, Fish. India, 1875-78, Pl. xx, fig. 4 (plate 29).

D. XIII-XIV; 13-18. A. III, 7-8. L. lat. 51-53. Tr. (10-11)+(20-21). G.R. 18-20. Depth 2¼-2½. Slaty grey with violet on head and bronze on body. Fins nearly black. A few irregular coppery spots on body. In young caudal is yellowish-white. 25 inches. Coastal waters and pearl banks.

438. GATERIN SCHOTAF (Forskål) GREY SWEETLIP

Sciaena abu-mgaterin schotaf Forskål, Descript. Animal., 1775, p. 51. *Plectorhinchus griseus. Diagramma griseum* Day, Fish. India, 1875-78, Pl. xxi, fig. 2 (plate 29).

D. XII-XIII; 18-23. A. III, 7-8. L. lat. 55-60. Tr. 12+20. Depth 2½-2¾. G.R. 16-18. Uniform grey or olive grey, sometimes with golden spots. The young are more olive with some sinuous narrow light blue lines. Fins nearly or quite black. Sometimes faint light cross-bars on back. 36 inches but seldom reaching more than 18 inches. Gulf of Manaar and trawling grounds.

Boraluva, Gal modha, Gobeya, Kobeya (S). Kallu kallewa, Tholen (T).

439. GATERIN CINCTUS (Schlegel) THREE-BANDED SWEETLIP

Diagramma cinctum Schlegel, Fauna Japonica, Pisces, 1843, p. 61. *Plectorhynchus cinctus. Diagramma cinctum* Day, Fish. India, 1875-78, Pl. xxi, fig. 1 (plate 29).

D. XII; 15. A. III, 7. L. lat. with 90-95 along above and 65-70 along below. Tr. 17+23. Depth 2½. G.R. 7+15. Slaty grey with three longitudinal brownish bands which may be broken into blotches. Numerous dark spots between two uppermost bands. Vertical fins with dark margins. Ventral blackish. Pectorals yellow. 24 inches. Gulf of Manaar.

440. GATERIN ORIENTALIS (Bloch) ORIENTAL SWEETLIP

Anthias orientalis Bloch, Nat. Ausländ. Fische, vii, 1793, p. 10. *Diagramma sibbaldi* or *orientale. Plectorhynchus orientalis* Weber and De Beaufort, Fish. Indo-Aust. Archipel., vii, 1936, fig. 83 (plate 29).

D. XIII-XIV; 17-18. A. III, 8. L. lat. 80, above along 100, below along 85. Tr. 13+23. Depth 3¾. Yellowish with several chestnut bands as wide or wider than interspaces. Bands may be broken into irregular patches and are better developed in the young. Dorsals whitish with large dark basal bands. Caudal with broad central band and broad dark tips. Pectoral black with pale margins. 16 inches. Coastal waters.

441. GATERIN LINEATUS (Linnaeus) YELLOW-BANDED SWEETLIP

Perca lineata Linnaeus, Syst. Nat., ed. 10, i, p. 293. *Diagramma lineatus* or *radja. Plectorhynchus lineatus* Bleeker, Atlas Ichth. Ind. Neerl., vii, 1873-76, Pl. cccvi, fig. 4 (plate 29).

D. XII-XIII; 18-21. A. III, 6-8. L. lat. 85-90. Tr. (13-14)+(24-25). G.R. 19-20. Depth 3·3-3·5. Yellowish-white with usually six dark brown horizontal bands of variable width and arrangement. Fins yellow with one or more dark blotches and broken continuations of body bands. Dorsal usually with dark edge. Light band along middle of anal fin. 14 inches. Coastal waters.

Deva boraluva, Ho bateya (S). Manjel manthanam (T).

442. GATERIN DIAGRAMMUS (Linnaeus) SILVER-BANDED SWEETLIP

Perca diagramma Linnaeus, Syst. Nat., ed. 10, i, 1758, p. 293. *Bodian, Diagramma* or *Plectorhynchus cuvieri. Plectorhynchus sebae* Bleeker, Atlas Ichth. Ind. Neerl., vii, 1873-76, Pl. cccv, fig. 3 (plate 29).

D. XII-XIII; 17-20. A. III, 7. L. lat. 93. Tr. 11+25. G.R. 20-21. Depth $3\text{-}3\frac{3}{4}$. Silvery with horizontal greyish or brownish bands which vary in number from 3 to 7 and vary greatly in width and arrangement. Fins yellowish marked with blotches and broken bands or spots, especially on soft dorsal, anal and caudal. 20 inches. Doubtfully distinct from preceding species. Coastal waters.

RIVER KINGFISH, JEWFISH, CROAKERS

Family SCIAENIDAE

Body elongate, robust, with small to moderate scales. Dorsal deeply notched before the soft portion. 2 weak anal spines. A fleshy process in pectoral axil. Caudal truncate to weakly emarginate. Teeth mostly small but some canines present. Air bladder with many branches. Moderate to large fishes mostly inhabiting shallow water. Capable of producing a croaking or drumming sound.

1.	A single barbel below chin	3
	No barbel below chin	2
2.	Distinct pores on chin	5
	No pores on chin	14
3.	Dorsal rays 23-26	*Sciaena dussumieri*
	Dorsal rays 28-31	4
4.	Barbel sub-equal to eye; anal base about 2 in base of soft dorsal	*Sciaena russelli*
	Barbel obsolete or less than pupil; anal base 4·5-5 in base of soft dorsal	*Sciaena macropterus*
5.	Dorsal rays 22-26	6
	Dorsal rays 26-30	10
	Dorsal rays 28-32	*Johnius dussumieri*
6.	Depth $3\text{-}3\frac{1}{2}$	7
	Depth $3\frac{1}{2}\text{-}4$	9
7.	Dorsal rays 22-23	*Johnius diacanthus*
	Dorsal rays 24-26	8
8.	No dark opercular blotch	*Johnius aneus*
	Diffuse dark blotch on opercle	*Johnius coibor*
9.	Body without black transverse bands	*Johnius osseus*
	Body with 4-5 black transverse bands	*Johnius maculatus*
10.	Depth $3\text{-}3\frac{1}{2}$	11
	Depth $3\frac{1}{2}\text{-}4$	12
11.	Inner row of enlarged mandibular teeth	*Johnius sina*
	No enlarged inner row of mandibular teeth	*Johnius carutta*
12.	A dark axillary blotch	*Johnius axillaris*
	No dark axillary blotch	13
13.	Lower gill-rakers 13	*Johnius soldado*
	Lower gill-rakers 14-15	*Johnius argentatus*
14.	Anal with 10-12 branched rays; body and fins with dark blotches	*Otolithus maculatus*
	Anal with 7-8 branched rays; body and fins uniform	15
15.	Gill-rakers 5+8; scales 63-66 along above lateral line	*Otolithus argenteus*
	Gill-rakers 9+16; scales 59-62 along above lateral line	*Otolithus ruber*

443. SCIAENA DUSSUMIERI (Valenciennes) DUSSUMIER'S JEWFISH

Umbrina dussumieri Cuvier and Valenciennes, Hist. Nat. Poiss., ix, 1833, p. 481. *Sciaena dussumieri* Bleeker, Atlas Ichth. Ind. Neerl., ix, 1877, Pl. ccclxxxvii, fig. 4 (plate 30).

D. X; I, 23-26. A. II, 7. L. lat. 50-52, tubules with 2 branches. Tr. (6-7)+(15-17). G.R. 4+(8-9), short. Depth $3\frac{1}{2}$-$4\frac{1}{2}$. Snout rounded. Barbel about half eye diameter. Anal base 4 times in dorsal base. Caudal rounded. Brownish with metallic reflections below. 4-5 ill-defined brownish stripes. Fins yellowish-brown. Spinous dorsal dark. 10 inches. Coastal waters.

444. SCIAENA RUSSELLI (Cuvier) RUSSELL'S JEWFISH

Umbrina russelii Cuvier and Valenciennes, Hist. Nat. Poiss., v, 1830, p. 178. *Sciaena indica. Sciaena russelli* Bleeker, Atlas Ichth. Ind. Neerl., ix, 1877, Pl. ccclxxxvi, fig. 2 (plate 30).

D. X; I, 28. A. II, 7-8. L. lat. ca. 50. Tr. (6-7)+(11-12). G.R. 6+10, very short. Depth 3-$3\frac{1}{2}$. Snout blunt. Barbel is a little less than eye diameter. Anal base about 2 in soft dorsal base. Caudal rounded in young but angularly pointed in adult. Grey, silvery below. A steel-blue opercular spot. Upper two-thirds of spinous dorsal black. Other fins yellow. 10 inches. Coastal waters.

445. SCIAENA MACROPTERUS (Bleeker) LARGE-FINNED JEWFISH

Umbrina macropterus Bleeker, Nat. Tijds. Ned. Ind., iv, 1853, p. 254. *Sciaena macropterus* Bleeker, Atlas Ichth. Ind. Neerl., ix, 1877, Pl. ccclxxxiv, fig. 5 (plate 30).

D. X; I, 30-31. A. II, 7-8. L. lat. 44-45. Tr. (4-5)+(14). G.R. 5+(7-8), short. Depth $3\frac{1}{4}$-$3\frac{1}{2}$. Snout obtuse. Barbel short, less than diameter of pupil or obsolete. Base of anal $4\frac{1}{2}$-5 in that of soft dorsal. Caudal rhomboid. Grey, silvery below and everywhere covered with minute black dots. Fins yellowish and similarly dotted. A dark mark on the opercle. 8 inches. Coastal waters.

446. JOHNIUS DUSSUMIERI (Cuvier) DUSSUMIER'S SILVER JEWFISH

Corvina dussumieri Cuvier and Valenciennes, Hist. Nat. Poiss., v, 1830, p. 119. *Sciaena glaucus* Day, Fish. India, 1875-78, Pl. xlvi, fig. 2 (plate 30).

D. X; I, 28-32. A. II, 8. L. lat. 50. Tr. 5+(10-12). G.R. (5-6)+(12-13). Depth $3\frac{1}{4}$ in length without caudal. Chin with 5 pores. Origin of anal below middle of soft dorsal. Caudal rounded. Lateral line tubules with 2 branches. Silvery. Back greenish grey. Diffuse blue blotch on opercles and a dark spot at base of pectorals. Spinous dorsal almost black and upper two thirds of soft dorsal spotted with dusky. Caudal and posterior half of pectorals similar. 8 inches, but attains greater size. Gulf of Manaar and trawling grounds.

447. JOHNIUS DIACANTHUS (Lacépède) TWO-SPINED JEWFISH

Lutjanus diacanthus Lacépède, Hist. Nat. Poiss., iv, 1802, p. 195. *Sciaena diacanthus* or *maculata. Corvina nigromaculata. Pseudosciaena diacanthus* Bleeker, Atlas Ichth. Ind. Neerl., ix, 1877, Pl. ccclxxxviii, fig. 2 (plate 28).

D. X-XI; I, 22-23. A. II, 7. L. lat. 56-67. Tr. (7-8)+(14-15). G.R. 6+9. Depth 3-4. Head bluntly pointed. Operculum with two weak spines. Chin with 6 pores. Base of anal nearly 4 in that of soft dorsal. Pectorals rounded. Caudal rhombic, more pointed in young. Lateral line tubules finely branched. Brownish grey above, silvery below. Head glossed with purple. Fins yellowish with black dots. Young fish with grey fins which have dark edges and dorsal with 2 rows of spots. 5 feet. Coastal waters, entering tidal waters.
Gobeya (S). Kopayen (T).

448. JOHNIUS ANEUS Bloch GREY-FIN JEWFISH

Johnius aneus Bloch, Naturg. Ausländ. Fische, vii, 1793, p. 135. *Sciaena aneus* Day, Fish. India, 1875-78, Pl. xlv, fig. 5 (plate 30).

D. X-XI; I, 22-26. A. II, 7-8. L. lat. 50-52. Tr. 8+18. G.R. 6+14. Depth 3-3½. Chin with 5 pores. Operculum with 2 flat weak spines. Anal originates below 12-14th dorsal ray; its base 4 in that of dorsal. Caudal rounded below and with pointed upper corner. Lateral line tubules branched. Silvery grey above, white below. Spinous dorsal tipped with black or dark grey. Soft dorsal greyish. Caudal tipped with grey. Other fins yellowish. 9 inches. Trawling grounds and Palk Straits.

449. JOHNIUS COIBOR (Hamilton-Buchanan) GANGES JEWFISH

Bola coibor Hamilton-Buchanan, Fish. Ganges, 1822, p. 78. *Sciaena albida* Day, Fish. India, 1875-78, Pl. xliv, fig. 4 (plate 30).

D. IX-X; I, 25-27. A. II, 7. L. lat. 53-55. Tr. 7+(16-17). Depth 3-4. Three pores below mandibular symphysis and 3 others behind. Snout blunt with 3 small pores on its top. Operculum with single broad flat spine. Lateral line tubules branched. Anal fin originates below 12th dorsal ray, its base 5 in that of soft dorsal. Caudal rounded in young but cuneate with age. Silvery with light streak along each row of scales. Spinous dorsal with black membrane in young or black outer edge in adult. Upper third of soft dorsal grey. A dark bluish mark on opercle, especially in young. Ventral, anal and caudal yellowish. 3 feet. Gulf of Manaar.

450. JOHNIUS OSSEUS (Day) BONY JEWFISH

Sciaena osseus Day, Fish. India, 1875-78, p. 193. *Bola osseus. Sciaena osseus* Day, *loc. cit.*, Pl. xlvi, fig. 3 (plate 30).

D. X; I, 25. A. II, 7. L. lat. 50. Tr. 6+15. G.R. 2+7. Depth 4¼. Snout obtuse. Operculum with 2 spines. Row of 5 pores across snout and 5 more along lower edge of mandible. Lateral line tubules with one short branch. Caudal wedge-shaped. Brownish-grey above, white beneath. Opercle bluish black. Upper half of spinous dorsal black. Outer edges of other fins grey. 7 inches. Coastal waters and Gulf of Manaar.

451. JOHNIUS MACULATUS Schneider BLACK-BANDED JEWFISH

Johnius maculatus Bloch and Schneider, Syst. Ichth., 1801, p. 75. *Sciaena maculata*.

D. X; I, 23-24. A. II, 7. L. lat. 45-48, along above 65, along below 58. Tr. 8+16. Depth 3½-4. Opercle with 2 obtuse spines. Transverse row of 4 pores across snout, 1 below

mandibular symphysis and 2 more on either side. Lateral line tubules with single branch. Base of anal 4 in that of soft dorsal. Caudal wedge-shaped in young, more obtuse in adults. Silvery grey tinted with gold. Back with 5 interrupted black bands. Upper two thirds of spinous dorsal black, less distinct with age. Caudal tinged with black. Other fins yellowish. 12 inches. Coastal waters.
Katala (S).

452. JOHNIUS SINA (Cuvier) DRAB JEWFISH

Corvina sina Cuvier and Valenciennes, Hist. Nat. Poiss., v, 1830, p. 122. *Pseudosciaena sina. Sciaena vogleri* Day, Fish. India, 1875-78, Pl. xlv, fig. 1 (plate 30).

D. X; I, 27-29. A. II, 7. L. lat. 48-50. Tr. (6-7)+(12-13). G.R. (4-8)+(9-18). Depth 3-3½. Chin with 6 pores. Lateral line tubules branched. Anal originates before middle of soft dorsal. Caudal rhomboidal with lower median rays longest. Brown above, silvery with gold reflections below. Spinous dorsal blackish. Other fins grey. Coastal waters.

453. JOHNIUS CARUTTA Bloch PURPLE JEWFISH

Johnius carutta Bloch, Naturg. Ausländ. Fische, vii, 1793, p. 133. *Sciaena carutta* Day, Fish. India, 1875-78, Pl. xliv, fig. 1 (plate 30).

D. X; I, 25-28. A. II, 7. L. lat. 50. Tr. (5-6)+18. G.R. 4+9. Depth 3½-4. Snout inflated, overhanging lower jaw. 5 pores on chin. A row of pores across snout, 5 along edge of free border. Caudal rhomboidal. Lateral line tubules with short branches. Upper part of body and head purplish, lower portions golden. Spinous dorsal dark. Other fins with grey edges. 12 inches. Coastal waters.

454. JOHNIUS AXILLARIS (Cuvier) BLACK-SPOT JEWFISH

Corvina axillaris Cuvier and Valenciennes, Hist. Nat. Poiss., v, 1830, p. 113. *Bola* or *Pseudosciaena axillaris. Sciaena axillaris* Day, Fish. India, 1875-78, Pl. xliii, fig. 6 (plate 30).

D. X; I, 28. A. II, 7-8. L. lat. 48-50. Tr. (5-6)+(9-10). G.R. 7+4. Depth equal to head, 3-7. Mouth terminal. No pores on top of snout. A median pore below mandibular symphysis followed by 2 slit-like pores on each side. Operculum with 2 flat spines. Lateral line tubules branched. Anal base 4 in that of soft dorsal. Caudal rounded. Silvery dashed with purple. A black axillary spot. Upper two thirds of spinous dorsal black and a black tinge along first portion of soft dorsal. Other fins grey. 10½ inches. Coastal waters.

455. JOHNIUS SOLDADO (Lacépède) GREEN-BACKED JEWFISH

Holocentrus soldado Lacépède, Hist. Nat. Poiss., iv, 1802, p. 344. *Corvina* or *Sciaena miles. Sciaena soldado* Ogilby, Mem. Qld. Mus., vi, 1918, Pl. xxiv (plate 30).

D. IX-X; I, 28-31. A. II, 7. L. lat. 50-55. Tr. (8-9)+(17-18). Depth subequal to head, 3-4. G.R. 7+13 rudimentary ones below. Mouth terminal. 5 pores on top of snout. Lateral line tubules strongly branched. Anal originates below 9-10th soft dorsal ray. Caudal wedge-shaped. Green on back, grey on sides and whitish below. Sometimes small brown spot in front of each dorsal ray. Outer edges of fins sometimes dark. Ventral always white. 24 inches. Coastal waters.

456. JOHNIUS ARGENTATUS (Houttuyn) SILVER JEWFISH

Sparus argentatus Houttuyn, Verh. Holland. Maatsch. Wet. Haarlem, xx, 1782, p. 319. *Sciaena bleekeri* Day, Fish. India, 1875-78, Pl. xlv, fig. 4 (plate 30).

D. X-XI; I, 24-28. L. lat. 60, along above 85-90, along below 70. Tr. (9-10)+18. G.R. 6+(14-15). Depth 4¾. Shoulder scale strongly serrated. 3 pores on top of snout. 5 pores on chin. Inner row of teeth in lower jaw much larger than others. Caudal cuneate. Lateral line tubules with simple bifurcations. Silvery grey on back, dull white below. A black spot in the axilla. Spinous dorsal grey with light centre band. Caudal dark on outer third. Other fins yellow. 27 inches. Coastal waters and Gulf of Manaar.

457. OTOLITHUS MACULATUS (Cuvier) SPOTTED JEWFISH

Otolithus maculatus Cuvier and Valenciennes, Hist. Nat. Poiss., v, 1830, p. 64; Bleeker, Atlas Ichth. Ind. Neerl., ix, 1877, Pl. ccclxxxiv, fig. 3 (plate 30).

D. IX-X; I, 30. A. II, 10-12. L. lat. 85, along above 95. Tr. (13-14)+(25-26). Depth 4½-5½. Head pointed. Three pores on snout. Operculum with 2 weak flat spines. Anal origin below middle of soft dorsal. Caudal wedge-shaped or rounded. Lateral line tubules simple. Grey above, golden below. 5-6 rows of black spots along body and on to caudal fin. Fins grey along their edges. 18 inches. Gulf of Manaar.

458. OTOLITHUS ARGENTEUS Cuvier SILVER-BANDED JEWFISH

Otolithus argenteus Cuvier and Valenciennes, Hist. Nat. Poiss., v, 1830, p. 62. *Otholithes argenteus* Ogilby, Mem. Qld. Mus., vi, 1918, Pl. xx (plate 30).

D. X; I, 27-31. A. II, 7. L. lat. 54, along above 80, along below 70-75. Tr. (7-8)+(10-12). Rather slender. Depth 4-4½. G.R. 5+8. Snout without pores. Operculum with 2 blunt spines. Anal base 5 in that of soft dorsal; its origin below 12th dorsal ray. Caudal cuneiform or in young, lanceolate. Silvery with 4 longitudinal bands along sides. A dark spot on opercle. Pectorals, ventrals and anal orange. Outer edge of dorsal grey. 2½ feet. Coastal waters and Gulf of Manaar.

459. OTOLITHUS RUBER (Schneider) ROSY JEWFISH

Johnius ruber Bloch and Schneider, Syst. Ichth., 1801, p. 75. *Otolithus ruber* or *rubra. Otolithus aureus* Fowler, Proc. Acad. Nat. Sci. Phila., 1929, fig. 2 (plate 30).

D. X; I, 28-31. A. II, 7. L. lat. 50-54, along above 60-65, along below 54. Tr. 7+17. G.R. 9+16. Depth 3½-4½. Operculum with 2 flat spines. Lateral line tubules branched. Anal originates behind middle of soft dorsal. Caudal wedge-shaped in young but round to truncate in older fish. Brownish-red shot with silver and sometimes golden below. Spinous dorsal with black edge. Soft dorsal and anal with grey margins. Other fins yellow. 2½ feet. Coastal waters.

Mundalava, Pannava (S).

RUDDER FISH, PIGFACE BREAM, SWEETLIPS

Family LETHRINIDAE*

Moderately compressed perch-like fish with a rather long scaleless snout. Mouth moderate with small teeth, a few of the front ones enlarged and caniniform. Palate normally not toothed. Dorsal always 10 spines and 8-9 rays. Scales moderate. Caudal forked.

1. Snout extremely long and pointed; maxillary reaches barely half way to eye .... *Lethrinella miniata*
   Snout not greatly elongated; maxillary reaches more than half way to eye 2
2. Black lateral blotch on body 3
   No black lateral blotch on body 6
3. Dark vertical band on cheek .... *Lethrinus variegatus*
   No dark vertical band on cheek 4
4. Dark lateral blotch faded in adults; each scale with a white spot .... *Lethrinus nebulosus*
   Dark lateral blotch persistent in adults 5
5. Small black blotch before middle of pectoral .... *Lethrinus frenatus*
   Large black blotch at pectoral end .... *Lethrinus rhodopterus*
   Black blotch behind pectoral end .... *Lethrinus reticulatus*
6. Blue streaks before eye .... *Lethrinus microdon*
   No blue streaks before eye 7
7. Body with longitudinal bands 8
   Body without longitudinal bands .... *Lethrinus mahsenoides*
8. Two pale longitudinal bands .... *Lethrinus ramak*
   Five or six reddish or yellow longitudinal bands .... *Lethrinus ornatus*

460. LETHRINELLA MINIATA (Schneider) LONG-NOSED PIGFACE BREAM

Sparus miniatus Bloch and Schneider, Syst. Ichth., 1801, p. 281. *Lethrinus rostratus. Lethrinus miniatus* Herre and Montalban, Philipp. Jour. Sci., xxxiii (4), 1927, Pl. i, fig. 3 (plate 31).

D. X; 8-9. A. III, 8. L. lat. 44-48. Tr. (5-6)+(17-19). G.R. 4+(5-8). Depth 3½-4. Snout very long and slender. Olive-brown, paler below. Dark brown blotch behind posterior upper angle of eye and narrow oblique dark bands from eye and opercle to upper jaw. Large blotch on opercle. Inside of mouth orange. Each scale with blue spot and 8 faint vertical bands on upper half of body. Fins with indistinct bands. 3 feet, usually about 12 inches. Pearl banks and trawling grounds.

Hota ula, Ura hota, Palu hakka (S). Thinan (T).

461. LETHRINUS VARIEGATUS Valenciennes VARIEGATED PIGFACE BREAM

Lethrinus variegatus Cuvier and Valenciennes, Hist. Nat. Poiss., vi, 1830, p. 287; Herre and Montalban, Philipp. Jour. Sci., xxxiii (4), 1927, Pl. iii, fig. 1 (plate 31).

D. X; 9. A. III, 8. L. lat. 43-46. Tr. (4-5)+(15-16). G.R. 5+7. Depth 4-4½. Maxilla reaches to below posterior nostril. Dull olive-brown. Young specimens with brownish band from snout through eye. Older fish have indistinct dark cross-bands. Fins yellowish and often with darker cross-bands. 15 inches. Coastal waters.

* *Lethrinus undulosus* is listed from the Gulf of Manaar by Pearson (1913), p. 2, but it is uncertain as to which species this refers.

462. LETHRINUS NEBULOSUS (Forskål) STARRY PIGFACE BREAM

Sciaena nebulosa Forskål, Descript. Animal., 1775, p. 52. *Lethrinus karwa* or *opercularis. Lethrinus nebulosus* Weber and De Beaufort, Fish. Indo-Aust. Archipel., vii, 1936, fig. 84 (plate 31).

D. X; 9. A. III, 8. L. lat. 45-48. Tr. (6-7)+(15-17). G.R. (5-6)+5. Depth 3. Body deepens and thickens with age. Mouth slightly oblique, jaws equal and corner of mouth below anterior nostril. Bluish bands from eye to posterior nostril and jaw. Preopercle spotted blue. Inside mouth orange. Opercular membrane yellow to red. Body olive with white spot on each scale. Young specimens with darkish spot between pectoral and lateral line. Fins yellowish; dorsal edged with reddish, others dark with narrow bluish bands. 30 inches. Coastal waters entering harbours, pearl banks and trawling grounds.
Mivatiya, Pulli mevativa (S). Vella meen, Pulli vella meen (T).

463. LETHRINUS FRENATUS Valenciennes BRIDLED PIGFACE BREAM

Lethrinus frenatus Cuvier and Valenciennes, Hist. Nat. Poiss., vi, 1830, p. 291. *Lethrinus cinereus. Lethrinus richardsoni* Herre and Montalban, Philipp. Jour. Sci., xxxiii (4), 1927, Pl. ii, fig. 2 (plate 31).

D. X; 9. A. III, 8. L. lat. 47-48. Tr. 6+17. G.R. 5+5. Depth 3-3¼. Olive brown with a quadrangular black blotch between pectoral fin and lateral line. Faint vertical cross-bars present but often broken into blotches. White or blue spots on scales of upper half of body. Dorsal with brown spots and red margin. Pectoral pink. Other fins dusky with indistinct bands. Inside mouth orange. 9 inches. Coastal waters.

464. LETHRINUS RHODOPTERUS Bleeker ROSE-FINNED PIGFACE BREAM

Lethrinus rhodopterus Bleeker, Nat. Tijds. Ned. Ind., iii, 1852, p. 65. (?) *Lethrinus erythrurus. Lethrinus harak* Herre and Montalban, Philipp. Jour. Sci., xxxiii (4), 1927, Pl. iii, fig. 2 (plate 31).

D. X; 9. A. III, 8. L. lat. 45-48. Tr. (5-6)+(14-16). G.R. 5+5. Depth 3¼-3½. Jaws equal with angle of mouth below vertical between nostrils. Greenish olive with oblong blackish blotch below lateral line opposite middle of dorsal fin. Fins yellowish, but dorsal tends to be rosy especially along margin. Caudal and anal olive with a rosy tinge. Inside mouth red. 20 inches. Trawling grounds.

465. LETHRINUS RETICULATUS Valenciennes RETICULATED PIGFACE BREAM

Lethrinus reticulatus Cuvier and Valenciennes, Hist. Nat. Poiss., vi, 1830, p. 298. *Lethrinus moensi* Herre and Montalban, Philipp. Jour. Sci., xxxiii (4), 1927, Pl. i, fig. 1 (plate 31).

D. X; 9. A. III, 8. L. lat. 45-47. Tr. 6+15. G.R. 5+5. Depth 3. Maxillary reaches opposite front nostril. Head reddish with 2-3 brown bands above eye. Body olive with 7-8 vertical diffuse cross-bars on back and sides. A dark blotch behind end of pectoral. Fins greyish. 12 inches. Coastal waters.

466. LETHRINUS MICRODON Valenciennes
SMALL-TOOTHED PIGFACE BREAM

Lethrinus microdon Cuvier and Valenciennes, Hist. Nat. Poiss., vi, 1830, p. 295. *Lethrinus rostratus*.

D. X; 9. A. III, 8. L. lat. 47. Tr. 5+16. Depth $3\frac{3}{4}$-4. Snout rather pointed. Uniform greenish. Muzzle clouded. Traces of bluish streaks in front of eye. Young specimens with indistinct cross-bars and a dark blotch between lateral line and middle of pectoral. Fins darkish. 12 inches. Trawling grounds.

467. LETHRINUS MAHSENOIDES Valenciennes
WHITE-SPOTTED PIGFACE BREAM

Lethrinus mahsenoides Cuvier and Valenciennes, Hist. Nat. Poiss., vi, 1830, p. 286. *Lethrinus mahsena. Lethrinus insulindicus* Herre and Montalban, Philipp. Jour. Sci., xxxiii (4), 1927, Pl. v, fig. 1 (plate 31).

D. X; 8-9. A. III, 8. L. lat. 48. Tr. 5+13. Depth $2\frac{3}{5}$. Pale green with a white spot at angle of every scale. Numerous white points on cheek. Membranous edge to opercle red. Vertical fins greenish. 8 inches. Coastal waters.

468. LETHRINUS RAMAK (Forskål) YELLOW-BANDED PIGFACE BREAM

Sciaena ramak Forskål, Descript. Animal., 1775, p. 52. *Lethrinus fasciatus. Lethrinus ramak* Günther, Jour. Mus. Godeffroy, ii (7), 1874, Pl. xlvi, fig. B (plate 31).

D. X; 9. A. III, 8. L. lat. 48. Tr. 6+16. G.R. 5+5. Depth equal to head. 3·4 in length. Brownish olive above, paler below. An indistinct broad axial band from upper gill-opening to middle of caudal base. Another narrower and shorter, from pectoral axil. Bands are yellowish formed by spots on scales. Fins brownish. 12 inches. Coastal waters and trawling grounds.
Hini gadeya (S). Vella meen (T).

469. LETHRINUS ORNATUS Valenciennes ORNATE PIGFACE BREAM

Lethrinus ornatus Valenciennes, Hist. Nat. Poiss., vi, 1830, p. 310. *Lutjanus ornatus. Lethrinus ornatus* Bleeker, Atlas Ichth. Ind. Neerl., viii, 1876-77, Pl. cccl, fig. 4 (plate 31).

D. X; 9-10. A. III, 8. L. lat. 47. Tr. (5-6)+15. Depth 3-$3\frac{1}{2}$. Lateral teeth rounded and molariform. G.R. (4-5)+5. Olive green with 6-7 yellowish horizontal bands. Membrane of opercle red. Caudal edged with red. Violet band across base of pectoral. Young specimens have dorsal and anal edged with red, and large irregular dark blotches on body. 18 inches. Pearl banks and trawling grounds.

LARGE-EYED BREAM

Family PENTAPODIDAE

Perch-like fishes with a noticeably large eye, a few knob-like gill-rakers and 10 dorsal spines. Caudal forked. Fine teeth in groups in both jaws and a few enlarged canines in front of each. No scales in interorbital area, preopercular flange, or on bases of soft dorsal and anal.

1. Teeth at side of jaw are all rounded molars .... *Monotaxis grandoculis*
   Teeth at side of jaw all conical 2
2. Lower canines normal .... *Gymnocranius griseus*
   Lower canines flare outwards .... *Gnathodentex aurolineatus*

470. MONOTAXIS GRANDOCULIS (Forskål)

ROUND-TOOTHED LARGE-EYED BREAM

Sciaena grandoculis Forskål, Descript. Animal., 1775, p. 53. *Sphaerodon heterodon* or *grandoculis. Monotaxis grandoculis* Weber and De Beaufort, Fish. Indo-Aust. Archip., vii, 1936, fig. 73 (plate 28).

D. X; 10. A. III, 9. L. lat. 45-47. Tr. (5-6)+(14-15). G.R. 6+5. Depth 3-3¼. No scales between eyes or on preopercular flange. 6 canines in upper, 4-6 in lower jaw, with 6-7 molars on each side of both jaws. Adults with very massive heads. Eye 2·4-3·3 in head in young, increasing to 1·2-1·5 in adults. Silvery grey to brown above, often with variously coloured spots about the body. Fins reddish. Lips yellow. 30 inches. Coastal reefs and trawling grounds.

Angana (S).

471. GYMNOCRANIUS GRISEUS (Schlegel)

NAKED-HEADED LARGE-EYED BREAM

Dentex griseus Schlegel, Fauna Japonica, Poiss., 1843, p. 72. *Dentex rivulatus, Gymnocranius robinsoni. Gymnocranius griseus* Weber and De Beaufort, Fish. Indo-Aust. Archip., vii, 1936, fig. 76 (plate 28).

D. X; 9-11. A. III, 9-10. L. lat. 47-50. Tr. 6+18. Depth 2·6-2·8 in young to 3·0-3·1 in adults. No scales on interorbital, preopercular flange, or on soft dorsal or anal. Snout normal in juveniles increasingly steep with age, developing a prominent bulge. Eye large in adults. 6 moderate canines in front of both jaws, conical teeth laterally and smaller teeth inside. G.R. 4+5. Juveniles silvery olive with some banding. Adults uniform silvery brown with band through eye. 20 inches. Coastal reefs and trawling grounds.

472. GNATHODENTEX AUROLINEATUS (Lacépède)

GOLD-LINED LARGE-EYED BREAM

Sparus aurolineatus Lacépède, Hist. Nat. Poiss., iv, 1803, p. 132. *Pentapodus aurolineatus. Pentapus aurolineatus* Bleeker, Atlas Ichth. Ind. Neerl., ix, 1877, Pl. cccxviii, fig. 3 (plate 28).

D. X; 10. A. III, 8-9. L. lat. 75. Tr. 6+20. G.R. 6+6. Depth 3·4. Eye 2·8 in head, equal to snout and interorbital. No scales on interorbital, preopercular flange, or on soft dorsal or anal. 4 close-set canines in upper jaw. An outward flaring canine on both sides of lower jaw and smaller pointed teeth between. Small sharp teeth laterally in jaws. Strong serrated bony ridge on maxilla. 18 inches. Coastal waters.

SILVER BREAMS

Family SPARIDAE

Perch-like fishes with well-developed fins. Dorsal with 10-13 stout spines and 10-15 soft rays. Anal with 3 spines. Snout never scaly and preopercular margin never serrate. A group of enlarged teeth in the front of each jaw followed by several rows of rounded molars. Carnivorous species inhabiting shallow waters.

1. Head between eyes wholly scaly .... *Argyrops spinifer*
   Head between eyes not scaly 2
2. One enlarged molar posteriorly in each jaw; 7 rows of scales above lateral line; 13 dorsal and 11 anal rays .... *Rhabdosargus sarba*
   No single greatly enlarged molar posteriorly; 4-5 rows of scales above lateral line; 12 dorsal and 9 anal rays 3
3. Head profile straight; lateral line highest under 6th-7th dorsal spines; body dusky without longitudinal streaks .... *Acanthopagrus berda*
   Head profile gibbous at eye; lateral line highest under 3rd-5th dorsal spines; body silvery with longitudinal streaks along the scale rows .... *Acanthopagrus latus*

473. ARGYROPS SPINIFER (Forskål) — LONG-SPINED RED BREAM

Sparus spinifer Forskål, Descript. Animal., 1775, p. 32. *Pagrus spinifer. Argyrops spinifer* Chevey, Mem. Inst. Oceanogr. Indochine, iv, 1932, Pl. xxxiv (plate 31).

D. XI-XII; 10-11. A. III, 8. L. lat. 49-53. Tr. 8+17. G.R. 9-10. Depth 1¾-2. First 2 dorsal spines short, 3rd-7th elongated; in young longer than whole fish, relatively shorter in adults. Snout bluntly convex. Scales present between eyes. Preopercular flange naked. Generally rosy or silvery with pink cross-bars and fins. Iris and opercular margin reddish. 24 inches. Trawling grounds.

Atissa, Mattiya, Seppiliya (S). Scheppili, Punanthalai (T).

474. RHABDOSARGUS SARBA (Forskål) — TARWHINE; SILVER BREAM

Sparus sarba Forskål, Descript. Animal., 1775, p. 31. *Chrysophrys sarba. Rhabdosargus sarba* Munro, Mem. Qld. Mus., xii (4), Pl. xxi (plate 31).

D. XI; 13-14. A. III, 11-12. L. lat. 55-70, normally 62-65. Tr. 7+(12-16). G.R. (5-7) +(6-10). No scales on preopercular flange. In front of upper jaw 4-6, in lower jaw 6-8 compressed incisors, followed by 5 and 4 rows of rounded molars respectively. Posterior molar in each jaw greatly enlarged. Silvery with golden longitudinal lines along scale rows. A yellow streak across belly. Caudal yellowish. Ventrals and anal dusky to silvery. 18 inches. Coastal waters.

475. ACANTHOPAGRUS BERDA (Forskål) — BLACK BREAM

Sparus berda Forskål, Descript. Animal., 1775, p. 32. *Chrysophrys berda* or *hasta. Mylio berda* Munro, Mem. Qld. Mus., xii (4), Pl. xviii (plate 31).

D. XI; 11-12. A. III, 8-9. L. lat. 46-54. Tr. 4+(11-12). G.R. (5-8)+(9-11). Six peg-like curved incisors in front of both jaws. Rounded molars in 4-5 series in upper and 3-4 series in lower jaw. Molars small anteriorly becoming progressively larger posteriorly. Usually black but specimens living in the sea much lighter to silvery. Young are olive with scale rows marked with longitudinal streaks but lack vertical cross-bars. Fins silvery to dusky. Pectoral axilla with black spot. 30 inches, normally much less. Coastal waters and lagoons.

Thiraliya (S).

476. ACANTHOPAGRUS LATUS (Houttuyn) JAPANESE SILVER BREAM

Sparus latus Houttuyn, Verh. Holl. Maatsch. Wet. Haarlem, 1782, p. 322. *Chrysophrys* or *Sparus datnia. Sparus latus* Tanaka, Fish. Japan, xv, 1914, Pl. lxxv, fig. 262 (plate 31).

D. XI; 11-12. A. III, 8-9. L. lat. 48-52. Tr. (4-5)+(11-14). G.R. (5-7)+(8-9). Teeth as in the preceding species. Silvery with golden longitudinal streaks along scale rows. Young with vertical cross-bars on back. Caudal and spinous dorsal with broad black margins. Pectoral axilla without a black spot. A dark area at origin of lateral line. Caudal yellow. Anal and ventrals yellowish. 18 inches. Coastal waters.

Atavara, Gal bateya, Mattivaya (S). Karrupu mattavan (T).

RED MULLET, SURMULLET, GOATFISH

Family MULLIDAE

Moderately elongate fishes with robust body. Mouth small with feeble teeth. 2 simple long barbels on chin. Scales large. Spinous dorsal with 6-8 spines, separated from soft dorsal. Anal short with 1-2 spines. Estuarine and shallow water species, living on bottom and using their feelers in the search for food. Mostly brilliantly coloured.

1. Teeth in both jaws on vomer and palatines 2
   No teeth on palate 4
2. Head and body with black spots .... *Upeneus tragula*
   Head and body without black spots 3
3. Large triangular reddish blotch at caudal base; dorsals each with 3 brownish horizontal bands; 6 dark bars across each caudal lobe .... *Upeneus taeniopterus*
   Bright yellow band from eye to caudal and another from pectoral axil to caudal; caudal uniform .... *Upeneus sulphureus*
   4-5 longitudinal yellow bands on sides; caudal with 4-6 oblique black bands on each lobe .... *Upeneus vittatus*
4. Teeth in jaws in several rows, small .... *Mulloidichthys auriflamma*
   Teeth in jaws uniserial, strong 5
5. One or more black blotches on side of body 6
   Body without black spots or blotches 10
6. Body with black cross-bars 7
   Body without black cross-bars 8
7. Body with 3 broad cross-bars; no yellow area before caudal peduncle .... *Parupeneus bifasciatus*
   Body with 4-5 black cross-bars; yellow area before caudal peduncle .... *Parupeneus trifasciatus*
8. Dark streak through eye to below soft dorsal fin 9
   Pale area below spinous dorsal fin .... *Parupeneus indicus*
9. Second dorsal spine rigid and pungent .... *Parupeneus macronemus*
   Second dorsal spine flexible, not pungent .... *Parupeneus barberinus*
10. Pale or white blotch behind soft dorsal .... *Pseudupeneus fraterculus*
    No pale blotch behind soft dorsal 11
11. Four blue horizontal lines on head; each body scale with a pale spot .... *Pseudupeneus luteus*
    Vermilion; 2 rows of scales above and 2 rows below lateral line with silvery spots .... *Pseudupeneus cinnabarinus*

477. UPENEUS TRAGULA Richardson BLACK-STRIPED GOATFISH

Upeneus tragula Richardson, Rept. 15th Meet. Brit. Assoc. Adv. Sci., 1846, p. 220. *Upeneoides tragula* Day, Fish. India, 1875-78, Pl. xxx, fig. 4 (plate 32).

D. VII-VIII; I, 7-8. A. I, 6-7. L. lat. 30-32. Tr. 2+6. G.R. 16-18. Depth 4½. Reddish or brownish above, lighter below. Head and body with brownish or blackish dots. A black lateral stripe from snout through eye to caudal. Spinous dorsal with black margin. Caudal with 5 oblique reddish or blackish bars. 10 inches. Pearl banks.

478. UPENEUS TAENIOPTERUS Cuvier — BARRED-FIN GOATFISH

Upeneus taeniopterus Cuvier and Valenciennes, Hist. Nat. Poiss., iii, 1829, p. 451. *Upeneoides taeniopterus.*

D. VII; I, 7. A. 7. L. lat. 38. Tr. 3+7. Depth 4¾. Reddish on back. Abdomen white. A large triangular red spot on the free portion of the tail. Spinous dorsal with 3 brownish longitudinal bands; soft dorsal similarly banded. Caudal with 6 oblique streaks across each lobe. 12 inches. Harbours.

479. UPENEUS SULPHUREUS Cuvier — YELLOW GOATFISH

Upeneus sulphureus Cuvier and Valenciennes, Hist. Nat. Poiss., iii, 1829, p. 450. *Upeneoides sulphureus. Upeneus sulphureus* Smith, Fishes Sth. Africa, 1949, Pl. xxviii, fig. 563 (plate 32).

D. VIII; I, 8. A. I, 6-7. L. lat. 33-37. Tr. 3+(7-8). G.R. 19-22. Depth 3¼-3⅞. Reddish-chestnut on back, becoming silvery on abdomen which is sometimes yellowish. Purplish blotch on opercle and subopercle. A bright yellow stripe from eye to caudal and one or more below and parallel to it. Spinous dorsal with black tip and 2 olive bars. Caudal reddish to yellow. Anal and ventrals yellow. 8 inches. Trawling grounds.

480. UPENEUS VITTATUS (Forskål) — YELLOW-STRIPED GOATFISH

Mullus vittatus Forskål, Descript. Animal., 1775, p. 31. *Upeneoides vittatus* Day, Fish. India, 1875-78, Pl. xxx, fig. 2 (plate 32).

D. VIII; I, 8. A. I, 6-7. L. lat. 33-38. Tr. 3+7. G.R. 16-20. Depth 4. Colours variable. Back chestnut. Golden below. 4-5 bright yellow longitudinal stripes along each side. First dorsal edged with black and with 2 dark bands. Pectoral pinkish. Upper caudal lobe crossed by 6 yellowish-brown bars and with a black tip. Lower caudal lobe with 3 such oblique bars and a white tip. 12 inches. Coastal waters, pearl banks and trawling grounds. Kaha iri nagareya (S). Manjel kithu nakharai (T).

481. MULLOIDICHTHYS AURIFLAMMA (Forskål) — GOLDEN-BANDED GOATFISH

Mullus auriflamma Forskål, Descript. Animal., 1775, p. 30. *Upeneus* or *Upeneoides zeylonicus, Mulloides ceylonicus* or *flavolineatus. Upeneoides flavolineatus* Day, Fish. India, 1875-78, Pl. xxx, fig. 6 (plate 32).

D. VII; I, 8. A. I-II, 6. L. lat. 35-38. Tr. 2+7. G.R. 20. Depth 4. Back reddish brown; sides white; belly tinged with yellow. Narrow yellow band from upper edge of eye to snout with a second band below it. A brilliant golden band from eye to middle of caudal base. Fins pinkish with yellow band along base of soft dorsal. 16 inches. Harbours.

482. PARUPENEUS BIFASCIATUS (Lacépède) TWO-BARRED GOATFISH

Mullus bifasciatus Lacépède, Hist. Nat. Poiss., iii, 1802, p. 383. *Upeneus bifasciatus* Günther, Jour. Mus. Godeffroy, ii (7), 1874, Pl. xliv, fig. A (plate 32).

D. VIII; I, 8. A. II, 6. L. lat. 28-31. Tr. 3+(6-7). G.R. 30. Depth 3-3⅙. Pale brown, darker on back. Body with broad dusky brown to black cross-bars, passing through spinous dorsal, soft dorsal and caudal peduncle. Spinous dorsal pale brown. Soft dorsal dark brown with 4-5 longitudinal pale brown bars. Anal similar with 6-7 such bands. Caudal dark brown. Coastal waters.

483. PARUPENEUS TRIFASCIATUS (Lacépède)

THREE-BARRED GOATFISH

Mullus trifasciatus Lacépède, Hist. Nat. Poiss., iii, 1802, p. 404. *Upeneus trifasciatus* Günther, Jour. Mus. Godeffroy, ii (7), 1874, Pl. xliv, fig. B (plate 32).

D. VIII; I, 8. A. I, 6-7. L. lat. 28-30. Tr. 2+6. G.R. 28. Depth 3¼-3½. Yellowish-brown, lighter below. A blackish-brown postocular blotch which may extend as a bar from snout through eye. Body with 4-5 blackish cross-bars, before spinous dorsal, under spinous dorsal, at dorsal interspace, under soft dorsal and on caudal peduncle, the latter being reduced to a rounded black spot. Spinous dorsal pale brown; soft dorsal similar but with black base and 2-3 brownish longitudinal stripes. Caudal pale brown. Ventrals yellowish. 9 inches. Coastal waters.

484. PARUPENEUS INDICUS (Shaw) INDIAN GOATFISH

Mullus indicus Shaw, General Zool., iv (2), 1803, p. 611. *Upeneus* or *Upeneoides indicus*. *Parupeneus indicus* Bleeker, Atlas Ichth. Ind. Neerl., ix, 1877, Pl. cccxciv, fig. 5 (plate 32).

D. VIII; I, 8-9. A. I-II, 6. L. lat. 27-30. Tr. 3+7. G.R. 17-20. Depth 3½-3⅗. Purplish red with large oval golden blotch below spinous dorsal, at level of lateral line. A purplish-black mark on caudal peduncle. Some light violet lines on upper part of head. A broad purple band from eye to snout. Cheeks with yellow and blue lines. Dorsal purplish streaked with blue. Yellow bands on anal. 16 inches. Trawling grounds and Gulf of Manaar.

Gal nagareya (S). Kal nakharai (T).

485. PARUPENEUS MACRONEMUS (Lacépède) LONG-RAYED GOATFISH

Mullus macronemus Lacépède, Hist. Nat. Poiss., iii, 1802, p. 400. *Upeneus macronemus* Day, Fish. India, 1875-78, Pl. xxxi, fig. 1 (plate 32).

D. VIII; I, 8. A. I, 6. L. lat. 28-30. Tr. 3+7. G.R. 24. Depth 3½. Black band from snout through eye to end of soft dorsal. A black blotch on caudal peduncle. Golden spots on body scales. Purplish stripe from eye to snout. Lower half of spinous dorsal purple. Soft dorsal with black basal area with 3 paler stripes above. Anal with 5-6 light longitudinal stripes. Blue spots on scales below lateral line. 14 inches. Coastal waters.

486. PARUPENEUS BARBERINUS (Lacépède)* ROSY GOATFISH

Mullus barberinus Lacépède, Hist. Nat. Poiss., iii, 1802, p. 283. *Upeneus barberinus. Parupeneus barberinus* Bleeker, Atlas Ichth. Ind. Neerl., ix, 1877, Pl. cccxciii, fig. 1 (plate 32).

D. VII; I, 8. A. I, 6. L. lat. 28-31. Tr. 2+7. G.R. 17-20. Depth 3⅗. A black band extends from eye along lateral line, and from below middle of soft dorsal above that line, terminating below end of soft dorsal. A round black spot on caudal peduncle. Some specimens with dark band along dorsal fins. General body colour pinkish. Dorsal fins grey, others pinkish. 20 inches. Coastal waters.

487. PSEUDUPENEUS FRATERCULUS (Valenciennes)

TWO-SPOT GOATFISH

Upeneus fraterculus Cuvier and Valenciennes, Hist. Nat. Poiss., vii, 1831, p. 524. *Upeneus dispilurus* Day, Fish. India, 1875-78, Pl. xxxi, fig. 3 (plate 32).

D. VIII; I, 8. A. I, 6. L. lat. 27-30. Tr. 3+7. G.R. 17-20. Depth 3-3½. Reddish with darker edges to the scales. A purplish band from eye to snout. The 2 scale rows above and 3 scale rows below lateral line with golden spots. A light golden area anteriorly on caudal peduncle often followed by a small or large black spot or saddle. 18 inches. Coastal waters.

488. PSEUDUPENEUS LUTEUS (Valenciennes)

GOLDEN-YELLOW GOATFISH

Upeneus luteus Cuvier and Valenciennes, Hist. Nat. Poiss., vii, 1831, p. 521. *Parupeneus luteus* Bleeker, Atlas Ichth. Ind. Neerl., ix, 1877, Pl. cccxciv, fig. 1 (plate 32).

D. VIII; I, 8-9. A. I, 6-7. L. lat. 30. Tr. 2+7. Depth 4. Rosy above, paler below. 4 bluish horizontal lines on head. Each body scale with a yellow spot. Fins rosy. Soft dorsal and anal with 3-4 yellowish longitudinal bands. Caudal yellowish posteriorly. 12 inches. Coastal waters and trawling grounds.

489. PSEUDUPENEUS CINNABARINUS (Cuvier) RED GOATFISH

Upeneus cinnabarinus Cuvier and Valenciennes, Hist. Nat. Poiss., iii, 1829, p. 475.

D. VIII; 9. A. 7. L. lat. 29. Tr. 3+7. Depth 4¼. Vermilion, darkest on back. Silvery central spot on each scale, forming 2 rows above and 2 rows below the lateral line. A large purple spot over opercle and subopercle. Dorsal and anal rays yellow and membranes reddish. Upper caudal lobe orange, lower one red. Barbels rosy. Coastal waters and trawling grounds.

SILVER BATFISH

Family MONODACTYLIDAE

Fishes with deep silvery compressed body, with fine scales extending over the fins. Mouth small with feeble teeth. Palatine teeth present. Fin spines reduced.

* Fowler, Bull. U.S. Nat. Mus. 100, xii, p. 278 includes Ceylon in the distribution of *P. barberinoides* (Bleeker) but this is apparently an error for *P. barberinus*.

490. MONODACTYLUS ARGENTEUS (Linnaeus) SILVER BATFISH

Chaetodon argenteus Linnaeus, Syst. Nat., ed. 10, 1758, p. 272. *Psettus argenteus* Richardson, Voy. Erebus and Terror, Ichth., 1844-48, Pl. xxxv, fig. 1 (plate 33).

D. VII-VIII; 28-30. A. III, 28-32. V. I, 2-4. L. lat. 50-60. G.R. 16-19. Depth $1\frac{1}{4}$-$1\frac{1}{2}$. Brilliant silvery all over. Young with 2 narrow vertical cross-bars of darker colour which fade with growth. 9 inches. Coastal waters and trawling grounds.

Kapu handha (S). Moolen (T).

SWEEPERS

Family PEMPHERIDAE

Body compressed. Scales large, deciduous, either ctenoid or cycloid. Eye relatively large, greater than interorbital. Mouth oblique with small teeth in bands in jaws and on palate. A single dorsal which is short and elevated. Anal long.

Teeth anteriorly in mandible directed outwards; distance from snout to dorsal origin subequal to depth of body; L.lat. with 64-66 tubules .... *Pempheris moluca*

No forward directed teeth in mandible; distance from snout to dorsal origin conspicuously less than body depth; L.lat. with 55 tubules .... *Pempheris vanicolensis*

491. PEMPHERIS MOLUCA Cuvier MOLUCCAN SWEEPER

Pempheris moluca Cuvier and Valenciennes, Hist. Nat., Poiss., vii, 1831, p. 304. *Pempheris mangula, mangula-kutti* or *macrolepidotus*. *Pempheris molucca* Day, Fish. India, 1875-78, Pl. xlii, fig. 2 (plate 32).

D. VI, 9. A. III, 41. L. lat. ca. 64-66. Depth 2·6-2·8. Head 4·1. Rostro-dorsal profile barely convex, upper rim of eye nearly level with it. Distance from snout to dorsal origin only slightly less than body depth. Silvery, brownish on back and coppery on flanks. Base of pectorals with brown or black patch. Vertical fins reddish. 6 inches. Gulf of Manaar.

Kimbissa (S).

492. PEMPHERIS VANICOLENSIS Cuvier VANIKORO SWEEPER

Pempheris vanicolensis Cuvier and Valenciennes, Hist. Nat. Poiss., vii, 1831, p. 305. *Pempheris mangula. Pempheris vanicolensis* Bleeker, Atlas Ichth. Ind. Neerl., ix, 1877, Pl. ccclxxxiii, fig. 6 (plate 32).

D. VI, 9. A. III, 39-43. L. lat. 55. Depth 2·6-2·8. Head 4·0-4·6. Distance of dorsal from tip of snout conspicuously less than body height. Teeth in jaws small, forming a rather narrow setiform band, the outer intermaxillary ones stronger than others. Brown, darker above, silvery below. Pectorals without basal darker patch. Dorsal with black tip. Anal and caudal with dusky margins. 7 inches. Coastal waters.

ARCHER FISH

Family TOXOTIDAE

Oblong compressed fishes with moderate sized ctenoid scales which extend on to head, soft dorsal, anal and caudal fins. Head flattened above with produced snout and protractile terminal mouth. Dorsal opposite anal, with 4-5 spines. Inhabits estuaries and enters fresh water. Renowned for its ability to shoot down insects some distance above the water surface, by rising and ejecting a drop of water from its mouth.

493. TOXOTES CHATAREUS (Hamilton-Buchanan) ARCHER FISH

Coius chatareus Hamilton-Buchanan, Fish. Ganges, 1822, p. 191. *Toxotes chatareus* Deraniyagala, Col. Atlas Vert. Ceylon, i, 1952, Pl. xxiii (plate 33).

D. V, 12-13. A. III, 16-18. P. 15. V. I, 5. L. lat. 33-35. Tr. 5+9. Depth 2·1. Head 2·7. Teeth on tongue granular, minute. Patches of granular teeth on vomer, palatines and pterygoids. Silvery with a yellowish tinge laterally. Back darker with some black blotches. Fins yellowish with dark margins to dorsal and anal. 8 inches. Fresh and brackish waters of coastal rivers and lagoons.
Dhimittha (S).

DRUMMERS

Family KYPHOSIDAE

Smoothly oval fishes with small soft scales covering head, body and fins. Mouth small, with an outer series of incisors and fine teeth in jaws, palate and on tongue. Small marine fishes inhabiting rocks and reefs.

Soft dorsal higher than spinous dorsal .... *Kyphosus cinerascens*
Soft dorsal lower than spinous dorsal .... *Kyphosus bigibbus*

494. KYPHOSUS CINERASCENS (Forskål) ASHEN DRUMMER

Sciaena cinerascens Forskål, Descript. Animal., 1775, p. 53. *Pimelepterus cinerascens* Bleeker, Atlas Ichth. Ind. Neerl., ix, 1877, Pl. ccclxiv, fig. 4 (plate 32).

D. XI; 12. A. III, 11-12. L. lat. 60. Tr. 10+18. G.R. 18-20. Depth 2¼. Olivaceous to silvery grey with a dark band between each row of scales. A silver band under the eye. Fins almost black. 18 inches. Coastal waters and Gulf of Manaar.

495. KYPHOSUS BIGIBBUS Lacépède STRIPED DRUMMER

Kyphosus bigibbus Lacépède, Hist. Nat. Poiss., iii, 1802, p. 115. *Kyphosus fuscus*. *Pimelepterus indicus* Klunzinger, Sitzb. Akad. Wiss. Wien., lxxx (1), 1879, Pl. vii (plate 32).

D. XI-XII; 12-14. A. III, 11-13. L. lat. 65. Tr. 12+20. G.R. 18-20. Depth 2¼. Dull, dusky or brownish, but occasionally golden, usually stripes along the scale rows. 30 inches. Coastal waters.

SPADEFISH

Family EPHIPPIDAE

Strongly compressed and elevated body, almost circular in shape and covered with moderately smooth scales which are smaller on head and extend on to vertical fins. Spinous dorsal differentiated from soft dorsal; first spine procumbent. Anal with 3 spines. Gill-membranes broadly united with isthmus. Mouth not protractile and maxilla not exposed. Pectorals short and rounded. A broad subocular shelf.

496. EPHIPPUS ORBIS (Bloch) SPADEFISH

Chaetodon orbis Bloch, Naturg. Ausländ. Fische, iii, 1787, p. 81. *Ephippus orbis* Herre and Montalban, Philipp. Jour. Sci., xxxiv (1), 1927, Pl. ii, fig. 2 (plate 33).

D. I, IX; 19-20. A. III, 15-16. L. lat. 35-38. Nearly circular in outline. Maxillary not quite reaching front of eye. G.R. 2+8. Pale green-brown, sides and belly silvery with leaden tinge. Fins pale yellow with dusky margins. 7½ inches. Gulf of Manaar, pearl banks and trawling grounds.

Hada (S). Thirali (T).

BATFISH

Family PLATACIDAE

Deep compressed fishes, almost circular in outline, and covered with small adherent scales. Spinous and soft dorsals confluent; no procumbent spine. Mouth not protractile and maxilla not exposed. Subocular shelf feeble. Anal with 3 spines. Gill-membranes broadly united with scaly isthmus. Body shape subject to considerable changes with age. Teeth tricuspid.

Cusps of teeth equal; anterior profile evenly convex; few teeth on vomer; 25-28 scales between arch of lateral line and dorsal .... *Platax teira*

Middle cusp of teeth larger than lateral cusps; anterior profile angular; no vomerine teeth; 20 scales between arch of lateral line and dorsal .... *Platax orbicularis*

497. PLATAX TEIRA (Forskål) LONG-FINNED BATFISH

Chaetodon teira Forskål, Descript. Animal., 1775, p. 60. *Platax teira* Bleeker, Atlas Ichth. Ind. Neerl., ix, 1877, Pl. ccclxxxii, fig. 1 (plate 33).

D. V; 30-33. A. III, 23-24. About 50 scales in horizontal line to caudal base. L. lat. 60-65. Tr. 25-30 above arch. Three cusps of teeth about equal in size. Soft dorsal and anal much produced and falcate in young, decreasing in size with age. Ventrals long and falcate. Grey or light brown with 3 broad cross-bars, passing through eye, pectoral and across soft dorsal and anal. 20 inches. Trawling grounds.

498. PLATAX ORBICULARIS (Forskål) ROUND BATFISH

Chaetodon orbicularis Forskål, Descript. Animal., 1775, p. 59. *Chaetodon vespertilio*, *Platax raynaldi*. *Platax vespertilio* Bleeker, Atlas Ichth. Ind. Neerl., ix, 1877, Pl. ccclxxx, fig. 3 (plate 33).

D. V; 34-38. A. III, 26-28. About 50 scales in horizontal line to caudal base, L. lat. 55. Tr. 20 above arch. The central cusp of teeth much longer and stronger than lateral cusps. Soft dorsal and anal less produced and barely falcate in young and becoming obtuse and rounded in older fish. Pectorals rounded. Banded as in preceding species but the first 2 cross-bars tend to be broader and the third more diffuse. Young have small widely spaced spots. 13 inches. Coastal waters.

Kottan thaleya, Kola handheya, Thaduva (S). Chellal (T).

SPOTTED BATFISH, SICKLE FISH

Family DREPANIDAE

Strongly compressed and elevated body, almost circular in shape with moderate scales. Spinous dorsal with strong spines, differentiated from soft dorsal; first spine procumbent. Anal with 3 spines. Gill-membranes united broadly with scaly isthmus. Mouth protractile and maxilla exposed distally. Pectorals falcate, elongate. No subocular shelf. Scales ciliated and medium-sized.

499. DREPANE PUNCTATA (Linnaeus) SPOTTED BATFISH

Chaetodon punctatus Linnaeus, Syst. Nat., ed. 10, 1758, p. 273. *Harpochirus punctatus* Bleeker, Atlas Ichth. Ind. Neerl., ix, 1877, Pl. ccclxv, fig. 4 (plate 33).

D. VIII-IX; 19-22. A. III, 17-19. L. lat. 50-55. Tr. 14-15 above. G.R. 10-11. Easily distinguished by the protractile mouth which when protruded projects downwards as a tube. Colour silvery with 4-10 series of dark spots arranged transversely and sometimes extending on to soft dorsal. In young the spots may be represented by darker cross-bars or may be entirely wanting. 16 inches. Coastal waters, Gulf of Manaar and trawling grounds.
Handeya, Pasinthiya (S). Painthi (T).

BUTTERFISH

Family SCATOPHAGIDAE

Deep, compressed angular body with firm skin and small ctenoid scales. Mouth small with bands of fine teeth in jaws but not on palate. Gill-membranes narrowly attached to isthmus. Anal with 4 spines. Spinous dorsal differentiated from soft dorsal; the first spine procumbent. Very young have bony plates on head and a large shoulder spine.

500. SCATOPHAGUS ARGUS (Linnaeus) SPOTTED BUTTERFISH

Chaetodon argus Linnaeus, Syst. Nat., ed. 12, 1766, p. 464. *Chaetodon atromaculatus*. *Scatophagus argus* Herre and Montalban, Philipp. Jour. Sci., xxxiv (1), Pl. ii, fig. 1 (plate 34).

D. XI; 16-18. A. IV, 14-15. Depth $1\frac{3}{4}$-2. L. lat. 95-120. Tr. 80 between dorsal and ventral fins. Rostro-dorsal profile strongly ascending and followed by a deep concavity above the eyes. Blue or greenish grey to dusky brown above with numerous large round brown spots which may extend on to soft dorsal. Belly silvery. Fins pink, yellowish to brown-grey. 12 inches. Estuarine waters and coastal lagoons.
Deva koraliya, Pulli ilathiva, Gal handheya, Vaharaliya (S). Ilethi, Pulli ilethi (T).

ANG ELFISH

Family POMACANTHIDAE

Compressed, fairly deep or elongate fishes with hard rough adherent scales; often with auxiliary scales, i.e. smaller ones over the normal scales. Mouth small with bristle-like teeth which are tricuspid. Preoperculum with serrated margin and strong spine at the angle. Fin spines well developed. No elongate scale in pelvic axil. Mostly small brilliantly coloured species that frequent coral reefs.

1. L.lat. 46-50; scales arranged in regular series 2
   L.lat. 70-90; scales arranged in irregular series 4
2. Interoperculum large; interorbital greater than eye 3
   Interoperculum small; interorbital equal or less than eye .... *Centropyge multispinis*
3. Lateral line complete; caudal rounded .... *Euxiphipops xanthometopon*
   Lateral line terminating at end of second dorsal; caudal truncate .... *Apolemichthys xanthurus*
4. Adults brightly coloured and strongly patterned with longitudinal or oblique stripes on sides 5
   Adults comparatively dull, young with vertical stripes .... *Pomacanthodes semicirculatus*
5. No auxiliary scales; 90 scales to caudal; deep brown or blue with 15-25 somewhat oblique yellow stripes on sides .... *Pomacanthodes imperator*
   Auxiliaries arranged in row around base of scales; 70 scales to caudal; deep yellow with about 6 strongly oblique curved blue stripes and a blue humeral ring .... *Pomacanthodes annularis*

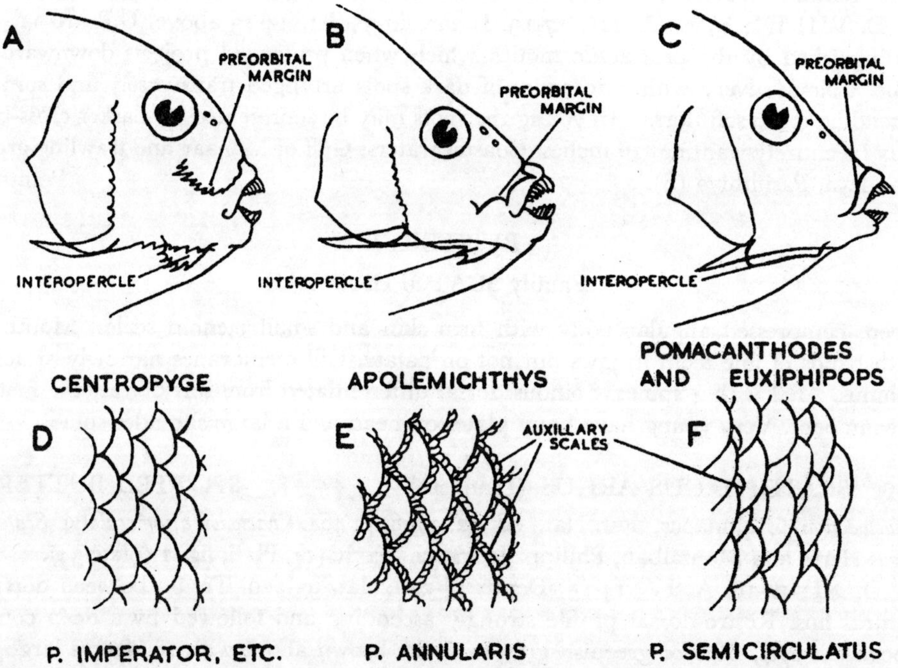

FIG. 12.

Explanation of characters of the Family Pomacanthidae (Angelfish). a-c.—Relative development of preorbital and interopercular bones in the various genera. d-f.—Types of scale arrangement, showing development of auxiliary scales.

501. CENTROPYGE MULTISPINIS (Playfair) SPINY ANGELFISH

Holacanthus multispinis Playfair, Fish. Zanzibar, 1866, p. 37, Pl. vi, fig. 4 (plate 33).

D. XIV; 15. A. III, 16. L. lat. 45-48, terminating at end of soft dorsal. Depth $1\frac{3}{4}$-$1\frac{5}{8}$. Head $3\frac{3}{4}$-4. Interoperculum small, curved, with 1-5 strong backwardly directed spines, the number varying on each side of the individual. Dark brown, numerous indistinct transverse streaks on side; a black patch larger than eye above pectoral. Dorsal and anal black; caudal black, narrowly edged with white. 4½ inches. Coastal waters.

502. EUXIPHIPOPS XANTHOMETOPON (Bleeker)
YELLOW-FACED ANGELFISH

Holacanthus xanthometopon Bleeker, Nat. Tijds. Ned. Ind., iv, 1853, p. 258; Atlas Ichth. Ind. Neerl., ix, 1877, Pl. ccclxxi, fig. 2 (plate 33).

D. XIV; 16-18. A. III, 16-18. L. lat. 46-50 in horizontal series. Depth 1·7-1·9. Head 2·4-3·5. Spinous dorsal much longer than soft dorsal. Caudal rounded. No auxiliary scales on trunk scales. Head blue with black-edged golden spots on opercle and cheek. Fine black lines on lips and chin. Interorbital with broad yellow band. Body violet with a brilliant blue spot on each scale and a yellow shoulder spot. Dorsal, caudal and pectorals yellow. A black spot at base of last 7 dorsal rays. Caudal with black edge. Ventrals and anal white with blue edge. 15 inches. Gulf of Manaar.

503. APOLEMICHTHYS XANTHURUS (Bennett)
YELLOW-BROWN ANGELFISH

Holacanthus xanthurus Bennett, Proc. Zool. Soc., 1832, p. 183; Day, Fish. India, 1875-78, Pl. xxix, fig. 2 (plate 33).

D. XIV; 19. A. III, 18. Depth 1⅔. Head 4. L. lat. 48, terminating at end of soft dorsal. Caudal truncate. Light brown, each scale on upper part of body with a pale edge, those below with white centres. Dorsal and anal rich brown. Pelvics and caudal yellow. 6 inches. Coastal waters.

Valgai kahi namba (S). Mawjel meen (T).

504. POMACANTHODES SEMICIRCULATUS (Cuvier) BLUE ANGELFISH

Holacanthus semicirculatus Cuvier and Valenciennes, Hist. Nat. Poiss., vii, 1831, p. 191. *Holocanthus alternans* and var. *meleagris*. *Acanthochaetodon semicirculatus* Bleeker, Atlas Ichth. Ind. Neerl., ix, 1877, Pl. ccclxx, fig. 5 (plate 33, A, young). *A. lepidolepis* Bleeker, *ibid.*, Pl. ccclxxii, fig. 1 (plate 33, B, adult).

D. XIII; 21-23. A. III, 20-21. L. lat. 75 in horizontal series. Depth 1·5-1·7. Head 3·3-4·0. Trunk scales moderate, unequal and without auxiliaries. Caudal rounded. Colour pattern develops with growth. Final adult stage is yellowish anteriorly and greyish posteriorly, the sides irregularly mottled with dark spots. Soft dorsal, anal and caudal blackish often with rows of white dots and blue margins. Head light or dark with blue vertical bars. Young are brilliant blue with semicircular shaped narrow bands of white or paler blue. 15 inches. Palk Straits.

505. POMACANTHODES IMPERATOR (Bloch) IMPERIAL ANGELFISH

Chaetodon imperator Bloch, Naturg. Ausländ. Fische, iii, 1787, p. 51. *Holacanthus* or *Holocentrum imperator*. *Acanthochaetodon imperator* Bleeker, Atlas Ichth. Ind. Neerl., ix, 1877, Pl. ccclxvii, fig. 2 (plate 33, B, adult). *A. nicobariensis* Bleeker, *ibid.*, Pl. ccclxv, fig. 1 (plate 33, A, young).

D. XIV; 19-21. A. III, 18-21. L. lat. about 90 in horizontal row. Depth in adult 1·6-1·8. Head 3·6-4·0. Trunk scales without true auxiliaries. Caudal convex. Colour pattern changes with growth. Young specimens (sometimes considered a separate variety, *nicobariensis*) are black with concentric streaks of white or blue, the last usually forming a complete white ring before the tail. Dorsal and anal fins with vermiculated lines. Adults purplish brown with black patch above pectoral extending on to chest and throat. From this run numerous oblique yellow bands towards soft dorsal, caudal and anal fins. Older fish with a brown band through eye. Paired fins yellow. 15 inches. Coastal rock pools, Gulf of Manaar and trawling grounds.
Kaka iri namba (S).

506. POMACANTHODES ANNULARIS (Bloch) RINGED ANGELFISH

Chaetodon annularis Bloch, Naturg. Ausländ. Fische, iii, 1787, p. 114. *Holacanthus annularis*. *Acanthochaetodon annularis* Bleeker, Atlas Ichth. Ind. Neerl., ix, 1877, Pl. ccclxx, fig. 2 (plate 33).

D. XIII; 21-22. A. III, 20-21. L. tr. about 70 in horizontal series. Depth 1·5-1·6. Head 3·5-4·0. Trunk scales moderate, rounded, with several auxiliaries forming a basal rim. Series somewhat irregular. Caudal rounded. Yellowish or greyish brown with 5-7 curved blue lines radiating obliquely from pectoral to soft dorsal, the last crossing caudal peduncle. Two similar horizontal lines across head, below and above eye. A blue irregular ring on shoulder. Caudal and pectorals yellow. Ventrals greyish. 15 inches. Gulf of Manaar and trawling grounds.

CORALFISH, BUTTERFLY FISH

Family CHAETODONTIDAE

Deep compressed fishes with moderate scales. Dorsal and anal well developed and scaly. Anal with 3-4 spines. Spinous dorsal without procumbent spine. Mouth protractile, small, with fine bristle-like teeth. Preoperculum without a strong spine in adults. Ventrals with an axillary scale. Small brilliantly coloured fishes living mostly around coral reefs.

1. Dorsal with 6 rather weak spines, gradually increasing in height posteriorly ...... *Parachaetodon ocellatus*
   Dorsal with 11-13 strong spines, the middle ones longest 2
2. Fourth dorsal spine much prolonged .... *Heniochus acuminatus*
   None of dorsal spines prolonged 3
3. All scales on body similar in size and shape; small to moderate and evenly rounded and arranged more or less regularly in horizontal rows 4
   Scales of two types on body; larger rhombic ones anteriorly and smaller rounded ones posteriorly; longitudinal rows of large ones indistinct and transverse rows of small ones very pronounced 8
4. Spinous dorsal nearly twice length of soft dorsal 5
   Spinous dorsal less than one and one half times length of soft dorsal 6
5. Dark ocular band present .... *Rhabdophorus trifasciatus*
   No dark ocular band .... *Rhabdophorus xanthocephalus*
6. Oblique lines of spots across body and clusters of smaller spots around dorsal, anal and caudal bases .... *Tetragonoptrus guttatissimus*
   Oblique lines across body; no clusters of small spots around fin bases 7

7. A white collar behind broad ocular band; no large dark spots on caudal peduncle ...... *Chaetodontops collaris*
   No white collar behind a narrow ocular band; two large dark spots on caudal peduncle .... *Chaetodontops melanotus*
8. Large round (usually ocellated) black spot over lateral line below posterior dorsal spines .... *Lepidochaetodon unimaculatus*
   No large round spot on back 9
9. Soft dorsal with long pointed filament and large black spot over posterior rays .... *Linophora auriga*
   Soft dorsal without long pointed filament and large black spot 10
10. Black ocular bar incomplete; a second short predorsal bar; no broad bands posteriorly on fins .... *Linophora mertensi*
    Black ocular bar complete; no second short predorsal bar; broad bars posteriorly across fins 11
11. Broad dark band along base of soft dorsal across caudal peduncle on to anal; 6 dark oblique streaks from operculum to dorsal, the lowest at right angles to another series directed obliquely downwards and backwards .... *Linophora vagabundus*
    No broad dark band along base of soft dorsal; sides checkered by oblique lines crossing one another .... *Linophora rafflesi*

507. PARACHAETODON OCELLATUS (Cuvier) OCELLATED CORALFISH

Platax ocellatus Cuvier and Valenciennes, Hist. Nat. Poiss., vii, 1831, p. 229. *Chaetodon oligacanthus*. *Parachaetodon ocellatus* Bleeker, Atlas Ichth. Ind. Neerl., ix, 1877, Pl. ccclxxvii, fig. 4 (plate 34).

D. VI; 28-30. A. III, 19-20. L. lat. 40. Tr. (15-16)+(20-21). G.R. 5+9. Depth 1¼. Head 3. Recognisable by the short dorsal fin which gradually increases in height posteriorly. Body yellowish. A narrow dark-edged band along dorsal profile from snout. A similar ocular band across head to breast. Body with three curved vertical cross-bars, the posterior with a black white-edged ocellus on the part crossing the soft dorsal fin. Caudal peduncle with a white-bordered vertical band. Fins otherwise yellowish. Younger specimens more silvery and ocellus more diffuse. 5½ inches. Coastal waters.

508. HENIOCHUS ACUMINATUS (Linnaeus) PENNANT CORALFISH

Chaetodon acuminatus Linnaeus, Syst. Nat., ed. 10, 1758, p. 272. *Heniochus macrolepidotus* Day, Fish. India, 1875-78, Pl. xxviii, fig. 3 (plate 34).

D. XI-XII; 23-28. A. III, 15-19. L. lat. 45. G.R. 11. Depth 1¼. Fourth dorsal spine and attached membrane greatly prolonged into a pennant-shaped appendage. Pearly white with dark purplish band over snout, another over eyes, a broad one from first 3 dorsal spines across side extending on to ventral fins and anal. A 4th band commences on hind part of spinous dorsal and extends on to posterior third of anal fin. Pectoral, soft portions of dorsal, anal and caudal fins bright yellow. 10 inches. Pearl banks, trawling grounds and coastal waters.

Gona, Manamaleya (S). Chudukan, Kadal madu, Puvali karuvel (T).

509. RHABDOPHORUS TRIFASCIATUS (Mungo Park) THREE-BANDED CORALFISH

Chaetodon trifasciatus Mungo Park, Trans. Linn. Soc., iii, 1797, p. 34. *Chaetodon vittatus* or *layardi*. *Chaetodon trifasciatus* Herre and Montalban, Philipp. Jour. Sci., xxxiv (1), 1927, Pl. xiii, fig. 1 (plate 34).

D. XII-XIII; 21-23. A. III, 19-20. L. lat. 38. G.R. 13. Depth $1\frac{1}{2}$. Head $4-4\frac{1}{4}$. Body yellowish. Head with three blackish vertical bands passing respectively through snout, eye and preopercular border. Body with about 15 narrow subhorizontal blackish streaks. A dark vertical band along base of soft dorsal and across to caudal peduncle. Soft dorsal yellow basally and orange distally. Anal similar with 2 dark bands bordering the orange area. Caudal peduncle orange, and divided from bluish rays by a vertical black band. 6 inches. Coastal waters and Gulf of Manaar.

510. RHABDOPHORUS XANTHOCEPHALUS (Bennett)

YELLOW-HEADED CORALFISH

Chaetodon xanthocephalus Bennett, Proc. Zool. Soc., ii, 1832, p. 182; Day, Fish. India, 1875-78, Pl. xxvi, fig. 4 (plate 34).

D. XIII-XIV; 25-27. A. III, 23-24. L. lat. ca. 40. Depth $1\frac{1}{2}$. Body silvery blue, cheeks and breast yellow. A dark band through eye, complete in young, disappearing with age. Six indistinct narrow dark subvertical streaks on sides. Dorsal and anal fins yellow in adults but brownish purple with orange margins in young. Adults with bluish edges and basal pink or red areas on these fins. Caudal bluish with yellow margins. Caudal peduncle brown in young and red in adult. Orange area on pectoral base and opercular margin. 8 inches. Coastal waters.

511. TETRAGONOPTRUS GUTTATISSIMUS (Bennett)

SPOTTED CORALFISH

Chaetodon guttatissimus Bennett, Proc. Zool. Soc., ii, 1832, p. 183; Day, Fish. India, 1875-78, Pl. xxvii, fig. 4 (plate 34).

D. XIII; 20-23. A. III, 17-20. L. lat. ca. 40. G.R. 12. Body yellowish, with dorsal areas and dorsal and anal fins basally grey-brown. A dark yellow-margined bar through eye. Body with numerous small dusky round spots scattered over sides and congregations of smaller ones around the bases of dorsal, caudal and anal fins. Soft dorsal with orange margin. Anal with yellow margin. Caudal yellowish basally and white distally, the colours separated by a dark crescent-shaped vertical bar. 5 inches. Coastal waters.

512. CHAETODONTOPS COLLARIS (Bloch)

WHITE-COLLARED CORALFISH

Chaetodon collare Bloch, Nat. Ausländ. Fische, iii, 1787, p. 116. *Tetragonoptrus collaris* Bleeker, Atlas Ichth. Ind. Neerl., ix, 1877, Pl. ccclxxiii, fig. 2 (plate 34).

D. XII; 25-27. A. III, 20-22. L. lat. ca. 30. Tr. 8+15. Depth $1\frac{1}{2}$. Head $3\frac{1}{2}$. Body brownish olive with yellow in centre of each scale. A blue-white band from before first dorsal across opercle to throat, a second across interorbital and a third across snout to chin. Dorsal and anal tinged with reddish violet. Soft dorsal margined with alternate bands of white, black, scarlet, black, white and black. Anal similarly banded with white, black and scarlet. Caudal scarlet basally and white distally, the colours separated by a black band. Ventrals black. 6 inches. Gulf of Manaar.

513. CHAETODONTOPS MELANOTUS (Bloch) TWO-SPOT CORALFISH

Chaetodon melanotus Bloch and Schneider, Syst. Ichth., 1801, p. 224; Herre and Montalban, Philipp. Jour. Sci., xxxiv (1), 1927, Pl. xi, fig. 2 (plate 34).

D. XII; 19-20. A. III, 17-18. L. lat. 35. G.R. 9. Depth 1⅗. Head 4. Body yellowish. A narrow dark ocular band from front of dorsal to chest. Upper part of body olivaceous. Sides marked with narrow dark oblique lines along the scale rows, passing from breast to dorsal fins. Dorsal and anal fins yellowish orange and with narrow black margins. Caudal yellowish-orange basally, white distally, the colours separated by a narrow vertical black bar. Caudal peduncle with two black blotches which may unite to form a complete bar. 7 inches. Gulf of Manaar.

514. LEPIDOCHAETODON UNIMACULATUS (Bloch)
ONE-SPOT CORALFISH

Chaetodon unimaculatus Bloch, Nat. Ausländ. Fische, iii, 1787, p. 75. *Tetragonoptrus unimaculatus* Bleeker, Atlas Ichth. Ind. Neerl., ix, 1877, Pl. ccclxxv, fig. 5 (plate 34).

D. XIII; 22-23. A. III, 19-20. L. lat. 38. G.R. 12. Depth 1½. Head 4. Yellowish. A moderate black ocular band from dorsal origin to chest. A large round black blotch on side, over lateral line and situated below last 4 spines and first 2 rays of dorsal. A black vertical bar across posterior borders of soft dorsal and anal and crossing the caudal peduncle. Dorsal otherwise yellow. Anal orange. Pectoral and caudal pale bluish white. 7 inches. Coastal waters.

515. LINOPHORA AURIGA (Forskål) FILAMENTED CORALFISH

Chaetodon auriga Forskål, Descript. Animal., 1775, p. 60. *Chaetodon setifer. Chaetodon auriga* Herre and Montalban, Philipp. Jour. Sci., xxxiv (1), Pl. vii, fig. 2 (plate 34).

D. XII-XIII; 23-26. A. III, 20-21. L. lat. 30. G.R. 13-14. Depth 1½. Head 3½. 5th, or 5th and 6th, or 6th and 7th dorsal rays filamentous and produced. Body whitish. A broad white-edged blackish ocular band from in front of dorsal to chest. A yellow band from level of eye on to first dorsal spines in postocular region. Body with dark narrow streaks passing obliquely upwards in anterior part and obliquely downwards in posterior part. Dorsal and anal fins broadly yellow. Soft dorsal edged with black and a large dark ocellus on centre of middle dorsal rays. Caudal basally yellowish with broad white margin and a greenish cross-bar. 9 inches. Coastal waters.

516. LINOPHORA MERTENSI (Cuvier) MERTEN'S CORALFISH

Chaetodon mertensii Cuvier and Valenciennes, Hist. Nat. Poiss., vii, 1831, p. 47. *Holacanthus xanthurus. Chaetodon chrysurus* Herre and Montalban, Philipp. Jour. Sci., xxxiv (1), 1927, Pl. viii, fig. 1 (plate 34).

D. XII-XIII; 21-22. A. III, 16-18. L. lat. 33. G.R. 12. Depth 2¼. Head 4⅓. Body pearly white with head and anterior parts lavender. An incomplete dark narrow ocular band. No other dark cross-bars. Anterior parts of body with narrow indistinct bent sub-vertical streaks. Ventrals and anterior parts of dorsal and anal fins white. Posterior parts of soft dorsal and anal broadly orange, the same colour extending across the caudal peduncle and posterior third of body. Soft dorsal and anal with yellow margins and inner dark line. Caudal blue basally followed by an orange crescent, a black edged narrow white bar and a dusky margin. 6 inches. Gulf of Manaar and coastal reefs.

517. LINOPHORA VAGABUNDA (Linnaeus) VAGABOND CORALFISH

Chaetodon vagabunda Linnaeus, Syst. Nat., ed. 10, 1758, p. 276. *Chaetodon pictus. Chaetodon vagabundus* Herre and Montalban, Philipp. Jour. Sci., xxxiv (1), Pl. v, fig. 2 (plate 34).

D. XII-XIII; 23-26. A. III, 20-22. L. lat. 35. G.R. 11. Depth 1½. Head 3½-4. General colour pearly white. A broad dark ocular band from before the dorsal, expanding on to chest. Body with series of dark narrow oblique streaks arranged as in the preceding species. A black broad band commences on posterior dorsal spines, passes across base of soft dorsal and caudal peduncle and terminates over the posterior anal rays. Dorsal and anal fins are broadly yellow. Dorsal has a marginal band of white with an inner one of black. Anal with black margin. Caudal peduncle yellow. Caudal white distally, yellow basally with 2 vertical black bars across the yellow area. 8 inches. Gulf of Manaar, trawling grounds and coastal reefs.

Gal lellah (S).

518. LINOPHORA RAFFLESI (Bennett) RAFFLES' CORALFISH

Chaetodon rafflesi Bennett, Life of Raffles, 1830, p. 689. *Chaetodon sebae. Chaetodon rafflesii* Günther, Jour. Mus. Godeffroy, ii (5), 1874, Pl. xxxv, fig. C (plate 34).

D. XIII; 22-23. A. III, 19-20. L. lat. 29-30. Tr. (6-7)+14. G.R. 4+9. Depth 1·6-1·7. Head 3. Yellowish green or brown with dark oblique streaks crossing each other in a lattice-work arrangement. The ascending ones are continued on to dorsal and curve posteriorly. Sometimes a large diffuse blotch on back. A broad white-edged ocular band complete to breast. Soft dorsal and caudal with broad intramarginal brown bands. Anal with dark intramarginal line. Juveniles with a rounded black ocellus on posterior half of soft dorsal, this developing into the broad band in adults. 5 inches. Coastal waters.

CICHLIDS

Family CICHLIDAE

Perch-like fishes with compressed oblong body covered in moderate-sized ctenoid scales. Dorsal spines numerous. Anal spines 3 or more. Teeth small, compressed, lobate and in one or more series in jaws. Bases of dorsal and anal fins not scaly. Lateral line interrupted. Fresh water species.

1. Anal fin with 3 spines .... *Tilapia mossambica*
   Anal fin with 12 or more spines 2
2. Dorsal rays 8-10; anal rays 8-9; dark blotches along sides .... *Etroplus maculatus*
   Dorsal rays 14-15; anal rays 11-12; transverse bands .... *Etroplus suratensis*

519. TILAPIA MOSSAMBICA (Peters) TILAPIA

Chromis mossambicus Peters, Monatsb. Akad. Wiss. Berlin, 1852, p. 681. *Tilapia mossambica* Jones and Sarojini, Jour. Bombay Nat. Hist. Soc., l) 3), 1952, fig. 7 (plate 36).

D. XV-XVI; 11-12. A. III, 10. L. lat. 30, interrupted. Tr. 4+12. Teeth small and in several series. Caudal subtruncate. Greenish-olive to golden, each scale with dark centre. Vertical fins blackish with indistinct white spots. Young specimens with an ovate black spot behind base of last dorsal spine. A native fresh water fish from southern Africa that has been recently introduced to fresh and brackish waters of Ceylon.

520. ETROPLUS MACULATUS (Bloch) SPOTTED ETROPLUS

Chaetodon maculatus Bloch, Ichth., 1785, Pl. 427, fig. 2. *Etroplus maculatus* Day, Fish. India, 1875-78, Pl. lxxxix, fig. 4 (plate 36).

D. XVII-XX; 8-10. A. XII-XV, 8-9. L. lat. 35. Tr. 6+19. Head $3\frac{1}{2}$-$3\frac{2}{3}$. Depth $2\frac{1}{4}$-$2\frac{1}{2}$. Yellowish with greenish back with 17 horizontal rows of golden spots and a few others on dorsal fin, abdomen and anal. Lateral line with three dusky blotches, middle being largest. Ventral and anal with black areas. 3 inches. Coastal fresh waters.
Ran koraliya, Raha koraliya, Ralliya. (S).

521. ETROPLUS SURATENSIS (Bloch) BANDED ETROPLUS

Chaetodon suratensis Bloch, Ichth., 1785, Pl. 217. *Etroplus suratensis* Day, Fish. India, 1875-78, Pl. lxxxix, fig. 6 (plate 36).

D. XVIII-XIX; 14-15. A. XII-XIII, 11-12. L. lat. 35-40. Tr. 6+17. Depth $2\frac{1}{4}$. Head 4. Light green with 8 transverse bands. Most of the scales above the lateral line with central pearly spot. Some irregular black spots on abdomen. Pectoral yellowish with black base. Other fins dark leaden colour. Young fish have a white-edged black ocellus from 4th to 10th dorsal ray. 12 inches. Fresh and brackish waters of coastal region.
Koraliya, Sethala (S). Sethel (T).

DEMOISELLES, ANEMONE FISH, PULLERS

Family POMACENTRIDAE

Short, deep, compressed body with large scales. Mouth small with compressed or conical teeth in 1-2 rows in each jaw. Gills reduced to $3\frac{1}{2}$. Anal spines 2. Lateral line interrupted, the anterior part consisting of tubes but often ending with scales perforated by pores only, while the second part on the caudal peduncle when present consists of pores only.

1. Scales small, 50-80 in a median lateral series 2
   Scales large, 27-36 in a median lateral series 4
2. Scales on head reaching line connecting hindborders of eyes; 19-25 rows of scales before dorsal .... *Amphiprion bicinctus*
   Scales on head not reaching line connecting hindborders of eyes; 12-16 rows of scales before dorsal ... 3
3. Second transverse white band bent backwards over superior part of soft dorsal .... *Amphiprion sebae*
   Second transverse white band with triangular prolongation anteriorly .... *Amphiprion percula*
4. Opercle, subopercle and interopercle finely serrated .... *Daya jerdoni*
   Opercle, subopercle and interopercle smooth 5
5. Teeth conical or villiform 6
   Teeth compressed and incisiform 8
6. Preopercle serrated 7
   Preopercle smooth .... *Chromis caeruleus*

7. Second and third dorsal spines subequal; no cross-bands .... *Dascyllus trimaculatus*
   Second dorsal spine two thirds length of third spine; 3 transverse dark cross-bands ...... *Dascyllus aruanus*
8. Lips swollen, fimbriate and curled back over the snout .... *Cheiloprion labiatus*
   Lips normal 9
9. Hindborder of preopercle serrated 10
   Hindborder of preopercle smooth 13
10. Twelve dorsal spines .... *Pomacentrus nigricans*
    Thirteen dorsal spines 11
11. Suborbital scaly .... *Pomacentrus cyanomos*
    Suborbital naked 12
12. A notch between preorbital and suborbital .... *Pomacentrus tripunctatus*
    No notch between preorbital and suborbital .... *Pomacentrus notophthalmus*
13. Twelve dorsal spines .... *Abudefduf lacrymatus*
    Thirteen dorsal spines 14
14. 3-4 scales between lateral line and scaly sheath of dorsal below last spine 15
    1½-2 scales between lateral line and scaly sheath of dorsal below last spine 16
15. Scales on head extending to interorbital space; inferior preopercle limb naked ... *Abudefduf septemfasciatus*
    Scales on head extending to nostrils; inferior preopercle limb scaly .... *Abudefduf saxatilis vaigiensis*
16. Two transverse rows of scales on preopercle .... *Abudefduf uniocellatus*
    Three transverse rows of scales on preopercle 17
17. Brown grey with 3 or 4 whitish cross-bands .... *Abudefduf xanthozona*
    Bluish or greenish grey, with a blue longitudinal streak above lateral line and 1-2 black patches on posterior part of dorsal .... *Abudefduf biocellatus*

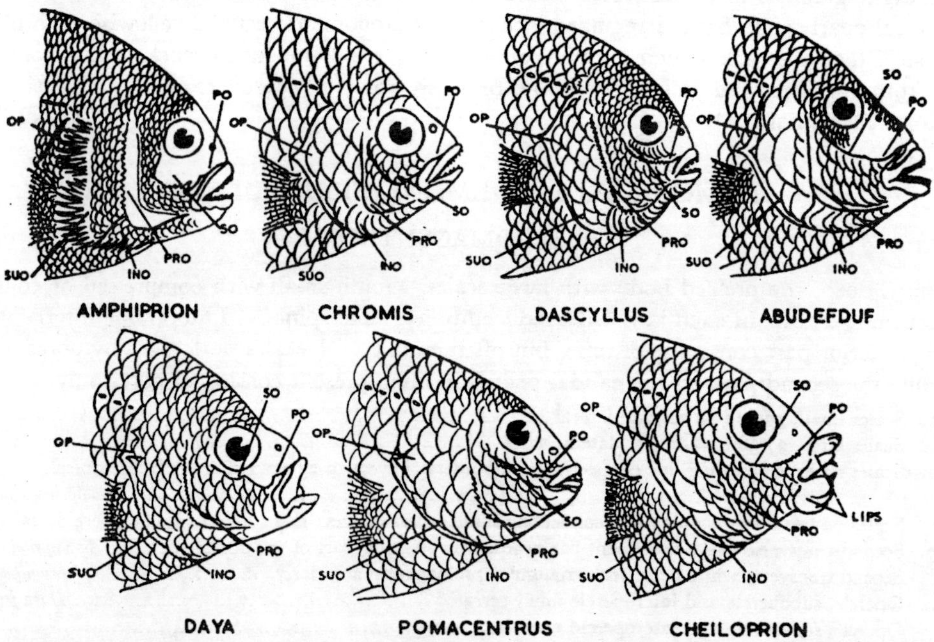

FIG. 13.

Principal differences between the seven genera of the Family Pomacentridae (Demoiselles, etc.), with particular reference to extent of scalation and nature of orbital and opercular bones. PO—Preorbital. SO—Suborbital. OP—Operculum. SUO—Suboperculum. INO—Interoperculum. PRO—Preoperculum.

522. AMPHIPRION BICINCTUS Rüppell THREE-BANDED ANEMONE FISH

Amphiprion bicinctus Rüppell, Atlas Reise Nordl. Afrika, 1828, p. 139. *Anthias* or *Amphiprion clarkii. Amphiprion polymnus. Prochilus polymnus* Bleeker, Atlas Ichth. Ind. Neerl., xi, 1877, Pl. cccc, fig. 7 (plate 35).

D. X; 15-17. A. II, 13-14. L. lat. 40 with 55 in horizontal series to caudal. Tr. 5+20. G.R. 11-12. Depth $1\frac{3}{4}$-2. Back purplish brown passing to pink on the sides and yellow below. Three bluish white bands across body, passing from in front of the dorsal [illegible] opercles, from the last 5 dorsal spines and across caudal peduncle. All fins yellowish. There is much variation in colour, especially in the darkening of the fins with age. Caudal may sometimes be black. 6 inches. Coastal waters.
Paydiya, Pol kitcha (S).

523. AMPHIPRION SEBAE Bleeker YELLOW-TAILED ANEMONE FISH

Amphiprion sebae Bleeker, Nat. Tijds. Ned. Ind., iv, 1853, p. 478. *Prochilus sebae* Bleeker, Atlas Ichth. Ind. Neerl., ix, 1877, Pl. cccc, fig. 9 (plate 35).

D. X-XI; 14-15. A. II, 13-14. L. lat. 34-40, with 49-52 in horizontal line to caudal. Tr. (5-7)+(18-19). G.R. 10. Depth 2·7-2·9. Brownish black with two milky cross-bands, one from nape to subopercle and the other from the last 3 spines and first 4 rays of dorsal to vent and curved back superiorly over the upper third or more of the soft dorsal. Base of pectoral, caudal and caudal peduncle canary yellow. $4\frac{1}{2}$ inches. Coral reefs, swimming amongst tentacles of sea anemones.

524. AMPHIPRION PERCULA (Lacépède) ORANGE ANEMONE FISH

Lutjanus percula Lacépède, Hist. Nat. Poiss., iv, 1802, p. 194. *Amphiprion percula* De Beaufort, Fish. Indo-Aust. Archip., viii, 1940, fig. 43 (plate 35).

D. X-XII; 14-15. A. II, 11-13. L. lat. 32-36, with 51-56 in horizontal series to caudal. Tr. (5-6)+(22-26). Depth 2·7-2·9. Bright orange to yellow with broad black-edged bluish-white cross-bands, first convex across hind part of head, second from middle of dorsal to vent with a triangular bulge projecting forwards, and the third across the caudal peduncle. All fins orange with broad white terminal and black subterminal bands. The extent of the bands varies amongst individuals. $3\frac{1}{2}$ inches. Symbiotic with anemones around coral reefs.

525. DAYA JERDONI (Day) GREEN PULLER

Pomacentrus jerdoni Day, Proc. Zool. Soc., 1873, p. 237. *Daya jerdoni* McCulloch, Mem. Qld. Mus., vii (3), 1921, Pl. xi, fig. 1 (plate 35).

D. XIII; 12-13. A. II, 12-14. L. lat. 21+12, with 30-34 in horizontal series to caudal. Tr. 4+10. Depth 3·9. Olive above and lighter below; 7 rows of light blue spots across gill-cover, one row along sub-orbitals, one over snout. A row of light lines along middle of scales on sides. Pectoral base with black spot. Fins dark. Base of caudal and anal barred with light lines. 5 inches. Trawling grounds.

526. CHROMIS CAERULEUS (Cuvier) BLUE PULLER

Heliases caeruleus Cuvier and Valenciennes, Hist. Nat. Poiss., v, 1830, p. 497. *Chromis lepidurus. Chromis caeruleus* De Beaufort, Fish. Indo-Aust. Archip., viii, 1940, fig. 52 (plate 35).

D. XII; 10. A. II, 10-11. L. lat. (14-16)+(8-10), with 26 in horizontal row to caudal. Tr. 1½+(7-8). Depth 2·8-3·2. Olive with a blue spot on each scale. A blue line from eye across preorbital. A blue spot at commencement of lateral line. Dorsal dotted with blue and with a dark edge. Anal dotted with yellow and with a dark margin. Ventrals green. Caudal with brown band along each lobe. A dark spot in pectoral axilla. 5 inches. Lives amongst branches of coral, along coastal reefs.

527. DASCYLLUS TRIMACULATUS (Rüppell) WHITE-SPOTTED PULLER

Pomacentrus trimaculatus Rüppell, Atlas Reise Nordl. Afrika, 1828, p. 39. *Pellochromis trimaculatus. Tetradrachmum trimaculatum* Bleeker, Atlas Ichth. Ind. Neerl., ix, 1877, Pl. ccccix, fig. 8 (plate 35).

D. XII; 15-16. A. II, 14-15. L. lat. (17-20)+8, with 28-29 in horizontal series to caudal. Tr. 3+11. G.R. 17. Depth 1·9-2·0. Body somewhat orbicular in shape. Light to dark brown with a darker hindborder to each scale. A blackish blotch at pectoral base. Young with a white blotch on nape and another below 7th-8th dorsal spines, the side one often remaining in adults. Fins brownish to very dark, especially ventrals. 5½ inches. Harbours and Gulf of Manaar.

528. DASCYLLUS ARUANUS (Linnaeus) BANDED PULLER

Chaetodon aruanus Linnaeus, Syst. Nat., ed. 10, 1758, p. 275. *Chaetodon araneus. Tetradrachmum aruanum. Tetradrachmum arcuatum* Bleeker, Atlas Ichth. Ind. Neerl., ix, 1877, Pl. ccccix, fig. 6 (plate 35).

D. XII; 12-13. A. II, 12. L. lat. 17+8, with 24-26 in horizontal row to caudal. Tr. 3+(9-10). G.R. 17. Pearly white on sides and above. Three vertical blackish bands, first through eye, second through first 3 dorsal spines to ventrals and third across soft dorsal, caudal peduncle and anal. Ventrals black. Caudal blackish basally. 3½ inches. Coastal reefs. Gal handha (S).

529. CHEILOPRION LABIATUS (Day) LARGE-LIPPED PULLER

Pomacentrus labiatus Day, Fish. India, 1875-78, p. 384, Pl. lxxxi, fig. 2 (plate 35).

D. XIII; 13. A. II, 12. L. lat. (17-18)+(8-9), with 26 in horizontal series to caudal. Tr. 2+9. Depth 2·3. Brown, lightest on chest and lower surface of head. Fins black. Blue spots on some scales of head and usually one on each scale in the row above the lateral line. Easily recognized by the large, fleshy and furrowed lips which fold back over snout and expose the needle-like teeth. 3½ inches. Coastal reefs.

530. POMACENTRUS NIGRICANS (Lacépède) DUSKY DEMOISELLE

Holocentrus nigricans Lacépède, Hist. Nat. Poiss., iv, 1803, p. 332. *Pomacentrus cyanospilus. Eupomacentrus albifasciatus* Bleeker, Atlas Ichth. Ind. Neerl., ix, 1877, Pl. cccciii, fig. 6 (plate 35).

D. XII; 14-16. A. II, 12-13. L. lat. (19-21)+(8-10), with 26-29 in horizontal series to caudal. Tr. (2-3)+(9-10). Light to dark brown with a white stripe from snout over preorbital and suborbital. Scales of preopercle and opercle each with a white spot; others on scales of lower part of body especially near anal. A black ocellus with white edge on base of posterior dorsal rays. Black blotch on pectoral base. Sometimes a broad white band from posterior part of dorsal to anal origin. Fins dark brown. 6 inches. Pearl banks.

531. POMACENTRUS CYANOMOS Bleeker VIOLET DEMOISELLE

Pomacentrus cyanomos Bleeker, Nat. Tijds. Ned. Ind., xi, 1856, p. 89; Atlas Ichth. Ind. Neerl., ix, 1877, Pl. ccccviii, fig. 3 (plate 35).

D. XIII; 11. A. II, 11. L. lat. 17+11, with 26 in horizontal series to caudal. Tr. 3+10. Depth 3·3. Silvery violet, being darker above. A dark brown blotch at origin of lateral line. A dark brown triangular patch at base of pectoral. Pectoral axilla brown. Dorsal, anal and ventrals dark violet. Pectorals dusky. 4 inches. Rock pools.

532. POMACENTRUS TRIPUNCTATUS Cuvier THREE-SPOT DEMOISELLE

Pomacentrus tripunctatus Cuvier and Valenciennes, Hist. Nat. Poiss., v, 1830, p. 421. *Pomacentrus bankanensis. Pomacentrus trilineatus* Bleeker, Atlas Ichth. Ind. Neerl., ix, 1877, Pl. ccccvi, fig. 1 (plate 35).

D. XIII-XIV; 14-15. A. II, 13-16. L. lat. (16-18)+(7-8), with 27-28 in horizontal series to caudal. Tr. 2+9. Depth 2·5-2·8. Dark or light brown with fins usually darker brown. Caudal sometimes yellow. Distal halves of dorsal and anal sometimes whitish. Anal sometimes with 1 or 2 whitish longitudinal stripes in young. Normally a blue-edged black spot at base of last dorsal rays; sometimes others at origin of lateral line and pectoral base. Young often with white or bluish lines across head and back and with an ocellus on the dorsal fin. 5¼ inches. Pearl banks.

533. POMACENTRUS NOTOPHTHALMUS Bleeker YELLOW DEMOISELLE

Pomacentrus notophthalmus Bleeker, Nat. Tijds. Ned. Ind., iv, 1853, p. 137. *Dischistodus notophthalmus* Bleeker, Atlas Ichth. Ind. Neerl., ix, 1877, Pl. ccci, fig. 4 (plate 35).

D. XIII; 13. A. II, 13. L. lat. 18+6, with 27 in horizontal series. Tr. (2-3)+9. Depth 2·8. Yellowish. Upper part of head and anterior portion of back dark brown. Anal region dark brown. Several white bands across head. Scales on opercle and above pectoral each with a white spot. A dark blotch at origin of the lateral line. Membranes of dorsal to 11th spine dark brown with white subterminal and black terminal edge. A dark blotch at base of 9th-11th spines, forming a white edged ocellus in young. Fins otherwise yellowish. 6 inches. Coastal waters.

534. ABUDEFDUF LACRYMATUS (Quoy and Gaimard)
WHITE-SPECKLED DEMOISELLE

Glyphisodon lacrymatus Quoy and Gaimard, Voy. Uranie et Physic., Zool., 1824, p. 388. *Glyphidodon lacrymatus* Bleeker, Atlas Ichth. Ind. Neerl., ix, 1877, Pl. cccciv, fig. 6 (plate 35).

D. XII; 16-17. A. II, 13. L. lat. 19+7, with 27 in horizontal series. Tr. 3+(9-10). Depth 2¼. Deep brown. Head and body with numerous small scattered light blue spots, more crowded on head below eye. Dorsal and anal brown with some of scales on sheaths bearing a white spot. Pectoral dusky. Ventrals black. Caudal and its peduncle yellowish to reddish. In young the white spots are relatively larger and there may be a black blotch on the spinous dorsal. 4 inches. Coastal waters.

535. ABUDEFDUF SEPTEMFASCIATUS (Cuvier) SEVEN-BAND DEMOISELLE

Glyphisodon septemfasciatus Cuvier and Valenciennes, Hist. Nat. Poiss., v, 1830, p. 463. *Glyphidodon septemfasciatus* Day, Fish. India, 1875-78, Pl. lxxxi, fig. 7 (plate 36).

D. XIII; 13. A. II, 12-13. L. lat. (21-22)+(7-9), with 28 in horizontal series. Tr. (3-4)+(11-12). Depth 3·9-4·3. Yellowish with darker hindborders to scales. Seven transverse dark bands. Spinous dorsal dark brown with lighter colour in middle. Soft dorsal, caudal and anal brown basally and yellow distally. Pectorals yellow with basal triangular black spot. Ventrals greyish basally and yellow distally. 7 inches. Rock pools.

536. ABUDEFDUF SAXATILIS VAIGIENSIS (Quoy and Gaimard) WAIGEU DEMOISELLE

Glyphisodon vaigiensis Quoy and Gaimard, Voy. Uranie et Physic., Zool., 1824, p. 391. *Chaetodon tyrrwhitti, Abudefduf sexfasciatus. Glyphidodon coelestinus* Day, Fish. India, 1875-78, Pl. lxxxiii, fig. 2 (plate 36).

D. XIII; 12-13. A. II, 12-13. L. lat. (20-23)+(0-9), with 28-29 in a horizontal series. Tr. (3-4)+11. Depth 2½. Bluish or brownish, being lighter below. Five vertical dark brown or blackish bars slightly less broad than interspaces. Dorsal and anal light or dark brown, the spinous dorsal sometimes quite black. A dark blotch sometimes present at base of posterior dorsal rays. Caudal brownish. Pectorals yellow with a triangular black basal spot. Ventrals grey. 7 inches. Rock pools.

537. ABUDEFDUF UNIOCELLATUS (Quoy and Gaimard) ONE-SPOT DEMOISELLE

Glyphisodon uniocellatus Quoy and Gaimard, Voy. Uranie et Physic., Zool., 1824, p. 393. *Glyphidodon assimilis. Abudefduf uniocellatus* Jordan and Seale, Bull. U.S. Bur. Fish., xxv, 1905, Pl. xliii, fig. 2 (plate 36).

D. XIII; 12. A. II, 13. L. lat. (16-17)+(8-9), with 26 in horizontal series. Tr. 2+9. Depth 2·7-2·8. Reddish with a deep bluish sheen over the scales. Belly yellow. Dorsal bluish grey with a lighter median band. Caudal, anal and ventrals mainly yellowish. Some small yellow spots on body scales. A black blotch, encircled with bluish-white sometimes present at base of median dorsal rays. 3 inches. Coastal waters.

538. ABUDEFDUF XANTHOZONA (Bleeker) THREE-BAND DEMOISELLE

Glyphisodon xanthozona Bleeker, Nat. Tijds. Ned. Ind., iv, 1853, p. 283, *Abudefduf albofasciatus. Abudefduf amabilis* Jordan and Seale, Bull. U.S. Bur. Fish., xxv, 1905, Pl. xlii, fig. 1 (plate 36).

D. XIII; 12-13. A. II, 11-12. L. lat. (17-20)+(7-9), with 27-28 in horizontal series. Depth 2·9-3·4. Brown or grey with three vertical silvery cross-bars, through opercle, middle of spinous dorsal and caudal peduncle respectively. A 4th band sometimes present between first two. Some light spots on opercular scales and pectoral base. Fins brown or grey. 3 inches. Coastal waters.

539. ABUDEFDUF BIOCELLATUS (Quoy and Gaimard)
TWO-SPOT DEMOISELLE

Glyphisodon biocellatus Quoy and Gaimard, Voy. Uranie et Physic., Zool., 1824, p. 389. *Chaetodon* or *Abudefduf brownriggii*, *Abudefduf modestus*, *Glyphidodon antjerius*. *Abudefduf antjerius* Jordan and Seale, Bull. U.S. Bur. Fish., xxv, 1905, Pl. xlii, fig. 2 (plate 36).

D. XIII; 12-14. A. II, 12-13. L. lat. (16-19)+(7-8), with 26-28 in horizontal series. Tr. 2+9. Depth 2·7-3·1. Young specimens are reddish brown with a blue central area on each scale and lighter to yellowish below. A light blue line runs from snout below dorsal profile to a large black, blue-edged ocellus on posterior dorsal spines. Two shorter blue lines pass through eye. A smaller blue-edged ocellus at upper edge of caudal peduncle. In older specimens the larger ocellus tends to disappear. The blue line may also disappear or break up in a row of spots. 4 inches. Coastal reefs.
Kaha batichcho, Kaha batikyah (S).

RAINBOW FISH, WRASSES*

Family LABRIDAE

Body deep and compressed to elongate and cylindrical. Mouth moderate and equipped with canine teeth, a posterior one sometimes projecting forwards in angle of the mouth. No teeth on palate. Lower pharyngeals united into a Y-shaped structure which bears granular teeth. Scales cycloid. Lateral line variable, sometimes in 2 parts. Fin spines mostly feeble. Mostly brilliantly coloured species. Mostly carnivores but some are herbivores.

1. 11-13 dorsal spines 2
   8-10 dorsal spines 5
2. Upper lip not covered when the mouth is closed; preorbital moderately deep 3
   Upper lip covered when the mouth is closed; preorbital very deep 4
3. Preopercular limb scaly .... *Lepidaplois diana*
   Preopercular limb naked .... *Lepidaplois bilunulatus*
4. Vertical fins with scaly bases .... *Choerodon anchorago*
   No scales along bases of fins .... *Xiphocheilus robustus*
5. Lateral line interrupted 6
   Lateral line continuous 12
6. Scales small; L.lat. with about 80 .... *Cymolutes lecluse*
   Scales large; L.lat. with about 20-30 7
7. 1st and 2nd spines of dorsal flexible .... *Inistius pavo*
   1st and 2nd spines of dorsal pungent and rigid 8

* Two nominal species *Julis finlaysoni* and *J. purpureo-lineatus* have also been described from Ceylon by Valenciennes and referred to *Stethojulis* and *Platyglossus* respectively in Günther's Catalogue, but since dropped from standard works.

8. 3rd anal spine longer than 2nd 9
2nd anal spine longer than 3rd .... *Pseudocheilinus hexataenia*
9. Dorsal with 10 spines .... *Cheilinus chlorurus*
Dorsal with 9 spines 10
10. Depth 2·8-3·3; 8 oblique black lines on head .... *Cheilinus diagramma*
Depth 2·2-2·6; no narrow oblique lines on head 11
11. Dorsal and anal with wide scale sheath; body with 4 dark spots and cross-bars .... *Cheilinus trilobatus*
Dorsal and anal with narrow scale sheath; 2 black lines from eye obliquely upwards .... *Cheilinus undulatus*
12. Two large flattened incisors bent forwards anteriorly in jaws .... *Anampses meleagrides*
No flattened incisors 13
13. Body elongate and cylindrical .... *Cheilio inermis*
Body less elongate and compressed 14
14. Dorsal with 8 spines 15
Dorsal with 9 spines 25
15. Snout produced and tubiform 16
Snout normal 18
16. Caudal truncate or rounded, the outer rays scarcely prolonged; vertical fins dark .... *Gomphosus varius*
Caudal semi-lunate, outer rays prolonged; vertical fins yellow 17
17. Violet or vivid green; no yellow blotch .... *Gomphosus coeruleus*
Green or brown; a wide yellow blotch from shoulder to behind pectoral .... *Gomphosus tricolor*
18. Head uniform in colour and without bands 19
Head with bands or bars of contrasting colour 20
19. Yellowish with 3-5 wide black vertical bars .... *Thalassoma janseni*
Greenish with a broad light vertical bar under 2nd-6th dorsal spines .... *Thalassoma melanoptera*
20. Opercle scaled above; body greenish or bluish 21
Head quite scaleless; body brown or yellowish 25
21. Black blotches on caudal peduncle and anterior rays of soft dorsal .... *Thalassoma cupido bipunctatum*
No black blotches in these positions 22
22. Six vertical dusky cross-bars on back .... *Thalassoma hardwicki*
No vertical dusky cross-bars 23
23. Longitudinal reddish or violet bands .... *Thalassoma purpurea*
No longitudinal bands on body 24
24. Each scale with a vertical red streak; dorsal and anal with red longitudinal bands; young with black spot at anterior of spinous dorsal .... *Thalassoma lunare*
Each scale with deeper vertical streak; dorsal and anal uniform; young with pale vertical cross-bar on shoulder .... *Thalassoma hebraicum*
25. Head not dark in contrast with body; a narrow blue line from lower edge of eye to pectoral .... *Thalassoma amblycephala*
Head darker in contrast to body; 4 or 5 broken green or brown bands through eye and across cheek .... *Thalassoma lutescens*
26. Cheeks scaly .... *Hemigymnus fasciatus*
Cheeks naked or with a few scales only 27
27. Large scales; L.lat. 25-30 28
Scales small; L.lat. 50-80 35
28. No canine teeth; scales on thorax larger than those on sides .... *Stethojulis axillaris*
A canine at angle of mouth; scales on thorax smaller than those on sides 29
29. A small patch of scales on upper part of opercle 30
Head naked except occiput 31
30. No scales behind or below eye .... *Guntheria scapularis*
Scales present behind and below eye .... *Hemitautoga centriquadrus*

31. Dorsal and anal with low scaly sheath; head pointed 32
Dorsal and anal without scaly sheath; head rounded 33
32. Purplish brown with red longitudinal bands; second dorsal with large ocellus ...... *Platyglossus notopsis*
Blackish green with undulating blue or green streaks on head; second dorsal without an ocellus .... *Platyglossus marginatus*
33. No streaks across cheek or around eye; 2 broad horizontal brown bars along sides ...... *Halichoeres hyrtli*
Red or violet streaks across cheeks and around eye; no horizontal bands on sides 34
34. An angular white-edged black blotch on middle of side .... *Halichoeres bimaculatus*
No large blotch on side .... *Halichoeres javanicus*
35. Tr. 8+35; grey with blue spots .... *Coris gaimardi*
Tr. 6+28; brown or green .... *Coris aygula*
Tr. 3+27; red with green spot on each scale .... *Coris halei*

540. LEPIDAPLOIS DIANA (Lacépède) DIANA'S WRASSE

Labrus diana Lacépède, Hist. Nat. Poiss., iii, 1802, p. 450. *Bodianus diana. Cossyphus diana* Day, Fish. India, 1875-78, Pl. lxxxvii, fig. 3 (plate 36).

D. XII; 9-10. A. III, 10-12. L. lat. 31. Tr. 5+11. Depth 3½. G.R. 12. Reddish with 3 large white spots between lateral line and dorsal fin, 2 below spines and one below the rays. Dark spots on scales from commencement of soft dorsal above lateral line. A black spot on posterior of lateral line. Fins reddish yellow usually with black blotch on ventrals and at bases of anterior and posterior anal rays. 9 inches. Coastal waters.

541. LEPIDAPLOIS BILUNULATUS (Lacépède) CRESCENT-BANDED WRASSE

Labrus bilunulatus Lacépède, Hist. Nat. Poiss., iii, 1802, p. 454. *Bodianus bilunulatus. Cossyphus bilunulatus* Bleeker, Atlas Ichth. Ind. Neerl., i, 1862, Pl. xxxviii, fig. 3 (plate 36).

D. XII; 10. A. III, 12. L. lat. 33-35. Tr. 6+(12-13). Depth 3·4-3·8 G.R. 6+12. Reddish with light or yellow stripes and a large black blotch below hind edge of soft dorsal and over anterior part of caudal peduncle. Two black lines on head, one from snout through eye, second from angle of mouth to subopercle. Fins reddish with a black spot between 1st and 3rd dorsal spines. 16 inches. Coastal waters.

542. CHOERODON ANCHORAGO (Bloch) PIG-TOOTHED WRASSE

Labrus anchorago Bloch, Naturg. Ausländ. Fische, v, 1791, p. 108. *Choerops anchorago* Day, Fish. India, 1875-78, Pl. lxxxii, fig. 5 (plate 36).

D. XIII; 7. A. III, 8-9. L. lat. 29. Tr. 3+(8-10). Depth 3. Both jaws with enlarged curved canines, inner pair longest in upper jaw and outer pair longest in lower jaw. Cheeks brownish yellow with scarlet spots. Back brown with a white vertical band below 4th-6th dorsal spines. A light band on caudal peduncle. Sometimes also 4 wide dark cross-bands. Spinous dorsal with 2 dark longitudinal bands. Soft dorsal with dusky area over anterior rays and extending on to body, and posteriorly yellow edged with orange. Other fins yellowish. 13 inches. Coastal waters.

543. XIPHOCHEILUS ROBUSTUS Günther SWORD-LIPPED WRASSE

Xiphochilus robustus Günther, Cat. Fish., iv, 1862, p. 98. *Choerodon robustus* Blegvad and Løppenthin, Fish. Iranian Gulf, 1944, Pl. ix, fig. 2 (plate 36).

D. XI-XII; 8-9. A. III, 10. L. lat. 29. Tr. 3+9. Depth 3·4. Outer canines of mandibles turned outwards. An obtuse osseus ridge surrounds the edges of the jaws in which the teeth are scarcely distinct. Yellowish red with a yellow band along basal half of anal and another along middle of dorsal fin. Both fins with a narrow basal blue line, the dorsals also with a margin of similar colour. Caudal yellowish with longitudinal violet streaks. 12 inches. Coastal waters.

544. CYMOLUTES LECLUSE (Quoy and Gaimard) BROWN-LINED WRASSE

Xyrichthys lecluse Quoy and Gaimard, Voy. Uranie et Physic., Zool., 1824, p. 284. *Cymolutes praetextatus* Day, Fish. India, 1875-78, Pl. xc, fig. 1 (plate 37).

D. IX; 12-13. A. III, 11-13. L. lat. (57-62)+(20-22). Tr. 4+20. G.R. 10. Depth 4-4⅓. Colour variable. Usually greenish with a dark blue-edged band across shoulder. Sometimes olive with narrow brownish lines across body. These are bent at an angle, the point directed backwards. Sometimes a round black spot below lateral line. Dorsal pale violet, sometimes with a black margin to spinous part. Anal rosy. Caudal orange with yellowish transverse lines. 8 inches. Coastal waters.

545. INISTIUS PAVO (Valenciennes) PEACOCK WRASSE

Xyrichthys pavo Cuvier and Valenciennes, Hist. Nat. Poiss., xiv, 1839, p. 61. (?) *Novacula rufa*. *Xyrichthys pavo* Bleeker, Atlas Ichth. Ind. Neerl., i, 1862, Pl. xxix, fig. 2 (plate 36).

D. II, VII; 12-13. A. III, 12-13. L. lat. 20+(5-6). Tr. 2+10. Depth 3·2-3·4. The first two dorsal spines are separated from the remainder of the fin and situated over the head. Pale blue with green cross-bands on back and sides. A large black white-edged ocellus sometimes on dorsal above 3rd and 4th cross-bands. Another may be present above the 7th or 8th lateral line scale. Bands fade with age but silvery patch between 2nd and 3rd bands remains. Dorsal pale bluish with darker oblique stripes. Soft dorsal with pale yellow margin. Anal yellowish with longitudinal submarginal dark band. 14 inches. Gulf of Manaar.

546. PSEUDOCHEILINUS HEXATAENIA (Bleeker) SIX-LINED WRASSE

Cheilinus hexataenia Bleeker, Act. Soc. Indo-Neerl., ii, 1857, p. 84. *Pseudocheilinus hexataenia* De Beaufort, Fish. Indo-Aust. Archip., viii, 1940, fig. 18 (plate 37).

D. IX; 11-12. A. III, 8-9. L. lat. (15-18)+(4-7). Tr. 2+7. G.R. 8. Depth 3·4-3·6. Golden to green above, pinkish below. Three longitudinal dull red bands bordered above and below by bright greenish blue stripes, forming in effect 6 prominent longitudinal stripes below the lateral line. 2 brown spots on chin and another on upper part of caudal peduncle. Fins with pastel shades of lemon, green, mauve and pink. 18 inches. Coastal reefs.

547. CHEILINUS CHLORURUS (Bloch) RED-SPOTTED GREEN WRASSE

Sparus chlorurus Bloch, Naturg. Ausländ. Fische, v, 1791, p. 24. *Cheilinus punctatus*. *Cheilinus chlorurus* Day, Fish. India, 1875-78, Pl. lxxxii, fig. 6 (plate 37).

D. X; 9-10. A. III, 7-9. L. lat. 15+(5-8). Tr. 2+(6-7). G.R. 5-6. Depth 2·8-3·3. Olive brown to greenish with numerous red spots on cheeks and around snout. These form lines and anastomosed figures. Fins greenish brown with anastomosing figuring as on head. Young specimens with brownish indistinct vertical bars. 12 inches. Coastal reefs. Muthu gireva, Pothy gireva (S). Kili meen (T).

548. CHEILINUS DIAGRAMMA (Lacépède) SCRIBBLED WRASSE

Labrus diagramma Lacépède, Hist. Nat. Poiss., iii, 1802, p. 448. *Cheilinus radiatus* Bleeker, Atlas Ichth. Ind. Neerl., i, 1862, Pl. xxvi, fig. 1 (plate 37).

D. IX; 10. A. III, 8. L. lat. (15-16)+(7-8). Tr. 2+6. G.R. 10. Depth 3·3-3·8. Olive green with 3 faint cross-saddles, one over nape, one below spinous dorsal and one below soft dorsal. Head with red stripes from lip through eye; also 8 oblique narrow black lines across lower parts of cheek. Dorsal and anal greenish with blotchy markings. Caudal marbled yellow with outer rays brown. 15 inches. Coastal reefs.

549. CHEILINUS TRILOBATUS Lacépède TRIPLE-TAILED WRASSE

Cheilinus trilobatus Lacépède, Hist. Nat. Poiss., iii, 1802, p. 529; Smith, Fish. Sth. Africa, 1949, Pl. lvi, fig. 816 (plate 37).

D. IX; 10-11. A. III, 8-9. L. lat. (14-16)+(7-8). Tr. 2+6. G.R. 5-6. Depth 2·8-3·4. Caudal rounded in young but central and outer rays prolonged to form 3 lobes in adults. Greenish brown with reddish brown vertical streaks on the body scales. Head with reddish stripes radiating from the anterior border of the eye. Red spots on cheeks, opercles and breast. Fins green with red-pink margins and some red spots. Ground colour often mottled. Some darker blotches on lower part of body in young. 18 inches. Coastal reefs.

550. CHEILINUS UNDULATUS Rüppell WAVY-LINED WRASSE

Cheilinus undulatus Rüppell, Neue Wirbelth., Fische, 1835, p. 20; Bleeker, Atlas Ichth. Ind. Neerl., i, 1862, Pl. xxvi, fig. 3 (plate 37).

D. IX; 10. A. III, 8. L. lat. (14-16)+(7-8). Tr. 2+6. Predorsal scales 6. G.R. 11-12. Mainly dull olive-green with mottlings. Each scale with a vertical purplish line. Wavy lines and streaks on head. Dorsal and anal with red and green bars and dots. Caudal slaty basally and yellow distally. Pectorals greenish. Pelvics light yellow. Attains 7 feet. Coastal waters and Gulf of Manaar.

551. ANAMPSES MELEAGRIDES Valenciennes SPECKLED WRASSE

Anampses meleagrides Cuvier and Valenciennes, Hist. Nat. Poiss., xiv, 1840, p. 12; Bleeker, Atlas Ichth. Ind. Neerl., i, 1862, Pl. xxiv, fig. 1 (plate 37).

D. IX; 12. A. III, 12. L. lat. 27-28, continuous but abruptly bent posteriorly. Tr. (3-4)+10. Depth 3·6. Scales beginning behind eye. G.R. 7+12. Brownish with head and body covered with white spots which are round on the head and elongated in short streaks on opercle and behind isthmus. On body there is a spot to each scale. Dorsal and anal brown with narrow white margin and a white spot or streak between each radial. A white-edged dark brown ocellus sometimes present between 10th and 11th rays of dorsal and anal. Caudal yellowish. 7½ inches. Coastal waters.

552. CHEILIO INERMIS (Forskål) SLENDER WRASSE

Labrus inermis Forskål, Descr. Animal., 1775, p. 34. *Cheilio inermis* Bleeker, Atlas Ichth. Ind. Neerl., i, 1862, Pl. xxxi, fig. 4 (plate 37).

D. IX; 13-14. A. III, 11-12. L. lat. 45-50. Tr. 5+(12-14). Depth 6·3-9·2. An elongate species. Mouth horizontal with lips which become thicker with age. G.R. 17-20. Colour varies from yellow to brilliant green. Sometimes a dark mid-lateral streak from opercular angle to caudal and across caudal also. Head sometimes reticulated or marbled with red or other colours. Dorsal sometimes with a meshwork of darker colour and anal with oblique streaks. 20 inches. Coastal reefs.

553. GOMPHOSUS VARIUS Lacépède OLIVE CLUB-NOSED WRASSE

Gomphosus varius Lacépède, Hist. Nat. Poiss., iii, 1802, p. 104. *Gomphosus fuscus* or *pectoralis*. *Gomphosus varius* Jordan and Evermann, Bull. U.S. Fish. Comm., xxiii (1), fig. 125 (plate 37).

D. VIII; 13. A. II-III, 10-12. L. lat. 26-30. Tr. 3+10. G.R. 8+14. Depth 4·0-4·8. Teeth in jaws mostly hooked, with anterior canines. Colour varying from olive to brownish-purple. Body scales with vertical streaks or dark spots. A dark brown or black band from snout through eye where it divides and crosses upper part of opercle. Pectoral axilla with small black spot. Dorsal brownish, often with large black blotch between 1st and 4th spines. Soft dorsal and anal with marginal darker bands and anal with another basally. Caudal dusky, deeper distally. 10 inches. Coastal waters.
Gireva, Koppara gireva (S).

554. GOMPHOSUS COERULEUS Lacépède BLUE CLUB-NOSED WRASSE

Gomphosus coeruleus Lacépède, Hist. Nat. Poiss., iii, 1802, p. 101. *Gomphosus viridis*. *Gomphosus coeruleus* Bleeker, Atlas Ichth. Ind. Neerl., i, 1862, Pl. xxi, fig. 5 (plate 37).

D. VIII; 13. A. III, 11. L. lat. 28. Tr. 3+10. Depth 4·5-4·7. G.R. 14. Usually deep blue or purplish but sometimes dominantly deep green. Dorsal and anal yellow but usually brownish purple basally, and last ray of both deep brown. Caudal yellow but base and outer rays purplish brown. Sometimes only a dark spot at tip of last anal and dorsal rays. Pectorals and ventrals also yellowish. Green specimens may have dark spots on each scale. 11 inches.
Nil talapath gireva (S).

555. GOMPHOSUS TRICOLOR Quoy and Gaimard
YELLOW CLUB-NOSED WRASSE

Gomphosus tricolor Quoy and Gaimard, Voy. Uranie et Physic., Zool., 1824, p. 280; Bleeker, Atlas Ichth. Ind. Neerl., i, 1862, Pl. xxi, fig. 6 (plate 37).

D. VIII; 13. A. III, 11-12. L. lat. 28-29. Tr. (3-4)+(9-10). G.R. 11+14. Depth 4·4. Outer caudal rays rather produced. Brownish or with a broad yellow band from shoulder to behind pectoral base. Vertical fins yellow. Pectoral brownish with dark spot at its base, and a narrow vertical bluish band parallel with its posterior border. 12 inches. Coastal waters.
Nil koppara gireva (S).

556. THALASSOMA JANSENI (Bleeker) JANSEN'S WRASSE

Julis jansenii Bleeker, Act. Soc. Sci. Ind. Neerl., i, 1856, p. 56; Atlas Ichth. Ind. Neerl., i, 1862, Pl. xxxiv, fig. 5 (plate 37).

D. VIII; 13. A. II-III, 11. L. lat. 26-27. Tr. 3+(9-10). G.R. 7+14. Depth 4·2-4·7. Caudal rounded in young, lobes produced in adults. Yellow with 3-5 wide black vertical bands from back down sides. In some specimens the black almost masks out the yellow. Both colours are continued on to the dorsal fin. A violet streak from behind opercle to base of pectoral. Pectoral, ventral and anal yellowish, the last with a black spot at end of last two rays. Caudal dusky with a dark spot in the middle of its base in young specimens. 7½ inches. Coastal reefs.

557. THALASSOMA MELANOPTERA (Günther) DUSKY-FINNED WRASSE

Julis melanoptera Günther, Cat. Fish., iv, 1862, p. 183.

D. VIII; 13. A. II, 11. L. lat. 27. Tr. 3+9. Depth ca. 4. Head uniform purplish brown. Body olive green, with a broad lighter transverse band below 2nd to 6th dorsal spines. Each scale within the light band with a shining vertical streak. Dorsal and anal dark violet, former with a dark spot anteriorly. Caudal light with upper and lower margins greenish. Pectoral with an oblique oblong black spot and axil black. An imperfectly known species based on a 5½ inch specimen which is thought to have come from Ceylon.

558. THALASSOMA CUPIDO BIPUNCTATUM Vasiliu TWO-SPOT WRASSE

Thalassoma cupido bipunctatum Vasiliu, Jena Z. Naturw., lxv, 1931, p. 349, Pl. x, fig. 11 (plate 37).

D. VIII; 13. A. III, 11. L. lat. 28. Tr. 2+(9-10). Depth 3·9. Brownish with darker head. A dark band from snout to eye where it bifurcates and continues to opercular border. Scales on back with vertical streaks. Dorsal hyaline with a median stripe which darkens to a black blotch between 2nd and 4th rays. Basal half of anal dark. Caudal light with a large black irregular blotch on the caudal peduncle. 8 inches. Coastal reefs.

559. THALASSOMA HARDWICKI (Bennett) HARDWICK'S WRASSE

Sparus hardwickii Bennett, Fish. Ceylon, 1828-30, p. 12. *Julis dorsalis. Thalassoma melanochir* Fowler, Jour. Acad. Nat. Sci. Phila. (2), xii, 1904, Pl. xx (plate 37).

D. VIII; 12-13. A. II, 10-11. L. lat. 27-28. Tr. 3+(10-11). G.R. 5+16. Depth 3·6-3·7. Greenish with silvery abdomen. Broad red bands radiate from eyes. Six vertical black bars cross upper part of body. A black spot in axilla. A red band along side to tail. Dorsal fin red with a black or deep blue longitudinal band. Anal yellow with a black spot anteriorly. Caudal yellow with a grey band distally. 8 inches, attaining 18 inches. Coastal waters. Mal gireva (S).

560. THALASSOMA PURPUREA (Forskål) RAINBOW WRASSE

Scarus purpureus Forskål, Descr. Animal., 1775, p. 27. *Scarus georgii. Julis purpurea* Day, Fish. India, 1875-78, Pl. lxxxvi, fig. 3 (plate 37).

D. VIII; 13. A. II-III, 11. L. lat. 27-30. Tr. 2+(9-11). G.R. 14. Depth 3·5-3·7. Green or blue, usually green above and blue below. Usually red or pink bands radiating from eye in adults. 2 to 4 reddish to purple horizontal bands from opercle to caudal base. Dorsal green or yellowish red with a median green horizontal band. Young with a black spot at origin of spinous dorsal and sometimes another on the anterior part of the soft dorsal. Anal green with a pink or dark basal band and sometimes an orange margin. Caudal bluish or pink with red, green, orange and yellow bands. Young with black spots in groups on sides. 18 inches. Coastal waters, amongst weeds and in rock crevices.
Lena gireva (S).

561. THALASSOMA LUNARE (Linnaeus) GREEN WRASSE

Labrus lunaris Linnaeus, Syst. Nat., ed. 10, 1758, p. 283. *Labrus zeylanicus*. *Julis porphyrocephala*. *Julis lunaris* Bleeker, Atlas Ichth. Ind. Neerl., i, 1862, Pl. xxxiii, fig. 5 (plate 38).

D. VIII; 13. A. III, 11. L. lat. 27-29. Tr. 3+(9-11). G.R. 10-12. Depth 4½. Caudal deeply lunate in adult. Body green, fading to blue on death. Each scale with a vertical red streak. Head green above and violet below with several oblique reddish bands through eye. Pectoral blue with oblong reddish violet spot. Dorsal red with blue and yellow margin. Anal similar. Both fins sometimes green basally. Caudal mainly yellow with outer rays reddish. Young with a black spot at base of dorsal rays, some light blotches along back and a black spot on caudal peduncle. 12 inches. Coastal waters.

562. THALASSOMA HEBRAICUM (Lacépède) CROSS-BANDED WRASSE

Labrus hebraicus Lacépède, Hist. Nat. Poiss., iii, 1802, p. 455. *Thalassoma hebraicum* Smith, Fish. Sth. Africa, 1949, Pl. liii, fig. 778 (plate 38).

D. VIII; 13. A. III, 11. L. lat. 27-28. Tr. (2-3)+(9-10). G.R. 13. Depth 4-4½. Greenish or bluish, each scale with a darker vertical streak. 3 wide violet-edged or blue bands across opercle, the lower ones extending to beneath the pectoral. A buff subvertical cross-bar from dorsal origin to pectoral axilla. Dorsal fin sometimes with an indistinct dark ocellus on its anterior part. Fins green or blue to match body; caudal with darker lunate band. 9 inches. Coastal waters.

563. THALASSOMA AMBLYCEPHALA (Bleeker)

BLUNT-HEADED WRASSE

Julis amblycephalus Bleeker, Nat. Tijds. Ned. Ind., xi, 1856, p. 83; Günther Jour. Mus. Godeffroy, viii (16), 1909, Pl. clii, fig. C (plate 38).

D. VIII; 13. A. III, 10-11. L. lat. 26-27. Tr. 3+11. G.R. 4+8. Depth 5·3. Upper two thirds of head and body olive brown, each scale having a vertical streak. Lower third of body white. A narrow blue dark-edged line from lower angle of eye to axilla and another across angle of preopercle. A broad triangular patch in pectoral axilla. Pectoral yellow with black tip. Dorsal dark anteriorly and soft portion with a narrow white edge. Caudal lobes dark olive. 4 inches. Coastal waters.

564. THALASSOMA LUTESCENS (Lay and Bennett) YELLOW WRASSE

Julis lutescens Lay and Bennett, Zool. Beechey's Voy., 1839, p. 65. *Thalassoma lutescens* Vasiliu, Jena Z. Naturw., lxv, 1931, Pl. x, fig. 12 (plate 38).

D. VIII; 13. A. II, 11. L. lat. 27. Tr. 4+9. Depth 3·4. G.R. 9+16. Yellow or brownish body with head and anterior portions of body grey or bluish. Head with about 5 broken greenish or brown bands, uppermost passing through eye. Dorsal yellowish with median grey or bluish band. Anal yellow with broad greyish blue band. Caudal yellow. Pectoral yellow with broad clouded terminal blotch. Young with black spot over 2nd and 3rd dorsal spines, and darker centres to body scales. 8 inches. Coastal waters.

565. HEMIGYMNUS FASCIATUS (Bloch) FIVE-BANDED WRASSE

Labrus fasciatus Bloch, Naturg. Ausländ. Fische, vi, 1792, p. 6. *Tautoga fasciata, Scarus quinquefasciatus. Hemigymnus fasciatus* Bleeker, Atlas Ichth. Ind. Neerl., i, 1862, Pl. xlvi, fig. 2 (plate 38).

D. XI; 11. A. III, 11. L. lat. 29. Tr. (4-5)+(10-14). G.R. 16-18. Depth 3·1-3·2. Creamy white above, pinkish below. Head greenish with a number of brownish bands radiating from lower border of eye. Body with about 5 brown broad cross-bars. Fins yellowish with the brown cross-bars continued on to them. 30 inches. Coastal waters. Panoo gireva (S).

566. STETHOJULIS AXILLARIS (Quoy and Gaimard) RED-SPOT RAINBOW FISH

Julis axillaris Quoy and Gaimard, Voy. Uranie et Physic., Zool., 1824, p. 272. *Cossyphus, Lepidaplois* or *Bodianus axillaris. Stethojulis bandanensis* Jordan and Seale, Bull. U.S. Bur. Fish., xxv, 1906, Pl. xlv, fig. 2 (plate 38).

D. IX; 11. A. II, 11-12. L. lat. 26-27. Tr. 3+9. G.R. 16-18. Depth 3·6-3·8. Caudal rounded. Brownish grey to olive green above, silvery below. Dorsal fin same colour as back and similarly freckled with small white dots. Cheeks golden, sometimes with 2 or 3 golden streaks from snout to opercular margin. A large triangular red spot in pectoral axil. Body sometimes with a pinkish longitudinal band from pectoral base to caudal. Pectoral, caudal and anal creamy. Lateral streak sometimes with 2 small blue-edged black ocelli on caudal peduncle. 5 inches. Coastal waters and trawling grounds.

567. GUNTHERIA SCAPULARIS (Bennett) BROWN-BANDED RAINBOW FISH

Julis scapularis Bennett, Proc. Comm. Zool. Soc., 1831, p. 167. *Julis ceylonicus, Platyglossus ceylonicus* or *scapularis. Halichoeres scapularis* Smith, Fish. Sth. Africa, 1949, Pl. lv, fig. 795 (plate 38).

D. IX; 11. A. III, 11. L. lat. 27-29. Tr. (2-3)+(8-9). G.R. 12. Depth 3·6-4·0. No scales on cheek, a few on opercle. Variable in colour and markings. Usually dominantly green with scales on back with blue borders. A brown or blackish band runs horizontally from upper angle of opercle to caudal peduncle, formed largely by large brown sopts along scale rows. Head usually green above and yellow below with several irregular bars of pink, red or brown, one extending through pectoral axil to belly. Dorsal and anal fins with horizontal bands of white, yellow or green, blue, and orange or red. Caudal yellow with lunate cross-bars of orange. 7 inches. Coastal waters.

568. HEMITAUTOGA CENTRIQUADRUS (Lacépède)

FOUR-SPOT WRASSE

Labrus centriquadrus Lacépède, Hist. Nat. Poiss., iii, 1802, p. 437. *Scarus* or *Julis decussatus*, *Platyglossus hortulanus*. *Halichoeres centriquadrus* Jordan and Seale, Bull. U.S. Bur. Fish., xxv, 1906, Pl. xlvi, fig. 3 (plate 38).

D. IX; 10-11. A. III, 11-12. L. lat. 26-28. Tr. (3-4)+9. G.R. 13-14. Depth 3·2-3·7. Caudal rounded to truncate. Colour variable, usually yellowish with back and centres of scales on sides green, or alternately creamy with brown on back and centres of scales. Head with 5 or 6 irregular bands of red or orange. Orange spots on scales above and below the pectoral origin. A dusky blotch preceded by a cream one on back under spinous dorsal. Dorsal creamy yellow with 3 or 4 rows of lighter spots and some blue or brown basally. Anal with 2 or 3 longitudinal bands of pink. Caudal yellow, sometimes with darker cross-bands. 12 inches. Coastal reefs and Gulf of Manaar.
Hembili gireva (S).

569. PLATYGLOSSUS NOTOPSIS (Valenciennes)

OCELLATED RAINBOW FISH

Julis notopsis Cuvier and Valenciennes, Hist. Nat. Poiss., xiii, 1839, p. 485. *Halichoeres notopsis* Bleeker, Atlas Ichth. Ind. Neerl., i, 1862, Pl. xli, fig. 2 (plate 38).

D. IX; 13. A. III, 12. L. lat. (3-4)+(9-11). Depth 4. Purplish brown with 4-5 reddish longitudinal bands. Two black ocelli on dorsal fin, the smaller between 1st and 2nd spines and larger between 3rd and 6th rays. One or two rows of light coloured round spots on dorsal behind large ocellus. Caudal with yellow basal band. Young darker than adults and with 3 white spots on back. 4 inches. Rock pools.

570. PLATYGLOSSUS MARGINATUS (Rüppell)

SPECKLED RAINBOW FISH

Halichoeres marginatus Rüppell, Neue Wirbelth., Fische, 1835, p. 16. *Platyglossus marginatus* Jordan and Seale, Bull. U.S. Bur. Fish., xxv, 1906, Pl. xlvi, fig. 1 (plate 38).

D. IX; 13-14. A. III, 12. L. lat. 27-28. Tr. (3-4)+(9-10). Depth 4·1-4·2. G.R. 6+11. Dusky green or bluish brown. Numerous reddish blue- or green-edged horizontal streaks following scale rows of body. Head and anterior part of body with undulating blue-edged green bands. Blue-edged streaks and spots on vertical fins and sometimes posterior part of body. Caudal with vertical crescent-shaped patch covered with spots. Basal half of pectoral dark or purplish. 6½ inches. Coastal reefs.

571. HALICHOERES HYRTLI (Bleeker) HYRTL'S RAINBOW FISH

Julis (*Halichoeres*) *hyrtlii* Bleeker, Act. Soc. Sci. Ind. Neerl., i, 1856, p. 60. *Halichoeres hyrtlii* Bleeker, Atlas Ichth. Ind. Neerl., i, 1862, Pl. xxxix, fig. 2 (plate 38).

D. IX; 11-12. A. III, 12. L. lat. 28-29. Tr. (2-3)+(9-10). Depth 4-4½. G.R. 8+11. Back olive, abdomen white. Two brownish-black longitudinal bands; the upper along back at base of dorsal fin; the lower from snout through eye to middle of base of caudal, where it ends in a black spot and its course on the opercle is interrupted by a red spot. Head with broad dark blue and narrow green vertical stripes behind eye. Dorsal anteriorly with 2 and posteriorly with 3 rows of spots, which reticulate to form bands and rings. Caudal similarly reticulated. Pectoral with dark spot at its base. 4½ inches. Pearl banks.

572. HALICHOERES BIMACULATUS Rüppell TWO-SPOT RAINBOW FISH

Halichoeres bimaculatus Rüppell, Neue Wirbelth., Fische, 1835, p. 10. *Platyglossus bimaculatus* Day, Fish. India, 1875-78, Pl. lxxxv, fig. 5 (plate 38).

D. IX; 10. A. III, 11. L. lat. 27. Tr. 4+9. Depth 4. Caudal truncated with outer rays slightly produced. Yellow with green around lower edge of eye and violet bands radiating from it. A broad green band from eye to caudal, and a row of light spots above it. A triangular black white-edged spot on 9th-10th lateral scale. Upper edge of dorsal yellow with narrow purple intramarginal line. Lower two thirds of dorsal red with 2 rows of blue spots edged with yellow. Anal with dark wide median band. Caudal vertically rivulated with green. Coastal waters.

573. HALICHOERES JAVANICUS (Bleeker) JAVANESE RAINBOW FISH

Julis (*Halichoeres*) *javanicus* Bleeker, Nat. Tijds. Ned. Ind., xiii, 1857, p. 341. *Platyglossus javanicus*. *Halichoeres javanicus* Bleeker, Atlas Ichth. Ind. Neerl., i, 1862, Pl. xl, fig. 3 (plate 38).

D. IX; 12. A. III, 12. L. lat. 26-28. Tr. 2+9. Depth 3½-4¼. Brownish red becoming silvery on abdomen. A vertical blue spot or band behind upper half of orbit; some oblique red streaks on head. A black spot at pectoral base. Dorsal reddish with 2-3 rows of round yellow spots. Caudal similar but the spots are less regularly disposed. Anal red. 4½ inches. Coastal waters.

574. CORIS GAIMARDI (Quoy and Gaimard) GAIMARD'S RAINBOW FISH

Julis gaimard Quoy and Gaimard, Voy. Uranie et Physic., 1824, p. 265. *Julis*, *Labrus* or *Coris formosus*. *Coris gaimard* Smith, Fish. Sth. Africa, 1949, Pl. lviii, fig. 809 (plate 38).

D. IX; 12-13. A. III, 12-13. L. lat. 70-80. Tr. (6-8)+(30-31). G.R. 12. Depth 3·8-4·2. Caudal rounded. Bluish-grey to mauve or brown with black or blue spots on the posterior part of the body. Head yellow with 2 oblique blue bands ascending towards dorsal. Dorsal and anal pinkish brown; former with a red margin and 2 narrow green lines within that band and black dots and elongate blotches basally on the membranes; latter with green margin and deeper green horizontal lines and dots along it and black dots basally on membrane. Caudal red basally and pale distally with blue vertical dividing line and basal

blue dots. Pectoral rosy or yellow. Sometimes a green area on body at posterior end of anal fin. 15 inches. Coastal waters.
Rathu gireva (S).

575. CORIS AYGULA Lacépède RED-THROATED RAINBOW FISH

Coris aygula Lacépède, Hist. Nat. Poiss., iii, 1802, p. 96. *Labrus aureomaculatus*, *Julis cingulum* or *rüppelli*, *Coris angulatum*. *Coris aygula* Day, Fish. India, 1875-78, Pl. lxxxviii, fig. 5 (plate 38).

D. IX; 12-13. A. III, 12. L. lat. 60-64. Tr. 4+(25-28), as well as 3 rows above which form the dorsal sheath. G.R. 7+(11-12). Depth $3\frac{1}{2}$. Adults develop a hump on the forehead. Body olive brown or greenish. Small blue or black spots on head and anterior part of body. A pale cross-bar near end of pectoral. Sometimes one or two red or orange saddles beneath dorsal fin. Fins brown with black spots and white margins. Young with two large black ocelli on dorsal fin. Adults with red angular bands across neck and body and fins variably green. 4 feet. Coastal waters.
Thik gireva (S).

576. CORIS HALEI Day HALY'S RAINBOW FISH

Coris halei Day, Supplement to Fishes of India, 1888, p. 803.

D. IX; 12. A. III, 12. L. lat. 75. Tr. 3+27. Depth $3\frac{3}{4}$. Vinous red barred by 11 purplish-grey stripes. Each scale with a spot of brilliant emerald green. Head orange with violet, red-bordered stripes radiating from eye; two of these uniting to form a band which descends from 4th dorsal spine past eye and mouth to subopercle. Broad red band on opercular margin. Dorsals red-grey. Anal red. Each with an orange border and covered with small blue spots. Caudal dark grey with large blue black-edged spots. Coastal waters and Gulf of Manaar.

PARROT FISHES

Family SCARIDAE

Body elongate and robust, with large scales of which several on the caudal base are enlarged. Small mouth with coalesced teeth either forming a beak-like palate or firmly joined to an interior plate. Caudal round in young but lobes elongated in adults. Lateral line bent or interrupted.

1. Teeth not fully united into a smooth beak .... *Cryptotomus spinidens*
   Teeth united forming a smooth beak 2
2. Two rows of scales on cheek, preopercular limb naked 7
   Three rows of scales on cheek 3
3. Inferior preopercular limb naked .... *Callyodon harid*
   Inferior preopercular limb with one or more scales 4
4. Body green with red margins or bases to scales 5
   Body bluish with darker margins to scales 6
5. Head red with undulating and anastomosing green lines; anal green .... *Callyodon fasciatus*
   Head green with yellow on lips, across to eye, radiating from eye and on preopercle; anal red .... *Callyodon blochi*

6. Head red with a blue band on lips and across to eye; pectoral with 13 rays .... *Callyodon ghobban*
Head blue-green with yellow bands on lips, across to eye and radiating around eye; pectoral with 12 rays .... *Callyodon dussumieri*
7. Teeth yellowish green .... *Callyodon oktodon*
Teeth white or pink 8
8. Lips narrow; pectoral with 13 rays .... *Callyodon erythrodon*
Lips broad; pectoral with 12 rays 9
9. Inner upper lip well developed; caudal with lobes produced and with transverse bands .... *Callyodon bataviensis*
Inner upper lip poorly developed; caudal rounded and light basally, dark distally ...... *Callyodon forsteri*

577. CRYPTOTOMUS SPINIDENS (Quoy and Gaimard) TOOTHED PARROT FISH

Scarus spinidens Quoy and Gaimard, Voy. Uranie et Physic., Zool., 1824, p. 289. *Callyodon carolinus. Callyodon spinidens* Bleeker, Atlas Ichth. Ind. Neerl., i, 1862, Pl. ii, fig. 2 (plate 39).

D. IX; 10-11. A. III, 9-10. L. lat. 24-25. Tr. 2+7. G.R. 7, short. Depth 3-3½. A single transverse row of scales on cheek and a few on opercle. Jaws with inner row of incisors posteriorly and an outer row of pointed ones anteriorly. Colour variable, usually green or brown mottled and speckled with darker. Sometimes red streaks on snout. Sometimes red centres to scales and white dots. Dorsal and anal with oblique brown lines or orange dots and a black spot between 1st and 2nd spines. Pectorals yellow with a black base. 12 inches. Coastal waters.

578. CALLYODON HARID (Forskål) RED-LINED PARROT FISH

Scarus harid Forskål, Descr. Animal., 1775, p. 30. *Pseudoscarus harid. Pseudoscarus mystax* Bleeker, Atlas Ichth. Ind. Neerl., i, 1862, Pl. x, fig. 1 (plate 39).

D. IX; 10. A. III, 9-10. L. lat. 17+9. Tr. 2+7. G.R. 17+37, fine and flexible. Olive green with 2 red streaks from forehead through eye; narrow red line including lips and passing towards pectoral base. Body scales with red margin and basal blue spot. Dorsal and anal red, edged with blue, and with several rows of blue or green spots. Upper rays of pectoral and outer caudal rays blue. Sometimes 2 vertical blue lines on caudal. 21 inches. Trawling grounds.

579. CALLYODON FASCIATUS (Valenciennes) RIVULATED PARROT FISH

Scarus fasciatus Cuvier and Valenciennes, Hist. Nat. Poiss., xiv, 1839, p. 222. *Pseudoscarus rivulatus* Bleeker, Atlas Ichth. Ind. Neerl., i, 1862, Pl. ix, fig. 3 (plate 39).

D. IX; 10. A. III, 9. L. lat. 18+7. Tr. 2+6. Depth 3-3½. G.R. 16+12, weak and flexible. Body green. Each scale with a reddish base. Snout and cheeks red with undulating and anastomosing green lines. A narrow green line along base and edge of dorsal and an intermediate row of spots. Anal green, lighter towards margin, which is edged with dark green. Caudal with green spots. 14 inches. Gulf of Manaar and trawling grounds.

580. CALLYODON BLOCHI (Valenciennes) BLOCH'S PARROT FISH

Scarus blochii Cuvier and Valenciennes, Hist. Nat. Poiss., xiv, 1839, p. 219. *Pseudoscarus chrysopoma. Pseudoscarus viridis* Bleeker, Atlas Ichth. Ind. Neerl., i, 1862, Pl. xvii, fig. 3 (plate 39).

D. IX; 10-11. A. III, 9. L. lat. 23-24. Tr. 2+7. G.R. 14+27. Depth 3¼. Green, scales with a reddish margin. Upper lip with one, lower lip with 2 cross-bands of yellow passing into a triangular spot between eye and angle of mouth; three short lines radiate from eye. Dorsal and anal rosy, a narrow band along their bases, and green margins. Caudal green. 12 inches. Trawling grounds.

581. CALLYODON GHOBBAN (Forskål) FLAME PARROT FISH

Scarus ghobban Forskål, Descr. Animal., 1775, p. 28. *Scarus pepo, Pseudoscarus pyrrhostethus* or *ghobban. Callyodon pyrrhostethus australianus* Paradice and Whitley, Mem. Qld. Mus., ix (1), 1927, Pl. xiv (plate 39).

D. IX; 10-11. A. III, 9. L. lat. 18+7. Tr. 2+(6-7). Depth 3½-3¾. Head reddish with white jaws. A blue cross-band on lips. A curved blue streak from angle of mouth to below eye. Scales on body with blue edges. Vertical fins red, with blue bases and margins. Caudal sometimes with blue spots. Blue spots may also be present on dorsal. 16 inches. Coastal waters and Gulf of Manaar.

582. CALLYODON DUSSUMIERI (Valenciennes)

DUSSUMIER'S PARROT FISH

Scarus dussumieri Cuvier and Valenciennes, Hist. Nat. Poiss., xiv, 1839, p. 252. *Pseudoscarus dussumieri* Bleeker, Atlas Ichth. Ind. Neerl., i, 1862, Pl. viii, fig. 1 (plate 39).

D. IX; 10-11. A. III, 9. L. lat. 19+6. Tr. 2+(6-7). G.R. 17+30. Depth 3¼. Caerulean blue with lower edge of body pale rose. Centre of every scale on blue portion bile green, as is also upper part of head. Yellow or blue bands around eye, across upper lip and behind lower lip. Dorsal and anal orange with a blue basal and second outer band. Caudal similar with outer rays and 3 vertical broken bars blue. Pectoral upper ray blue, next 4 orange and rest hyaline. Outer ventral ray blue. 16 inches. Gulf of Manaar and trawling grounds. Labu gireva, Nil gireva (S).

583. CALLYODON OKTODON (Bleeker) EIGHT-TOOTHED PARROT FISH

Pseudoscarus oktodon Bleeker, Versl. Akad. Amsterdam, xii, 1861, p. 235; Atlas Ichth. Ind. Neerl., i, 1862, Pl. xiii, fig. 2 (plate 39).

D. IX; 10. A. III, 9. L. lat. 18+7. Tr. 2+6. Depth 3¼. Two canines posteriorly in both upper and lower jaws. Brown with lighter snout. A dark blue band above upper lip continued to eye. Lower lip with blue band followed by a second crossing chin and an indistinct one on interopercle. Dorsal dark brown with a dark bluish edge and a median dark band on its posterior part. Anal red with dark blue edge and a dark median band. Caudal brown with a vertical blue band. Teeth yellowish green. 13 inches. Gulf of Manaar.

584. CALLYODON ERYTHRODON (Valenciennes)
RED-TOOTHED PARROT FISH

Scarus erythrodon Cuvier and Valenciennes, Hist. Nat. Poiss., xiv, 1839, p. 255. *Pseudoscarus erythrodon. Pseudoscarus sumbawensis* Bleeker, Atlas Ichth. Ind. Neerl., i, 1862, Pl. xv, fig. 4 (plate 39).

D. IX; 10-11. A. III, 9. L. lat. 25. Tr. 2+6. G.R. 12+27. Depth 3·3-3·7. Teeth pink. A posterior canine may be present. Purplish brown, edges of scales darker than centres. Scales on caudal peduncle yellowish. Thoracic region with violet tint. Snout greenish-yellow. Vertical fins brown, dorsal with a narrow black margin. Pectoral hyaline. 11 inches. Coastal waters.

585. CALLYODON BATAVIENSIS (Bleeker) BATAVIAN PARROT FISH

Scarus bataviensis Bleeker, Nat. Tijds. Ned. Ind., xiii, 1857, p. 342. *Pseudoscarus bataviensis* Bleeker, Atlas Ichth. Ind. Neerl., i, 1862, Pl. xii, fig. 3 (plate 39).

D. IX; 10. A. III, 9. L. lat. (17-18)+7. Tr. 2+7. Depth 3·4-3·7. Teeth dirty yellow; one or two canines posteriorly in both jaws. Head indian red, becoming more dilute over belly. Cheeks and opercle with golden green. Eye surrounded by emerald green which passes down in stripes to upper lip and chin. Body olive green, paler below. Dorsal banded with cobalt, red, green, red and cobalt. Pectoral orange. Anal with 3 bands of cobalt separated by bands of red. Caudal red with outer ray and 3 vertical bands blue. Scales red basally. 14 inches. Gulf of Manaar.

586. CALLYODON FORSTERI (Valenciennes) FORSTER'S PARROT FISH

Scarus forsteri Cuvier and Valenciennes, Hist. Nat. Poiss., xiv, 1839, p. 275. *Pseudoscarus capistratoides* Bleeker, Atlas Ichth. Ind. Neerl., i, 1862, Pl. vi, fig. 2 (plate 39).

D. IX; 10-11. A. III, 9. L. lat. (17-18)+(8-10). Tr. 2+6. G.R. (15-25)+30. Depth 3·3-3·6. Teeth yellowish, largely covered by broad lips. Dark brown with paler margins to scales. Pale yellow or greenish band across each lip and another from their unison to eye. Similar bands from nostrils to eye, and sometimes other short ones behind eye. Dorsal and anal dark brown with light edge. Caudal light basally and darker distally, olive, sometimes crossed by dark vertical band. Pectoral yellow with black axillary spot. 12 inches. Gulf of Manaar.

CURLYFINS

Family CIRRHITIDAE

Body moderately elongate and compressed, with moderate scales. Fin spines well developed. Caudal never forked. Small mouth with fine pointed teeth in jaws and vomer, and on palatines in some genera. Lower pectoral rays simple and thickened.

Palatine teeth present; posterior border of preopercle strongly denticulated; large scales on interopercle .... *Cirrhitichthys aureus*

Palatine teeth absent; posterior border of preopercle strongly crenulated; small scales on interopercle .... *Paracirrhites forsteri*

587. CIRRHITICHTHYS AUREUS (Schlegel) GOLDEN CURLYFIN

Cirrhites aureus Temminck and Schlegel, Fauna Japonica, Poiss., 1843, p. 15; Day, Fish. India, 1875-78, Pl. xxxv, fig. 5 (plate 39).

D. X; 12-13. A. III, 6-7. P. 7+(6-7 unbranched). L. lat. 43. Tr. 4+12. Depth 3·3. Villiform teeth on vomer and palatines as well as jaws, the later having outer rows enlarged. Rosy with lighter longitudinal lines and an ill-defined blotch below soft dorsal and sometimes 2 others under spinous dorsal. Dorsal and caudal lightly banded, the latter with red spots. 4 inches. Gulf of Manaar.

588. PARACIRRHITES FORSTERI (Bloch) FORSTER'S CURLYFIN

Grammistes forsteri Bloch and Schneider, Syst. Ichth., 1801, p. 191. *Serranus tankervillae*. *Paracirrhites forsteri* Bleeker, Atlas Ichth. Ind. Neerl., viii, 1876-77, Pl. cccxlix, fig. 5 (plate 39).

D. X; 11. A. III, 6. P. 7+7, or 8+6. L. lat. 45-46. Tr. 5+(10-12). Depth 3½. Palate without teeth. Villiform teeth on vomer and in jaws, upper series more complete than lower, with anterior canines and enlarged outer series. Reddish. Head, chest and base of pectoral with black spots. Broad dark band broken into black or brown blotches under soft dorsal and on caudal peduncle. Broad yellow mid-lateral band. A black band across soft dorsal. Front edge of anal and outer rays of caudal black. Rocky shores around coast. 18 inches.

Thik kossa (S).

GRUB FISH

Family PARAPERCIDAE

Body elongate, robust and subcylindrical. Scales small, large mouth with thick lips with teeth in jaws and palate. Dorsal fin divided into a low short spinous portion and a long high soft portion. Caudal rounded in young, lobes produced in adults. Lateral line simple.

1. Dorsal with 4 spines .... *Parapercis quadrispinosa*
   Dorsal with 5 spines 2
2. Pink with red cross-bars reaching to ventral profile; 17-18 scales below lateral line ... *Parapercis pulchella*
   White with red or brown blotches above lateral line only, these with black spots on them; 14-16 scales below lateral line .... *Parapercis punctata*

589. PARAPERCIS QUADRISPINOSA (Weber) FOUR-SPINED GRUB FISH

Percis quadrispinosa Weber, Siboga Exped. Fische, 1913, p. 519. *Parapercis tetracanthus* Günther, Jour. Mus. Godeffroy, iv (11), 1876, Pl. xciii, fig. B (plate 40).

D. IV; 21. A. 17. L. lat. 64. Depth 7·8. Brownish above, lighter below. 10 dark brown cross-bars on sides, extending on to belly. Head with large black spots above. Lips light with 3 vertical brown bands. Two blotches on cheeks and continued as transverse bands on throat and interopercle. Another blotch on opercle. Dorsals with basal row of black spots and soft dorsal with a subterminal row of smaller ones. Anal with a subterminal row also. Caudal with black spots around periphery. 7 inches. Coastal waters.

590. PARAPERCIS PULCHELLA (Schlegel) ROSY GRUB FISH

Percis pulchella Temminck and Schlegel, Fauna Japonica, Poiss., 1843, p. 24; Day, Fish. India, 1875-78, Pl. lviii, fig. 2 (plate 40).

D. V; 21-22. A. I, 16-17. L. lat. 60-62. Tr. 5+(17-18). Depth 6½. Reddish with 6 or more darker red cross-bars which extend on to belly. Head spotted with dark purple marks above. Three vertical blue bands across sub- and interopercles and sometimes a fourth in front and below pectoral. Dorsals yellow, black basally and a row of spots along upper half of soft part. Anal yellow with 5 rows of canary yellow spots and a red margin. Caudal yellow in middle, red below and with fine vertical bars above. 5 inches. Pearl banks and trawling grounds.

591. PARAPERCIS PUNCTATA (Cuvier) SPOTTED GRUB FISH

Percis punctata Cuvier and Valenciennes, Hist. Nat. Poiss., iii, 1829, p. 264. *Percis millepunctata. Percis punctata* Day, Fish. India, 1875-78, Pl. lviii, fig. 1 (plate 40).

D. V; 21-22. A. II, 17-18. L. lat. 58-60. Tr. 5+(14-16). Depth 6½. White with reddish cross-bars and brownish blotches and black spots over them, restricted to upper half of body. A golden line from below eye to base of dorsal and another from centre of eye to snout. Spinous dorsal with or without a row of black spots; soft dorsal with 3 rows. Caudal with 2 grey bands and sometimes an ocellus on peduncle. Anal yellow. 5 inches. Coastal waters.

SMELT

Family BEMBROPIDAE

Elongate body with moderate scales. Head depressed with large mouth and projecting lower jaw. Fine teeth on jaws, vomer and palatines.

592. BEMBROPS CAUDIMACULATA Steindachner BIG-EYE SMELT

Bembrops caudimaculata Steindachner, Sitzb. Akad. Wiss. Wien, lxxiv, 1877, p. 212; Alcock, Illust. Zool. Investigator, vii, 1900, Pl. xxxi, fig. 2 (plate 51).

D. VI; 14-15. A. 16. L. lat. 42-48. Tr. 2+9. G.R. 15-16. Depth about 6. Fine teeth in a knob each side of apex of upper jaw. Mainly brown-yellow. Vague dusky spots along body, one more definite at upper caudal base. 10 inches. Gulf of Manaar in 143 fathoms.

SAND EELS

Family TRICHONOTIDAE

Elongate body with dorsal fin along most of the back, rays and spines not distinguishable and anterior ones filamentous and produced. Mouth large but teeth feeble. Burrow in sand.

593. TRICHONOTUS SETIGER Bloch FILAMENTED SAND EEL

Trichonotus setiger Bloch and Schneider, Syst. Ichth., 1801, p. 179. *Taeniolabrus filamentosus* Steindachner, Sitzb. Akad. Wiss. Wien, lv (1), Pl. iii, fig. 5 (plate 52).

D. 46-47. A. 36. L. lat. 60. Tr. 4+5. Depth 16½-17¼. Teeth pointed, those of lower jaw hooked. Anterior rays of dorsal in males much prolonged and filamentous. Ventral rays also elongated. Brown with a black spot at hindborder of each lateral line scale. Dorsal, posterior part of caudal with many small black-edged ocelli in rows. Sometimes others on head, body and anal. 7½ inches. Gulf of Manaar, pearl banks.

STAR GAZERS

Family URANOSCOPIDAE

Body elongate and robust with thick skin. Head bony and heavy. Eyes entirely dorsal. Large mouth which is vertical and contains small teeth on palate and jaws.

594. ICHTHYOSCOPUS LEBECK (Bloch) STAR GAZER

Uranoscopus lebeck Bloch and Schneider, Syst. Ichth., 1801, p. 47. *Anema inerme. Ichthyoscopus inermis. Uranoscopus inermis* Valenciennes, Illus. Poiss. Cuvier's Regne Anim., 1843, Pl. xvii, fig. 3 (plate 42).

D. III-IV; 16-18. A. 16-19. P. 18. Depth 4-4¼. Eyes anterior and entirely dorsal. Lips with numerous branched papillae. Posterior edge of opercle fringed and an elongate angular flap with papillae behind shoulder. Lateral line close to dorsal base. Canary yellow with brown markings on sides enclosing white blotches. Caudal yellow brown. A dark bar across pectoral. Estuarine, burying in müd with only eyes exposed. 24 inches.

Mudhu bassa (S). Nillum karaiyan, Vaggan (T).

Suborder BLENNIOIDEI

Perciform fishes with ventral fins jugular or mental (if present) and consisting of 1-5 rays. Body moderate to very elongate with well developed scales or naked, never fully scaled.

Body naked; ventrals with 2-4 simple rays .... *Family* BLENNIIDAE, p. 200
Body scaled; ventrals with 1 spine and 3-4 articulated rays .... *Family* CLINIDAE, p. 206

BLENNIES

Family BLENNIIDAE

Body elongate, smooth and without scales. Dorsal fin along most of the back, the spines feeble and barely distinguishable from rays. Anal spines absent or feeble, reduced in females and sometimes with spongy excrescences in males. Pelvics far forward. Mouth small; close-set teeth, fixed or movable and sometimes large lateral canines. Lateral line absent or much reduced. Small carnivorous fishes living around reefs and in tide pools.

1. Gill-membranes united with isthmus restricting opening to sides or to a small pore; teeth in jaws, fixed, and with an enormous lateral canine posteriorly in lower jaw 2
   Gill-membranes free from isthmus; teeth small, implanted in lips and freely movable 5
2. D.40, A.27; gill-opening extends downwards in front of pectoral fin .... *Aspidontus striatus*
   D.30, A.18-20; gill-opening entirely above base of pectoral fin 3
3. First 3 dorsal rays higher than rest and separated by a notch .... *Petroscirtes lienardi*
   Dorsal continuous, nearly the same height throughout 4
4. White bands along sides; depth 4·25 .... *Dasson breviceps*
   Black band along sides; depth 5·5 .... *Dasson cyprinoides*
5. A band of more than 30 tentacles transversely across nape; males with dendritic masses on anal spines .... *Cirripectus leopardus*
   Not more than 2 tentacles on either side of nape; males without dendritic masses on anal spines 6
6. Vomerine teeth present 7
   Vomerine teeth absent 8
7. Six pairs of dark vertical bands on sides not obscured by dark spots; nuchal cirrhus with 5-6 secondary filaments .... *Entomacrodus epalzeochilos*
   Body bands largely obscured by several rows of prominent black spots on sides; nuchal cirrhus simple and filiform .... *Entomacrodus striatus*
8. Lateral line absent; body depth 7·5-8·5; caudal rays unbranched .... *Alticus saliens*
   Lateral line conspicuous, at least anteriorly; body depth 2·3-5·8; inner 9 caudal rays deeply branched ... 9
9. Ventrals with 2-4 rays; nasal tentacles short; margin of dorsal fin notched 12
   Ventrals with 2 rays; nasal tentacles long or margin of dorsal entire 10
10. No orbital or nuchal tentacles; outer rays of caudal produced in males .... *Ecsenius frontalis*
    Orbital and nuchal tentacles present; no caudal rays produced 11
11. Dorsal fin not distinctly notched; no crest on head .... *Salarias fasciatus*
    Dorsal fin distinctly notched; a longitudinal crest on head, at least in males .... *Salarias alboguttatus*
12. D. XII; 17-18. A. II, 18-20 13
    D. XIII-XIV; 19-23. A. II, 21-28 14
13. Males with crest; maxilla extends well beyond eye .... *Istiblennius unicolor*
    No crest in either sex; maxilla does not extend beyond posterior edge of eye ...... *Istiblennius tanegasimae*
14. Ventrals with 2 rays 15
    Ventrals with 3 rays .... *Istiblennius edentulus*
    Ventrals with 4 rays .... *Istiblennius andersoni*
15. Supraorbital tentacle branched; body with 8-10 horizontal black lines .... *Istiblennius lineatus*
    Supraorbital tentacle simple; body with irregularly shaped broad vertical cross-bars .... *Istiblennius enosimae*

595. ASPIDONTUS STRIATUS (Day)

STRIATED SABRE-TOOTHED BLENNY

Petroscirtes striatus Day, Proc. Zool. Soc., 1888, p. 262.

D. 40. A. 27. V. 3. Depth 6. Exceedingly large curved canines in lower jaw; smaller ones in upper. A single row of 14 teeth between canines in both jaws. Dorsal continuous, nearly same height throughout, and $\frac{3}{8}$ height of body. Ten broad vertical bands from base of dorsal to anal, separated by very narrow light lines. Dorsal and anal black edged and membranes with fine brown spots. Caudal light. $1\frac{1}{2}$ inches. Rock pools.

596. PETROSCIRTES LIENARDI Day

LIENARD'S BLENNY

Petroscirtes lienardi Day, Fish. India, 1875-78, p. 327, Pl. lxix, fig. 8 (plate 40).

D. 3, 10, 17. A. 18-20. V. 3. Depth $4\frac{1}{4}$. A short fringed orbital tentacle, another on nostril, several on cheeks and opercles and below symphysis of lower jaw. First 3 dorsal rays higher than and separated by a notch from the remainder. Light brown or stone, vertically banded and mottled with darker brown or grey. Dorsal light brown, marbled and spotted darker, and upper $\frac{2}{3}$ of first portion nearly black. Caudal yellow with dark spots. Anal with oblique rows of spots and a dark outer edge. Pectoral yellow internally with 2 basal black spots. Adam's Bridge.

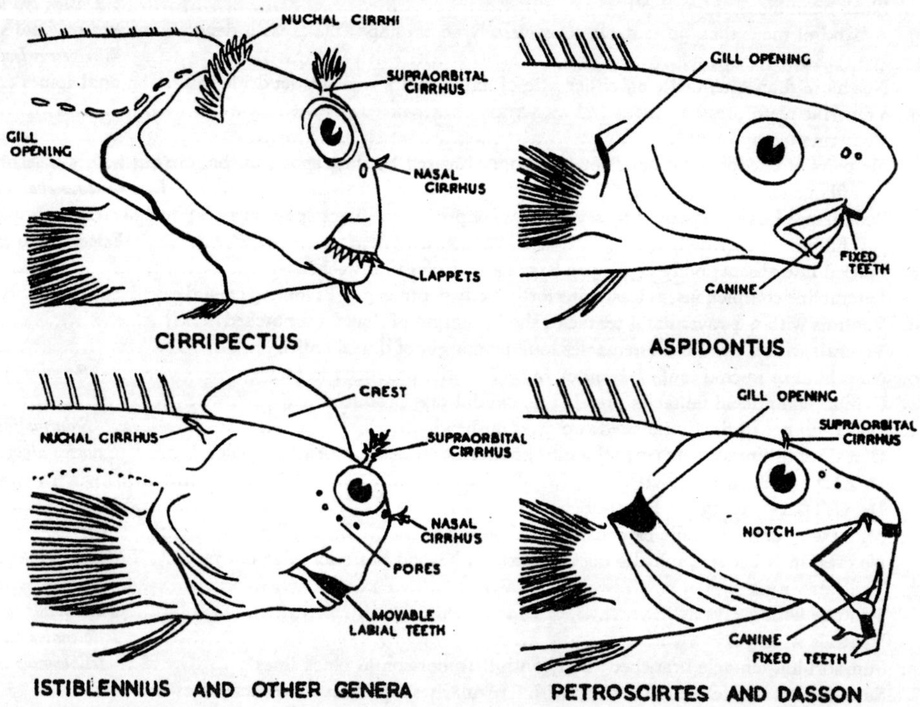

FIG. 14.
Principal differences between the various genera of the Family Blenniidae (Blennies), with particular reference to extent of the gill-opening, arrangement of teeth and tentacles.

597. DASSON BREVICEPS (Valenciennes) BLACK-BANDED BLENNY

Blennechis breviceps Cuvier and Valenciennes, Hist. Nat. Poiss., xi, 1836, p. 283.

D. 30. A. 19-20. V. 3. Depth $5\frac{1}{2}$. Small orbital tentacle. First dorsal commences above gill-opening. A black band along side of body. Dorsal fin dotted with black. Caudal yellowish. Adam's Bridge and pearl banks.

598. DASSON CYPRINOIDES (Valenciennes) WHITE-BANDED BLENNY

Blennechis cyprinoides Cuvier and Valenciennes, Hist. Nat. Poiss., xi, 1836, p. 286.

D. 30. A. 19-20. V. 3. Depth $4\frac{1}{4}$. Some irregularly placed open pores on head. Dorsal commences over hind border of opercle and does not quite reach caudal. Caudal truncate.

A white band along side and another below base of dorsal fin. Dorsal with dark basal band and 1-2 rows brown spots. Pectoral with basal brown spot. Anal brown with 2-3 rows of dark spots. Caudal yellow. Seven vertical bands along upper half of back. Coastal waters.

599. CIRRIPECTUS LEOPARDUS (Day) FRINGED LEOPARD BLENNY

Salarias leopardus Day, Proc. Zool. Soc., 1869, p. 518. *Salarias brevis. Blennius leopardus* Day, Fish. India, 1875-78, Pl. lxviii, fig. 6 (plate 40).

D. XII; 12-13. A. II, 14-15. V. I, 4. Depth $2\frac{1}{2}$-4 increasing with age. Transverse row of 30-36 tentacles across nape. Prominent branched supraorbital tentacle. Lips crenulate, upper with 24 lappets. Dorsal deeply notched. Anal lower than dorsal. Pale with small dark spots on all fins, head, throat and body. Body spots grouped in 5 or 10, and groups arranged in vertical bands. Coastal reefs.

600. ENTOMACRODUS EPALZEOCHILOS (Bleeker) TUFTED BLENNY

Salarias epalzeocheilos Bleeker, Nat. Tijds. Ned. Ind., xix, 1859, p. 333. *Salarias marmoratus* and (?) *frenatus*.

D. XIII; 15. A. II, 16. V. I, 4. Depth 4·8-5·0. Both sexes without crest. Nasal tentacle comprising 5-7 filaments on a common stalk. Supraorbital tentacle longer than eye and bearing 5-6 filaments. Upper lip strongly crenulate. Dentary with posterior canine. Dorsal deeply notched. Dark brown to tan with diffuse remnants of 6 paired vertical bars. 30-40 narrow vertical, alternately dark and light bands across upper lip and snout. Also small brown dots on head. Dorsal irregularly mottled with light and dark oblique lines or blotches. Anal without markings. Caudal with 2-3 diffuse rows of spots. Rock pools.

601. ENTOMACRODUS STRIATUS (Valenciennes) STREAKED BLENNY

Salarias striatus Cuvier and Valenciennes, Hist. Nat. Poiss., xi, 1836, p. 309. *Salarias marmoratus* Günther, Jour. Mus. Godeffroy, iv (13), 1877, Pl. cxvi, fig. B (plate 40).

D. XIII; 15-16. A. II, 16-17. V. I, 4. Depth 6·0-6·4. Both sexes without crest. Nasal tentacle comprising 4-7 filaments on a common stalk. Supraorbital tentacle shorter than eye and bears a few short filaments. Dorsal deeply notched. Dentary with single strong curved canine. Sandy with faint vertical body bands which are largely obscured by rows of black spots, some as large as pupil. Dusky spot behind eye. Head with brown specks but no vertical bands. Dorsals with oblique dark lines. Caudal with 4 broad dark bands. Coastal waters.

602. ALTICUS SALIENS (Forster) JUMPING BLENNY

Blennius saliens Forster, Itin., ii, 1788, p. 343. *Salarias tridactylus* Day, Fish. India, 1875-78, Pl. lxx, fig. 3 (plate 40).

D. XIV; 21-23. A. II, 24-27. V. I, 4. Depth 7·6-8·5. Crest well developed in male, triangular. Nasal tentacle simple, small and pointed. Supraorbital tentacle less than eye, with 2-3 simple filaments on each side. Both lips strongly crenulate. Males with prominent canine on dentary fitting into cavity in roof of mouth when latter is closed. Dorsal not deeply notched. Body dark with darker fins. Crest with dark spots which extend to opercle.

10-15 light vertical streaks on body. Narrow oblique lines on dorsal and 11 light spots along its base. Anal dark with light margin. Coastal waters.

603. ECSENIUS FRONTALIS (Valenciennes) TWO-COLOUR BLENNY

Salarias frontalis Cuvier and Valenciennes, Hist. Nat. Poiss., xi, 1836, p. 328. *Salarias furcatus* Johnstone, Suppl. Rept. Ceylon Pearl Oyster Fish., xv, 1904, Pl. i, fig. 4 (plate 40).

D. XII; 16-17. A. II, 18-20. V. I, 3. Depth 5·0-5·2. Nasal tentacle double. Forehead slightly projecting. Upper labial teeth more than 100 and independently movable. Dentary with 45-50 more or less fixed teeth. Dorsal with notch only in young. In large males outer rays of caudal are prolonged. Anterior part of body dark brown or red, sharply divided from posterior which is yellow. The lighter area is restricted to caudal peduncle in females. 2 inches. Pearl banks.

604. SALARIAS FASCIATUS (Bloch) YELLOW-BANDED BLENNY

Blennius fasciatus Bloch, Naturg. Ausländ. Fische, ii, 1786, p. 110. *Salarias fasciatus* Günther, Jour. Mus. Godeffroy, iv (13), 1877, Pl. cxv, fig. G (plate 40).

D. XII; 18-20. A. II, 19-21. V. I, 2. Depth 4·3-4·8. L. lat. curving down behind pectoral and ending there. No crest in either sex. Nasal tentacle bearing about 12 filaments. Supraorbital tentacle bushy and less than half eye. Nuchal tentacle bushy with 30 filaments. No canines. First few dorsal rays in adult males elongated. Light brown, banded and spotted with yellow and blue and sometimes blue spots along base of dorsal. Throat and chest with light coloured bands and white spots. Males and females similar. Coastal waters.

605. SALARIAS ALBOGUTTATUS Kner WHITE-SPOTTED BLENNY

Salarias alboguttatus Kner, Sitzb. Akad. Wiss. Wien, lvi, 1867, p. 724; Günther, Jour. Mus. Godeffroy, iv (13), 1877, Pl. cxviii, fig. B (plate 40).

D. XII; 18. A. II, 19-20. V. I, 2. Depth 5·3. Head with a low crest. Nasal and supra-orbital tentacles. Small posterior canines in lower jaw. Dorsal deeply notched. Light brown with 8 pairs of well marked vertical bands as wide as interspaces. An oval blue spot behind eye. Many small dark-edged white spots on head and 2-3 rows in lower posterior half of body. Spinous dorsal and pectoral brown. Soft dorsal white. Both dorsals with oblique dark bands. Anal light with dark outer edge. 2½ inches. Pearl banks.

606. ISTIBLENNIUS UNICOLOR (Rüppell) OLIVE BLENNY

Salarias unicolor Rüppell, Neue Wirbelth., Fische, 1835, p. 136; Day, Fish. India, 1875-78, Pl. lxx, fig. 6 (plate 40).

D. XII; 17-18. A. 18-19. V. I, 2. Depth 5-5½. Males with well-developed crest and a fringed supraorbital tentacle, 2-3 times diameter of eye. Females without crest and with a supraorbital tentacle subequal to eye. Nasal tentacle short. No canines. A very deep notch between spinous and soft dorsals. Olive brown. Males with blackish blotches on sides and dorsal, and 2-3 rows blue spots on body and head. Females marbled with brown lines isolating lighter patches, and upper ⅔ of body dotted with blue. Fins yellowish. Caudal with 6 vertical bands. Pectoral with basal brown band. Rock pools.

623. ZANCLUS CORNUTUS (Linnaeus) MOORISH IDOL

Chaetodon cornutus Linnaeus, Syst. Nat., ed. 10, 1758, p. 273. *Zanclus cornutus* Bleeker, Atlas Ichth. Ind. Neerl., ix, 1877, Pl. ccclxvi, fig. 2 (plate 42).

D. VII; 38-41. A. III, 32-35. Depth $1\frac{1}{4}$. First dorsal spine short, 3rd in young up to twice total length, shorter with age. A horn-like projection above eye increases with age. Yellow with 2 broad black or brown cross-bars. The anterior band is broader inferiorly and includes eye. Bands with narrow blue lines along them. Caudal dark with yellow margin and narrow blue submarginal line. 9 inches. Gulf of Manaar.

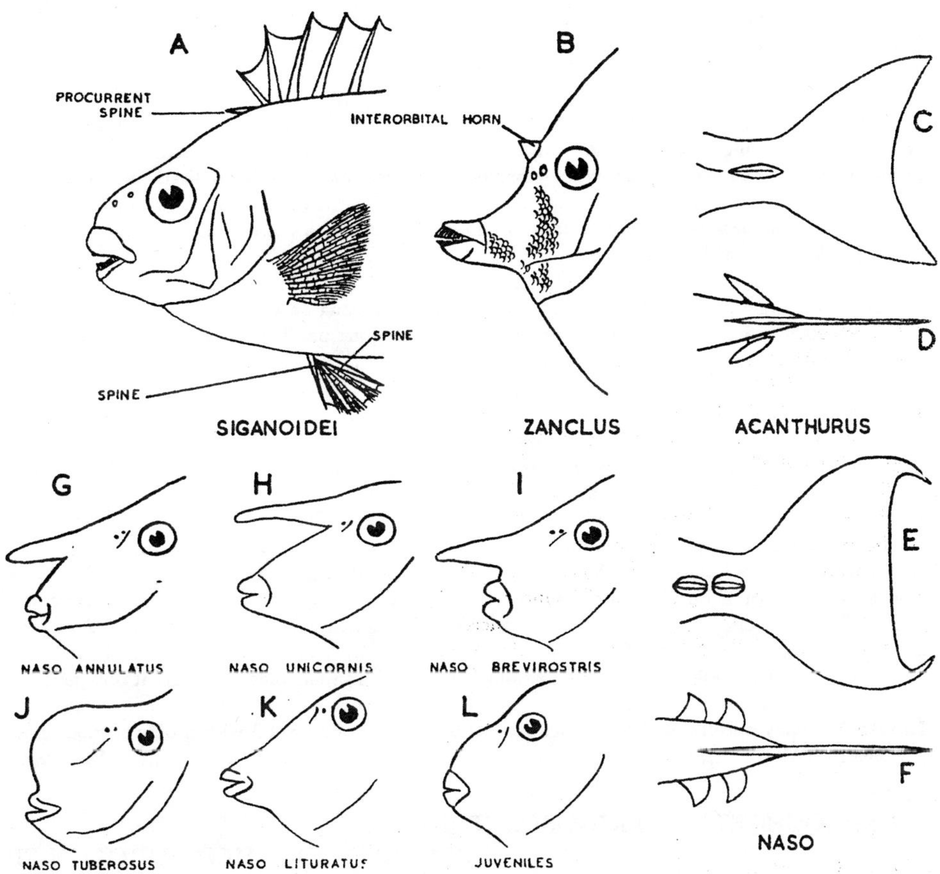

Fig. 15.

Explanation of characters of the Suborders Siganoidei (Spinefeet) and Acanthuroidei (Tang, Surgeon-fish). a.—External anatomy of *Siganus* showing dorsal and ventral fin spines. b.—Head of *Zanclus* showing bristle-like teeth and interorbital horn. c-d.—Tail of Surgeon-fish showing movable lateral spines. e-f.—Tail of Unicorn-fish showing lateral bony bucklers. g-l.—Profiles of species of *Naso* showing extent of development of the horn and its position relative to the mouth.

TANG, UNICORN-FISH, SURGEON-FISH

Family ACANTHURIDAE

Moderately elongate and deep fishes, rather compressed. Skin thick and leathery and covered with small rough scales. Mouth small with incisors. Base of tail with one or more bony plates or spines.

1. Caudal peduncle with one erectile spine on each side; anal with 3 spines 2
   Caudal peduncle with 1-2 immovable bony bucklers, each keeled or with a rigid spine; anal with 2 spines 11
2. Teeth movable, setiform .... *Ctenochaetus strigosus*
   Teeth fixed, broad and lobate 3
3. Dorsal with 7-9 spines 4
   Dorsal with 3-5 spines 10
4. A light-coloured ring around mouth .... *Acanthurus leucosternon*
   No light-coloured ring around mouth 5
5. Dorsal with 22, anal with 19-20 rays; 5 narrow transverse dark bands on head and body .... *Acanthurus triostegus*
   Dorsal with 24-29, anal with 23-27 rays; no transverse bands 6
6. Body with longitudinal bands or lines 7
   Body without longitudinal bands or lines 8
7. Light blue bands bordered by black, radiating from eye and along back .... *Acanthurus lineatus*
   Head and body with numerous narrow yellow and blue longitudinal lines .... *Acanthurus bleekeri*
8. Shoulder with a round black-edged yellow mark .... *Acanthurus pyroferus*
   Shoulder with a dark brown bar .... *Acanthurus gahm*
   Shoulder without any mark 9
9. Corner of mouth nearer hindborder of preopercle than eye .... *Acanthurus matoides*
   Corner of mouth equidistant between hindborder of preopercle and eye .... *Acanthurus weberi*
10. Black, head and shoulders reticulated with grey; caudal yellow .... *Zebrasoma xanthurus*
    Brown, olive or yellow, with minute bluish dots or fine undulating lines .... *Zebrasoma flavescens*
11. Adults with horn-like protuberance on forehead 12
    Adults with a convex hump on forehead .... *Naso tuberosus*
    Adults with no horn or hump on forehead .... *Naso lituratus*
12. Snout below horn long; caudal lobes produced into filaments .... *Naso unicornis*
    Snout below horn short; caudal lobes not produced 13
13. Head 4·4; body brownish; horn directed obliquely down with profile of snout under it oblique .... *Naso annulatus*
    Head 4·8-5; body purplish with indistinct vertical bands; horn horizontal with profile of snout under it nearly vertical .... *Naso brevirostris*

624. CTENOCHAETUS STRIGOSUS (Bennett)

SLENDER-TOOTHED SURGEON-FISH

Acanthurus strigosus Bennett, Zool. Jour., iv, 1828, p. 41. *Acanthurus ctenodon. Ctenochaetus strigosus* Herre, Philipp. Jour. Sci., xxxiv (4), 1927, Pl. xv, fig. 2 (plate 41).

D. VIII-IX; 26-29. A. III, 25. L. lat. 115. Tr. (11-12)+(39-40). Depth 2·3-2·7. Teeth in jaws in single series, movable, setiform, slender and somewhat curved. They are expanded at the top, flattened and serrated. Deep reddish-brown or yellow with narrow horizontal blue lines. Numerous dull red spots on head. Dorsal and anal fins similar to body, caudal with a light vertical band. 11 inches. Coastal waters, trawling grounds and pearl banks.

Iri dhattha, Iri orava (S). Ora, Thettan (T).

625. ACANTHURUS LEUCOSTERNON Bennett

WHITE-BREASTED SURGEON-FISH

Acanthurus leucosternon Bennett, Proc. Zool. Soc., 1832, p. 183; De Beaufort, Fish. Indo-Aust. Archip., ix, 1951, fig. 26 (plate 41).

D. IX; 28-29. A. III, 26-27. L. lat. 192-209. Tr. 17+93. Depth 2·4. 4-5 large truncate and lobate teeth in each side of both jaws. Bluish-grey, lighter on caudal peduncle. Head dark brown or black with patch on chest and ring around mouth white. Broad bluish band on shoulder. A wide black crescentic band across caudal margin joined to a narrower one which follows outer rays and across base. Fins greyish. 6½ inches. Coastal waters.

626. ACANTHURUS TRIOSTEGUS (Linnaeus) BARRED SURGEON-FISH

Chaetodon triostegus Linnaeus, Syst. Nat., ed. 10, 1758, p. 274. *Acanthurus hirudo. Hepatus triostegus. Acanthurus triostegus* Herre, Philipp. Jour. Sci., xxxiv (4), 1927, Pl. i, fig. 1 (plate 41).

D. IX; 22-23. A. III, 19-20. L. lat. 150-160. Tr. (17-18)+(72-77). Depth 2·3. Teeth deeply lobate, 6-8 in each side of both jaws. Greenish-brown above, white below, with a dark brown undulating stripe from chin to anal. Six dark vertical cross-bars, the first obliquely through eye and last across caudal peduncle in form of two spots divided by the lateral spine. 10 inches. Coastal waters.

Kara hamava, Kara paremuva (S).

627. ACANTHURUS LINEATUS (Linnaeus) BLUE-LINED SURGEON-FISH

Chaetodon lineatus Linnaeus, Syst. Nat., ed. 10, 1758, p. 274. *Acanthurus vittatus, Hepatus lineatus. Acanthurus lineatus* Herre, Philipp. Jour. Sci., xxxiv (4), 1927, Pl. xiv, fig. 1 (plate 41).

D. IX; 25-26. A. III, 26-27. L. lat. 110. Tr. 8+30. Depth 2·6-2·9. Teeth large, lobate and truncate, 6-7 in each side of both jaws. Body canary yellow with 9-10 more or less oblique blue bands from head to back and caudal fin. Lower third of body reddish-grey. 2-3 blue bands pass down in front of eye to snout, and about 7 more cross cheeks to join those on body. Others radiate from behind eye to those on body. Dorsal and anal dark with grey margin. Pectoral and ventral with outer ray blue. Caudal with semilunate blue band in middle and a blue margin. 11 inches. Coastal waters, Gulf of Manaar and trawling grounds.

Seveya (S).

628. ACANTHURUS BLEEKERI Günther

BLEEKER'S LINED SURGEON-FISH

Acanthurus bleekeri Günther, Cat. Fish., iii, 1861, p. 335. *Acanthurus mota, mata* or *aurolineatus. Acanthurus bleekeri* Herre, Philipp. Jour. Sci., xxxiv (4), 1927, Pl. xiv, fig. 2 (plate 41).

D. IX; 25-26. A. III, 24-25. L. lat. 200. Tr. 18+70. Depth 2·6-2·7. Teeth lobate, tapering to base, edge rounded, 8-9 in each side of both jaws. Yellow or brown in general tint but with yellow horizontal lines along body, about half width of blue interspaces. A narrow light band over caudal peduncle. Dorsal and anal blue with several longitudinal yellow bands as on body. 15 inches. Gulf of Manaar and trawling grounds.

629. ACANTHURUS PYROFERUS Kittlitz YELLOW SPOT SURGEON-FISH

Acanthurus pyroferus Kittlitz, Mus. Senckenb., i, 1834, p. 193. *Acanthurus tennentii. Acanthurus pyroferus* Kittlitz, *loc. cit.*, Pl. xii, fig. 2 (plate 41).

D. IX; 25. A. III, 22. Depth 2½. 8 lobate incisors on each side of both jaws. About 14 rows of scales between lateral line and base of last dorsal spine. Brown. Caudal with a broad white posterior edge and a dark base. Shoulder with an oblong blotch of yellow encircled by a thin black ring. Caudal spine with a backwardly directed process. Coastal waters.

630. ACANTHURUS GAHM (Forskål) DUSKY SURGEON-FISH

Chaetodon gahhm Forskål, Descr. Animal., 1775, p. 64. *Hepatus nigricans. Acanthurus mata* Day, Fish. India, 1875-78, Pl. xlviii, fig. 1 (plate 41).

D. IX; 26. A. III, 23. L. lat. 143-147. Tr. (13-14)+(65-80). Depth 2·7-3·0. Teeth slender and strongly lobate, 8 on each side of both jaws. Chocolate brown. A blackish postocular bar extends a short distance behind shoulder. Fins all deep brown with dusky tinge. Hind caudal edge narrowly white. 14 inches. Coastal waters.
Kalu orava (S).

631. ACANTHURUS MATOIDES Valenciennes
WHITE-TAILED SURGEON-FISH

Acanthurus matoides Cuvier and Valenciennes, Hist. Nat. Poiss., x, 1835, p. 204. *Acanthurus melas, Hepatus matoides* or *fuliginosus. Acanthurus matoides* Herre, Philipp. Jour. Sci., xxxiv (4), 1927, Pl. xvi, fig. 3 (plate 41).

D. IX; 25-26. A. III, 23-25. L. lat. 170-210. Tr. (16-20)+(60-80). Depth 2·3-2·7. Teeth slender, strongly lobate on tips and sides, 8-9 in each side of both jaws. Dark brown, rarely with slight indications of narrow light bands across cheeks and along body. Caudal with a white band across its base and on caudal peduncle, fading with age. Dorsal and anal dark brown with traces of light bands on soft portions. Outer half of pectoral yellow-green. 2 feet. Coastal waters, Gulf of Manaar and trawling grounds.
Orava, Peela (S).

632. ACANTHURUS WEBERI (Ahl) WEBER'S SURGEON-FISH

Hepatus weberi Ahl, Mitt. Zool. Mus. Berlin, xi (1), 1923, p. 37. *Hepatus* or *Acanthurus nigrofuscus. Acanthurus rubropunctatus* Rüppell, Atlas Reise Nordl. Afrika, 1828, Pl. xv, fig. 1 (plate 42).

D. IX; 25-26. A. III, 23-25. L. lat. 130-150. Tr. 18+72. Depth 2·3-2·4. 8-9 lobate teeth in each side of both jaws. Closely resembles *A. bleekeri* except for the lack of markings. Brown on body and fins. Colour of dorsal and anal darker than that of body and these fins may have some narrow light longitudinal bands. Caudal brown and finely edged with white posteriorly. 5 inches. Coastal waters.

633. ZEBRASOMA XANTHURUS (Blyth)
DUSKY SAIL-FINNED SURGEON-FISH

Acanthurus xanthurus Blyth, Prodr. Fauna Ceylon, Kelaart, Append., 1852, p. 50; Playfair, Fishes of Zanzibar, 1866, Pl. viii, fig. 4 (plate 42).

D. V; 24-25. A. III, 20-21. Depth 2½. A single row of 10 lobate incisors on each side of both jaws. Longest dorsal ray ⅘ of body depth. Scales on head and body very rough. Body blackish, caudal canary yellow. Head and shoulders reticulated with grey. End of pectoral stained yellow. 8 inches.
Raja orava (S).

634. ZEBRASOMA FLAVESCENS (Bennett)
YELLOW SAIL-FINNED SURGEON-FISH

Acanthurus flavescens Bennett, Zool. Jour., iv, 1828, p. 40. *Acanthurus rhombeus. Acanthurus flavescens* Günther, Jour. Mus. Godeffroy, ii (9), 1875, Pl. lxxvi (plate 42).

D. V; 24-25. A. III, 19-20. Depth 1·7-2·0. 12 lobate teeth in each side of both jaws. Scales very rough and semi-ossified. Lateral spines on caudal peduncle preceded by scales bearing tufts of spines. Usually uniformly dark brown. Older fish with head and body covered by minute grey dots which form streaks giving the appearance of wavy lines. Ventral fins black. 8½ inches. Coastal waters.

635. NASO TUBEROSUS Lacépède HUMP-NOSED UNICORN-FISH

Naso tuberosus Lacépède, Hist. Nat. Poiss., iii, 1802, p. 105. *Naseus tuberosus. Naseus punctulatus* Steindachner, Sitzb. Akad. Wiss. Wien, lxx (1), 1874, Pl. i (plate 42).

D. V; 27-29. A. II, 26-27. Depth 2¼-2¾ without tail. A bump on the snout scarcely developed in young and projecting beyond mouth in large specimens. Caudal peduncle with 2 bony bucklers, each with a broad, strongly compressed recurved spine. Back olive brown, lighter below. Back and upper sides with dark small round spots. Dorsal and anal dark brown with white margin and 2 rows black spots. Caudal with white edge. 2 feet. Coastal waters, Gulf of Manaar and trawling grounds.

636. NASO LITURATUS (Schneider) SMOOTH-HEADED UNICORN-FISH

Acanthurus lituratus Bloch and Schneider, Syst. Ichth., 1801, p. 216. *Naseus lituratus. Naso lituratus* Herre, Philipp. Jour. Sci., xxxiv (4), 1927, Pl. xvi, fig. 1 (plate 42).

D. VI; 28-29. A. II, 28-29. Depth 2·5-2·8. Upper profile of snout very oblique but without frontal horn or bump. 14 teeth with round tips in each side of both jaws. Caudal lunate, lobes filamentous with age. Scales rough. Dull uniform brown, paler below. A yellow or orange band from eye to corner of mouth and upward along lower preopercular edge. Interorbital and postorbital region sometimes orange. Caudal bucklers yellow and with recurved spines. Dorsals and anal black with creamy areas anteriorly. Caudal brown with white border. 20 inches. Coastal waters.

637. NASO UNICORNIS (Forskål) LONG-SNOUTED UNICORN-FISH

Chaetodon unicornis Forskål, Descript. Animal., 1775, p. 63. *Naseus unicornis. Naso unicornis* Herre, Philipp. Jour. Sci., xxxiv (4), 1927, Pl. vi, fig. 1 (plate 42).

D. V-VI; 28-31. A. II, 26-30. Depth 2·8-3·1. Degree of development of the horn varies with age from a small bump over eye in young to a long straight horn in older fish. The snout below horn is always long and pointed. Older specimens with caudal lobes produced into filaments. Grey-brown above, tinged with yellow on abdomen. Young stages with dark band on nape and about 6 series of dark spots on body. Dorsal and anal with basal dark spots, oblique bars and longitudinal orange stripes and outer blue edge. Caudal yellowish. 22 inches. Coastal waters and trawling grounds.

638. NASO ANNULATUS (Quoy and Gaimard) BANDED UNICORN-FISH

Priodon annulatus Quoy and Gaimard, Voy. Uranie et Physic., Zool., 1825, p. 377. *Naso marginatus* Herre, Philipp. Jour. Sci., xxxiv (4), 1927, Pl. vi, fig. 2 (plate 42).

D. V; 28. A. II, 28-29. Depth 2·4-2·8. Development of horn proceeds with age from a small protuberance in young. The horn is directed slightly downwards. Caudal peduncle with 2 keeled bucklers in adults. Caudal slightly emarginate. Slaty brown or greyish. Sides with many short dusky or bluish transverse parallel lines. Upper and lower boundaries of body and head with numerous close-set brown spots. Dorsal and anal uniform brown. Caudal with dusky base. 20 inches. Coastal waters.

639. NASO BREVIROSTRIS (Valenciennes) SHORT-NOSED UNICORN-FISH

Naseus brevirostris Cuvier and Valenciennes, Hist. Nat. Poiss., x, 1835, p. 277. *Naso brevirostris* Herre, Philipp. Jour. Sci., xxxiv (4), 1927, Pl. v, fig. 3 (plate 42).

D. VI; 28-29. A. II, 28-29. Depth 2·6-2·8. Frontal horn horizontal and level with middle of eye, increasing in size with age and represented by a bump in the young. Caudal spines 2, knob-like in young and blade-like in older fish. Caudal truncate or slightly emarginate. Greyish with numerous blue transverse spots or lines on posterior half of body. Caudal with a light edge. Dorsals and anal dark with narrow white margin and submarginal black band. 17 inches. Coastal waters and Gulf of Manaar.

Suborder TRICHIUROIDEI

Perciform fishes with maxillaries fixed to non-protractile premaxillaries. Bases of caudal fin rays not overlapping the hypural bone. Pectoral fins placed low.

Much compressed, elongate and band-like with tail tapering to a point; no caudal fin; anal reduced to numerous small spines .... *Family* TRICHIURIDAE, p. 217

Oblong compressed body with forked caudal fin; finlets behind dorsal and anal .... *Family* GEMPYLIDAE, p. 217

RIBBON-FISH

Family TRICHIURIDAE

Elongate, highly-compressed band-like body without scales, silvery. Dorsal and anal fins feeble but long and with numerous rays. Body tapers behind to a point. No caudal fin. Marine fishes entering estuaries.

First anal spine well-developed, more than half diameter of eye .... *Trichiurus savala*
First anal spine minute, not longer than following ones .... *Trichiurus haumela*

640. TRICHIURUS SAVALA Cuvier — SMALL-HEADED RIBBON-FISH

Trichiurus savala Cuvier, Regne Animal, ed. 2, 1829, p. 219; Day, Fish. India, 1875-78, Pl. xlvii, fig. 4 (plate 13).

D. 105-134. A. 72-80. Ventrals absent. Depth 14·8-19·8. Head 7·4-10·5. Eye 6·7-10·0., 2·0-3·5 in snout. Fangs in both jaws barbed. First anal spine more than ½ diameter of eye, longer than others. Anal originates below 40th-43rd dorsal ray. Silvery. Head brown above. Dorsal and pectorals yellowish with a rather broad black terminal border. 3 feet. Coastal waters, trawling grounds and Gulf of Manaar.
Savaleya (S). Savalai (T).

641. TRICHIURUS HAUMELA (Forskål) — LARGE-HEADED RIBBON-FISH

Clupea haumela Forskål, Descript. Animal., 1775, p. 72. *Trichiurus haumela* Fowler, Jour. Acad. Nat. Sci. Phila., xii, 1904, Pl. vii, lower (plate 13).

D. 127-133. A. 75-80. Ventrals absent. Depth 12·9-15·8. Head 6·8-7·6. Eye 5·8-6·6, 2·0-2·25 in snout. Fangs in both jaws barbed. First anal spine small, less than diameter of pupil and subequal to other spines. Anal originates below 40th dorsal ray. Silvery. Dorsals and pectorals light yellow, the dorsal blackish anteriorly and along its upper border. Pectorals sometimes dusky. 3 feet or more. Coastal waters and trawling grounds.

SNAKE MACKEREL

Family GEMPYLIDAE

Elongate compressed body with minute scales. Large mouth with powerful fangs. Soft dorsal shorter than spinous portion. Dorsal and anal with or without detached finlets. Caudal forked. Lateral line sometimes multiple.

642. REXEA PROMETHEOIDES (Bleeker) — SNAKE MACKEREL

Thyrsites prometheoides Bleeker, Acta Soc. Sci. Indo-Neerl., i, 1856, p. 42. *Promethichthys solandri*. *Thyrsites bengalensis* Alcock, Illust. Zool. Investigator, iii, 1895, Pl. xv, fig. 10 (plate 13).

D. XVIII; II, 13; 2. A. II, 11; 2. P. 15. V. I. Depth 7. Eye large, 4 in head. Lower jaw prominent, the 2 anterior canines outside of mouth. Caudal deeply forked. Silvery with brown back. Fins hyaline. A black blotch between 1st and 3rd dorsal spines. 7 inches. Deep waters off the coast.

Suborder SCOMBROIDEI

Perciform fish with maxillaries fixed to non-protractile premaxillaries forming a pointed beak. Caudal fin rays overlapping the hypural bone.

1. Snout not produced into a spear 2
   Snout produced as a spear 3
2. Moderately large scales covering the whole body; 5-6 detached finlets ... *Family* SCOMBRIDAE, p. 218
   Scales restricted to pectoral region forming a corselet; 6-9 detached finlets ... *Family* THUNNIDAE, p. 218
   Body apparently naked and without corselet; 7-10 detached finlets .... *Family* SCOMBEROMORIDAE, p. 220
3. Ventrals absent .... *Family* XIPHIIDAE, p. 222
   Ventrals spear-like .... *Family* HISTIOPHORIDAE, p. 222

MACKEREL

Family SCOMBRIDAE

Fusiform body covered with small or relatively large scales which are not restricted to a specialized corselet, but larger on cheeks and below pectoral origin. Eyes with well developed adipose lid. Teeth in jaws small. Two slight ridges on each side of tail but no wide keel. 5-7 detached finlets behind dorsal and anal fins.

643. RASTRELLIGER KANAGURTA (Cuvier) RAKE-GILLED MACKEREL
Scomber kanagurta Cuvier, Regne Animal, ii, 1817, p. 313. *Scomber microlepidotus* and (?) *japonicus*. *Scomber kanagurta* Fowler, Jour. Acad. Nat. Sci. Phila. (2), xii, 1904, Pl. xii, fig. 1 (plate 43).

D. X; 12; 5. A. 12; 5. P. 2, 17. Depth 4·1-4·7. Head longer than high. A single series of very small teeth in jaws, almost obsolete in adults. No teeth on palate. About 35-38 gill-rakers on lower arm of first arch, longest equal to distance from pupil to snout, projecting well into the mouth cavity. Back green, silvery below. A row of 16 spots along back near base of dorsal. Dorsals yellow with black tips. Caudal and pectoral also yellow but anal and ventrals dusky. Large specimens may have longitudinal lines along back. 15 inches. Coastal waters, Gulf of Manaar and trawling grounds.
Kumbalava, Maha kara bolla (S). Ailai, Karung kuluttan, Kumbala (T).

TUNA

Family THUNNIDAE

Mackerel-like fishes usually with a more robust body and slender caudal peduncle. Scales minute or apparently absent except in pectoral region where they form a specialized band or corselet. Each side of the caudal peduncle is a wide keel. Dorsal and anal fins followed by 6-9 detached finlets.

1. Spinous dorsal and soft dorsal separated .... *Auxis thazard*
   Spinous dorsal and soft dorsal close together 2
2. Body naked except for corselet 3
   Body with perceptible scales other than in corselet 4

3. Palate without teeth; origin of anal below end of soft dorsal; longitudinal bands extending to belly .... *Katsuwonus pelamis*
   Palatine teeth present; origin of anal slightly behind that of soft dorsal; no longitudinal bands on belly but several black blotches on breast .... *Euthynnus affinis*
4. Second dorsal and anal falcate, at least as long as head .... *Neothunnus macropterus*
   Second dorsal and anal normal, second dorsal not higher than spinous dorsal .... *Kishinoella tonggol*

644. AUXIS THAZARD (Lacépède) FRIGATE MACKEREL

Scomber thazard Lacépède, Hist. Nat. Poiss., iii, 1802, p. 9. *Auxis hira* Kishinouye, Jour. Coll. Agric. Tokyo Univ., viii (3), 1923, Pl. xxxi, fig. 55 (plate 43).

D. X; 11; 6-9. A. 14; 6-8. P. 23. Depth 4. Teeth in jaws minute, absent from palate. Second dorsal very low, almost three times its base distant from spinous dorsal. Anal similar to second dorsal. About 30 slender gill-rakers. Bluish brown above with irregular markings of blue or green on back. Silvery below. Fins brownish. 24 inches. Coastal waters.

645. KATSUWONUS PELAMIS (Linnaeus) STRIPED TUNA, SKIPJACK

Scomber pelamis Linnaeus, Syst. Nat., ed. 1, 1758, p. 297. *Thynnus pelamys*, *Euthynnus pelamis*. *Katsuwonus pelamis* Shapiro, SCAP Nat. Resources Rept. 104, 1948, fig. 5 (plate 43).

D. XV; II, 12-14; 8. A. III, 12; 7. P. 2, 24. Depth 3¾. About 50 small conical teeth in each side of both jaws. No teeth on vomer and palate. Corselet well-developed, the scales on its anterior part, especially those on back, larger and rhombic. Lateral line with sharp curve under soft dorsal. G.R. 35. Back dark blue with indistinct light transverse marks. Belly and sides silvery with 4-6 dark longitudinal bands. Anal finlets white, other fins and finlets dark. 36 inches. Visit coastal waters in large schools which follow the anchovies.

Alagoduva, Baleya (S).

646. EUTHYNNUS AFFINIS (Cantor) MACKEREL TUNA

Thynnus affinis Cantor, Jour. Asiatic Soc. Bengal, xviii, 1850, p. 1088. *Thynnus thunnina*. *Euthynnus alletteratus* Serventy, C.S.I.R. (Aust.) Pamph., 104, 1941. Pl. iii (plate 43).

D. XV; II, 11; 8. A. II, 12; 7. P. 2, 24. Depth 3·9-4·8. A single series of slightly curved conical teeth in both jaws, those in lower jaw stronger. Vomer and palatines both with minute teeth. Hindborder of corselet formed by a line from base of soft dorsal to behind ventral tips, with a strong emargination along lateral line to below 8th dorsal spine. G.R. 24. Bluish, deeper on back and silvery below. A number of dark oblique broken lines in area above lateral line behind corselet. Several, usually about 3 black blotches in area between pectoral and ventrals. Fins bluish. Ventral with a dark blotch. 2 feet. Coastal waters and Gulf of Manaar.

Atavalla, Ragoduva, Sureya (S). Shurai (T).

647. NEOTHUNNUS MACROPTERUS (Schlegel) YELLOW-FIN TUNA

Thynnus macropterus Schlegel, Fauna Japonica, Pisces, 1824, p. 98. *Thunnus* or *Germo macropterus*. *Thunnus (Neothunnus) argentivittatus* Deraniyagala, Col. Atlas Vert. Ceylon, i, 1952, Pl. xxv (plate 43).

D. XIII-XIV; 14; 9. A. 14-15; 8-9. Depth 3·6-3·7. Teeth in single series in both jaws and others on vomer and palate. Soft dorsal and anal long and pointed especially in older fish, at least as long as head. Posterior part of corselet indistinct. Body greyish, blue black above, silvery below. Belly and sides with small silvery elongate spots with dark edge. Fins mostly bright yellow with dark tips. 6 feet. Coastal waters and occasionally trawling grounds.

As geddi kelawalla, Pihatu kelawalla, Kelawalla, Hovalla (S). Kelavalai (T).

648. KISHINOELLA TONGGOL (Bleeker) BLUE-FIN TUNA

Thynnus tonggol Bleeker, Nat. Tijds. Ned. Ind., i, 1851, p. 356. *Thynnus* or *Neothunnus rarus*. *Kishinoella tonggol* Roughley, Fish and Fisheries Aust., 1951, Pl. xlvi (plate 43).

D. XII-XIII; II, 12; 8-9. A. II, 12; 8-9. P. 2, 27. Depth 4·3. Rather large conical teeth in a single series in jaws. Patches of small teeth on vomer and palate. Corselet with 2 posterior emarginations. Second dorsal and anal not prolonged, about ½ body depth. Greyish-blue to black above, sides silvery with elongated spots in 4 or 5 longitudinal rows. Dorsals, pectorals and ventrals blackish. Second dorsal and anal washed with yellow. Finlets similar. 2½ feet. Coastal waters.

SPANISH MACKEREL, SIER, WAHOO

Family SCOMBEROMORIDAE

Elongate mackerel-like fishes without perceptible scales or corselet. Jaws with large, sharp flattened teeth. Caudal peduncle with a medium sized lateral keel and a lesser ridge above and below. Lateral line simple or branched, not multiple.

1. Spinous dorsal with 24-27 radials; jaws beak-like; gill-rakers absent; lateral line with long branches .... *Acanthocybium solandri*

   Spinous dorsal with 15-17 radials; jaws not beak-like; a few gill-rakers at least; lateral line never with long branches 2

2. Lower limb of first gill-arch with 3 rudimentary rakers; body with numerous vertical narrow cross-bars .... *Cybium commersoni*

   Lower limb of first gill-arch with more than 3 well-developed rakers; body without vertical cross-bars 3

3. Sides with longitudinal streaks; second dorsal originates midway between eye and caudal base .... *Indocybium lineolatum*

   Sides with round spots; second dorsal originates midway between snout tip and caudal base .... *Indocybium guttatum*

649. ACANTHOCYBIUM SOLANDRI (Cuvier) WAHOO

Cybium solandri Cuvier and Valenciennes, Hist. Nat. Poiss., viii, 1831, p. 192. *Acanthocybium sara*. *Acanthocybium solandri* Kishinouye, Jour. Coll. Agric., Tokyo Univ., viii (3), 1923, Pl. xx, fig. 31 (plate 43).

D. XXV-XXVII; I, 6; 10. A. II, 11; 8. Body elongate and fusiform. Depth 7. About 50 serrated triangular teeth in each jaw. Jaws produced and beak-like. Lateral line arched

anteriorly and with long straight branches at right angles throughout its length. Gill-rakers absent but gill-lamellae reticulated. Back dark slaty blue, belly light silvery grey. Sides with about 30 silvery grey double transverse cross-bars. Spinous dorsal blue. Soft dorsal blackish. 6 feet or more. Coastal waters.
Hera maha, Pirimi thora, Thal thora (S). Savara (T).

650. CYBIUM COMMERSONI (Lacépède) BARRED SPANISH MACKEREL

Scomber commerson Lacépède, Hist. Nat. Poiss., ii, 1800, p. 598. *Scomberomorus commersoni. Cybium commerson* Munro, Mem. Qld. Mus., xii (2), 1943, Pl. vi, fig. B (plate 43).

D. XVI; III, 13-14; 9-10. A. 17; 9-10. P. 2, 20-21. Depth 4·8-6·2. Teeth minutely serrated. G.R. rudimentary, about 3 on lower limb of first gill-arch. Lateral line with a sharp bend behind soft dorsal. Dark blue or greenish above, silvery grey on belly and sides with numerous wavy narrow greyish vertical cross-bars which in young may be reduced to blotches. Cross-bars are restricted mainly to the lower half of the body. First dorsal black, at least anteriorly in young. Other fins blue-grey. 3 feet, or occasionally larger. Coastal reefs, trawling grounds and Gulf of Manaar.
Ahin thora, Bambāra thora, Anjilava (S). Arekula (T).

651. INDOCYBIUM LINEOLATUM (Cuvier)
STREAKED SPANISH MACKEREL

Cybium lineolatum Cuvier and Valenciennes, Hist. Nat. Poiss., viii, 1831, p. 170. *Scomberomorus lineolatus. Cybium, Sawara* or *Scomberomorus interruptus. Scomberomorus leopardus* Smith, Fish. Sth. Africa, 1949, Pl. lxiv, fig. 841 (plate 43).

D. XV-XVII; V, 13-15; 9-10. A. IV, 16; 9-10. P. 2, 22. Depth 6-6½. Strong compressed pointed teeth in jaws, those in lower largest. G.R. 2+9, short. Lateral line almost straight. Blue grey with purplish reflections above and grey to white below. About 4 lateral longitudinal rows of yellowish brown elongate spots, resembling interrupted bars. Spinous dorsal white, anteriorly black. Fins brownish. 26 inches. Coastal reefs, trawling grounds and Gulf of Manaar.
Alu thora, Ganu thora, Sekkal thora, Haru mas thora (S). Pulli arakula (T).

652. INDOCYBIUM GUTTATUM (Bloch and Schneider)
SPOTTED SPANISH MACKEREL

Scomber guttatus Bloch and Schneider, Syst. Ichth., 1801, p. 23. *Sawara* or *Scomberomorus guttatus. Cybium guttatum* Day, Fish. India, 1875-78, Pl. lvi, fig. 4 (plate 43).

D. XV-XVI; IV, 14-16; 8-9. A. IV, 16; 8-9. P. 20-23. Depth 4·7-5·6. Moderate, compressed teeth in both jaws, those in lower longer. G.R. (1-2)+(8-10). Lateral line almost straight. Dorsally dark, sides silvery with 3 rows of almost circular spots as large as pupil of eye. The band crosses the lateral line and originates under pectoral fin. Spinous dorsal black anteriorly and white basally on posterior third. Other fins dark. 2 feet. Coastal waters, Gulf of Manaar and trawling grounds.
Alu thora, Genu thora, Sekkal thora, Aru maha, Anjilava (S).
Nei arekula, Anjila (T).

SWORDFISH

Family XIPHIIDAE

Elongate robust body, rough in young and smooth and scaleless in adults. Adults with 2 dorsal fins, the second small and situated well back. Caudal peduncle with a keel. Upper jaw strongly produced into a powerful sword. Pelvics absent. Laminae of gill-arch united into a single unit.

653. XIPHIAS GLADIUS Linnaeus BROADBILL SWORDFISH

Xiphias gladius Linnaeus, Syst. Nat., ed. 10, 1758; Deraniyagala, Col. Atlas Vert. Ceylon, i, 1952, Pl. xxix (plate 43).

D. 40; 4. A. 18; 4. P. 2, 14. Depth 6. Sword from front of eye 3 in total length, flattened. Grey or bronze purple above, lighter below. Attains 15 feet and well over 1000 lb. Deep water beyond the continental shelf.
Kadu koppara (S).

SAILFISH, SPEARFISH, MARLIN

Family HISTIOPHORIDAE

Elongate robust body, with fine lancet-like scales in skin. Dorsal fin high and continuous, or lower and almost continuous with a separate hind portion. 2 anal fins. Ventrals with either 1, or 2-3 rays. 2 keels each side of caudal peduncle. Upper jaw produced into a rounded spear, never as long as the flattened sword of the preceding family.

1. Dorsal much higher than body depth .... *Histiophorus gladius*
   Dorsal lower than body depth 2
2. Middle dorsal rays equal at least half height of anterior lobe .... *Tetrapturus brevirostris*
   Middle dorsal rays shorter than half height of anterior lobe 3
3. 12-20 pale cross-bars; spear very long; head with spear 2·4 in length .... *Makaira mitsukurii*
   Cross-bars absent or obscure; spear short; length of spear from tip to eye 1·5 rest of head .... *Makaira indica*

654. HISTIOPHORUS GLADIUS (Broussonnet)
INDIAN OR BANANA SAILFISH

Scomber gladius Broussonnet, Mem. Acad. Sci., 1786, p. 454. *Histiophorus immaculatus*. *Istiophorus gladius* McCulloch, Rec. Aust. Mus., xiii (4), 1921, Pl. xxiv, fig. 1 (plate 43).

D. XLII-XLVIII; 6-7. A. X-XI; 6-7. Spinous dorsal much elevated and separated from second dorsal by a series of 6-12 short stubby spines. Pectoral slightly longer than head from tip of lower jaw. Ventrals much longer than pectorals, with 3 rays. Spear very long, from tip to eye $3\frac{1}{2}$ times remainder of head. Blue on body and fins. Body with traces of vertical broken bars. Dorsal blue, basally purplish with numerous black spots. 9-10 feet and over 230 lb. Coastal waters.
Thalapatha (S). Myl meen (T).

655. TETRAPTURUS BREVIROSTRIS (Playfair) SHORT-NOSED SWORDFISH

Histiophorus brevirostris Playfair, Fishes Zanzibar, 1866, p. 53. *Tetrapturus tenuirostratus* or *acutirostratus*, Deraniyagala, Col. Atlas. Vert. Ceylon, i, 1952, Pl. xxvii (plate 43).

D. IV, 32-35; I, 6. A. III, 10; I, 6. P. 19. Moderately stout. Depth of body 5·6. Median dorsal spines ⅔ height of longest. No short spines between the two dorsals. Length of spear from tip to tip of lower jaw exceeds length from tip of lower jaw to eye. Ventrals shorter than pectorals. Back bluish black. Dorsal deep blue. Sides and belly silvery white. Caudal dark. 10 feet. Coastal waters.
Habara (S).

656. MAKAIRA MITSUKURII (Jordan and Snyder) STRIPED MARLIN

Tetrapturus mitsukurii Jordan and Snyder, Jour. Coll. Sci. Univ., Tokyo, xv (2), 1901, p. 303. *Makaira mitsukurii* Walford, Pacif. Marine Game Fish., 1937, Pl. xli (plate 43).

D. III, 38-44; 6. A. 14; 7. Pectorals 1½ times length of ventrals which equal height of dorsal lobe. Head with spear 4·2 in body to base of caudal. Caudal very long, lobes 1¼ times pectoral. Colour blue-black with about 17 pale bluish bars on sides reaching to belly. Dorsal black, usually unspotted. Pectorals dark grey edged with black. Grows to 780 lb. or more. Coastal waters.
Seraman koppara (S).

657. MAKAIRA INDICA (Cuvier) INDIAN SPEARFISH; JOO-HOO

Tetrapturus indicus Cuvier and Valenciennes, Hist. Nat. Poiss., viii, 1831, p. 286. *Makaira indica* Deraniyagala, Col. Atlas. Vert. Ceylon, i, 1952, Pl. xxviii (plate 43).

D. III; 33-35; I, 6. A. III, 10-11; I, 6. Height of dorsal lobe less than length of ventral fin, 1·3 to 1·0 in body depth. Dorsal fin low posteriorly. Spear long, 5 in body with caudal. Uniform bluish to dark grey. Dorsal blue with a few dark spots. Caudal and pectorals dark. Up to 2½ cwt. Coastal waters.
Ahin koppara, Kalu koppara, Dhappara, Ghappara Kopparan, Kopparaikulla (T).
Kapparava, Makara, Sapparava (S).

Suborder STROMATEOIDEI

Perciform fishes with pelvic bones very loosely attached to pectoral arch. Ventrals if present, thoracic or subthoracic. Ventrals I, 5. Posterior part of pharynx with lateral sacs which are internally provided with papillae supported on bone which bears teeth. Scales normal, cycloid.

POMFRETS

Family STROMATEIDAE

Ovate, compressed body covered with moderately small scales. Head mostly scaly. Lateral line simple and complete. Moderately small mouth with weak jaws and a single series of slender pointed teeth. Dorsal with a distinct spinous portion but spines more or less wholly embedded in the skin.

1. Lateral line not keeled 2
   Posterior portion of lateral line keeled .... *Parastromateus niger*
2. Posterior part of dorsal comprising rays of equal length; caudal deeply forked .... *Pampus argenteus*
   Posterior part of dorsal comprising rays of gradually decreasing height; caudal emarginate
   .... *Pampus chinensis*

658. PARASTROMATEUS NIGER (Bloch) BROWN POMFRET

Stromateus niger Bloch, Ichth., 1788, p. 422. *Stromateus fiatola. Stromateus niger* Day, Fish. India, 1875-78, Pl. liii, fig. 4 (plate 42).

D. IV; I, 41-46. A. II, 35-40. P. 23-24. L. lat. ca. 100. Depth 1·9-2·4. Eye with narrow adipose lid. Maxillary reaches to front border of eye. Teeth almost absent in adults, in young are needle-like, inserted on a ridge. Spinous dorsal embedded in skin. Ventrals well-developed in young, absent in adults. Caudal deeply forked. Deep brown or grey with blue reflections; cheeks, opercles and abdomen paler. Dorsal and anal grey-brown with black margins. Young with dorsal and anal black and caudal yellow with 3 brown cross-bars. 2 feet. Occurs in shoals off coast, Gulf of Manaar and trawling grounds.

659. PAMPUS ARGENTEUS (Euphrasen) SILVER POMFRET

Stromateus argenteus Euphrasen, Kongl. Vetensk. Acad. Handl. Stockholm, ix, 1788, p. 49. *Stromateus cinereus* Day, Fish. India, 1875-78, Pl. liii, fig. 3 (plate 42).

D. V-IX; I, 37-43. A. I, 38-43. P. 23-26. Depth 1·8-2·1. Eye with feeble adipose lid. Narrow row of villiform teeth in jaws. Rays of last $\frac{7}{10}$ of dorsal, low, slightly less than eye, equal in length. Origin of anal behind that of dorsal. Anal lobe longer than dorsal lobe. Caudal deeply forked. Upper surfaces neutral grey with purplish sheen. Sides of head and body silvery, fading to white on abdomen. Everywhere covered in small black dots. Dark spot on upper part of opercle. Dorsal and anal grey and similarly dotted, outer half darker. Other fins yellowish. 12 inches. Coastal waters, Gulf of Manaar and trawling grounds.

Sudhu vaouvaleya (S). Vella vavel (T).

660. PAMPUS CHINENSIS (Euphrasen) CHINESE POMFRET

Stromateus chinensis Euphrasen, Kongl. Vetensk. Acad. Handl. Stockholm, ix, 1788, p. 53. *Stromateus sinensis* Day, Fish. India, 1875-78, Pl. li, C, fig. 6 (plate 42).

D. 43-50; A. 39-42. P. 23-25. Depth 1·4-1·6. Eye with narrow adipose lid. A row of slender truncate teeth in jaws, those of upper smaller. Dorsal rays rapidly decrease in length after 16th. Anal similarly concave along its border. Caudal emarginate in young, more slightly forked in adults. Brownish grey with metallic reflections, lighter and more silvery towards the abdomen. Body dotted all over brown, larger points silvery in centre. Fins silver grey with black margin. Young grey with stellate spots and fins nearly black. 8 inches. Coastal waters, trawling grounds and Palk Straits.

Kalu vaouvaleya, Kiri vaouvalaya (S). Karappu vavel (T).

PLATE I

Cat Sharks, Tawny Sharks, Man-eating Sharks and Dog Sharks.

PLATE 2

Sharks, Saw-fishes, Skates, Rays and Electric Rays

PLATE 3

Rays, Stingrays, Eagle Rays, Cow Rays and Devil Rays.

PLATE 4

Sardines, Herrings and Shads.

Plate 5

Sprats, Anchovies and Naked Heads.

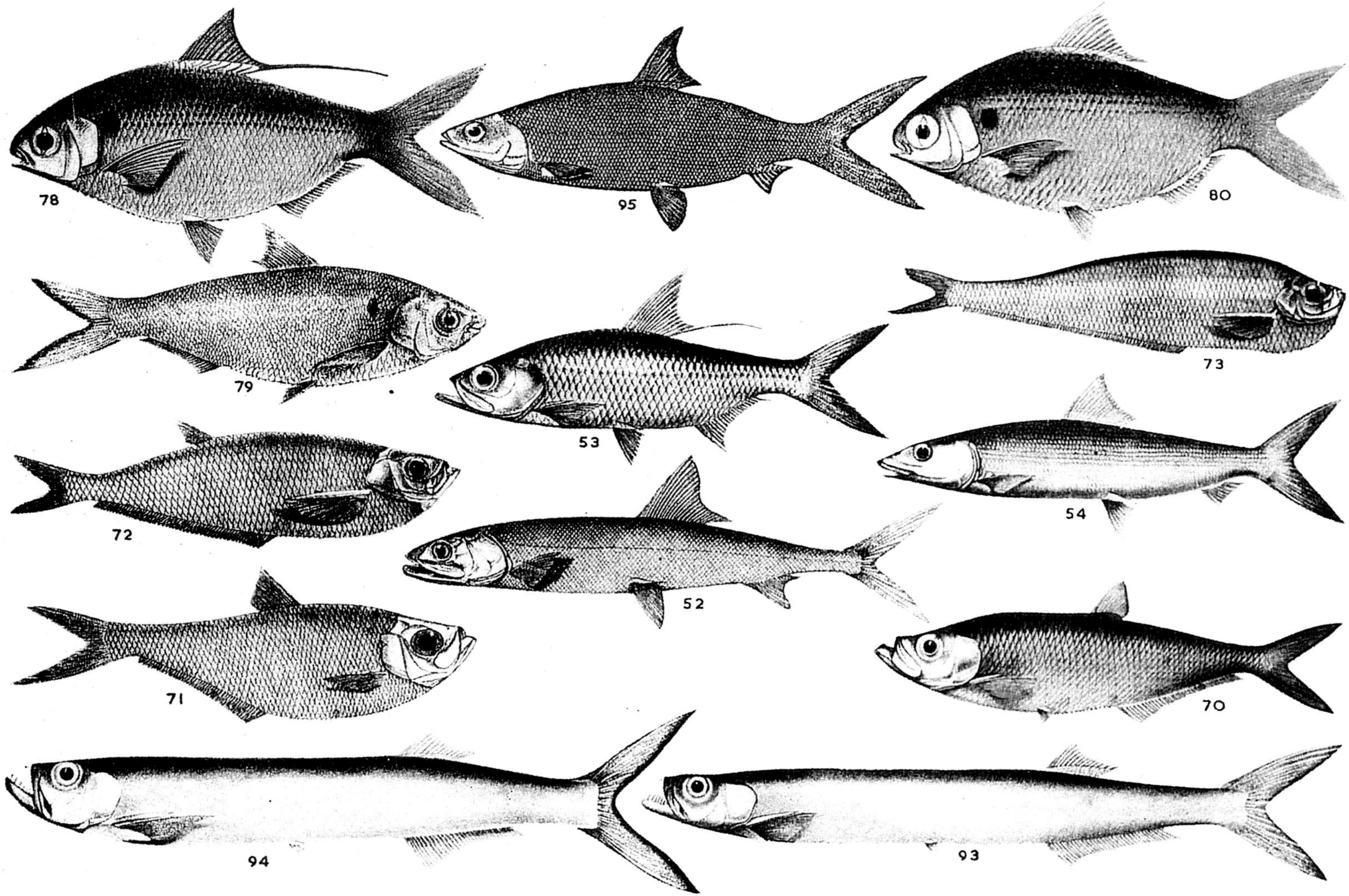

PLATE 6

Tarpon, Lady Fish, Shads, Gizzard Shads, Herrings and Wolf Herrings.

PLATE 7

Carps, Carplets, Barbs, Rasboras, Labeos and Stone Suckers.

PLATE 8

Carplets and Barbs.

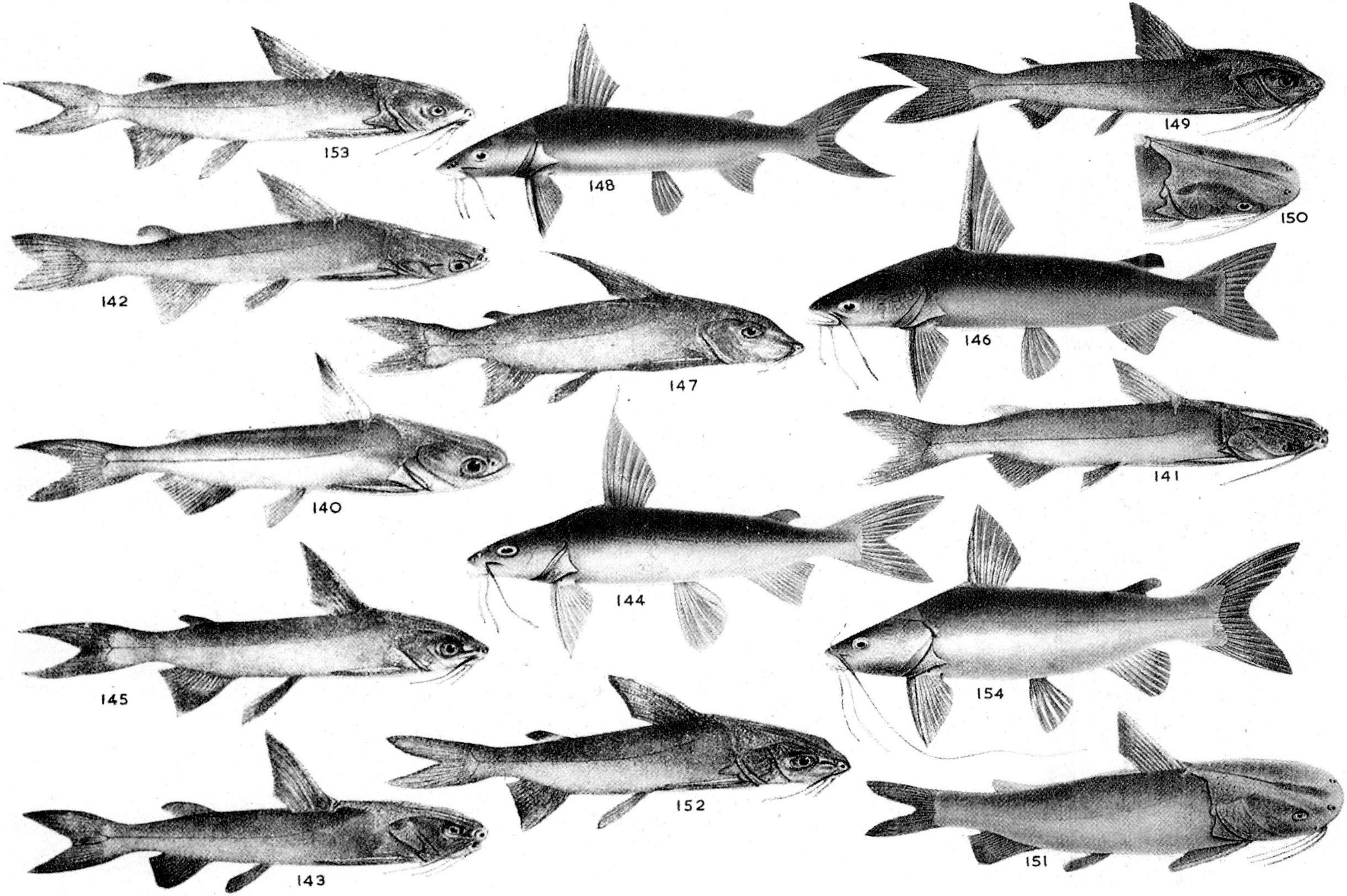

PLATE 9

Salmon Cat-fish, Estuarine Cat-fish and Fresh Water Cat-fish.

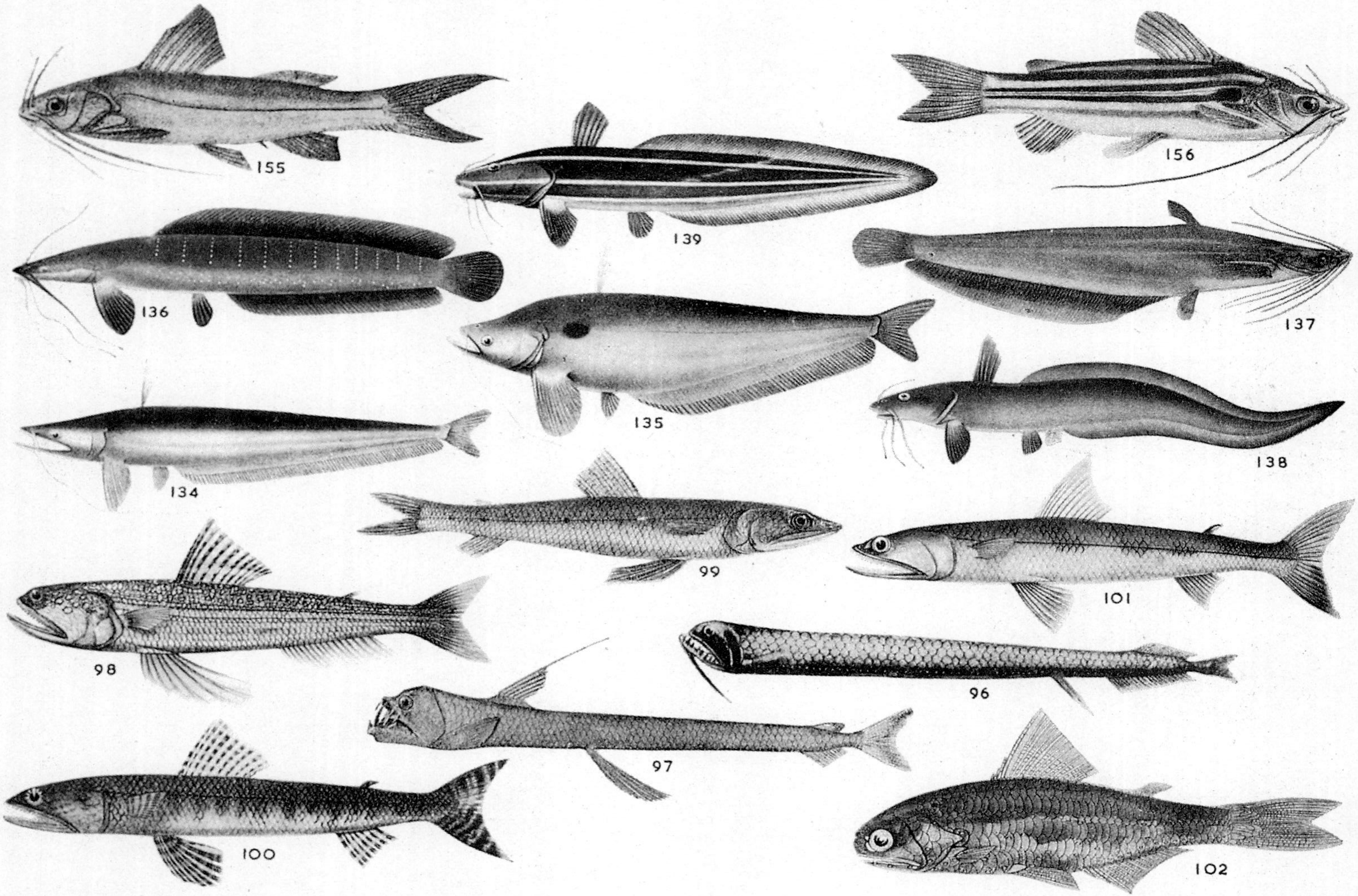

PLATE 10

Boa-fish, Fang-fish, Lizard-fish, Lantern Fish, Fresh Water Cat-fish and Cat-fish Eels.

PLATE II

Fresh Water Eels, Moray Eels, Conger Eels and Snake Eels.

PLATE 12
Fresh Water Eels, Moray Eels, Thread Eels, Conger Eels, Worm Eels, Snake Eels and Shore Eels.

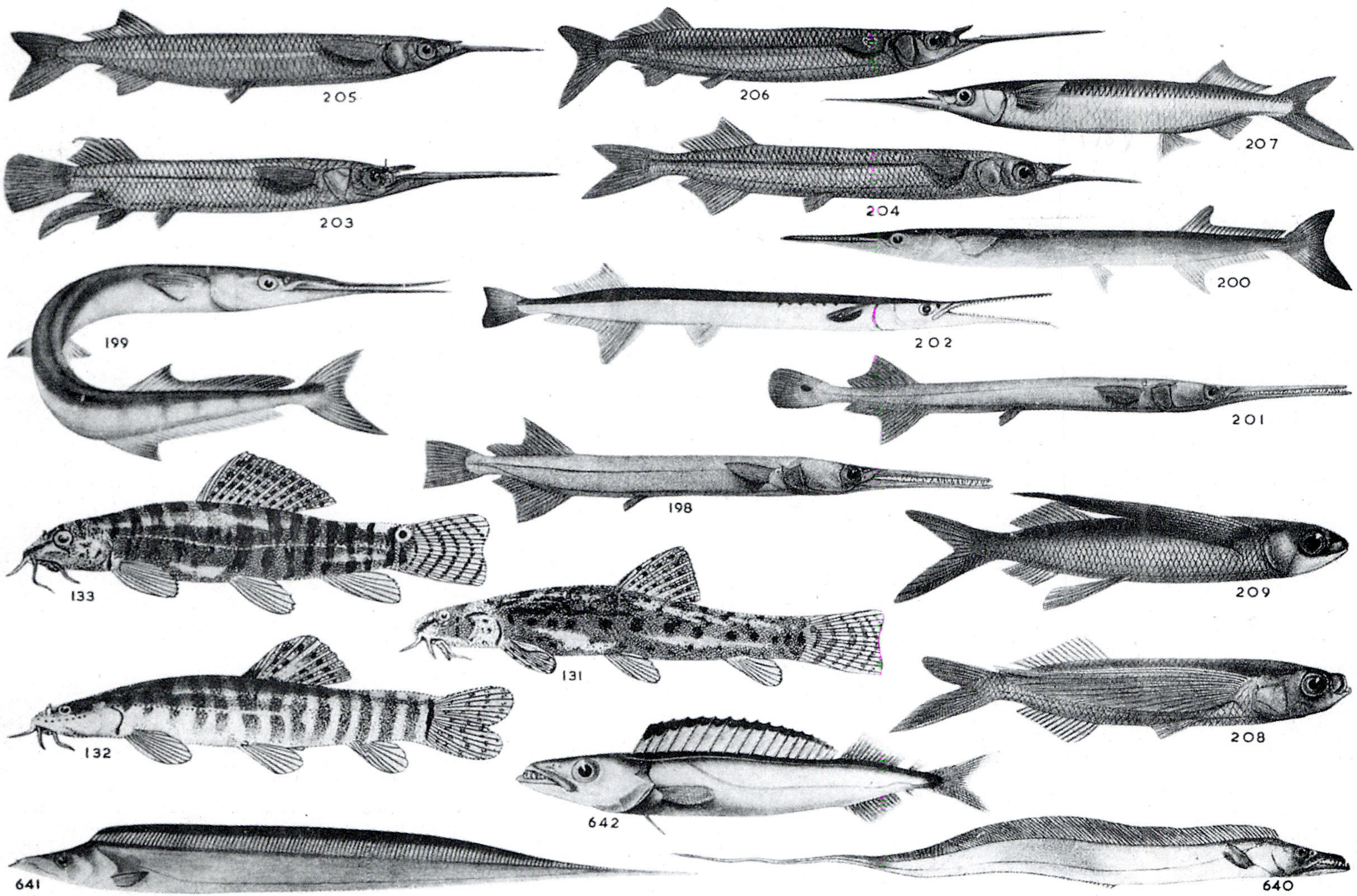

PLATE 13

Garfish, Needle-fish, Half-beaks, Flying Fish, Loaches, Ribbon-fish and Snake Mackerel.

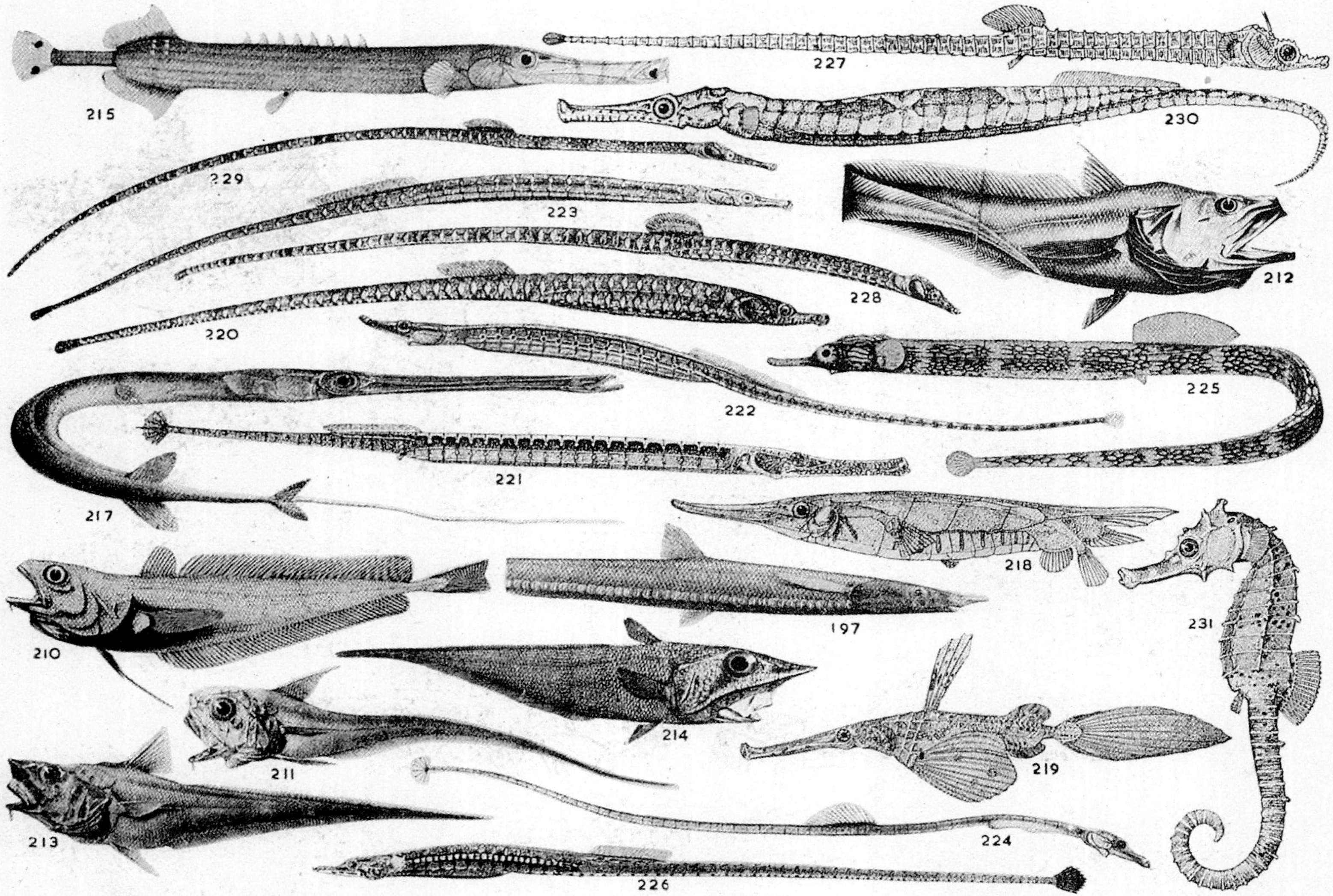

PLATE 14

Lizard-fish, Cod, Rat-tails ;Flute Mouths, Razor Fish, Pipe Fish and Sea Horses.

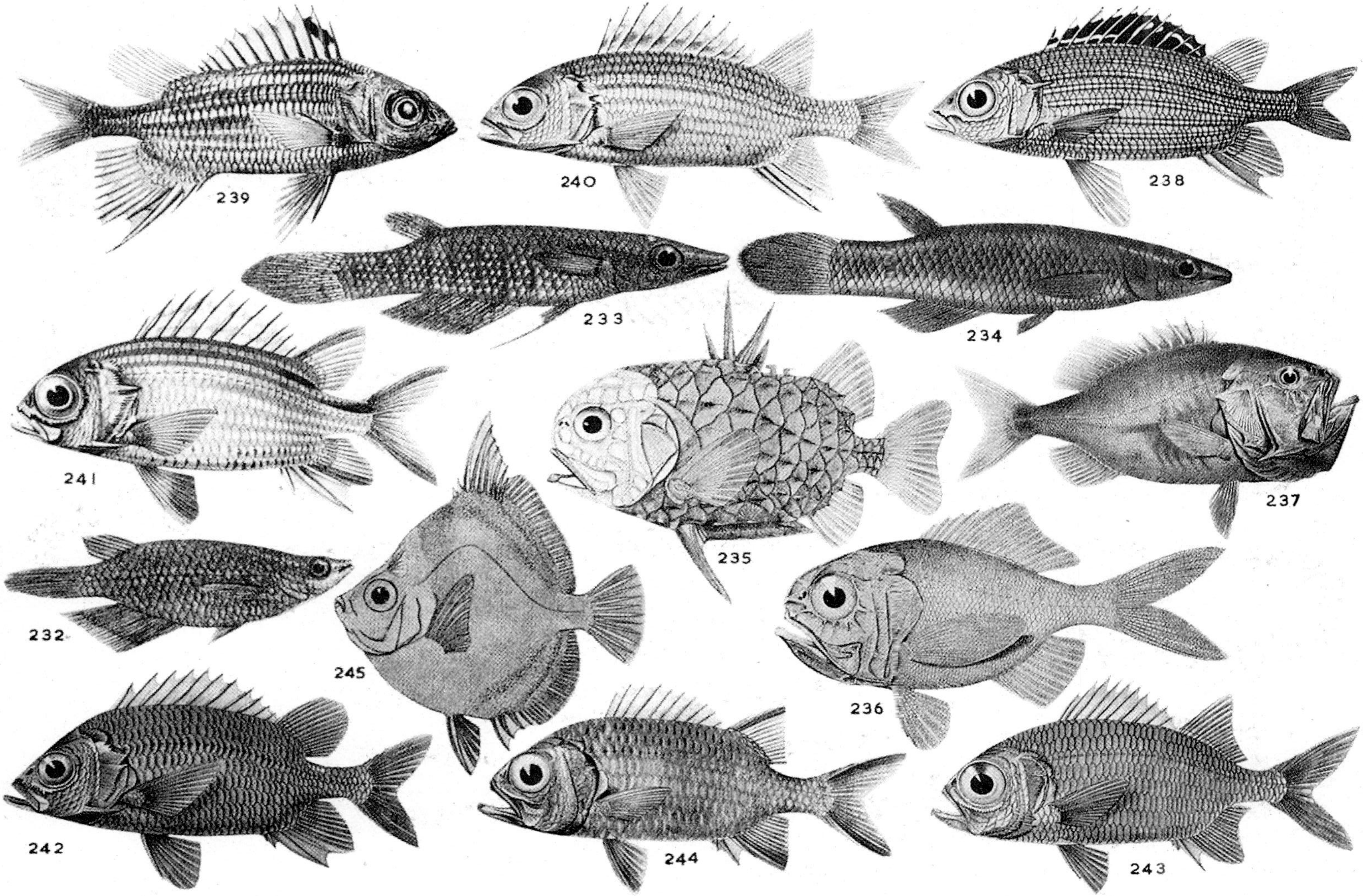

Plate 15

Top-minnows, Knight Fish ,Red-fish, Soldier Fish and Boar Fish.

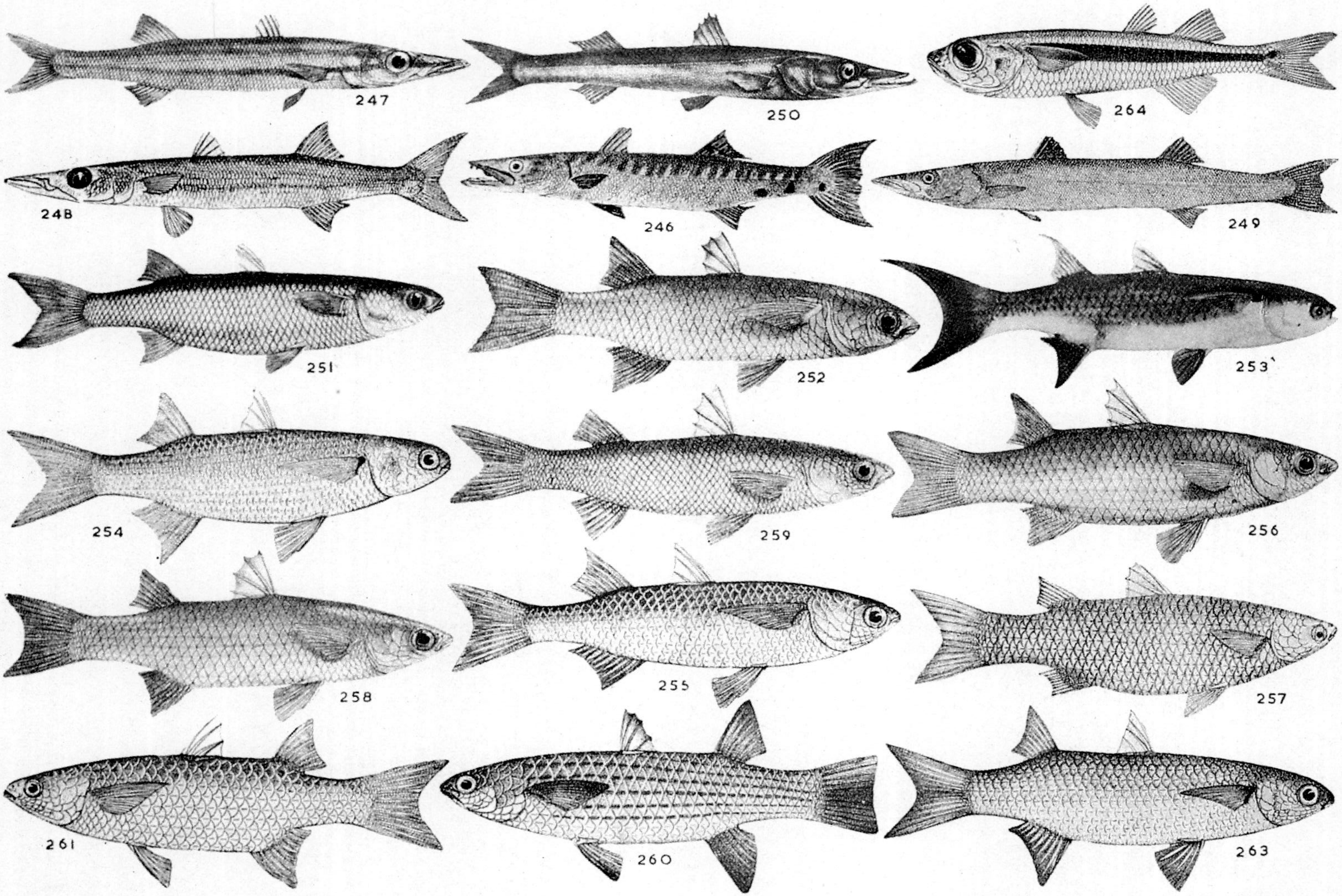

PLATE 16

Sea Pike, Grey Mullet and Hardyheads.

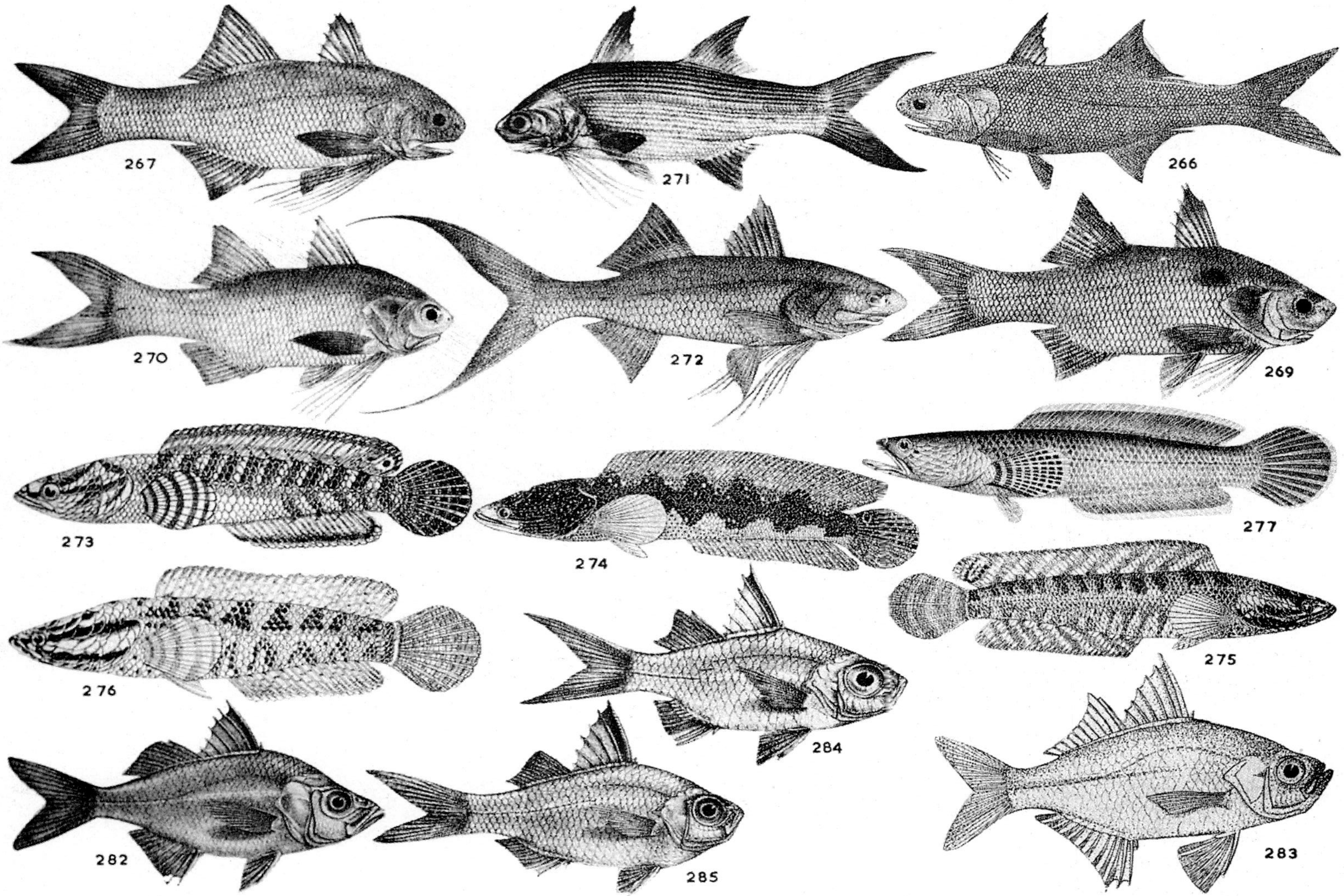

PLATE 17

Tassel Fish, Snake-heads and Glassy Perchlets.

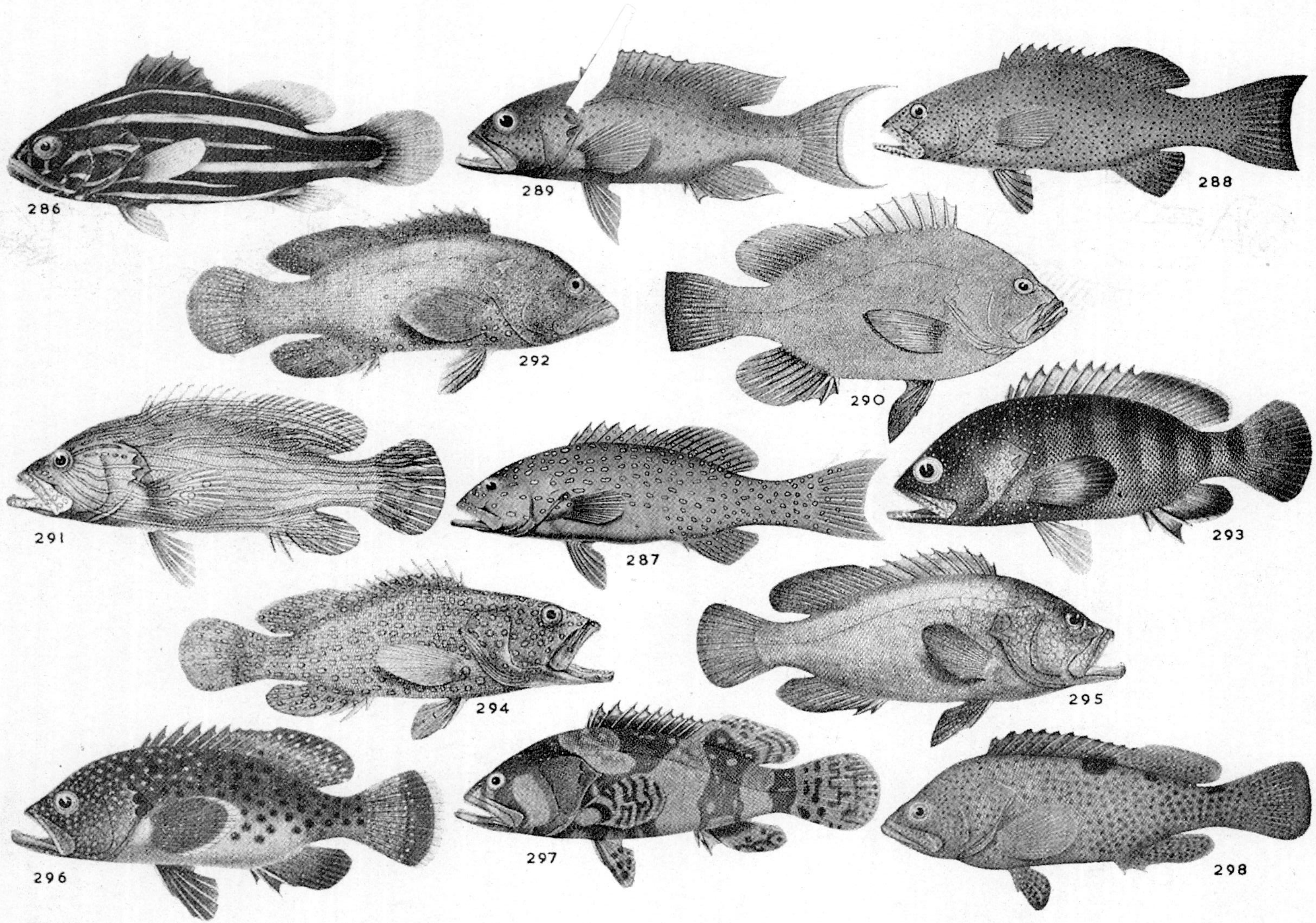

PLATE 18

Perch, Coral-trout, Rock-cod, Coral-cod and Groupers.

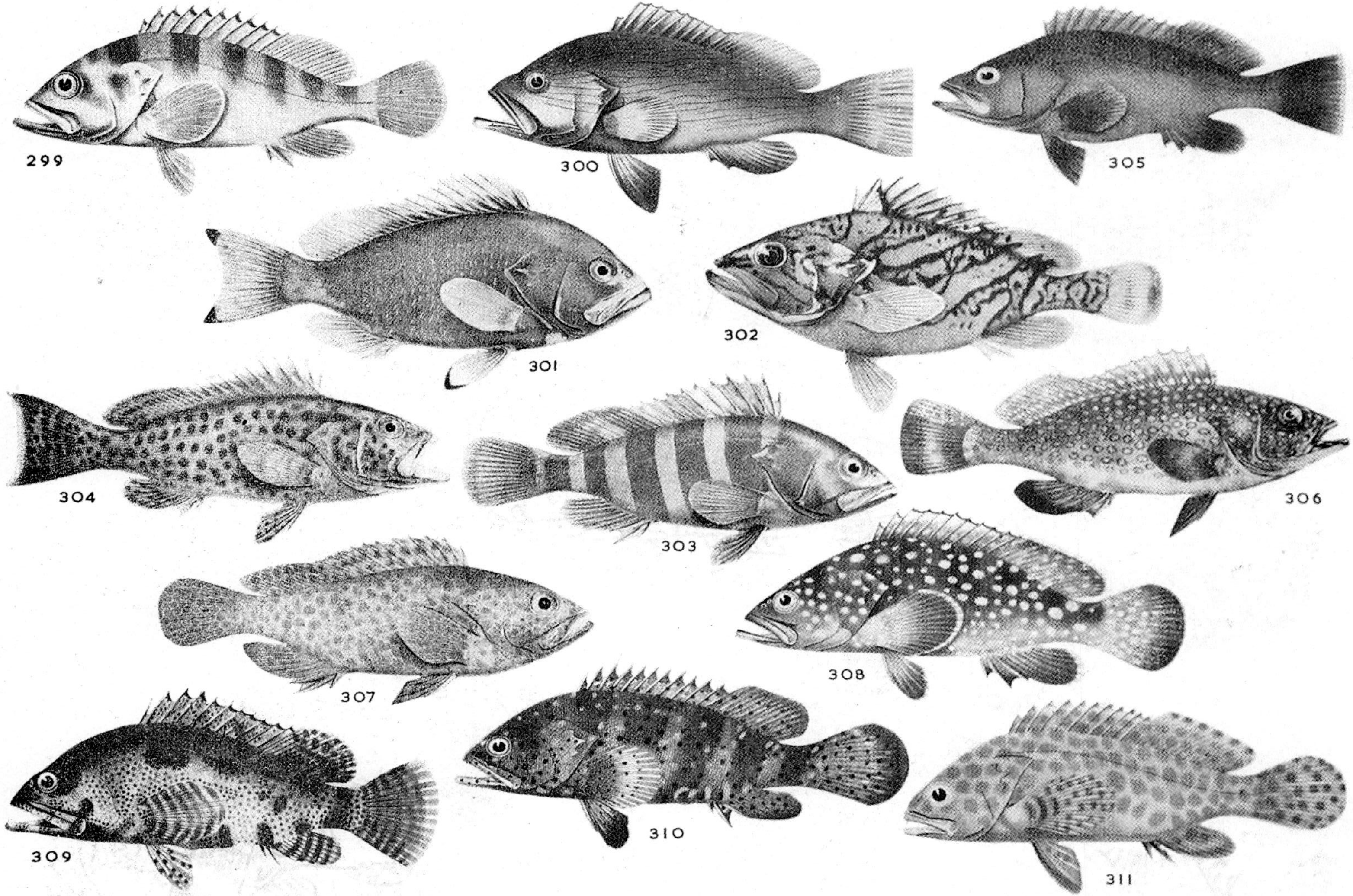

PLATE 19

Coral-cod, Reef-cod and Groupers.

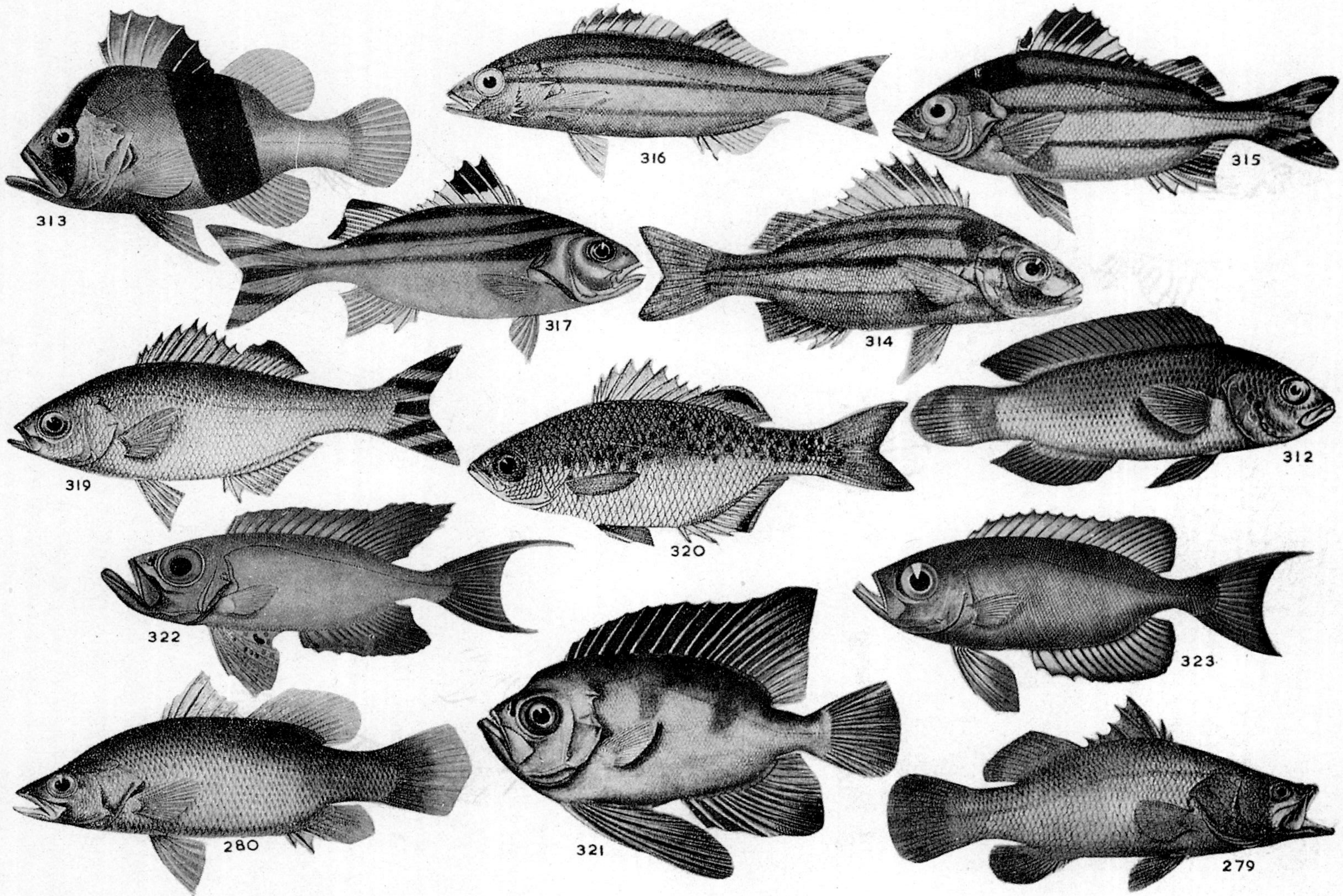

PLATE 20

Dottybacks, Perch, Grunters, Flagtails and Bullseyes.

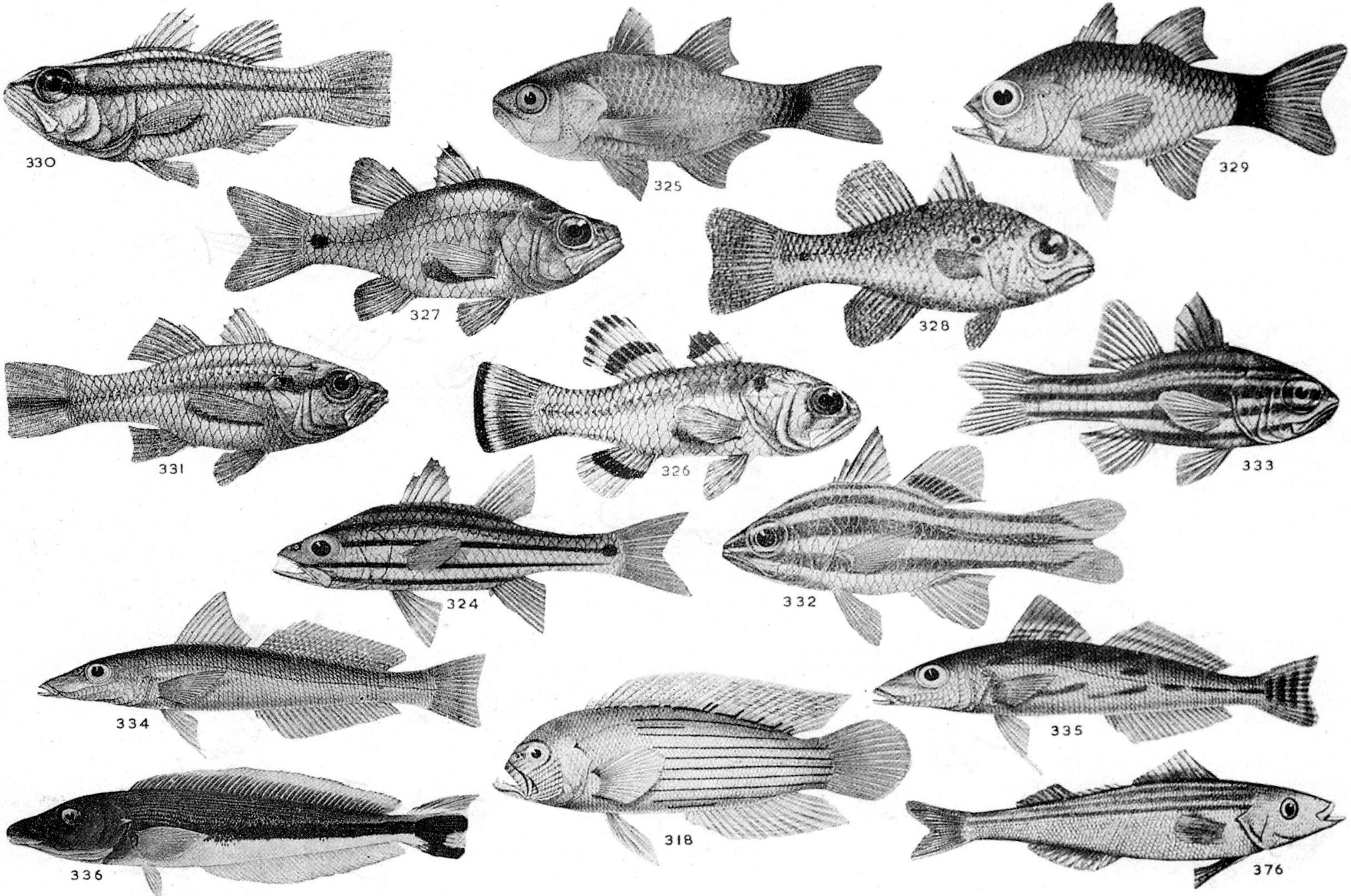

Plate 21

Round-heads, Cardinal Fish, Whiting and Redbait.

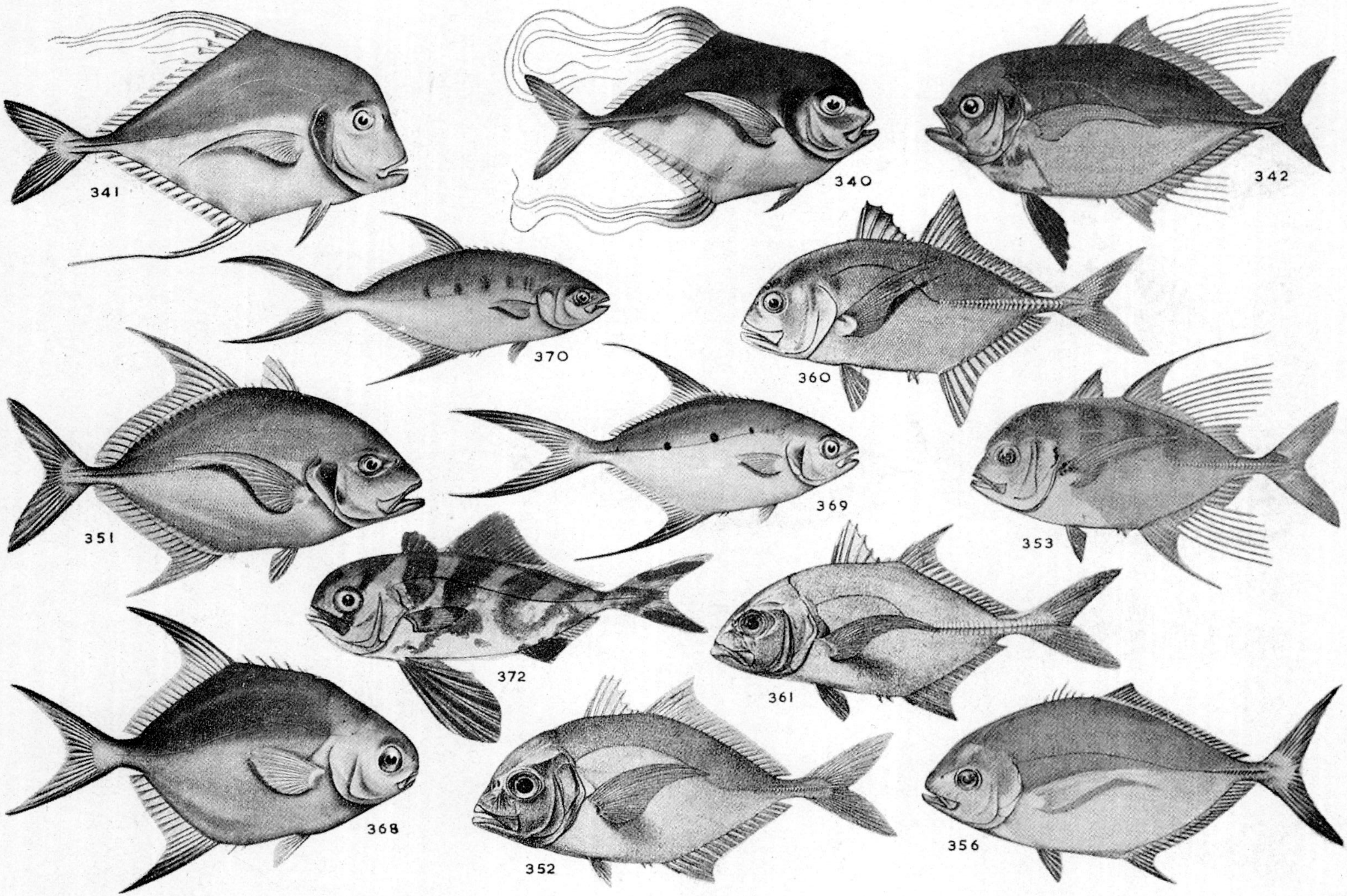

PLATE 22

Trevally, Kingfish and Darts.

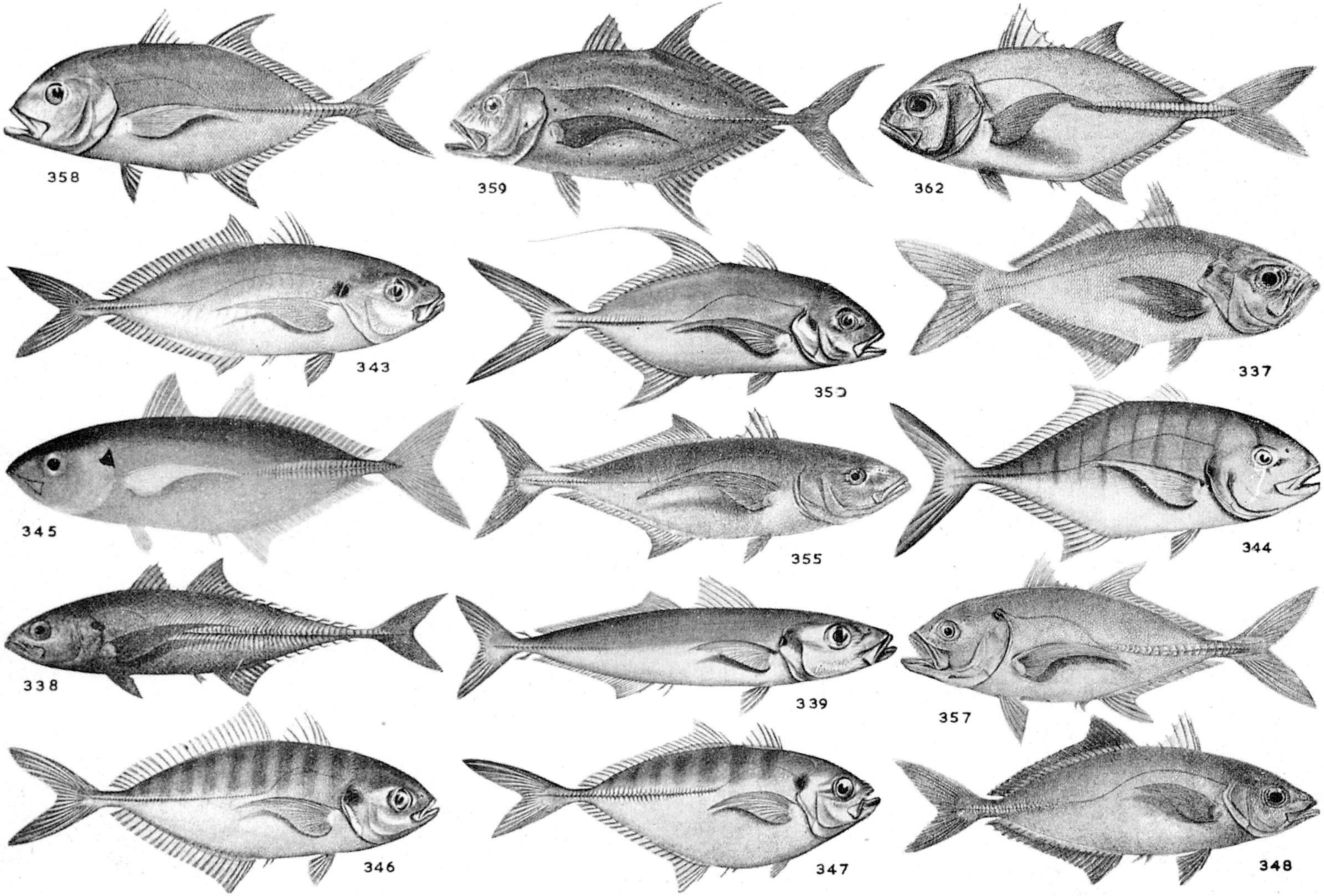

Plate 23
Whitefish, Trevally and Scad.

PLATE 24

Dolphin, Pilot Fish, Moonfish, Leather-skins, Kingfish, Jobfish, Bass and Fusiliers.

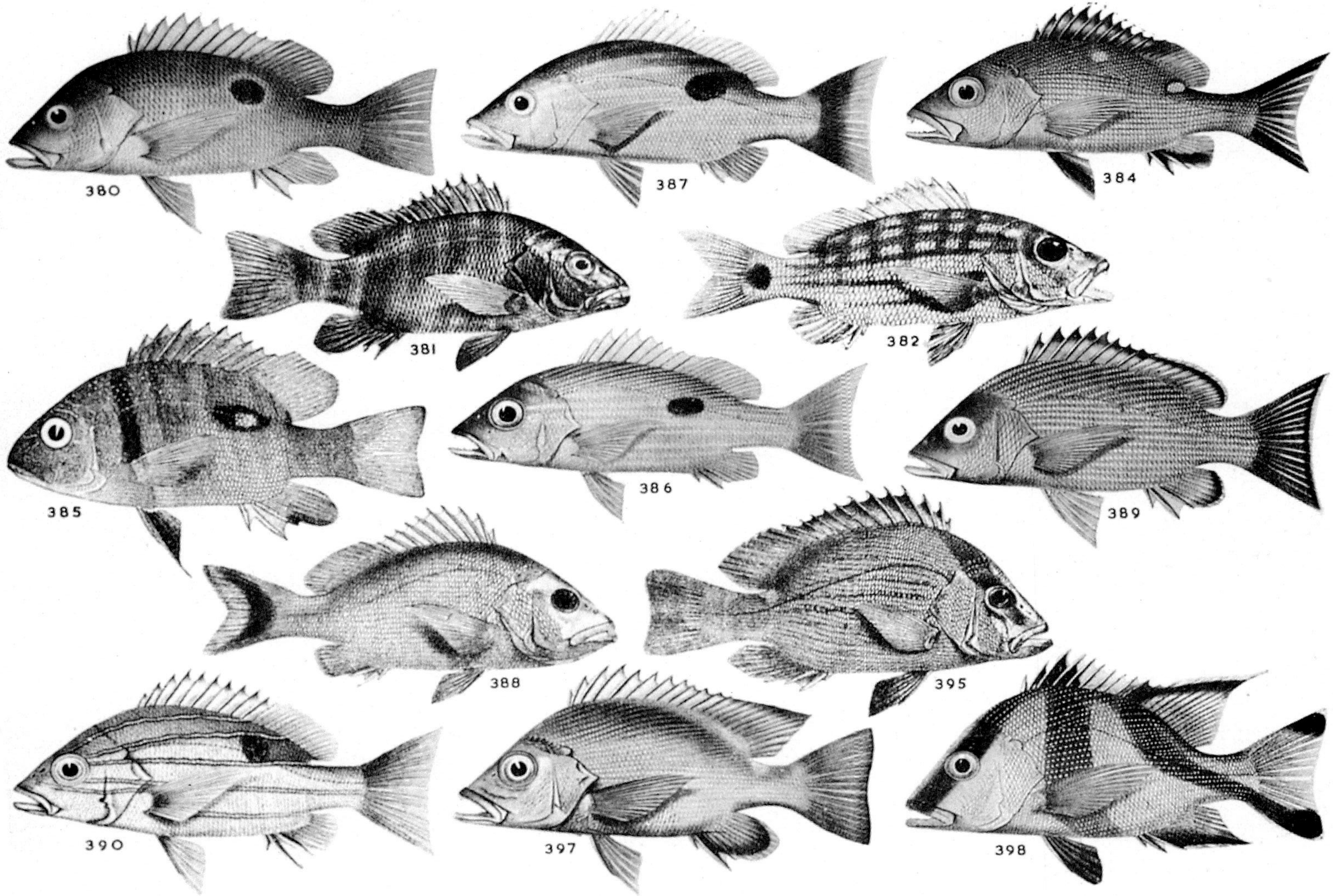

Plate 25

Moses Perch, Sea Perch, Snappers and Bass.

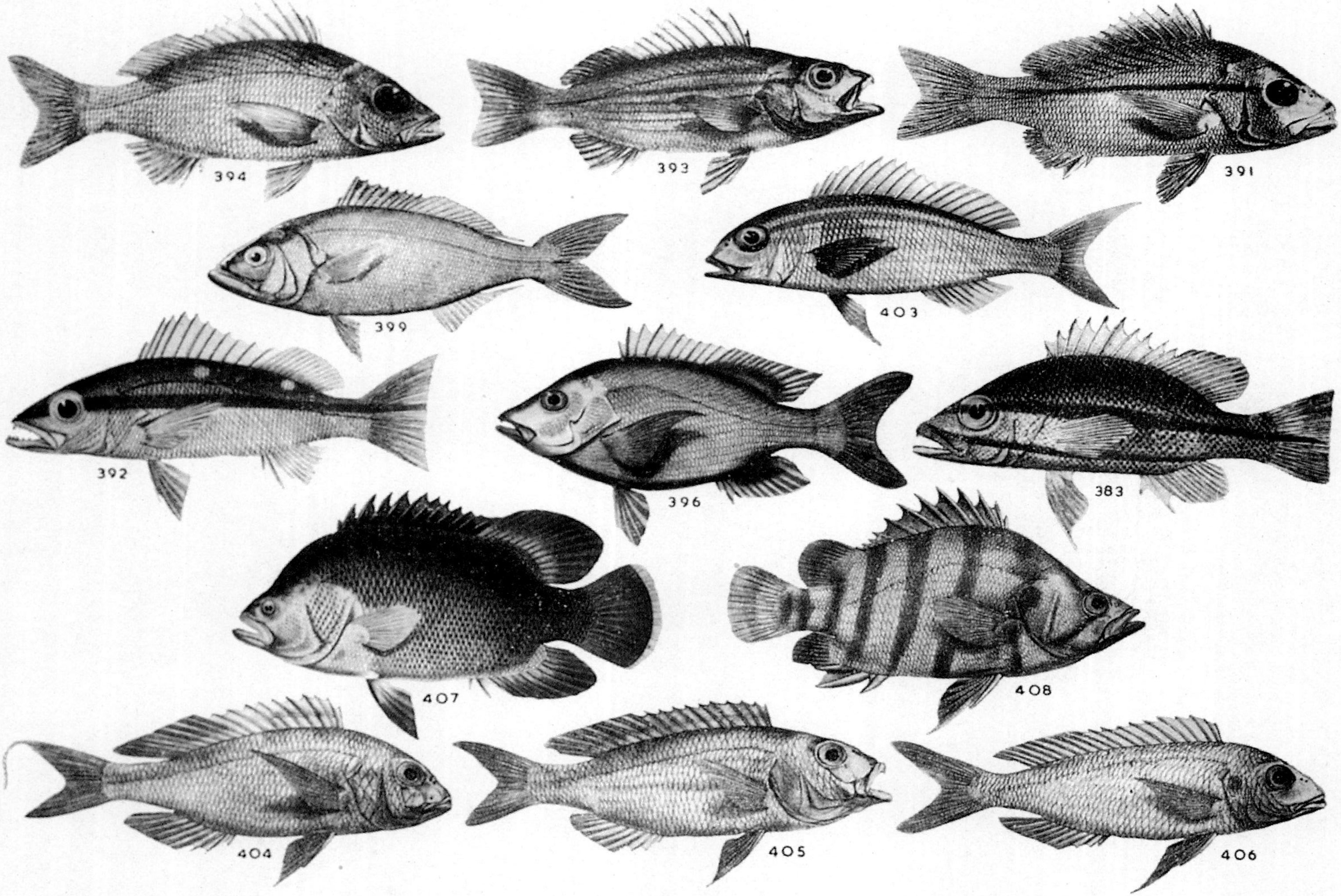

Plate 26

Snappers, Threadfin-bream and Triple-tails.

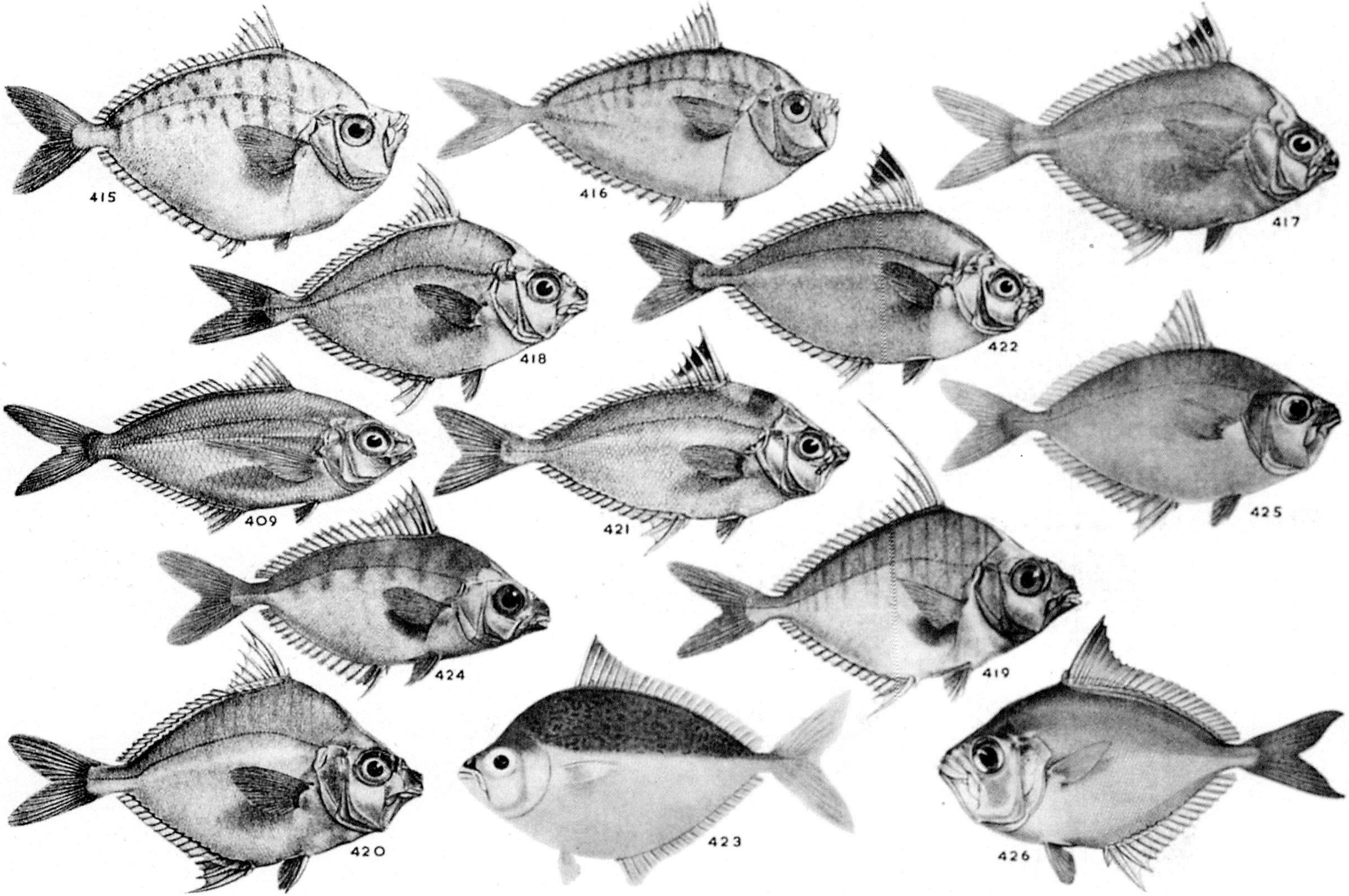

PLATE 27
Silver-Biddies and Pony Fish.

PLATE 28

Silver-biddies, Monocle-bream, Large-eyed Bream and Jewfish.

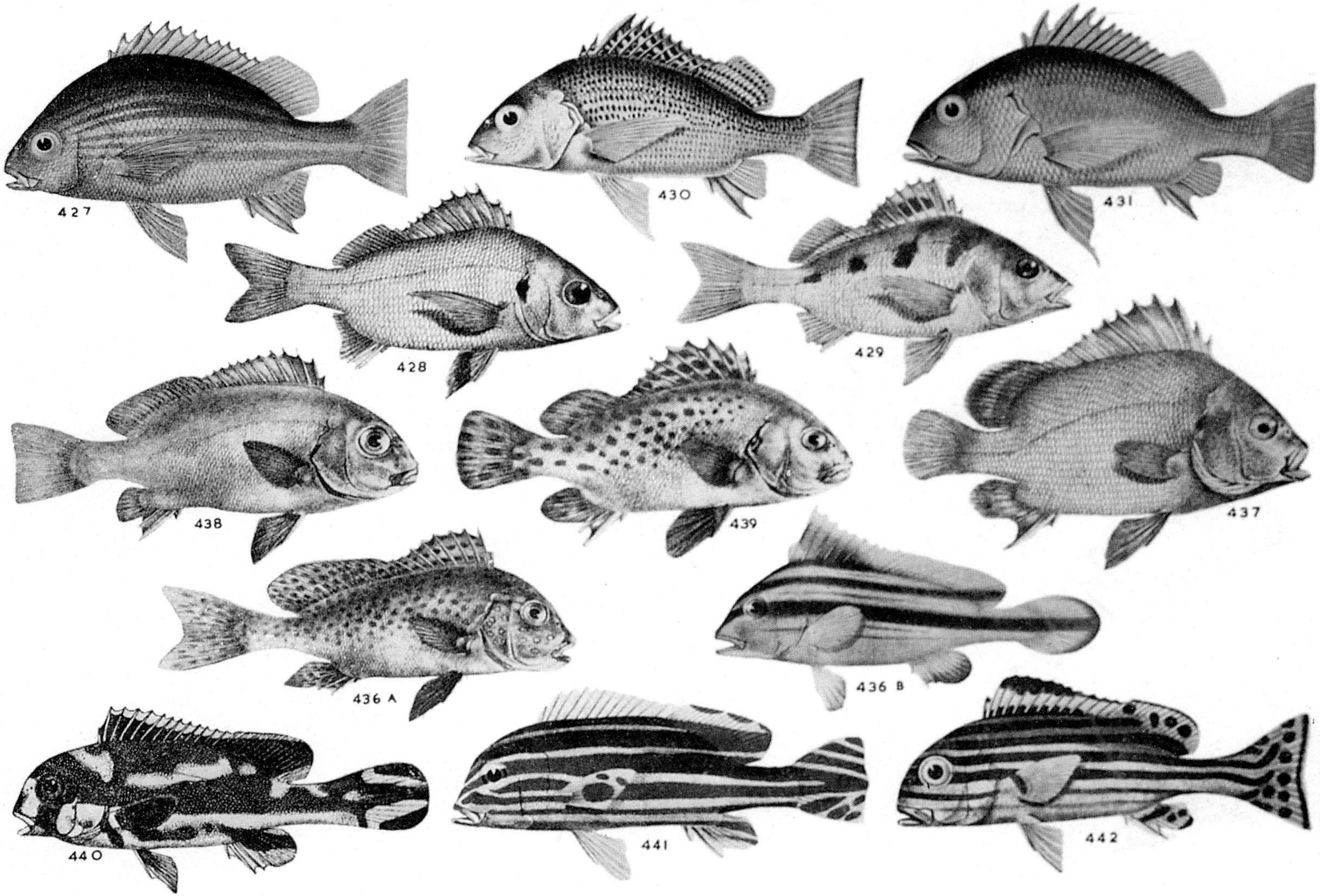

PLATE 29

Grunters and Sweetlips.

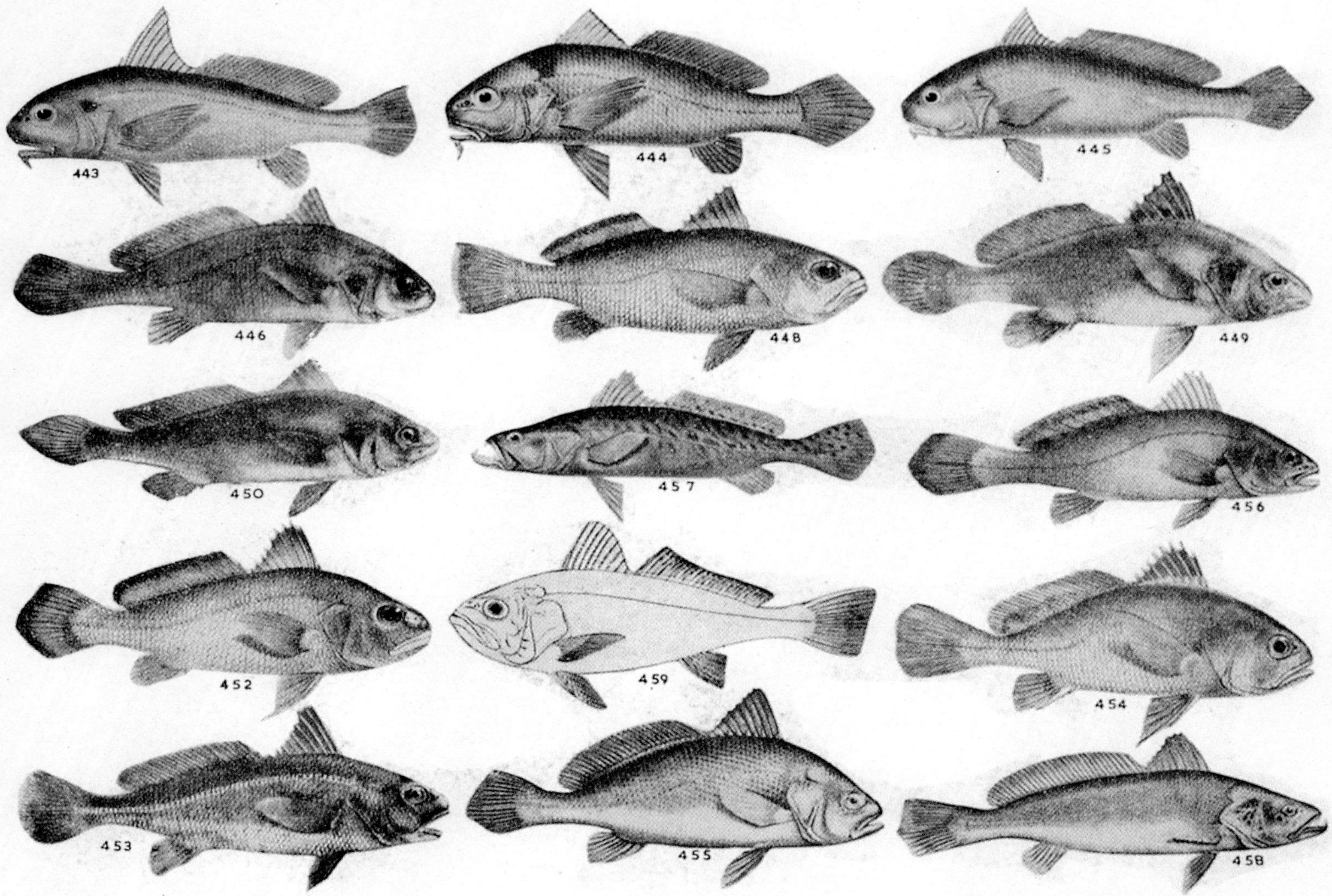

Plate 30

Jewish and Croakers.

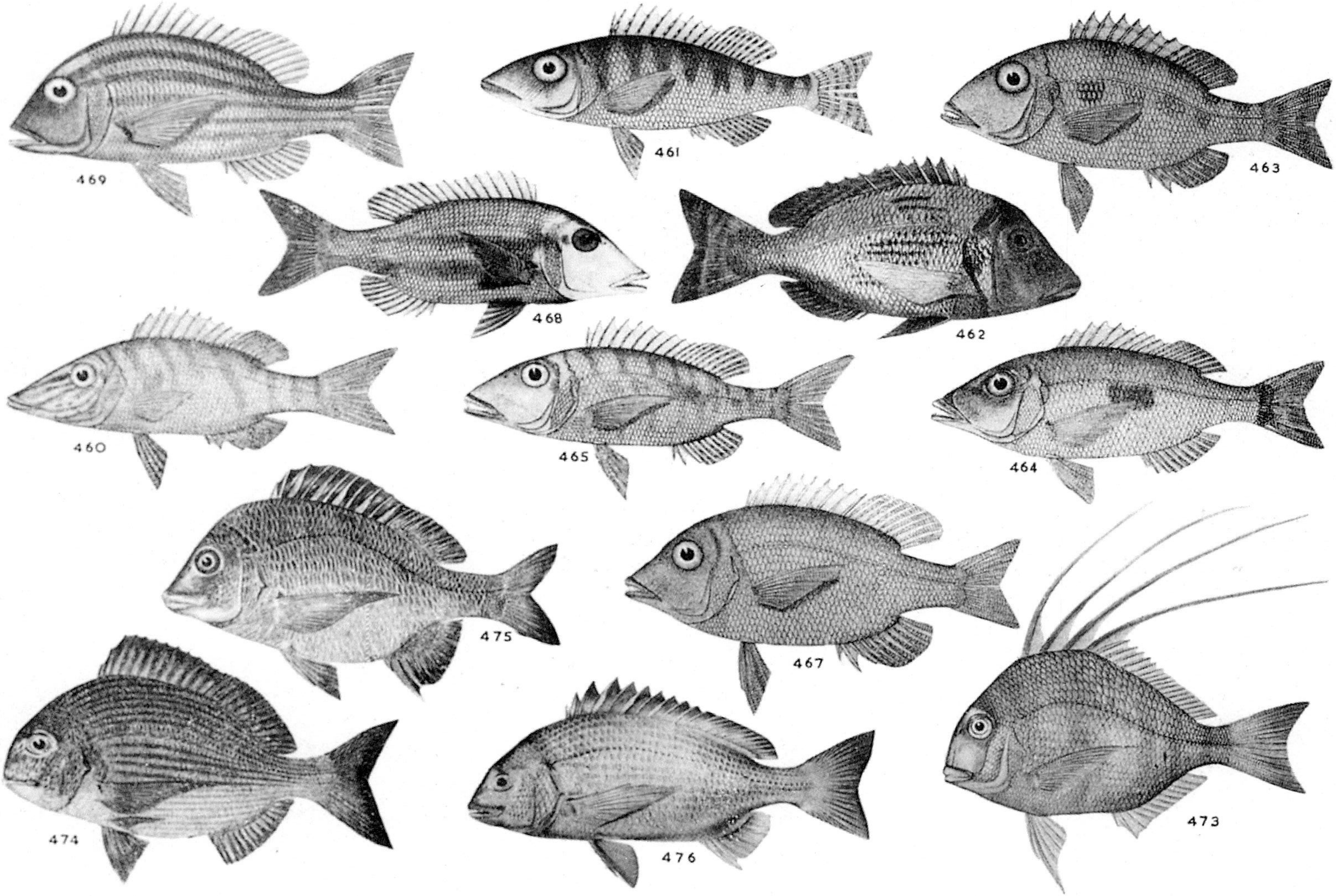

PLATE 31

Pigface Bream, Red Bream and Silver Bream.

PLATE 32
Goatfish, Sweepers and Drummers.

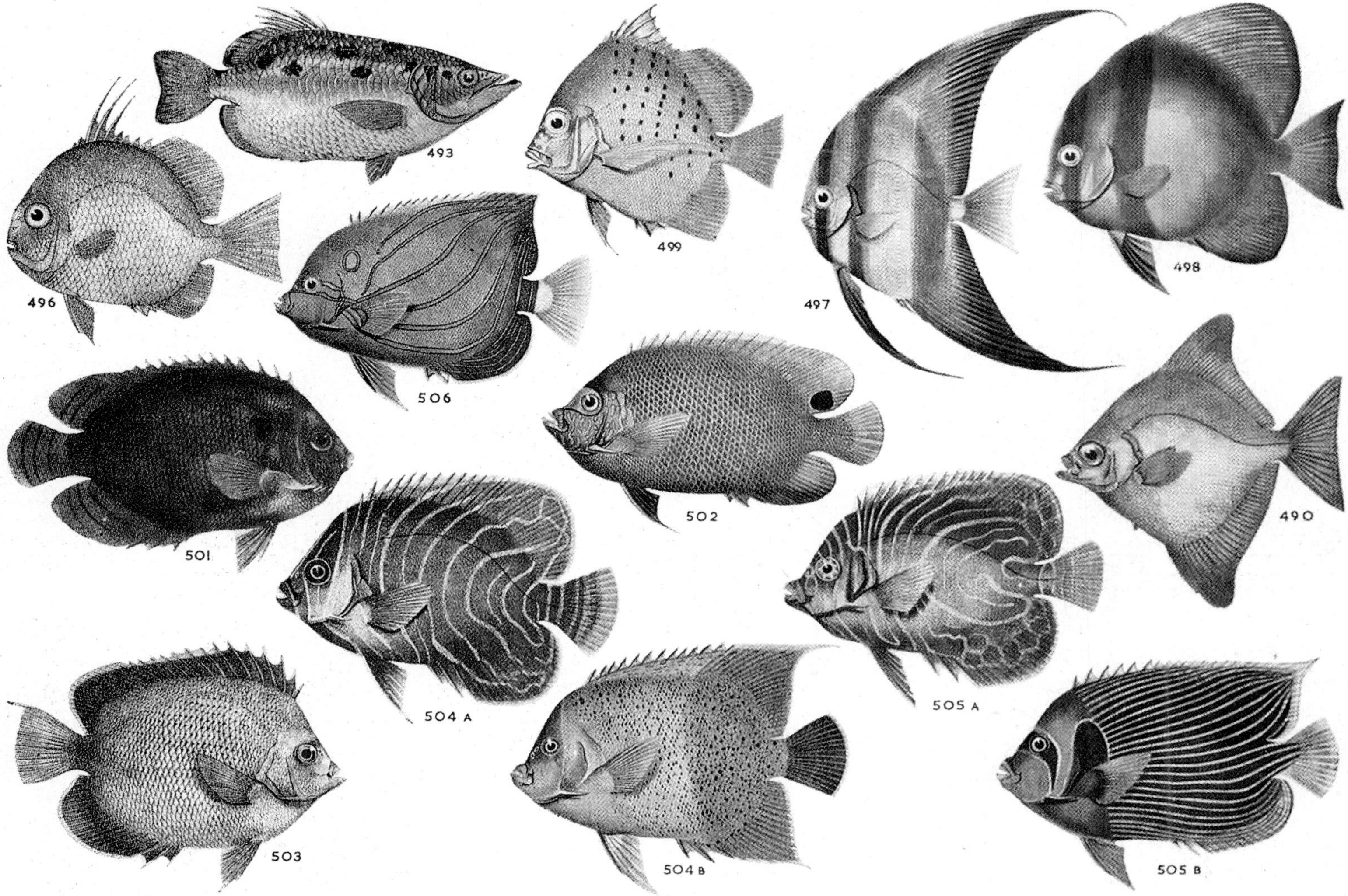

PLATE 33

Archer Fish, Batfish, Spadefish and Angelfish.

PLATE 34

Butterfish, Coral Fish and Butterfly Fish.

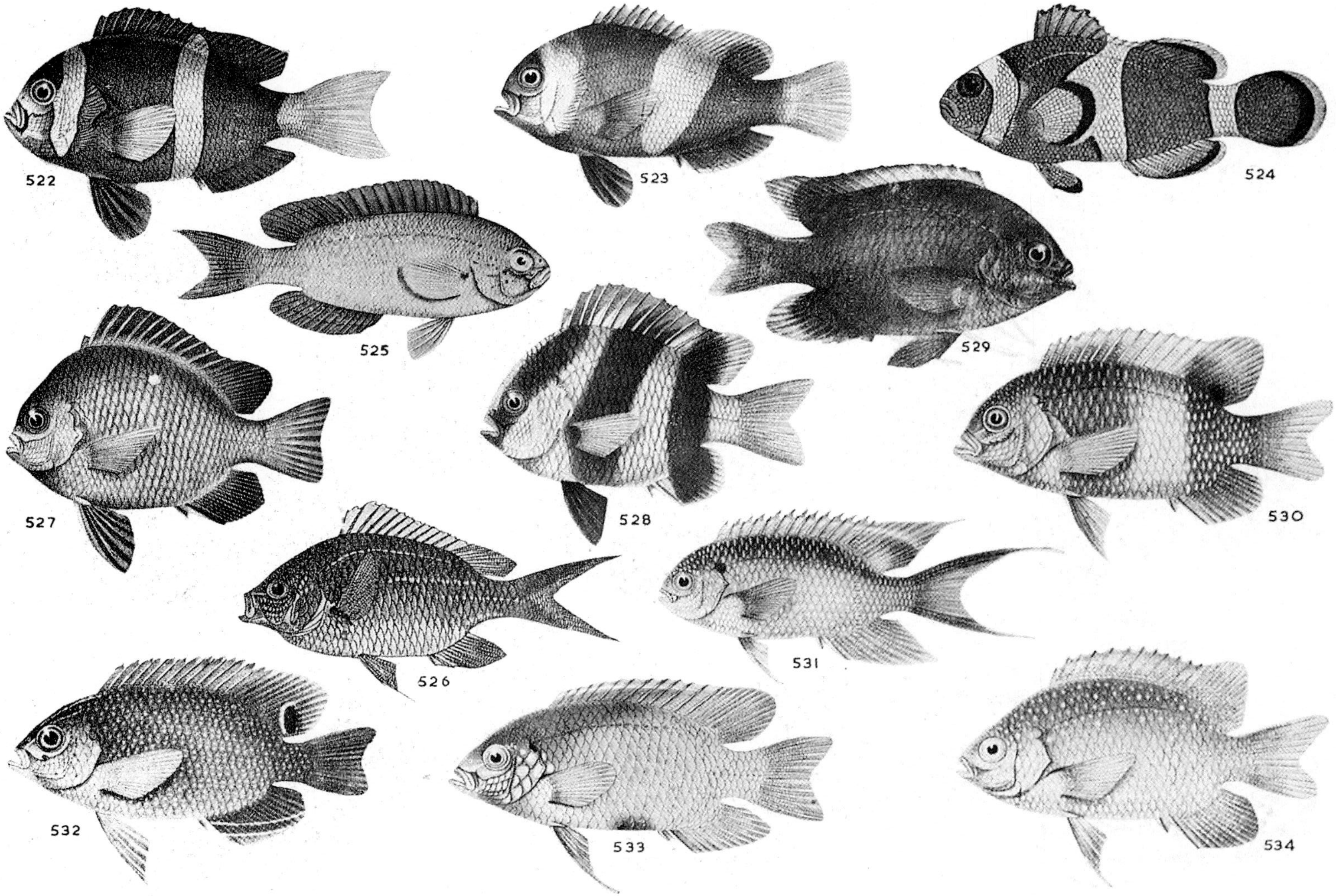

PLATE 35

Demoiselles, Anemone Fish and Pullers.

PLATE 36

Tilapia, Etroplus, Demoiselles and Wrasses.

PLATE 37
Wrasses.

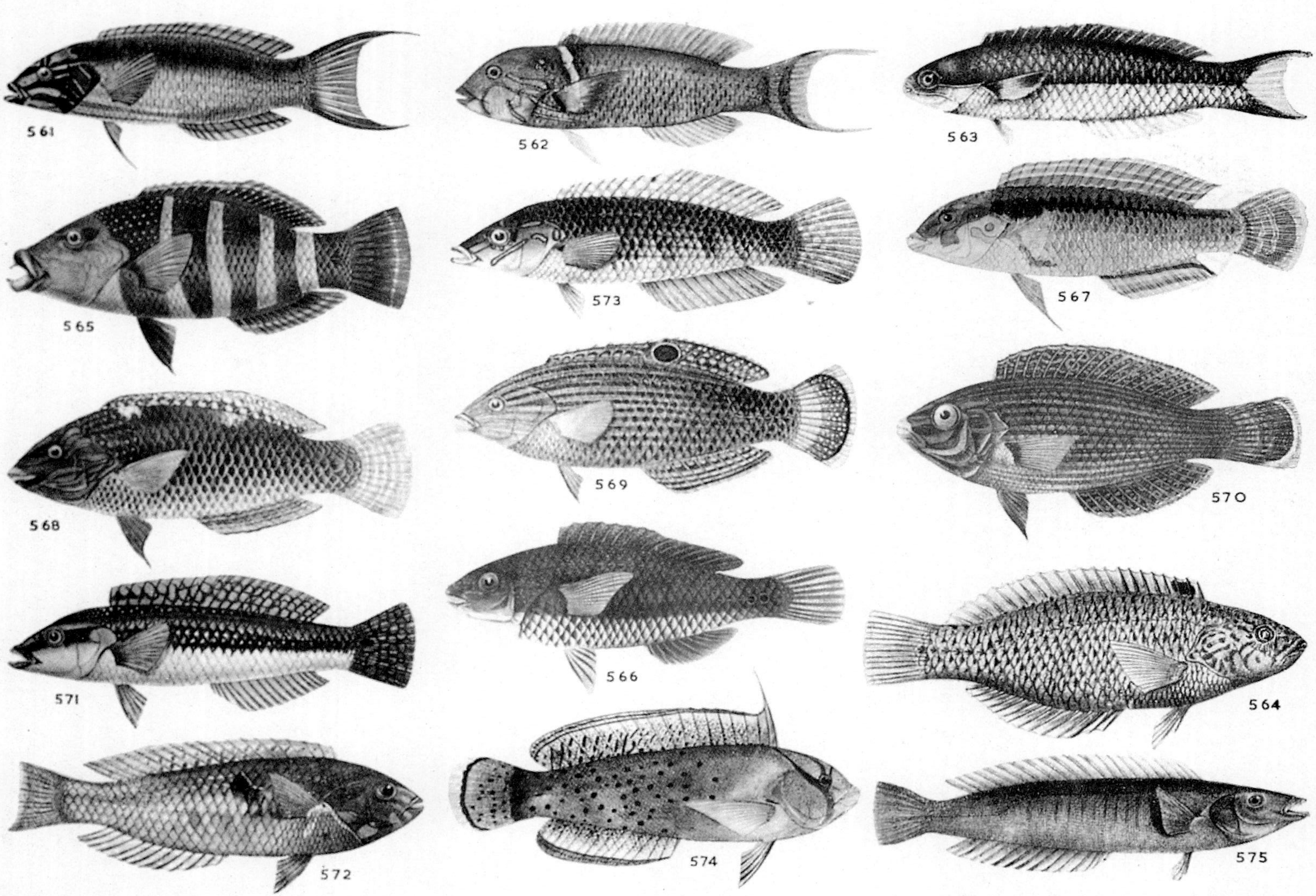

PLATE 38

Wrasses and Rainbow Fish.

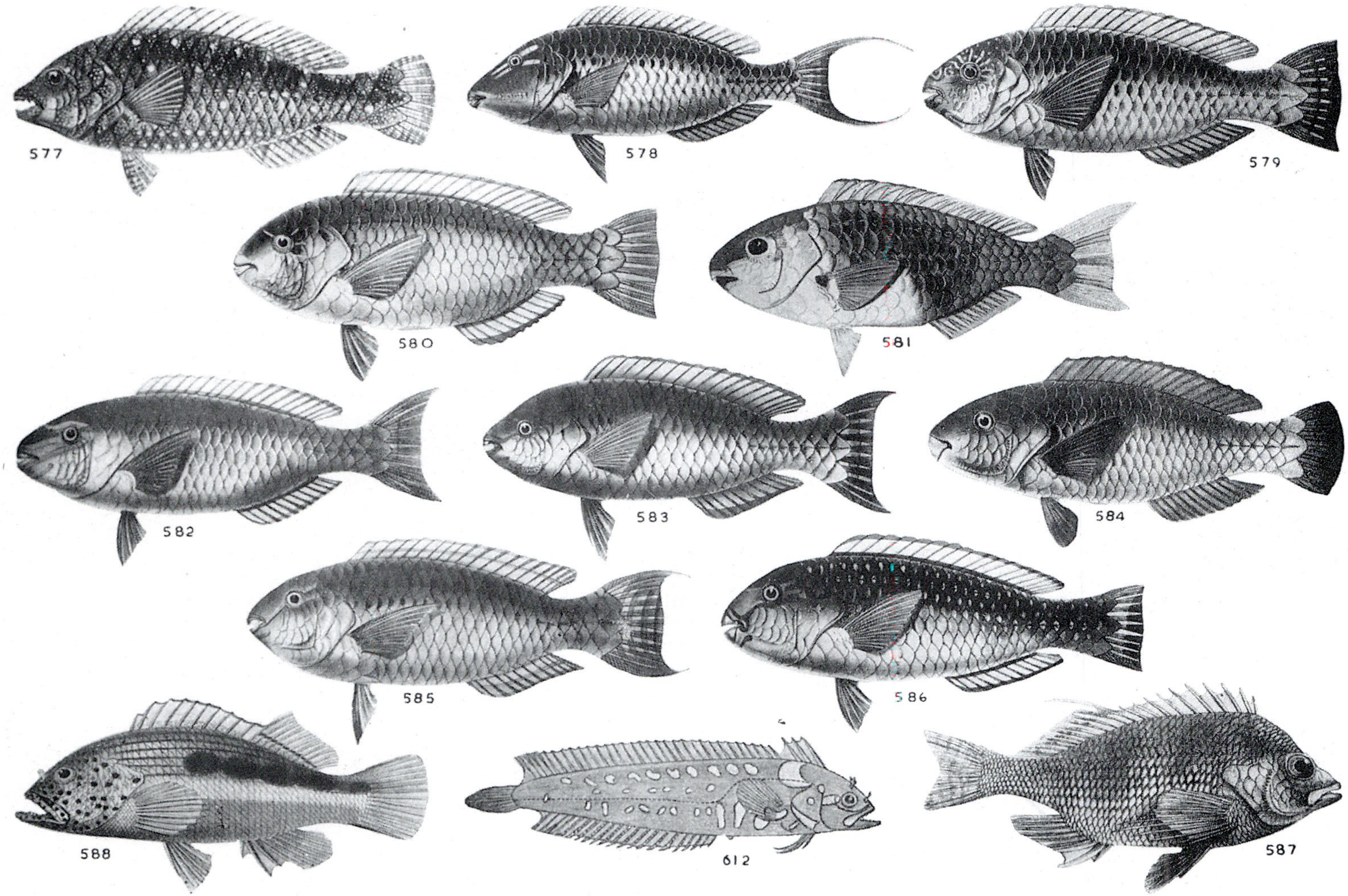

Plate 39

Parrot Fish, Curlyfins and Weedfish.

PLATE 40
Grub Fish and Blennies.

PLATE 41

Spinefeet, Rabbitfish and Surgeon-fish.

PLATE 42

Surgeon-fish Unicorn-fish, Moorish Idol, Stargazers and Pomfrets

PLATE 43

Mackerel, Spanish Mackerel, Tuna, Wahoo, Swordfish, Spearfish, Sailfish and Marlin.

Plate 44

Climbing Perch, Bubble-nest Builders, Gourami, Hump-heads and Gudgeons

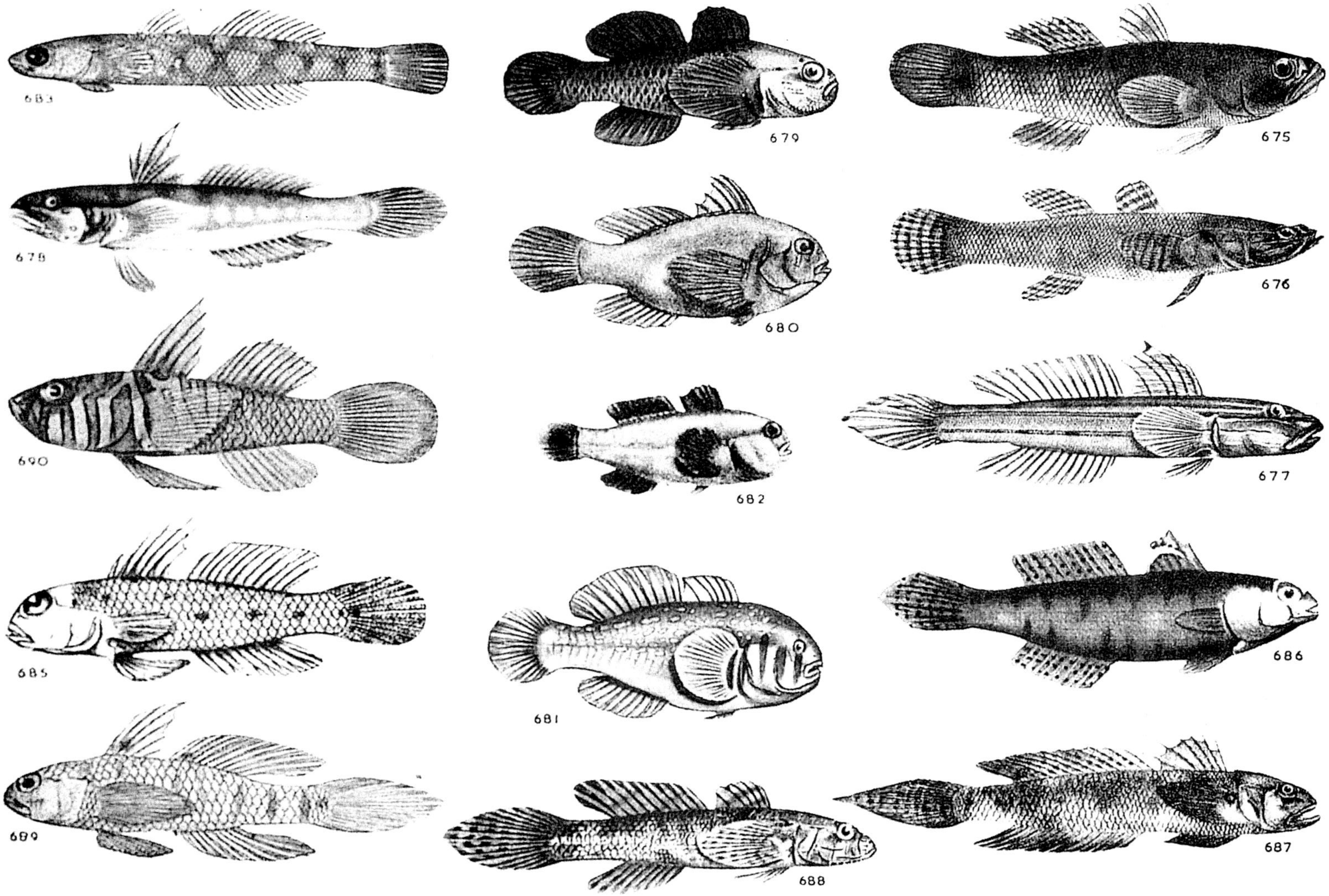

PLATE 45

Gudgeons and Gobies.

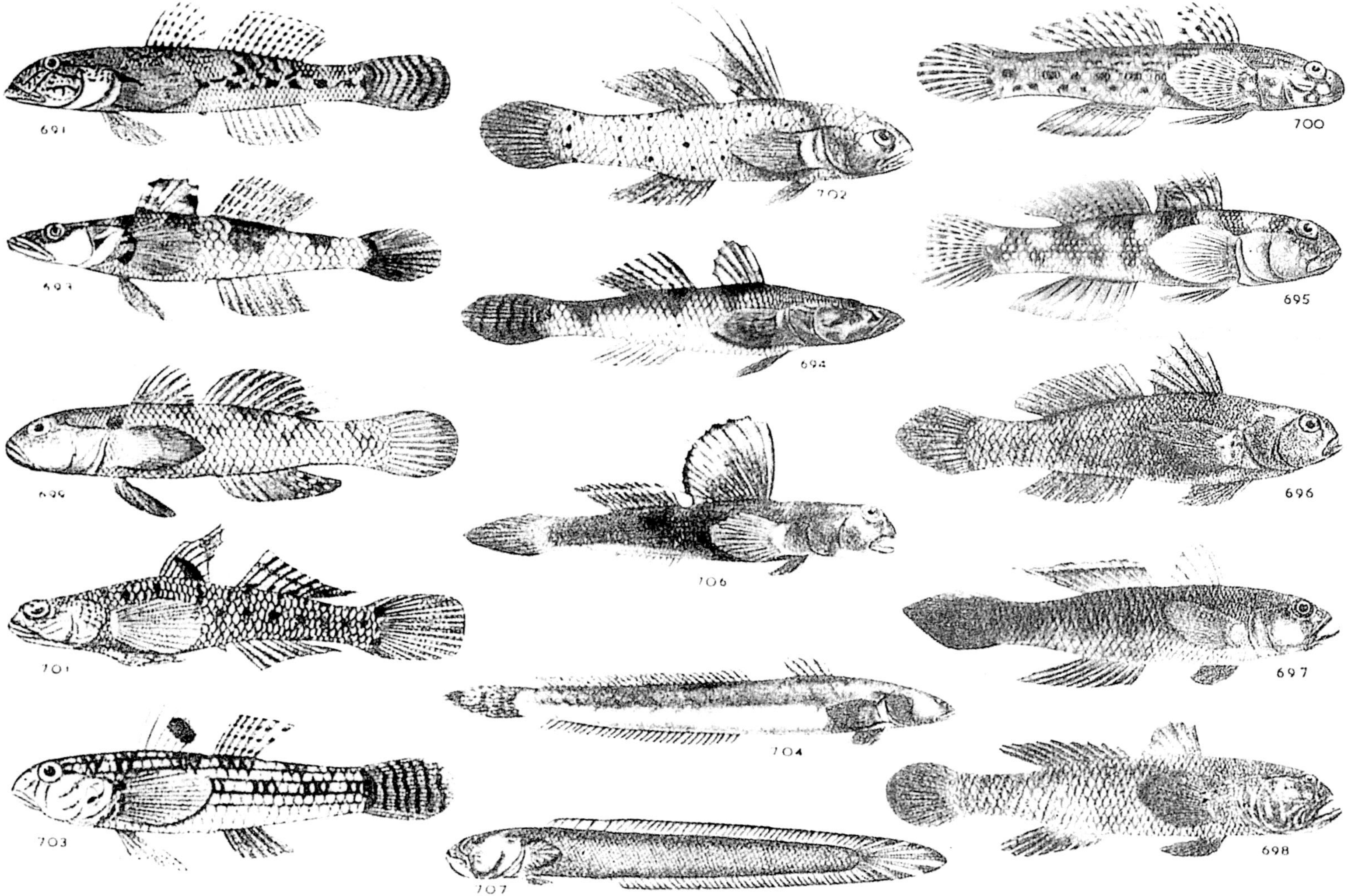

Plate 46

Gobies, Burrowing Gobies and Mud Skippers.

PLATE 47

Fire Fish and Stingfish.

PLATE 48

Stingfish, Velvet Fish, Goblinfish and Flatheads.

Plate 49

Halibut and Flounders.

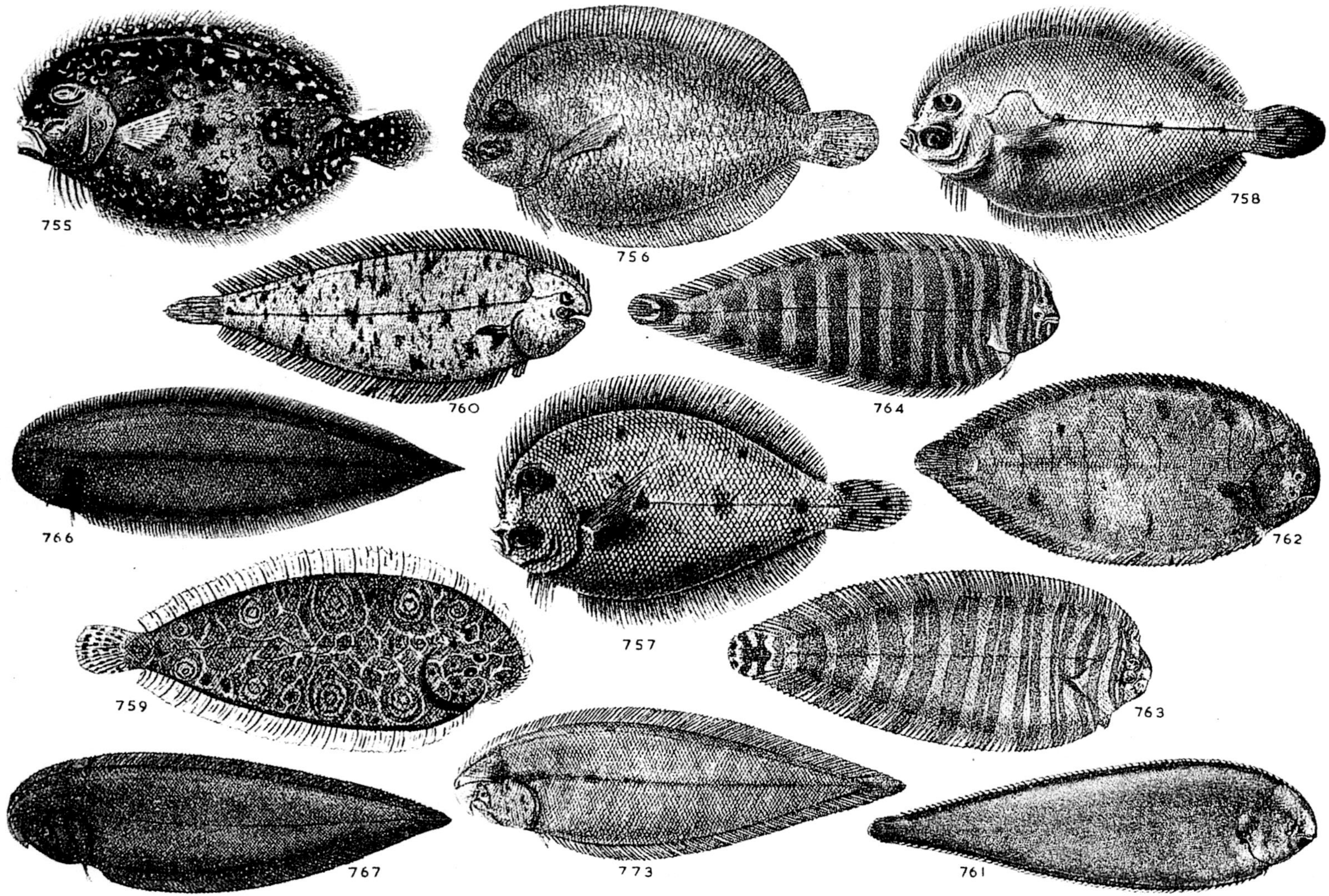

Plate 50

Flounders, Soles and Tongue-soles.

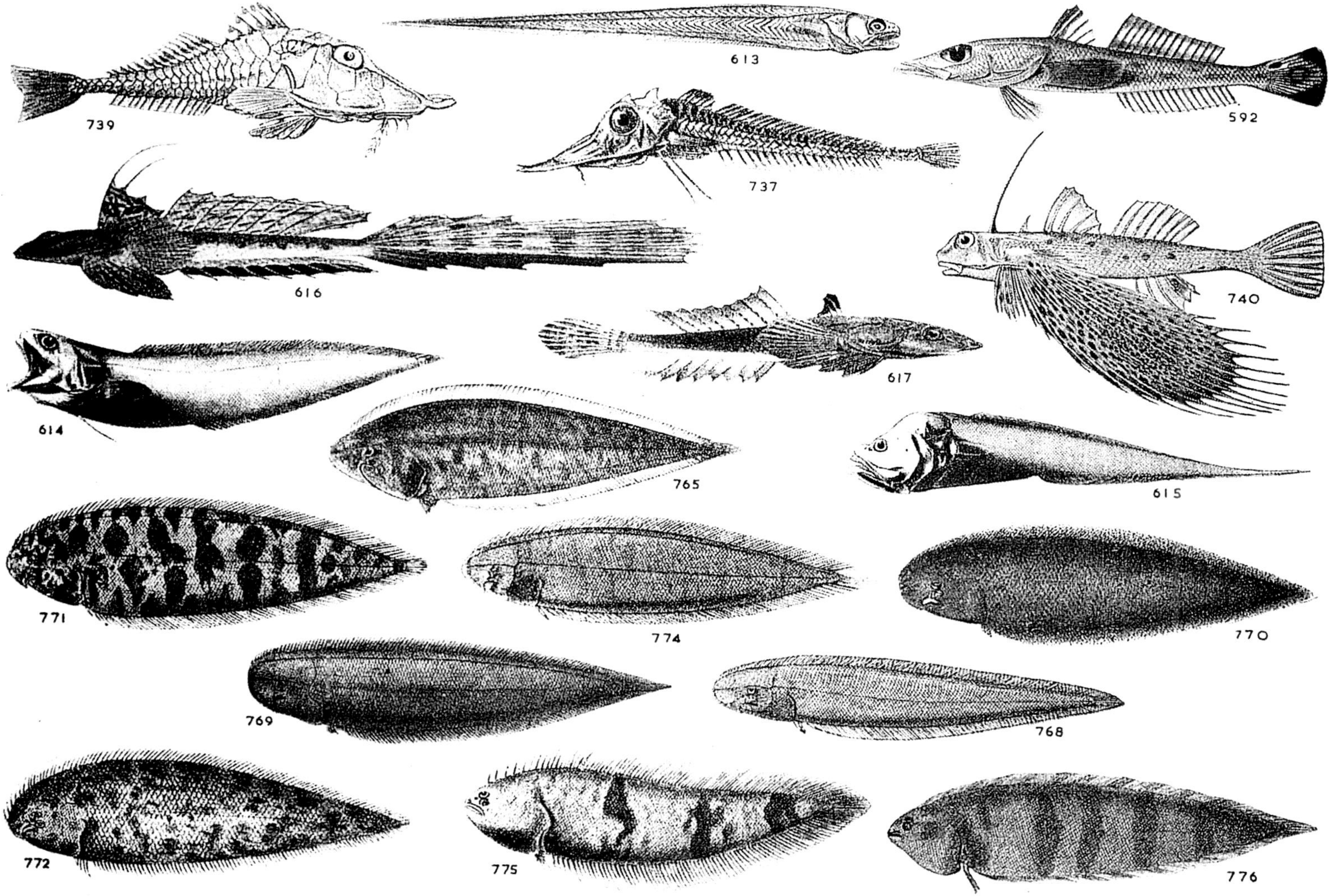

PLATE 51

Sea Robins, Smelt, Pearl Fish, Blindfish, Dragonets, Flying Gurnards and Tongue-soles.

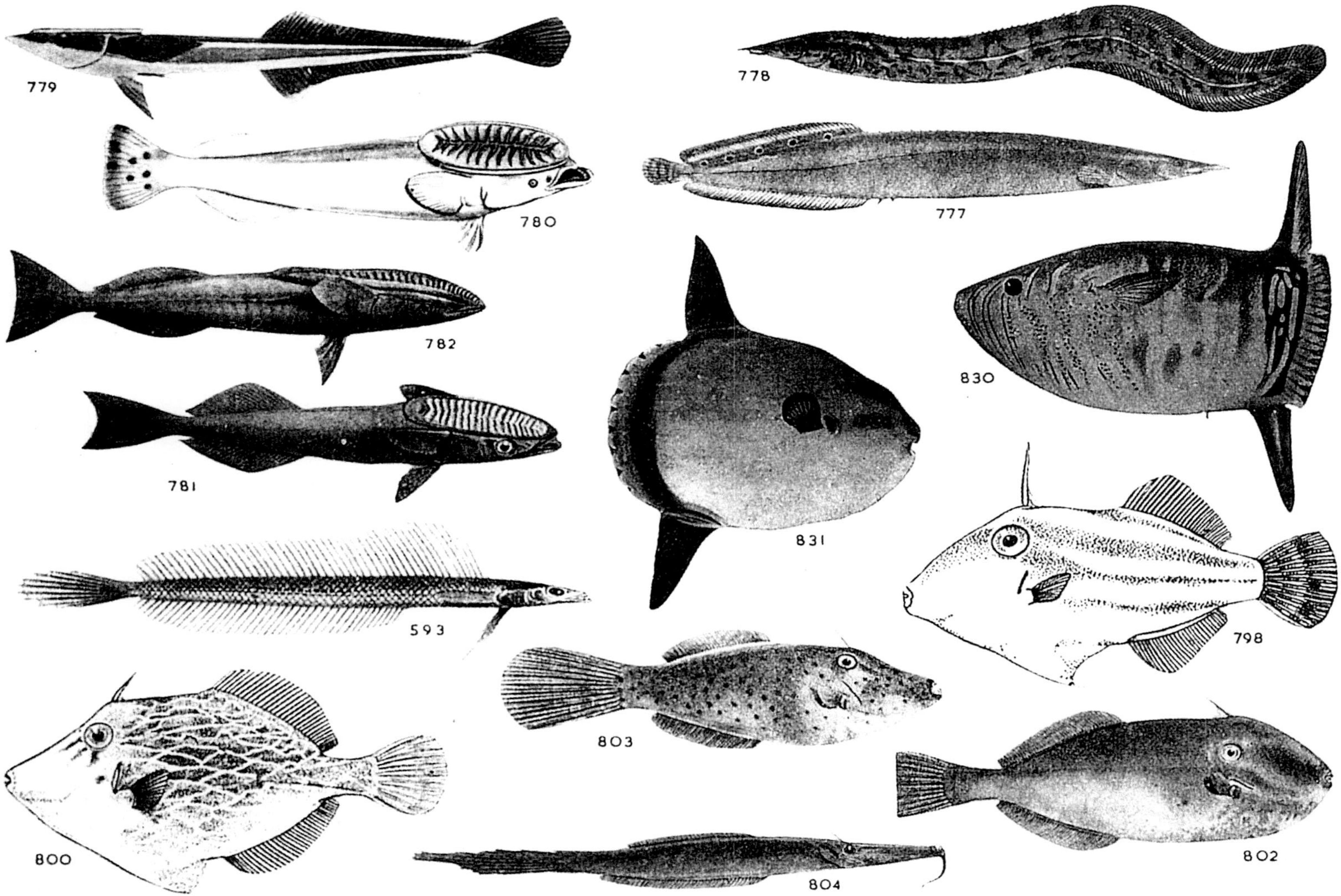

PLATE 52

Sand Eels, Spiny Eels, Suckerfish, Remoras, Leather-jackets and Sunfish.

PLATE 53

Leather-jackets and File-fish.

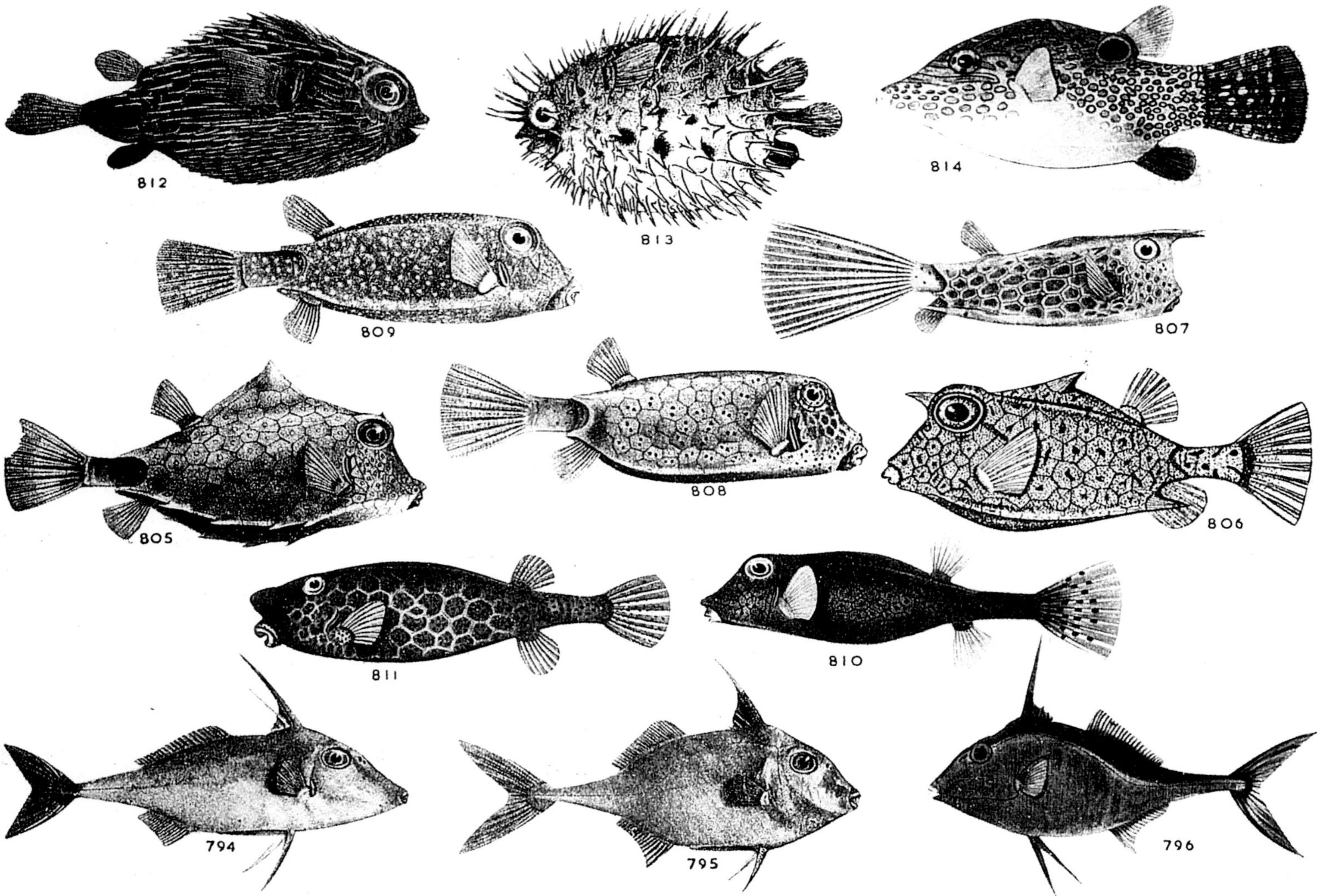

Plate 54

Tripod-fish, Box-fish, Cow-fish, Porcupine-fish and Tobys.

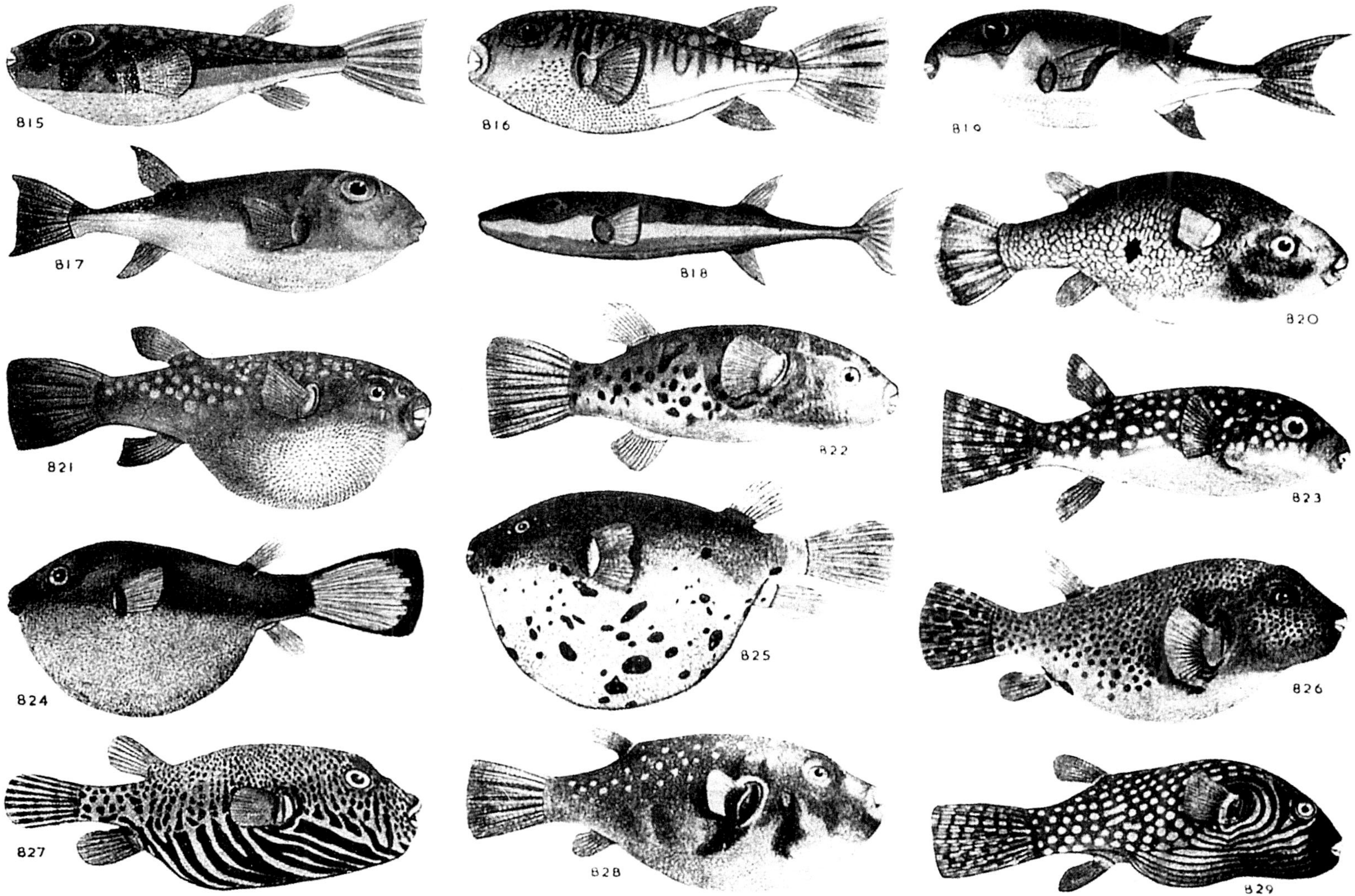

PLATE 55

Puffer-fish, Blowfish and Toad-fish.

PLATE 56

Frogfish, Monkfish, Handfish, Angler-fish, Fishing-frogs and Dragon-fish.

CLIMBING PERCH, BUBBLE-NEST BUILDERS

Suborder ANABANTOIDEI

Family ANABANTIDAE

Perciform fishes with a labyrinthic suprabranchial organ formed by expansion of the epibranchial of the first gill-arch. Air-bladder divided posteriorly. Ventrals I, 1-5, thoracic, the spine sometimes rudimentary. Dorsal and anal usually with spines. Scales ctenoid. Fresh water and estuarine species. Several introduced to Ceylon.

1. Dorsal longer than anal 2
   Dorsal shorter than anal 5
2. Palatine teeth present; anal originates below 8th or 9th dorsal spine; outer rays of ventrals normal .... *Anabas testudineus*
   Palatine teeth absent; anal originates only slightly behind dorsal origin; outer ray of ventrals produced and filamentous 3
3. Ventrals reduced to a single long filamentous ray; lateral line absent or vestigial .... *Trichogaster pectoralis*
   Ventrals with one spine and 5 rays, the outer being bifid and filamentous; lateral line complete but interrupted 4
4. Anal with 10-12 rays; mouth reaches orbit; teeth in jaws .... *Belontia signata*
   Anal with 17-19 rays; mouth not reaching orbit; teeth on lips .... *Helostoma temmincki*
5. Lateral line complete and continuous; anal with 20 or more rays .... *Osphronemus goramy*
   Lateral line reduced to a few pores anteriorly; anal with less than 15 rays 6
6. Dorsal inserted approximately above pectoral origin .... *Macropodus cupanus*
   Dorsal inserted far behind pectoral origin .... *Malpulutta kretseri*

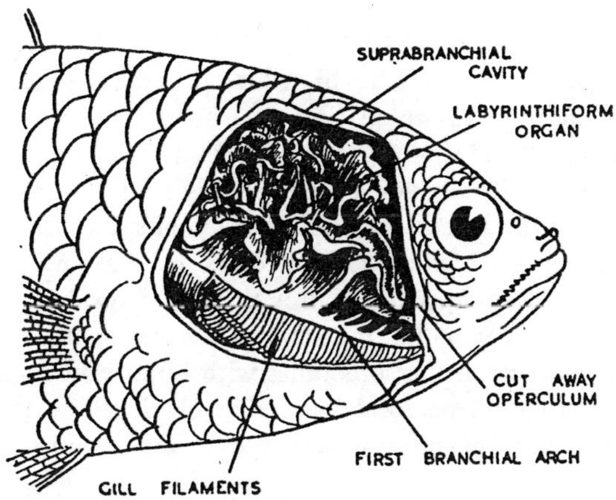

FIG. 16.

Dissection of gill-chamber of *Anabas* showing the development of a labyrinthiform organ from the first gill-arch as an accessory organ for aerial respiration.

661. ANABAS TESTUDINEUS (Bloch) CLIMBING PERCH

Anthias testudineus Bloch, Naturg. Ausländ. Fische, vii, 1795, p. 121. *Anabas oligolepis, scandens* or *testudineus kavaiya*. *Anabas testudineus* Deraniyagala, Ceylon Jour. Sci (B), xv (2), 1929, Pl. xxviii (plate 44).

D. XV-XVII; 9-10. A. IX-XI, 9-11. L. lat. (15-19)+(10-15). Tr. 5+(9-11). Predorsals 15. Preventrals 14-15. Depth $2\frac{1}{2}$-$3\frac{1}{2}$. Lateral line interrupted. Preorbital, interopercle, subopercle and opercle strongly serrated. (Entire in young.) Teeth small, conical and fixed. Suprabranchial organ consists of 2-6 osseous, perforated, fan-shaped lamellae. Light to dark green above, greenish-yellow to orange below. About 10 indistinct olive cross-bars. Black spot on caudal peduncle and sometimes another behind opercle. Dorsal pale green. Paired fins pinkish. Eye orange. Young have large ocellus on caudal peduncle. 6 inches. Capable of living out of water for short periods and crossing land, usually at night. Fresh waters of low-country, in tanks, ponds, streams and lakes.

Kavaiya, Pol kavaiya (S). Kavaiyan (T).

662. TRICHOGASTER PECTORALIS (Regan) BUBBLE-NEST BUILDER

Trichopodus pectoralis Regan, Proc. Zool. Soc., 1909 (1910), p. 784. *Trichopodus trichopterus* Chevey, Mem. Inst. Oceanogr. Indochine, iv, 1932, Pl. xxxiii (plate 44).

D. VII; 10-11. A. IX-XII; 33-38. L. lat. 55-63, in longitudinal series above. Soft dorsal in adult males, when flexed, reaches beyond caudal base; in females falls short of caudal base. Anal almost confluent with caudal. First ventral ray elongate and filamentous, reaching to end of caudal. Brownish with darker subvertical brown bands, about equal to interspaces. Soft anal with 2-3 longitudinal dark bands. No round black spot on middle of side or on caudal peduncle. 10 inches. Recently introduced to fresh waters in Ceylon, a native of Thailand, French Indo-China and the Malay Peninsula.

663. BELONTIA SIGNATA (Günther) LEAF LATES

Polyacanthus signatus Günther, Cat. Fish., iii, 1861, p. 379; Deraniyagala, Ceylon Jour. Sci. (B), xv (2), 1929, Pl. xxix (plate 44).

D. XVI-XVIII; 8-10. A. XIV-XVI; 10-12. P. 10-12. L. lat. (2-3)+(10-15)+(2-3)+(13-16). Tr. 4+(10-11). Predorsals 20-22. Preventrals 12-14. Ventral fin with outer ray elongated into 2 filaments which reach to about 11th anal spine. Caudal truncate in young but rounded with filamentous median rays in older fish. 2 rows of fixed conical teeth behind a single outer row of enlarged teeth in both jaws. Suprabranchial organ consists of 2 osseus lamellae invested in vascular mucous membrane; with 1-2 perforations. Olive dorsally and green on sides and belly, sometimes with blue reflections. A black spot under soft dorsal base and another on pectoral base. Fins olive yellow to reddish orange. Eye yellow. 5 inches. Fresh water hill-streams, and ponds, tanks and streams in low-country.

Pulutta, Kola modeya (S).

664. HELOSTOMA TEMMINCKI Cuvier KISSING GOURAMI

Helostoma temmincki Cuvier and Valenciennies, Hist. Nat. Poiss., vii, 1831, p. 342; Weber and De Beaufort, Fish. Indo-Aust. Archipel., iv, 1922, fig 88 (plate 44).

D. XVI-XVIII; 13-16. A. XIII-XV; 17-19. P. II, 11. L. lat. 44-48. Tr. 8+(15-16). Depth 2½, body strongly oval. Mouth protractile, lips thick and provided with some series of small movable black-tipped teeth. Preorbital serrated. Ventrals I, 5; first ray bifid and a little prolonged. Olive or greyish above, much lighter below with dark longitudinal stripes along the scale rows. 12 inches. Introduced recently to fresh waters in Ceylon. A native of Java, Borneo, Sumatra, Malay Peninsula and Central Thailand.

665. OSPHRONEMUS GORAMY Lacépède GOURAMI

Osphronemus goramy Lacépède, Hist. Nat. Poiss., iii, 1802, p. 116. *Osphronemus olfax* Bleeker, Atlas Ichth. Ind. Neerl., ix, 1877, Pl. cccxcv, fig. 6 (plate 44).

D. XII-XIII; 11-13. A. IX-XI; 19-21. P. 2, 13. L. lat. 30-33. Tr. (5-6)+(13-14). Head in young acute, obtuse in adults. Ventrals I, 5; outer ray single, elongated into a filament which extends to beyond the caudal fin. Preorbital serrated. Bands of small conical fixed teeth in jaws; exterior series largest. Lateral line almost straight. Brown or slaty above, silvery or yellowish below. Young with about 8 dark transverse cross-bands. A dark spot with yellow or silvery edge above soft anal and a black spot on pectoral base. Introduced into fresh waters of Ceylon in 1909 from Java. A native of the East Indian islands. 24 inches.
Goramy, Gurami (S).

666. MACROPODUS CUPANUS (Cuvier) PALMYRA-FIBRE FISH

Polyacanthus cupanus Cuvier and Valenciennes, Hist. Nat. Poiss., vii, 1831, p. 357. *Macropodus cupanus* Deraniyagala, Ceylon Jour. Sci (B), xv (2), 1929, Pl. xxxi (plate 44).

D. XIII-XV, 5-7. A. XVII-XX, 10-13. P. 11. V. I, 5. L. lat. (9-13)+(17-20). Tr. (4-5)+7. Predorsals 19-21. Preventrals 8-9. Soft dorsal and anal pointed. Caudal pointed with median rays elongated. Teeth small, conical and fixed, in many series in jaws only. Lateral line interrupted and rudimentary. Outer ray of ventrals produced into a single moderately long filament. Dark olive to green, darker above. Sometimes a brown stripe from eye to corner of opercle and brown spots on head. Elongated ventral rays red. Fins otherwise pale green. Soft dorsal and caudal with rows of black spots. A black blotch on caudal base. 1¾ inches. Fresh water ponds and ditches of low-country.
Tal kossa, Tal kadaya, Tal padda (S).

667. MALPULUTTA KRETSERI Deraniyagala DE KRETSER'S FISH

Malpulutta kretseri Deraniyagala, Ceylon Jour. Sci. (B), xx (3), 1937, p. 352; Col. Atlas Vert. Ceylon, i, 1952, Pl. xxxi (plate 44).

D. VIII-X; 6. A. XVI-XVII, 9-11. P. 12. V. I, 5. L. lat. 29-30. Tr. 9-10. Mouth protractile. Posterior angle and lower limb of preopercle serrated. Teeth in jaws conical, fixed and pluriserial. First ventral ray single, produced, as long as head less snout. Caudal in adult including the 2 central filamentous rays, about half body length to caudal base. Posterior few rays of dorsal and anal filamentous and produced. Lateral line with only 6 perforated scales. Cinnamon brown, lighter below. 3 blackish bands from eye to opercle.

2 rows of blackish blotches along sides. Fins yellowish with black spots and blue margins. Produced filamentous rays also blue. 1½ inches. Small fresh water ponds adjacent to streams.

Suborder KURTOIDEI

Perciform fishes with enlarged ribs which enclose the air-bladder. Males have a peculiar denticulated crest on the supraoccipital. One dorsal fin.

HUMP HEADS

Family KURTIDAE

Oblong, much compressed body with small cycloid scales. Head naked except preopercle and opercle. Lateral line developed for a short stretch behind head. Mouth large and protractile. Maxillary completely excluded from border of the mouth. Opercular bones thin and paper-like. Teeth in villiform bands in jaws and smaller ones on vomer and palatines. Males with prominent hook on occiput comprising skin covering the supraoccipital crest. Anus close behind ventrals.

668. KURTUS INDICUS Bloch — INDIAN HUMP HEAD

Kurtus indicus Bloch, Naturg. Ausländ. Fische, ii, 1786, p. 122; Day, Fish. India, 1875-78, Pl. xlii, fig. 1 (plate 44).

D. V; II, 12-13. A. II, 31-32. P. 18. V. I, 5. Depth 3·1-3·3. Preopercle with 4 spines near its angle. Operculum thin and paper-like. Caudal deeply forked. Lateral line ceasing opposite end of soft dorsal. Silvery shot with steel blue or lilac. Back with fine black dots which form a round black spot behind occiput. 4½ inches. Coastal waters, Palk Straits.

Suborder GOBIOIDEI

Body normally robust, usually rather elongate, sometimes little compressed or depressed. Usually scaly but sometimes naked. Dorsals divided or nearly so. Pelvics thoracic, with feeble spine and 4-5 rays, sometimes separate, often partly or completely united to form a sucking disc. Carnivorous, often sluggish fishes of shallow waters from marine to fresh water conditions, especially estuaries and tide pools.

1. Pelvics entirely separate .... *Family* ELEOTRIDAE, p. 228
   Pelvics fully united .... *Family* GOBIIDAE, p. 232
   Pelvics united at base only 2
2. Two dorsal fins .... *Family* PERIOPHTHALMIDAE, p. 242
   One dorsal fin .... *Family* TRYPAUCHENIDAE, p. 243

GUDGEONS

Family ELEOTRIDAE

Elongate body with small or moderate scales. Pelvics fully separate. Two dorsal fins, separated or connected at their bases, first usually with 6 feeble spines. Mouth moderate, teeth feeble, usually none on palate.

1. Teeth present in vomer; an ocellus on base of caudal fin .... *Bostrichthys sinensis*
   No teeth on vomer 2
2. Margin of preopercle with 3-5 spines directed backwards; head compressed ... *Asterropteryx semipunctatus*
   A single downwards curved strong spine at angle of preopercle; head subcylindrical or depressed
   .... *Eleotris fusca*
   Preopercle without spines 3
3. Bony crests on head in interorbital space 4
   No bony crests on head 5
4. Head flat, pointed; lower jaw prominent .... *Butis butis*
   Head short, obtuse; jaws subequal .... *Prionobutis koilomatodon*
5. Ventral fins long, narrow and with fringed rays; L.lat. less than 30 .... *Eviota gymnocephala*
   Ventral fins normal; L.lat. more than 30 6
6. Scales moderate, 30-40 in L.lat. .... *Ophiocara porocephala*
   Scales small, more than 50 in L.lat. 7
7. Sides of head scaled .... *Bunaka gyrinoides*
   Sides of head naked 8
8. Depth 8-9; cheek, opercle and dorsal fins with pink longitudinal bands .... *Eleotriodes muralis*
   Depth 7; cheeks with violet-edged blue spots and first dorsal with violet spots ...... *Eleotriodes sexguttatus*

669. BOSTRICHTHYS SINENSIS (Lacépède)

CHINESE OCELLATED GUDGEON

Bosirychus sinensis Lacépède, Hist. Nat. Poiss., v, 1801, p. 183. *Bostrichthys sinensis* Herre, Monogr. Bur. Sci. Manila, xxiii, 1927, Pl. iii, fig. 3 (plate 44).

D. VI; I, 9-10. A. I, 8-9. P. 16-17. L. lat. 100. Tr. 30. Depth 7½-8½. Body elongate, cylindrical anteriorly. Head depressed and snout pointed. Lower jaw slightly prominent. Maxilla extends to posterior margin of eye. Teeth in outer row distant; inner row small, mixed with larger ones. Tongue rounded. Head scaled. Caudal rounded. Green above, yellow below. Back and sides with oblong spots and transverse violet streaks. Fins yellow. Dorsal and anal with longitudinal streaks. Caudal base with orange ocellus, bordered with black. 4½ inches. Marine.

670. ASTERROPTERYX SEMIPUNCTATUS Rüppell

STAR-FINNED GUDGEON

Asterropteryx semipunctatus Rüppell, Atlas Reise N. Afrika, Fische, 1828, p. 138; Ogilby, Mem. Qld. Mus., iii, 1915, Pl. xxix, fig. 2 (plate 44).

D. VI; I, 9-11. A. I, 8-10. P. 17. L. lat. 24. Tr. 8. Predorsals 7. Depth 4½ in length. Scales ctenoid. Scaled on head behind eyes, on cheek and opercle. Teeth in several rows, outer enlarged. A canine on each side of lower jaw. Caudal rounded. Brown. Each scale except those on back, with a bluish spot. Irregular brownish cross-bars and blotches. Unpaired fins with blue dots. 2 inches. Coral reefs.

671. ELEOTRIS FUSCA (Schneider) BROWN GUDGEON

Poecilia fusca, Bloch and Schneider, Syst. Ichth., 1801, p. 453. *Eleotris fusca* Herre, Monogr. Bur. Sci. Philipp., xxiii, 1927, Pl. ii, fig. 1 (plate 44).

D. VI; I, 8. A. I, 8. P. 15-18. L. lat. 60-65. Tr. 16-19. Predorsals 50. Depth 5-6 in length. Scales cycloid before first dorsal, ctenoid posteriorly. Head depressed, scaled above between and behind eyes, on cheek and opercle, cycloid. Lower jaw prominent with teeth in several rows, outer enlarged. No canines. Dorsals separate. Caudal oblong. Blackish-green above, blackish-orange below. Scales with dark spot forming longitudinal lines. Fins violet with orange rays. Spinous dorsal with 2-3 dark bands. Second dorsal, pectorals and caudal with dark stripes. 6½ inches. Brackish to fresh waters, including hot springs.
Bin thalla, Bin tholla, Puvak badilla (S).

672. BUTIS BUTIS (Hamilton-Buchanan) FLAT-HEADED GUDGEON

Cheilodipterus butis Hamilton-Buchanan, Fish. Ganges, 1822, p. 57. *Butis amboinensis* McCulloch and Ogilby, Rec. Aust. Mus., xii (10), 1919, Pl. xxxvi, fig. 4 (plate 44).

D. VI; I, 8. A. I, 8. P. 18-21. L. lat. 30. Tr. 9-10. Predorsals 20. Depth 6-8 in length. Body cylindrical anteriorly, compressed posteriorly and with ctenoid scales. Head flat, pointed, with ctenoid scales on snout, between nostrils, on cheek and opercles. Interopercles naked. Bony crests on interorbital and around eyes. Lower jaw prominent. Teeth in several rows, in both jaws outer enlarged. Blackish, lighter below. With irregular cross-bands and reddish spots. Spinous dorsal blackish with red tip. Other fins dusky to violet with orange or red rays. Dorsal, anal and caudal bordered with red. Pectoral with 1-2 black spots at base. 6 inches. Coastal lagoons and estuaries.
Vaneya (S).

673. PRIONOBUTIS KOILOMATODON (Bleeker) SAW-TOOTHED GUDGEON

Eleotris koilomatodon Bleeker, Verh. Bat. Gen., xxii, 1849, p. 21. *Prionobutis koilomatodon* Herre, Monogr. Bur. Sci. Philipp., xxiii, 1927, Pl. iv, fig. 2 (plate 44).

D. VI; I, 8. A. I, 8. P. 19-21. L. lat. 30. Tr. 7-10. Predorsals 12-14. Depth 4⅔-5 in length. Body cylindrical anteriorly, compressed posteriorly. Scales ctenoid. Head convex with scales behind eye, upper part of cheek and opercle. Orbital crest denticulated, 2 crests on each side of snout. Teeth in several rows, outer enlarged. Dusky green to olive, lighter below. Body scales often with a red spot. Spinous dorsal with dusky tip and pink posterior margin. Pectorals red-orange with black spot on base. Other fins dusky. Anal and caudal bordered with red. 3¼ inches. Harbours.

674. EVIOTA GYMNOCEPHALA Weber NAKED-HEADED GUDGEON

Eviota gymnocephalus Weber, Siboga Exped., lvii, Fische, 1913, p. 452, fig. 87 (plate 44).

D. VI; I, 9. A. I, 8. P. 15. L. lat. 25. Tr. 7. Depth 5½ in total length. Body elongate, compressed. Scales ctenoid. Head naked. Teeth in several rows, outer enlarged. Dorsal fins nearly connected at base. Ventral fins long and narrow, with rays fringed. Caudal rounded. Yellowish, sometimes with about 8 transverse dark blotches. Head and nape with blotches. Anal dusky. 1 inch. Pearl banks.

675. OPHIOCARA POROCEPHALA (Valenciennes)

PORE-HEADED GUDGEON

Eleotris porocephala Cuvier and Valenciennes, Hist. Nat. Poiss., xii, 1837, p. 237; Day, Fish. India, 1875-78, Pl. lxvii, fig. 1 (plate 45).

D. VI; I, 8-9. A. I, 6-7. P. 15. L. lat. 38-40. Tr. 11-13. Predorsals 24-26. Body anteriorly cylindrical, posteriorly compressed, depth 5-6. Head depressed. Snout pointed. Teeth in many rows, outer enlarged. In lower jaw some inner teeth also enlarged. Mucous canals around eye and on cheek. Open pores on snout, interorbital and preopercle. Head totally scaled. Scales of head, breast and belly cycloid to ctenoid, those on body ctenoid. Caudal rounded. Dark green to olive above, lighter below. Body with dark longitudinal stripes, becoming indistinct with age. Fin membranes dusky to violet with orange rays. Soft dorsal, anal and caudal with reddish-orange border. Head, body, soft dorsal and anal with orange spots. Young with 2-5 silvery cross-bars. 12 inches. Brackish water.

676. BUNAKA GYRINOIDES (Bleeker) ROUND-BODIED GUDGEON

Eleotris gyrinoides Bleeker, Nat. Tijds. Ned. Ind., iv, 1853, p. 272. *Eleotris canarensis* Day, Fish. India, 1875-78, Pl. lxix, fig. 2 (plate 45).

D. VI; I, 8. A. I, 8. P. 18-19. L. lat. 60. Tr. 16-17. Predorsals 40. Depth 5-6½. Body elongate, cylindrical anteriorly and compressed posteriorly. Scales ctenoid. Head depressed, scaled above behind eyes, on cheek and opercle. Teeth in several rows, outer and inner rows not or hardly enlarged; no canines. Dorsal fins separate. Blackish-green above, orange-green below. Scales with dark spot forming longitudinal streaks. Fins orange, rays with dark spots forming bands. Black spot on pectoral base. 13 inches. Harbours.

677. ELEOTRIODES MURALIS (Valenciennes) PAINTED GUDGEON

Eleotris muralis Cuvier and Valenciennes, Hist. Nat. Poiss., xii, 1837, p. 253. *Valenciennea muralis* McCulloch & Ogilby, Rec. Aust. Mus., xii (10), 1919, Pl. xxxvii, fig. 4 (plate 45).

D. VI; I, 12. A. I, 12. P. 19-20. L. lat. 85-95. Tr. 25. Body compressed, 7-9. Head obtuse, compressed. Conical teeth. Upper jaw with one row and 3-5 canines in front. Lower jaw with 2 rows anteriorly and one row laterally; outer row with 5-6 enlarged teeth and a curved canine; inner row with 3-7 enlarged teeth. 3-4 pores on preopercular margin, and others on interorbital, behind eye and on snout. Head naked. Body scales ctenoid. Caudal rounded. Greenish-red above, pink to pearly below. Body with 3 longitudinal red stripes, lowest from axil to caudal. Fins yellow or orange. Spinous dorsal with red oblique stripes and black spot at tip. Soft dorsal with 2 red stripes basally and transverse oblong red spots distally. Pectoral base with red spot. Anal with basal red stripe. Caudal with irregular red ocelli. Inside mouth black. 6½ inches. Harbours and estuaries.

678. ELEOTRIODES SEXGUTTATUS (Valenciennes)

OCELLATED GUDGEON

Eleotris sexguttata Cuvier and Valenciennes, Hist. Nat. Poiss., xii, 1837, p. 254. *Valenciennea violifera* Jordan and Seale, Bull. U.S. Bur. Fish., xxv, 1906, Pl. lii, fig. 2 (plate 45).

D. VI; I, 12. A. I, 11-12. P. 19-20. L. lat. 85-90. Tr. 22-23. Body elongate, compressed. Scales ctenoid. Head obtuse, naked. Teeth in upper jaw in single row, in front of each side 3-6 caninoid teeth. Lower jaw with 2 rows in front and 1 row laterally; 5-6 enlarged anteriorly in both rows. Dorsal fins subcontinuous. Reddish-green above, pinkish below. Three rows blue spots across cheek to pectoral base. Red longitudinal streak along lower part of body. Fins yellow, orange or red. Fins with violet spots or red streaks. Base of anal with 8-10 orange ocelli bordered with violet. Caudal with similar ocelli. $5\frac{1}{2}$ inches. Harbours and estuaries.

GOBIES

Family GOBIIDAE

Pelvic fins united to form a feeble sucking disc. Body typically scaly, sometimes naked. Mostly with 2 dorsal fins, the first with 5-6 feeble spines. Mouth usually large with small canine teeth.

1. Body deeply oval, strongly compressed 2
   Body elongate 5
2. With large scales on body .... *Paragobiodon echinocephalus*
   Body without scales, covered by thick, granulous and slimy skin 3
3. In lower jaw on each side 1-2 canines; spinous dorsal lower than soft dorsal, foremost ray shortest ... 4
   In lower jaw on each side 3-4 canines; dorsal fins equal in height, foremost ray of first dorsal longest .... *Gobiodon citrinus*
4. Gill-opening extends below lower rays of pectoral; head deeper than long; upper part of opercle with brown spot .... *Gobiodon erythrospilus*
   Gill-opening does not extend below rays of pectoral; head longer than deep; upper part of opercle without brown spot .... *Gobiodon quinquestrigatus*
5. Teeth in lower jaw in one row 6
   Teeth in lower jaw in several rows 8
6. Second dorsal fin elongate; teeth in upper jaw fixed; teeth in lower jaw horizontal 24
   Second dorsal fin not elongated; teeth in upper jaw movable and bent; teeth in lower jaw not horizontal 7
7. Predorsal scales 11-12; no interorbital dark band; fins reddish .... *Sicyopterus gymnauchen*
   Predorsal scales 19; a dark interorbital band; fins with dark bands .... *Sicyopterus halei*
8. First ray of both dorsal fins strong and bony; preopercle armed with some teeth behind .... *Oplopomus oplopomus*
   First ray of both dorsal fins not strong and bony; preopercle unarmed 9
9. Caudal fin long, pointed, lanceolate, longer than head 10
   Caudal fin short, obtuse, rounded or oblong, shorter than head 14
10. Fleshy flaps on inner edge of shoulder girdle .... *Stenogobius malabaricus*
    No fleshy flaps on inner edge of shoulder girdle 11
11. Teeth in upper jaw in one row 12
    Teeth in upper jaw in several rows 13
12. No tentacle on eye, but a black triangular spot in upper part of iris .... *Oxyurichthys microlepis*
    A tentacle at upper margin of eye .... *Oxyurichthys tentacularis*
13. Nape naked; head depressed and with ridges of papillae .... *Callogobius hasseltii*
    Nape partly scaled; head compressed and without ridges of papillae .... *Oligolepis acutipennis*
14. Basal membrane of ventral fins weak or absent .... *Zonogobius semidoliatus*
    Basal membrane of ventral fins normally developed 15
15. Tongue emarginate or bilobate 16
    Tongue rounded or truncate 22

16. Fleshy flaps on inner edge of shoulder girdle .... *Awaous grammepomus*
    No such fleshy flaps 17
17. Gill-openings continued forward below, isthmus narrow 18
    Gill-openings not continued forward below, isthmus wide 19
    Gill-openings continued forward below, membranes united across isthmus .... *Cottogobius bilobatus*
18. Tr. 7-8; two ocelli on first dorsal; iris with process in pupil .... *Glossogobius biocellatus*
    Tr. 9-14; first dorsal immaculate or with one spot; iris without process in pupil ...... *Glossogobius giuris*
19. Body subcylindrical anteriorly, compressed posteriorly; head depressed .... *Bathygobius fuscus*
    Body compressed; head slightly compressed 20
20. L.lat. with 40-42 scales; Tr. 14 .... *Acentrogobius griseus*
    L.lat. with 35-36 scales; Tr. 10 .... *Acentrogobius viridipunctatus*
    L.lat. with 26-32 scales; Tr. 7-10 21
21. Predorsal scales about 25 .... *Acentrogobius masoni*
    Predorsal scales 17-20 .... *Acentrogobius caninus*
    Predorsal scales less than 14 .... *Acentrogobius ornatus*
22. Foremost scale in median line behind eye enlarged 23
    Foremost scale in median line not enlarged .... *Mugilogobius valigouva*
23. Lower jaw prominent; teeth of inner row of lower jaw laterally enlarged ...... *Stigmatogobius sadanundio*
    Upper jaw prominent; teeth of inner row of lower jaw not laterally enlarged ... *Stigmatogobius javanicus*
24. Canine teeth in both jaws; scales minute, more than 200 in longitudinal series .... *Pseudapocryptes lanceolatus*
    Canine teeth in lower jaw only; scales moderate, about 75 in longitudinal series .... *Parapocryptes macrolepis*

679. PARAGOBIODON ECHINOCEPHALUS (Rüppell) PRICKLY-HEADED GOBY

Gobius echinocephalus Rüppell, Atlas Reise Nordl. Afrika, 1828, p. 136. *Paragobiodon echinocephalus* McCulloch and Ogilby, Rec. Aust. Mus., xii (10), 1919, Pl. xxxiv, fig. 1 (plate 45).

D. VI; I, 9-10. A. I, 8-9. P. 18-22. L. lat. 22-27. Tr. 9-11. Depth 3⅓-4. Body oblong, compressed. Head obtuse. Open pores on front of head, around eyes, on opercle and preopercle. Teeth multiserial, outer row enlarged. An erect canine on each side of symphysis of lower jaw. Scales ctenoid on body. Head and body to 2nd-6th dorsal spine naked. Head orange-red to yellow. Body similar, more usually dusky or black. Iris orange-green to violet. Fins darker than body. 1½ inches. Marine, in rock pools, etc.

680. GOBIODON CITRINUS (Rüppell) LEMON GOBY

Gobius citrinus Rüppell, Neue Wirbelth., Fische, 1838, p. 139. *Gobiodon citrinus* Day, Fish. India, 1875-78, Pl. lxiv, fig. 2 (plate 45).

D. VI; I, 10. A. I, 9. P. 19-21. Body oblong, compressed. Depth 3-3¼. Head obtuse. Tubiform pores around margin of preoperculum. Lower border of preoperculum and mandible with small papillae. Teeth in both jaws in several rows, outermost enlarged. 3-4 canines on each side of symphysis of lower jaw. No scales but covered with a thick, granulous, slimy skin. Body and fins black to orange. Iris violet to green. Upper part of opercle with dark spot. 4 blue transverse bands on anterior part of body. Bases of soft dorsal and anal with longitudinal blue band bordered with black or violet. 2 inches. Reefs and rock pools.

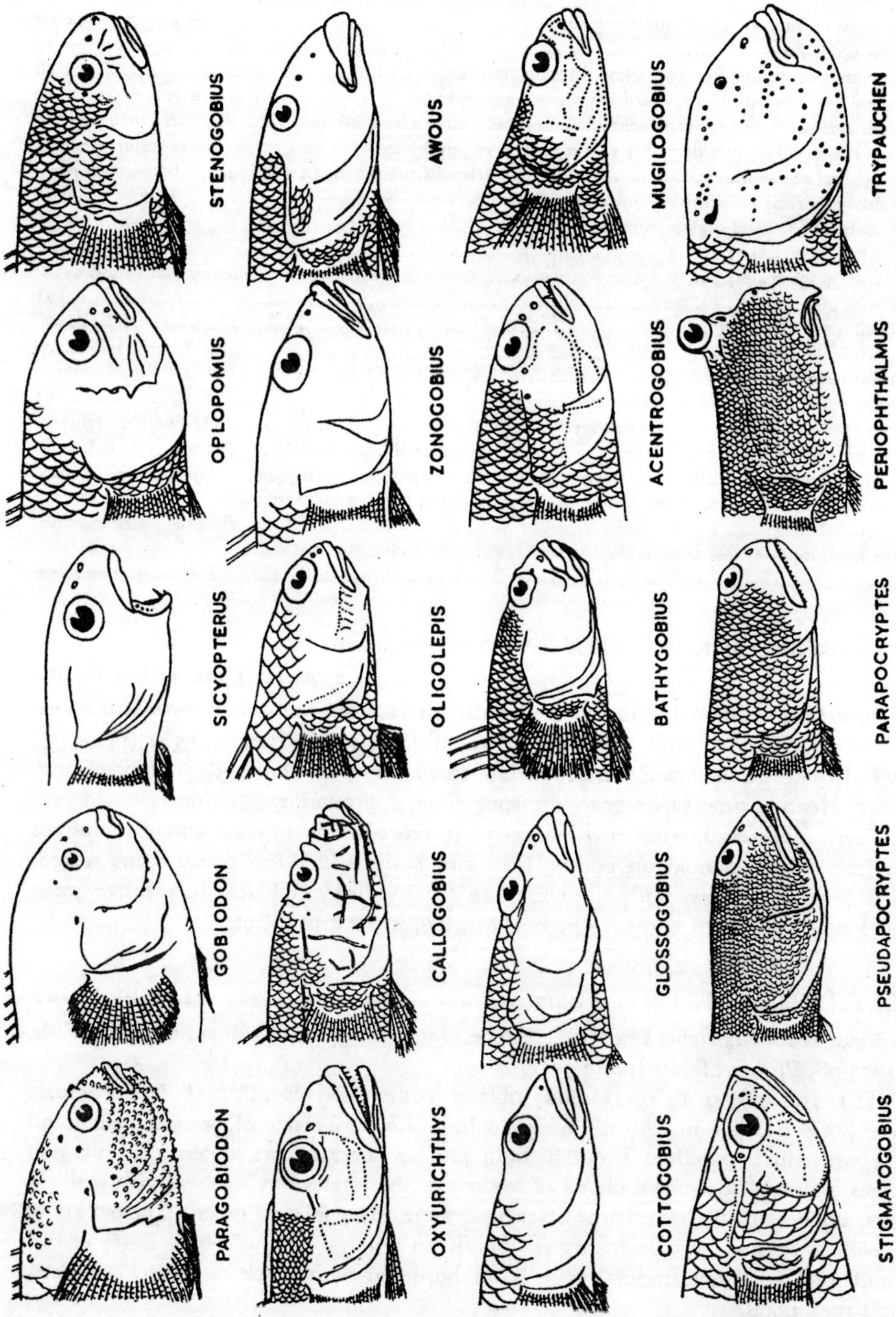

FIG. 17.
Principal differences between the twenty genera of the family Gobiidae (Gobies), with particular reference to extent of scalation, development of mucous canals, pores and papillae.

681. GOBIODON ERYTHROSPILUS Bleeker RED-SPOTTED GOBY

Gobiodon erythrospilus Bleeker, Arch. Neerl. Sci. Nat., x, 1875, p. 122. *Gobiodon verticalis* McCulloch and Ogilby, Rec. Aust. Mus., xii (10), 1919, Pl. xxxii, fig. 2 (plate 45).

D. VI; I, 10-11. A. I, 9. P. 19-21. Body oblong, compressed. Head obtuse. Pores and papillae as in preceding species. Teeth in both jaws in several rows, outermost enlarged. A canine tooth on each side of symphysis of lower jaw. Scales absent and replaced by a thick, granulous, slimy skin. Yellow to orange. Iris green. Head and anterior part of body with purple transverse bands. 7 longitudinal rows of purple spots on body. Vertical fins with violet margins. Dark spot on upper part of opercle. 2 inches. Coastal waters and pearl banks, reefs.

682. GOBIODON QUINQUESTRIGATUS (Valenciennes)
FIVE-STREAKED GOBY

Gobius quinquestrigatus Cuvier and Valenciennes, Hist. Nat. Poiss., xii, 1837, p. 134. *Gobiodon ceramensis* Günther, Jour. Mus. Godeffroy, vi (11), 1877, Pl. cix, fig. D (plate 45).

D. VI; I, 10. A. I, 8-9. P. 17-19. Body oblong, compressed. Depth 4-4$\frac{3}{5}$. Head obtuse. Pores and papillae as in preceding species. Teeth in several rows in both jaws, outermost enlarged. 1-2 canines on each side of symphysis of lower jaw. Scales absent and replaced by a thick, slimy, granulous skin. Black to blackish-orange. Iris greenish. No black spot on upper part of opercle. Fins reddish or blackish. Head and anterior part of body with bluish transverse lines. 2 inches. Coastal reefs.

683. SICYOPTERUS GYMNAUCHEN (Bleeker) NAKED-HEADED GOBY

Sicydium gymnauchen Bleeker, Act. Soc. Sci. Indo-Neerl., iii, 1858, p. 11. *Sicyopterus tauae* Jordan and Seale, Bull. U.S. Bur. Fish., xxv, 1906, fig. 96 (plate 45).

D. VI; I, 9-10. A. I, 10. P. 10. L. lat. 52-55 of which 12-15 are cycloid and remainder ctenoid. Tr. 14-15. Predorsals 11-12. Body cylindrical anteriorly, compressed posteriorly. Depth 7-8. Cheeks naked. Scales deciduous, cycloid anteriorly and ctenoid posteriorly. Caudal emarginate. Lower lip with numerous slender horizontal teeth. Upper jaw with single row of movable teeth; lower with row of widely set conical teeth and a pair of canines behind symphysis. Green above, pearly below. Body with blackish transverse stripes, narrower than interspaces, which may be diffuse or wanting. Fins reddish. Anal with black spots. 1$\frac{1}{2}$ inches. Harbours.
Kongani, Hin kongani, Kos kongani, Pol kongani (S).

684. SICYOPTERUS HALEI (Day) HALY'S GOBY

Sicydium halei Day, Fishes India Supplement, 1888, p. 794.

D. VI; 12. A. 11. P. 19. L. lat. 65. Tr. 16. Depth 7. Body subcylindrical. Upper jaw longer than lower. Teeth in upper jaw comprising an external movable row directed almost horizontally and an inner row of longer curved pointed ones. A large recurved canine on each side of symphysis of lower jaw, followed by a single row of teeth smaller than in upper jaw. Spines of first dorsal ending in filamentous prolongations. Scales all ctenoid. Greenish brown. A black interorbital band continued through eye to angle of mouth. Some dark vertical bands on body. Dark outer edge to ventral, anal, and caudal. 3 inches. Coastal waters.

685. OPLOPOMUS OPLOPOMUS (Valenciennes) ARMOURED GOBY

Gobius oplopomus Cuvier and Valenciennes, Hist. Nat. Poiss., xii, 1837, p. 66. *Hoplopomus acanthistius* Regan, Trans. Linn. Soc., xii, 1908, Pl. xxix, fig. 3 (plate 45).

D. VI; I, 9-10. A. I, 10. P. 18. L. lat. 30. Tr. 7-8. Predorsals 6-12. Body elongate, compressed. Depth 5½. Outer row of teeth in upper jaw enlarged. In lower jaw, outer row are enlarged, extending only half way along jaw, last tooth on each side a canine. Preopercle with 2 marginal serrations. Head naked except for a few scales behind eye. Body scales ctenoid but cycloid on nape and breast. Pectorals and caudal obtuse. Reddish-green above, pearly red below. Cheek and opercle with pearly spots and streaks bordered with violet. Laterally a row of dusky spots. Fins yellow-orange. Spinous dorsal with oblique violet streak. Second dorsal with small ocelli. Ventral with black posterior. Black spots on caudal. 3¼ inches. Harbours.

686. STENOGOBIUS MALABARICUS (Day) MALABAR GOBY

Gobius malabaricus Day, Proc. Zool. Soc., 1865, p. 27. *Chonophorus malabaricus. Gobius malabaricus* Day, Fishes of Malabar, 1865, Pl. vii, fig. 2 (plate 45).

D. VI; I, 10. A. I, 10. L. lat. 50. Tr. 9-10. Body elongate, compressed. Depth 5¼. Teeth in both jaws in several rows, outer enlarged, pointed. No canines. Some mucous canals longitudinally and transversely across cheek. An open pore on each side of interorbital space. Fleshy flaps on inner edge of shoulder girdle. Body scales ctenoid. Caudal lanceolate, a little longer than head. Pale brown with irregular dusky bands on back and sides. A dark band descends from eye. Brown blotches on head. Prominent crescentic mark on spinous dorsal. Other vertical fins brown with rows of darker spots.

687. OXYURICHTHYS MICROLEPIS (Bleeker) SMALL-SCALED GOBY

Gobius microlepis Bleeker, Verh. Bat. Gen., xxii, p. 35. *Gobius cristatus* Day, Fish. India, 1875-78, Pl. lxii, fig. 8 (plate 45).

D. VI; I, 12. A. I, 13. P. 20-22. L. lat. 55. Tr. 14. Predorsals 17. Body elongate, compressed. Depth 7½-9. Breast and belly scaled. Scales ctenoid posteriorly and cycloid anteriorly. Head scaled above eye, naked on cheek and opercle. Teeth in upper jaw in one row; in lower jaw in 2-3 rows, inner row enlarged anteriorly. Longitudinal mucous canals over cheek. Inner edge of shoulder girdle without fleshy flaps. Caudal pointed, about twice head. Green above, pearly-red below. Back and sides with large violet blotches. Vertical fins red with blue spots on membrane, and 2 transverse rows of dark spots between rays. Pectoral orange bordered with yellow and violet. Caudal bordered with violet below and with black spots above. 5½ inches. Harbours.

687a. OXYURICHTHYS TENTACULARIS (Valenciennes) TENTACLED GOBY

Gobius tentacularis Cuvier and Valenciennes, Hist. Nat. Poiss., xii, 1837, p. 128.

D. VI; I, 12. A. I, 13-14. P. 19-22. L. lat. 52-65. Tr. 14-16. Predorsals 17. Body elongate, compressed. Depth 7-7½. Head scaled behind eyes but median line of head and nape naked. Scales on head and anterior part of body cycloid, others ctenoid. Upper jaw with 20 curved teeth on each side; lower jaws with 2-3 rows, innermost enlarged. A short tentacle at upper margin of eye. A low dermal crest on nape. Mucous canals on cheek. First dorsal higher than body, anterior radials filiform. Caudal lanceolate, about twice head length. Reddish green above, pearly below. Scales of sides and back with marginal red spot. An oblong dark spot below eye. Dorsals with rows of reddish-violet spots. Pectoral orange. Ventral and anal violet. Caudal yellowish, violet below and with oblique reddish streaks. 6 inches. Coastal waters.

688. CALLOGOBIUS HASSELTII (Bleeker) VAN HASSELT'S GOBY

Eleotris hasseltii Bleeker, Nat. Tijds. Ned. Ind., i, p. 253. *Macgregorella indica. Callogobius hasseltii* var. *mucosus* McCulloch and Ogilby, Rec. Aust. Mus., xii (10), 1919, Pl. xxxii, fig. 4 (plate 45).

D. VI; I, 9-10. A. I, 7-8. P. 16-18. L. lat. 40-45. Tr. 15-16. Predorsals 20. Body elongate, cylindrical anteriorly and compressed posteriorly. Depth 7½-8. Head pointed, depressed. Lower jaw prominent. Head, cheek and opercle with rows of cutaneous papillae and folds. Cycloid scales on cheek, opercle and behind eyes. Scales enlarged posteriorly, ctenoid. Caudal lanceolate, longer than head. Teeth in both jaws in several rows, outer of which enlarged. No canines. Shoulder girdle without fieshy flaps. Reddish green above, orange below. Head and body clouded with black. Body with 3-4 indistinct transverse bands. Fins yellow, with transverse stripes. 3 inches. Gulf of Manaar. Enters brackish water.

689. OLIGOLEPIS ACUTIPENNIS (Valenciennes) POINTED-FINNED GOBY

Gobius acutipennis Cuvier and Valenciennes, Hist. Nat. Poiss., xii, 1837, p. 80. *Rhinogobius ocyurus* Jordan and Seale, Bull. U.S. Bur. Fish., xxvi, 1907, fig. 14 (plate 45).

D. VI; I, 10-11. A. I, 10-11. P. 20-21. L. lat. 27-30. Tr. 7-8. Body elongate, compressed. Depth 5-6. Teeth anteriorly in several rows; in upper jaw outer row enlarged anteriorly; in lower jaw inner row enlarged. Open pores above and behind eyes. A bifurcating mucous canal from posterior nostril across cheek. Head and nape naked. Belly scales cycloid, others ctenoid. Caudal lanceolate, longer than head. Yellowish. An oblique violet streak from eye to behind maxillary. 5 large dark spots alternating with short dark transverse stripes on sides. Back with transverse bars. Fins reddish; membranes of dorsal and caudal with rows of dark spots. Dark spot on pectoral base. 4½ inches. Adam's Bridge.

690. ZONOGOBIUS SEMIDOLIATUS (Valenciennes) GIRDLED GOBY

Gobius semidoliatus Cuvier and Valenciennes, Hist. Nat. Poiss., xii, 1837, p. 67. *Glossogobius semidoliatus. Zonogobius semidoliatus* Jordan and Seale, Bull. U.S. Bur. Fish., xxv, 1906, fig. 86 (plate 45).

D. VI; I, 8-9. A. I, 7-8. P. 16-17. L. lat. 27. Tr. 9-10. Body elongate, compressed. Depth 4. Both jaws with several rows of teeth, outer enlarged; upper with short inner row of enlarged teeth anteriorly; lower with complete inner row of enlarged teeth and some lateral canines. Head naked. Body scales ctenoid. Spinous dorsal high. Caudal rounded. Brown with 7-8 broad pale cross-bars, edged with dark brown on head and anterior part of body. Fins uniform. 1½ inches. Coastal waters and pearl banks.

Gal huna (S). Kal oluvai (T).

691. AWAOUS GRAMMEPOMUS (Bleeker) SCRIBBLED GOBY

Gobius grammepomus Bleeker, Verh. Bat. Gen., xxii, 1849, p. 33. *Awaous macropterus* or *personatus*. *Chonophorus melanocephalus* Herre, Monogr. Philipp. Bur. Sci., xxiii, 1927, Pl. xvii, fig. 1 (plate 46).

D. VI; I, 10. A. I, 10. P. 15-17. L. lat. 50-52. Tr. 13-15. Predorsals 13-15. Body elongate, anteriorly cylindrical, posteriorly compressed. Depth 5½-7½. Teeth in lower jaw equal, those in upper jaw a little enlarged in outer series. Tongue bilobate. Scales on head, breast and belly cycloid; body ctenoid. Caudal obtusely truncate, a little shorter than head. Olive green, lighter below. Head violet to olive green. 2 oblique blackish streaks from eye to maxillary. Back and sides with irregular blackish streaks. Caudal with 7-9 blackish streaks. Ventral and caudal often blackish-violet. 5½ inches. Coastal waters.

Vala potha, Vali honda, Vali kanaya, Vali thiruva, Vel kalandha, Thalangoiya (S).

692. COTTOGOBIUS BILOBATUS Koumans TWO-LOBED GOBY

Cottogobius bilobatus Koumans, Mem. Indian Mus., xiii (3), 1941, p. 253.

D. VI; I, 7-8. A. I, 7-8. L. lat. 26-27. Tr. 6-7. Predorsal 10 or naked. Body elongate, anteriorly cylindrical, posteriorly compressed. Depth 7. Head naked or few ctenoid scales behind eyes. Breast naked. Body scales ctenoid. Upper jaw with many rows of minute teeth, outer row enlarged canines. Lower jaw with many rows, outer enlarged, inner row laterally enlarged to curved canines, and 1-2 strong curved symphysial canines. Tongue bilobate. 2-3 longitudinal mucous canals on cheek. Lowermost pectoral rays unbranched. Brownish. 3 faint bands radiating from eye. Whole body with minute dark spots. Spinous dorsal with a few dark spots. Soft dorsal with an irregular black blotch posteriorly. 1¼ inches. Pearl banks.

693. GLOSSOGOBIUS BIOCELLATUS (Valenciennes) TWO-SPOT GOBY

Gobius biocellatus Cuvier and Valenciennes, Hist. Nat. Poiss., xii, 1837, p. 73. *Glossogobius vaisiganis*. *Glossogobius biocellatus* Herre, Monogr. Philipp. Bur. Sci., xxiii, 1927, Pl. xii, fig. 3 (plate 46).

D. VI; I, 9. A. I, 8. P. 17-19. L. lat. 28-32. Tr. 7-8. Predorsals 14-17. Body elongate, anteriorly cylindrical, posteriorly compressed. Depth 6-8. Snout pointed. Teeth in many rows, some rows caninoid in maxilla; inner and outer rows caninoid, enlarged in lower jaw, outer row strongly curved in front. 3-5 longitudinal mucous canals on cheek. Pectoral and caudal rounded. Greenish to blackish with shiny green spots and 5-6 larger blotches

in one or two alternating rows on upper parts of sides. Fins blackish to orange. Spinous dorsal with blackish ocellus between 1st and 2nd rays and another between 5th and 6th rays. 4½ inches. Coastal waters and pearl banks.
Kuduppuva, Mul batta (S).

694. GLOSSOGOBIUS GIURIS (Hamilton-Buchanan) BAR-EYED GOBY

Gobius giuris Hamilton-Buchanan, Fishes Ganges, 1822, p. 366; Day, Fish. India, 1875-78, Pl. lxvi, fig. 1 (plate 46).

D. VI; I, 8-9. A. I, 7-8. P. 17-21. L. lat. 28-36. Tr. 8-14. Predorsals 12-30. Body elongate, anteriorly cylindrical, posteriorly compressed. Depth 6-8½. Head pointed, depressed. Teeth of outer and inner rows enlarged, and caninoid in front. Tongue bilobate. Scales on upper part of head, upper parts of cheek and opercle, breast and belly cycloid. Other scales ctenoid. Caudal obtuse, as long as head. Olive to dusky green above, lighter below. Head with dark to violet spots laterally. Body with 2 alternating rows of 4-6 dark blotches. Fins yellow-green with spots and dark margins. 14 inches. Coastal waters and harbours, entering fresh waters.
Weligowa, Gal katta (S). Uluvai (T).

695. BATHYGOBIUS FUSCUS (Rüppell) BROWN GOBY

Gobius fuscus Rüppell, Atlas Reise Nordl. Afrika, Fische, 1828, p. 137. *Gobius albopunctatus*, *Gobius elegans*. *Mapo fuscus* McCulloch and Ogilby, Rec. Aust. Mus., xii (10), 1919, Pl. xxxiii, fig. 3 (plate 46).

D. VI; I, 9-10. A. I, 8-9. P. 19-20. L. lat. 38-40. Tr. 11-13. Predorsals 24 or less. Body elongate, anteriorly cylindrical, posteriorly compressed. Depth 6-6½. Head depressed. Teeth in several rows; in upper jaw, outer row enlarged; lower jaw with outer row enlarged in anterior half and inner row enlarged laterally. Tongue bilobate. Longitudinal rows of mucous canals over cheek. Scales of head, nape, breast and belly cycloid; remainder ctenoid. Pectoral rounded with silk-like upper rays. Caudal obtuse, about as long as head. Olive above, lighter below. Head and body with irregular dark spots, often in one or two alternating rows. Fins dark to light, with or without markings, some rays spotted. 5 inches. Coastal waters.

696. ACENTROGOBIUS GRISEUS (Day) GREY GOBY

Gobius griseus Day, Fish. India, 1875-78, p. 285, Pl. lxiii, fig. 3 (plate 46).

D. VI; I, 10. A. I, 9. P. 17. L. lat. 42. Tr. 14. Predorsals more than 20. Body elongate, compressed. Depth 4¾-5½. Head compressed. Snout obtuse. Teeth in several rows in both jaws; outer row enlarged and anterior ones canines in upper jaw; outer row enlarged and posterior recurved canines in lower jaw. Longitudinal mucous canals on cheek. 2 open pores in interorbital. Gill-openings continued far forwards below. Body scales ctenoid, those before spinous dorsal cycloid. Head scaled above behind eyes with cycloid scales. Rays of spinous dorsal filiform, longer than height of body, 2nd longest. Caudal wedge-shaped. Olive with bands and many distinct brown or black spots. Pectoral and surrounding area finely dotted with chestnut. Spinous dorsal base yellow with 3-4 horizontal brown bands. Soft dorsal yellow with dark margin. Ventrals, anal and caudal with dark margins. 3½ inches. Estuarine.

697. ACENTROGOBIUS VIRIDIPUNCTATUS (Valenciennes)
GREEN-SPOTTED GOBY

Gobius viridipunctatus Cuvier and Valenciennes, Hist. Nat. Poiss., xii, 1837, p. 62; Day, Fish. India, 1875-78, Pl. lix, fig. 5 (plate 46).

D. VI; I, 10. A. I, 9. P. 18-20. L. lat. 35-36. Tr. 10. Predorsals 30. Body elongate, anteriorly subcylindrical, posteriorly compressed. Depth 5-7. Head slightly compressed. Teeth filiform, in many rows, outer enlarged with anterior canines. 2-4 canines on each side of lower jaw. Longitudinal mucous canals on snout, cheeks and radial ones around eye. Head scaled above behind eye and on upper parts of preopercle and opercle. These and those on breast and belly cycloid, others ctenoid. Caudal oblong. Blackish-green, laterally dotted with dark spots and a longitudinal row of larger ones along middle of sides. Fins yellowish-orange. Spinous dorsal with black stripe basally. Soft dorsal with 2 dark bands. Base and margin of anal blackish. Ventrals blackish. Caudal with black spots. 6½ inches. Coastal reefs.

698. ACENTROGOBIUS MASONI (Day) MASON'S GOBY

Gobius masoni Day, Proc. Zool. Soc., 1873, p. 107; Fish. India, 1875-78, Pl. lxi, fig. 6 (plate 46).

D. VI; I, 10-11. A. I, 9. P. 19. L. lat. 28. Tr. 9. Predorsals 25. Body elongate, compressed. Depth 6. Head compressed. Teeth in several rows; outer row enlarged in upper jaw; outer row enlarged, extends half way along jaw followed by a recurved canine in lower jaw. Pair of small broad barbels behind chin. Head scaled above behind eyes and on upper part of opercle. These and those on breast and belly cycloid; rest are ctenoid. Olive with blue spots on nape and behind pectoral. Some black spots on sides. Dorsals, anal, ventral and caudal blackish. Pectoral yellow with dark margin. 4 inches. Gulf of Manaar.

699. ACENTROGOBIUS CANINUS (Valenciennes) DOG-TOOTHED GOBY

Gobius caninus Cuvier and Valenciennes, Hist. Nat. Poiss., xii, 1837, p. 86. *Coryphopterus bernadoui* Jordan and Starks, Proc. U.S. Nat. Mus., xxviii, 1905, fig. 9 (plate 46).

D. VI; I, 9. A. I, 9. P. 18-19. L. lat. 30. Tr. 9. Predorsals 17-20. Body elongate, anteriorly a little compressed but more so posteriorly. Depth 5-6½. Head obtuse. Longitudinal mucous canals on cheek. An open pore in interorbital and another behind each eye. Outer row of teeth in upper jaw enlarged. 2 canines on each side of lower jaw. Head scaled above behind eye and on upper part of opercle with cycloid scales. Scales on breast and belly also cycloid, others ctenoid. 2nd and 3rd dorsal rays often filiform. Caudal obtuse. Olive above, lighter below. Head with blue or pearly spots. Body with alternating rows of large black spots. Blue spot on shoulder. Fins violet. Soft dorsal with 3 rows of dark spots bordered with blue. Caudal with black spots. 5¼ inches. Harbours.

700. ACENTROGOBIUS ORNATUS (Rüppell) ORNATE GOBY

Gobius ornatus Rüppell, Atlas Reise Nordl. Afrika, Fische, 1828, p. 135; McCulloch and Ogilby, Rec. Aust. Mus., xii (10), 1919, Pl. xxxiii, fig. 2 (plate 46).

D. VI; I, 10-11. A. I, 8-9. P. 19-20. L. lat. 28. Tr. 8-9. Predorsals 10-12. Body elongate, anteriorly cylindrical, posteriorly compressed. Depth 4½-5¼. 3-4 canine teeth in outer row of upper jaw. Lower jaw with no teeth enlarged in outer row, except 1 weak canine on each side in old specimens. 2-3 longitudinal mucous canals on cheek. No interorbital or postorbital pores. Head scaled above behind eyes with cycloid scales. Cheek and opercle naked. Breast scales cycloid, others ctenoid. Green to olive above, lighter below. Head laterally with small yellow spots and 3 purple transverse streaks. Back and upper part of head with irregular rows of violet spots. Ventral part of body with larger violet spots in 2 longitudinal rows. Dorsals red to yellow with oblong violet spots alternating with oblique yellow streaks. Ventrals orange with violet tip. Anal yellow with dark border, with 2-4 purple streaks basally. Caudal orange with dark and yellow spots. 4 inches. Coastal reefs.

701. MUGILOGOBIUS VALIGOUVA (Deraniyagala) MULLET-HEADED GOBY

Vaimosa valigouva Deraniyagala, Ceylon Jour. Sci. (B), xix, 1936, p. 219, fig. 1 (plate 46).

D. VI-VII; I, 7-8. A. I, 7-8. P. 16-17. L. lat. 34-35. Tr. 15. Predorsals 16-18. Body elongate, slightly depressed anteriorly and slightly compressed posteriorly. Depth 4¼. Teeth in 4 subequal rows. 2-3 longitudinal mucous canals on cheek. Body scales ctenoid, smaller anteriorly. Caudal subtruncate. Brown with a lateral row of 9-11 spots connected by diffuse dark bands to as many dorsal spots. Spinous dorsal with an oblique dark band. Soft dorsal with 3 black spots basally and a diffuse dark median band. Caudal with 2 round black spots basally. ¾ inch. Estuarine.

702. STIGMATOGOBIUS SADANUNDIO (Hamilton-Buchanan) SPOTTED GOBY

Gobius sadanundio Hamilton-Buchanan, Fish. Ganges, 1822, p. 52; Day, Fish. India, 1875-78, Pl. lxiii, fig. 10 (plate 46).

D. VI; I, 7. A. I, 8. P. 18-19. L. lat. 28. Tr. 8-9. Predorsals 8-9. Body elongate, compressed. Depth 4⅝-5. Head subcylindrical. Teeth in several rows, outer a little enlarged. Lower jaw with inner row enlarged, a pair of symphysial canines and some lateral canines. Short mucous canals radiating from eye. Head scaled above behind eye and on opercles, some enlarged. Head and nape scales cycloid, body scales ctenoid. 3rd and 4th dorsal rays prolonged. Yellow with a number of black spots in one or more lines longitudinally along the sides. A few black spots at insertion of caudal rays. Spinous dorsal with black blotches between 4th and 6th rays. Soft dorsal and anal with 3-4 dark spots surrounded by a narrow white ring. Caudal with small black spots. 3½ inches. Harbours and estuarine waters.

703. STIGMATOGOBIUS JAVANICUS (Bleeker) JAVANESE GOBY

Gobius javanicus Bleeker, Nat. Tijds. Ned. Ind., xi, 1856, p. 88. *Gobius melanostictus. Vaimosa tesselata* Herre, Monogr. Bur. Sci. Manila, xxiii, 1927, Pl. xii, fig. 1 (plate 46).

D. VI; I, 6-7. A. I, 6-7. P. 15. L. lat. 25-27. Tr. 7. Predorsals 6-8. Body elongate, anteriorly cylindrical, posteriorly compressed. Depth 5. Head subcylindrical. Teeth anteriorly in several rows, outer row enlarged. Lower jaw with inner row not enlarged. No canines. 2 mucous canals longitudinally across cheek. Some open pores around eyes. Head scaled above behind eye and on opercle. Head, breast and belly scales cycloid, others ctenoid. Green above, reddish-green below. Head with a dark stripe from eye to maxillary and dark oblong spots on cheek. Body with numerous minute spots, and 2-3 longitudinal rows of 5 blotches. 2 spots at caudal base and a predorsal dark oblique stripe. Spinous dorsal with dark spot between 3rd and 5th ray. Pectoral base with 2 spots. 2¼ inches. Gulf of Manaar.

704. PSEUDAPOCRYPTES LANCEOLATUS (Bloch and Schneider)
POINTED-TAILED GOBY

Eleotris lanceolatus Bloch and Schneider, Syst. Ichth., 1801, p. 67. *Apocryptes dentatus. Apocryptes lanceolatus* Day, Fish. India, 1875-78, Pl. lxiv, fig. 5 (plate 46).

D. V; I, 30. A. I, 29. P. 17-20. More than 200 minute scales in a longitudinal series. Very elongate body, cylindrical anteriorly and compressed posteriorly. Depth 9. Head subcylindrical. 24 pointed teeth in both jaws. In upper jaw they are larger in front; in lower jaw horizontal and pointed and a pair of symphysial canines. Caudal pointed. Brownish-grey with brown markings on back and cheeks. Belly yellowish. Dorsals, pectorals and caudal greenish. Dorsal and caudal rays with brown spots. Ventral and anal yellowish. 8 inches. Coastal waters.

705. PARAPOCRYPTES MACROLEPIS (Bleeker) LARGE-SCALED GOBY

Apocryptes macrolepis Bleeker, Nat. Tijds. Ned. Ind., ii, 1851, p. 66.

D. VI; I, 28. A. I, 28. P. 20. L. lat. 75. Body very elongate, anteriorly subcylindrical, posteriorly compressed. Depth 11. Upper jaw with 6-8 caninoid teeth anteriorly and 18-24 smaller ones laterally, lower jaw with 18-20 subhorizontal teeth and a pair of canines behind symphysis. Body scales larger posteriorly. Caudal lanceolate. Green above, silvery to yellowish below. Snout black. Back and sides with smoky reddish blotches. Dorsal fins violet with 4-6 oblique rows of oblong black blotches. Pectoral green. Ventral yellow. Anal reddish. Caudal with green rays and dusky violet membrane. 9 inches. Harbours.

MUD SKIPPERS

Family PERIOPHTHALMIDAE

Gobioid fishes with very elongate body. Head truncate, scaled above and on cheek and opercle. Eyes prominent above dorsal profile, lower eyelid well developed. Teeth in 1-2 rows in upper jaw and 1 row in lower jaw. Scales cycloid to weakly ctenoid. Pectorals with a muscular base and bent at insertion of rays for use as a locomotory organ on dry land. Ventrals united totally or partially. Gill-openings not very wide.

706. PERIOPHTHALMUS KOELREUTERI* (Pallas) MUD SKIPPER

Gobius koelreuteri Pallas, Spicilegia Zool., viii, 1770, p. 8. *Periophthalmus koelreuteri* vars. *koelreuteri* and *papilio*. *Periophthalmus koelreuteri* Day, Fish. India, 1875-78, Pl. lxiv, fig. 8 (plate 46).

D. 10-17; I, 11. A. I, 10-11. P. 13. L. lat. 70-90. Tr. 2+(12-13). Predorsals 35. Depth 6-7·3. Teeth in front caninoid, laterally smaller. Dorsals separate. Inner rays of ventral united only at base. Caudal obliquely truncate. Brownish to blackish-brown. Irregular transverse bands on body. Opercle with light dots on lower half. First dorsal light or dark brown with white margin, a dark submarginal band and a narrower white one below. Soft dorsal with broad white margin, dark brown submarginal band and dark stripes below. Anal white. 6 inches. Estuarine, coming ashore onto mud flats.

Dhiya huna (S). Thetthi koruvai (T).

BURROWING GOBIES

Family TRYPAUCHENIDAE

Gobioid fishes with elongate compressed body, with small isolated scales. A single dorsal fin along most of back, confluent with caudal. Pelvics united at base only, not forming a disc. Mouth oblique with small sharp teeth. A cavity in head above gill-cavity, opening into an external pit. No gill-rakers. Gill-membranes fused with isthmus.

707. TRYPAUCHEN VAGINA (Bloch and Schneider) BURROWING GOBY

Gobius vagina Bloch and Schneider, Syst. Ichth., 1801, p. 73. *Trypauchen vagina* Herre, Monogr. Bur. Sci. Manila, xxiii, 1927, Pl. xxvi, fig. 3 (plate 46).

D. VI; 40-49. A. I, 39-46. P. 15-16. L. lat. 80-115. Tr. 21. Depth 11½. Eyes very small and covered with skin. Mouth very oblique, lower jaw prominent. Maxilla extends behind eye. Teeth in many rows, pointed; outer row enlarged and caninoid. Head, nape and breast scaled. Body scales cycloid. Dorsal continuous. Reddish. 8½ inches. Lives under surface of mud. Palk Straits.

Suborder COTTOIDEI

Perciform fishes characterized by a bony ridge across the cheek, the end articulating with the front edge of the preopercle. Head usually armoured with bony plates, sometimes completely encased, and with few or many spinous projections. Body naked, wholly or partly scaled, covered in velvet-like villi or in bony scutes.

1. Head not markedly depressed, deeper than broad 2
   Head greatly depressed, much broader than deep 5
2. Gill-membranes free from isthmus 3
   Gill-membranes united with isthmus 4

* *Periophthalmus vulgaris ceylonensis* Eggert is also listed from Ceylon.

3. Body wholly or partly scaled, or if covered with villi the first 3 or 4 spines of dorsal are not separated from the rest of the fin .... *Family* SCORPAENIDAE, p. 244
   Body covered with villi instead of scales; the first 3 or 4 spines of the dorsal wholly or partly separated from the remainder of the fin .... *Family* APLOACTIDAE, p. 249
4. Body suborbicular, covered with minute villi; dorsal with 7-8 spines; ventrals rudimentary .... *Family* CARACANTHIDAE, p. 249
   Body elongate, without scales; dorsal with 9-13 spines; ventrals large ... *Family* SYNANCEIDAE, p. 250
5. Head not encased in complete bony armour; body with normal scales *Family* PLATYCEPHALIDAE, p. 251
   Head completely encased in a bony armour with anterior horns; body covered with bony scutes, each with a lateral spine .... *Family* PERISTEDIIDAE, p. 253

STINGFISH, SCORPION FISH, GOBLIN FISH, FIRE FISH

Family SCORPAENIDAE

Body moderately elongate, not much compressed. Scales normal. Head distinctly armoured with bony plates and characteristically spiny. Below the eye is a bony ridge joined to the preopercle. Fin spines well developed. Ventrals close together, 1+5. Gill-openings wide and membranes not united with isthmus.

1. Origin of dorsal above or just behind eye 2
   Origin of dorsal far behind eye, at or behind nape 3
2. Nearly black .... *Gymnapistes niger*
   Grey-brown with black patch on dorsal .... *Gymnapistes dracaena*
3. Pectoral rays much elongated; 12-14 venomous dorsal spines, usually elongate and largely free 4
   Dorsal spines and pectoral rays not elongate or greatly produced 7
4. Pectorals with several upper rays divided, their tips little or not at all free and not extending to caudal .... *Dendrochirus zebra*
   Pectoral rays all simple, the upper ones largely free of the membrane which is deeply and obliquely notched 5
5. Interorbital space without scales .... *Pterois volitans*
   Interorbital space scaly 6
6. Caudal without spots .... *Pterois russelli*
   Caudal spotted .... *Pterois miles*
7. Dorsal with 13 spines .... *Scorpaenodes guamensis*
   Dorsal with 12 spines 8
8. No palatine teeth 9
   Palatine teeth present 12
9. Pectoral reaches anal origin; orbital tentacles longer than eye .... *Scorpaenopsis roseus*
   Pectoral not reaching anal origin; orbital tentacles absent or shorter than eye 10
10. Orbital tentacle, if present, small; 9-11 scale rows above lateral line .... *Scorpaenopsis cirrhosus*
    Orbital tentacle well-developed; 7-8 scales above lateral line 11
11. 22 scales above lateral line .... *Scorpaenopsis leonina*
    18 scales above lateral line .... *Scorpaenopsis novae-guineae*
12. No depression or groove at nape .... *Sebastapistes strongia*
    A quadrilateral pit or depression at nape 13
13. An orbital tentacle present 14
    No orbital tentacle 15
14. Pectoral reaches beyond anal origin; membrane of dorsal deeply notched .... *Parascorpaena erostris*
    Pectoral falls far short of anal origin; membrane of dorsal not deeply notched .... *Parascorpaena bleekeri*
15. Tr. (7-8)+16; pectoral does not reach anal origin .... *Parascorpaena picta*
    Tr. 9+19; pectoral reaches to dorsal origin .... *Parascorpaena armata*

708. GYMNAPISTES NIGER (Cuvier) BLACK SMOOTH STINGFISH

Apistus niger Cuvier and Valenciennes, Hist. Nat. Poiss., iv, 1829, p. 415. *Gymnapistus niger* Day, Fish. India, 1875-78, Pl. xxxvii, fig. 5 (plate 47).

D. XII-XIII; 8. A. III, 5-7. P. I, 10-11. First 3 or 4 dorsal spines more widely separated than others. Scales absent except in the form of roughnesses here and there in the skin. Depth 3. Dorsal originates over posterior third of eye. Brownish black but pink or white underneath mottled with reddish-brown. All fins with white margin. Caudal yellowish or reddish with broad black submarginal band. 3 inches. Pearl banks. Is said to inhabit fresh waters.

709. GYMNAPISTES DRACAENA (Cuvier) BROWN SMOOTH STINGFISH

Apistus dracaena Cuvier and Valenciennes, Hist. Nat. Poiss., iv, 1829, p. 403. *Gymnapistus dracaena* Day, Fish. India, 1875-78, Pl. xxxviii, fig. 1 (plate 47).

D. XI-XII; 7-8. A. III, 5-6. P. 13. Scales rudimentary. Depth 3. Dorsal originates over middle of eye. Pectoral reaches to above middle rays of anal. Ventral reaches anal origin. Grey-brown with a black blotch on the dorsal fin between 3rd or 4th and 7th or 9th spines. Posterior three quarters of soft dorsal black. Pectoral black. Outer half of ventrals and anal also black. Caudal yellowish white with some small brown spots. 3 inches. Gulf of Manaar and coastal waters.

710. DENDROCHIRUS ZEBRA (Quoy and Gaimard) ZEBRA FIRE FISH

Pterois zebra Quoy and Gaimard, Voy. Uranie et Physic., Zool., 1825, p. 329. *Pseudomonopterus* (*Dendrochirus*) *zebra* Bleeker, Atlas Ichth. Ind. Neerl., ix, 1877, Pl. ccccxi, fig. 1 (plate 47).

D. XIII; 10-11. A. III, 6-7. P. I, 7+8, or II, 8+7. Scales ctenoid, 48-55 in horizontal rows above L. lat. Tr. 8-9 above. Depth 3½. Orbital tentacle long, more than half head. Dorsal spines posteriorly as high as body. Upper pectoral rays branched, lower simple and mostly free of the membrane at tips only. Pectoral does not reach to caudal base. Red or rosy with reddish-black cross-bars, 2-3 on head and 6-7 on trunk. Often 2-3 black bars across eye. Soft dorsal, caudal and anal with 5 or more rows of dark brown spots on hyaline membrane. Pectoral with several rows of semilunate or angular spots. 7 inches. Gulf of Manaar.

711. PTEROIS VOLITANS (Linnaeus) WINGED FIRE FISH

Gasterosteus volitans Linnaeus, Syst. Nat., ed. 10, 1758, p. 296. *Scorpaena volitans*. *Pterois volitans* Day, Fish. India, 1875-78, Pl. xxxvii, fig. 1 (plate 47).

D. XII, I, 10-11. A. II-III, 6-8. P. 14-16. Scales cycloid, 90-95 in horizontal series above L. lat. Tr. 12-13 above. Depth 3-3·3. Interorbital space deeply concave, it and the nape scaleless. Orbital tentacle very long, about ½ head. Dorsal spines 1-1½ depth of body, the interspinous membrane deeply emarginate. Pectoral reaches to caudal base or beyond, with membrane between upper 4 rays deeply cleft. Reddish with vertical brown bands

and narrower pale borders. 3-4 broad stripes radiating from eye. Soft dorsal, anal and caudal with rows of dark brown spots. Spots on ventral and pectoral larger and paler. 13 inches. Coral reefs around the coast.
Gon maha (S).

712. PTEROIS RUSSELLI Bennett — RUSSELL'S FIRE FISH

Pterois russellii Bennett, Proc. Zool. Soc., i, 1831, p. 128; Day, Fish. India, 1875-78, Pl. xxxvi, fig. 5 (plate 47).

D. XII, I, 11-12. A. III, 7-8. P. 13. Scales cycloid, 70-80 in horizontal series above L. lat. Tr. 8-10 above. Depth 4. Interorbital space deeply concave, it and occiput scaled. Orbital tentacle very small or absent. Dorsal spines much higher than body, with membranes deeply incised. Pectoral in young reaches tip of caudal or beyond, in older fish only to caudal base; its membrane, especially between upper 4 rays deeply incised. Ventrals extend as far as anal rays. Reddish with 14-15 broad transverse brown bars on head and body. A black spot on shoulder. Soft dorsal, caudal and anal roseate, without spots. Ventral membranes violet, rays with yellow spots. Pectorals variegated with blackish spots. 12 inches. Gulf of Manaar.

713. PTEROIS MILES (Bennett) — MILES' FIRE FISH

Scorpaena miles Bennett, Fish. Ceylon, 1828, p. 9. *Pterois miles* Day, Fish. India, 1875-78, Pl. xxxvii, fig. 2 (plate 47).

D. XII; I, 10-11. A. III, 6-7. P. 14. 92 scales in horizontal series above L. lat. Tr. 12 above. Depth 3½. Interorbital space moderately convex, it and nape with scales. Orbital tentacle short. Dorsal spines as high as body with membrane deeply incised. Pectoral with membrane most deeply cleft between first 3 rays, reaching at least to end of dorsal base. Ventral reaches anal. Red with many dark vertical bands which are wider than interspaces. Interspaces narrower anteriorly. Head with bands mostly radiating from eye. Two S-shaped bands on chest. Dorsal spines with 5-6 dark bands. Soft dorsal, caudal and anal with small black spots. Pectoral with large black blotches on a lighter ground. Ventral black, also spotted. 14 inches. Coastal reefs, Gulf of Manaar and trawling grounds.
Ratoo gini maha, Gini maha (S). Saval meen (T).

714. SCORPAENODES GUAMENSIS (Quoy and Gaimard) — GUAM STINGING FISH

Scorpaena guamensis Quoy and Gaimard, Voy. Uranie et Physic., Zool., 1824, p. 326. *Scorpaenopsis guamensis*. *Scorpaena guamensis* Günther, Jour. Mus. Godeffroy, ii (7), 1874, Pl. lvi, fig. B (plate 47).

D. XII-XIII; I, 8-9, (sometimes as few as 5). A. III, 5-6. P. I-II, (6-7)+(11-12). 38-42 (+3) scales in horizontal series above L. lat. Tr. 5+(11-12). L. lat. 22-26. G.R. 6+8. Depth 3½. Interorbital concave. No groove beneath eyes. No orbital tentacle. Dorsal spines ½ body depth. Scales on cheeks and opercle. Rich dark reddish to dusky brown with

3-4 darker cross-bars and marbled with brown spots. Caudal peduncle with dusky brown bar followed by a white bar on caudal base. Deep red spot on opercle. Fins mottled reddish-brown. 4 inches. Coral reefs under and around coral heads.
Gon maha, Thumba kossa (S).

715. SCORPAENOPSIS ROSEUS (Day) ROSY STINGFISH

Scorpaena rosea Day, Proc. Zool. Soc., 1867, p. 703. *Scorpaenopsis rosea* Day, Fish. India, 1875-78, Pl. xxxvi, fig. 4 (plate 47).

D. XI; I, 9-10. A. III, 5. P. 6+12. L. lat. 43. Tr. 7+18. Depth 3. Eyes elevated, with groove below. Interorbital deeply concave. Broad fleshy orbital tentacle, longer than eye. Other tentacles on snout, angle of mouth and under the chin. Lower 12 rays of pectoral unbranched and with minute fleshy appendages. Caudal truncate. Fleshy appendages also on body and on anterior dorsal spines. Rosy, marbled with greyish. 1-2 irregular vertical grey bands on caudal. Dorsal, anal and ventrals also banded. Pectorals with numerous dark spots. 6 inches. Coastal waters.
Lunu kara kossa (S).

716. SCORPAENOPSIS CIRRHOSUS (Thunberg) HAIRY STINGFISH

Perca cirrhosa Thunberg, Nya Handl. Act. Holmae, xiv, 1793, p. 193. *Scorpaenopsis oxycephala. Scorpaenopsis cirrhosus* Bleeker, Atlas Ichth. Ind. Neerl., ix, 1877, Pl. ccccxiii, fig. 6 (plate 47).

D. XI-XII; I, 9-10. A. III, 5. P. (5-7)+(11-12). G.R. 10-12. 50-60 scales in longitudinal series above L. lat. Tr. 9-11 above. Depth 3·3. Interorbital deeply concave. A deep groove under eye. Sometimes a small orbital tentacle; others about the head and a large one above the angle of the mouth. Also other short tentacles scattered over body. Dorsal spines about ½ body depth. Ventral reaches anus. Caudal truncate. Clouded rose with dusky brown mottling. Lower parts paler. All fins transversely and obliquely striped with black and light grey. 10 inches. Coastal reefs and pearl banks.

717. SCORPAENOPSIS LEONINA (Richardson) LION-LIKE STINGFISH

Scorpaena leonina Richardson, Rept. 15th meet. Brit. Assn. Adv. Sci., 1846, p. 216. *Scorpaenopsis oxycephala* Day, Fish. India, 1875-78, Pl. xxxvi, fig. 3 (plate 47).

D. XI; I, 10. A. III, 5. P. 6+12. 52 scales in longitudinal series above lateral line. Tr. 7+22. Depth 4¼. Interorbital deeply concave. Deep groove in front of eye continued as a shallow one below. Tentacles on orbit, snout, preopercular margin, angle of mouth, on lateral line and scattered over body scales. Caudal truncate. Scales on upper parts of preopercle and opercle. Reddish clouded with brown and a few blackish spots. Spinous dorsal with dark transverse band. Many blotches on fins. Caudal posteriorly with a dark band. Anal with 3 narrow horizontal bands. Pectoral and ventral spotted. Gulf of Manaar.

718. SCORPAENOPSIS NOVAE-GUINEAE (Cuvier) NEW GUINEA STINGFISH

Scorpaena novae-guineae Cuvier and Valenciennes, Hist. Nat. Poiss., iv, 1829, p. 235. *Scorpaenopsis novae-guineae* Bleeker, Atlas Ichth. Ind. Neerl., ix, 1877, Pl. ccccxvii, fig. 6 (plate 47).

D. XII; 9-10. A. III, 5-6. P. I, (4-6)+(11-13). G.R. 4+10. Depth 2·6 without caudal. 44-45 scales in longitudinal series above lateral line. Tr. 7+18. Eyes prominent. Deep groove below eye. Pectoral and ventral fins reach as far as anus. Many tentacles and threads on head and trunk, especially lower jaw, preopercular margin and on dorsal and pectoral fins. Clouded rose marbled with dusky brown and 2-3 broad diagonal blackish bands. A basal dusky band on caudal followed by a pale band and 2 black stripes. Spinous dorsal with black band between 2nd and 5th spines. Soft dorsal and anal with dark diagonal bands. 4½ inches. Pearl banks.

719. SEBASTAPISTES STRONGIA (Cuvier) BROWN-BANDED STINGFISH

Scorpaena strongia Cuvier and Valenciennes, Hist. Nat. Poiss., iv, 1829, p. 323. *Sebastichthys strongia. Sebastichthys cyanostigma* Bleeker, Atlas Ichth. Ind. Neerl., ix, 1877, Pl. ccccxvi, fig. 1 (plate 47).

D. XI-XII; I, 9. A. III, 5. P. 19. L. lat. 45. Tr. 5+16. Depth more than 3. Supra-orbital ridge spinous. Suborbital ridge also spiny. 2 tentacles above orbit and others on head. Pectoral with lower 8 rays with free extremities. Tentacles along lateral line. Brownish, banded darker, the first band passing through eye. A large brown spot on opercle. Fins irregularly banded with lines of dots. Coastal waters.

720. PARASCORPAENA EROSTRIS (Alcock) DEEP WATER STINGFISH

Scorpaena erostris Alcock, Jour. Asiatic Soc. Bengal, lxv (2), 1896, p. 302; Illust. Zool. Investigator, Fish., v, 1898, Pl. xviii, fig. 6 (plate 47).

D. XI; I, 9. A. III, 5. L. lat. 40. Opercles and cheeks densely scaly. Lateral line with a few long tentacles. Pectorals reach anal, only a few of the rays branched. Supraorbital margin toothed. A tentacle above eye. Mottled purple above, light reddish below. Fins light red. A dark purple blot between 7th and 10th dorsal spines, and a fainter streak of the same colour between 2nd and 6th dorsal rays. 3 inches. Coastal waters in 34 fathoms.

721. PARASCORPAENA BLEEKERI (Day) BLEEKER'S STINGFISH

Scorpaena bleekeri Day, Fish. India, 1875-78, p. 747. *Scorpaena haplodactylus* Day, *loc. cit.*, Pl. xxxvi, fig. 2 (plate 48).

D. XI; I, 9. A. III, 5. P. 5+12. L. lat. 26, with about 42-44 in longitudinal series below. Tr. 6+22. Depth 3·3. Interorbital deeply concave. No groove below eyes. A small orbital tentacle, and others on head. A strong spine on shoulder girdle. A few fleshy tentacles along lateral line. Brownish black, banded and marbled with darker. Anal with broad band along its basal half. Caudal with basal band and two cross-bars and a narrow white margin. Dorsal with brown blotches. 3 inches. Gulf of Manaar.

722. PARASCORPAENA PICTA (Cuvier) PAINTED STINGFISH

Scorpaena picta Cuvier and Valenciennes, Hist. Nat. Poiss., iv, 1829, p. 321. *Scorpaena polyprion. Parascorpaena picta* Bleeker, Atlas Ichth. Ind. Neerl., ix, 1877, Pl. ccccxii, fig. 5 (plate 48).

D. XII; 10. A. III, 5. P. I, (4-5)+(11-12). Scales cycloid, 45 in longitudinal series above L. lat. Tr. (7-8)+16. Depth 3. No orbital tentacles. Interorbital deeply concave. Skinny flaps and tentacles along lateral line, on sides and head. Brown or greenish brown, marbled with darker brown. Soft dorsal and anal with 2-3 oblique bars. Pectoral and caudal with several cross-bars. Pectoral axil with irregular yellow spots. 6½ inches. Coastal waters.

723. PARASCORPAENA ARMATA (Sauvage) SPINY STINGFISH

Scorpaena armata Sauvage, Bull. Nouv. Arch. Mus., Paris, ix, 1873, p. 49, Pl. vi, fig. 1 (plate 48).

D. XI; I, 9. A. III, 5. P. 6+12. L. lat. 35. Tr. 9+19. Depth 3. Interorbital very concave, with a quadrangular fossa. A deep groove below eye. Opercle with 2 spines. Pectorals and ventrals equal, extending slightly behind anus. Roseate brown with darker blotches and a few yellowish-white spots. Head spotted with black. Fins yellowish-brown, spotted with darker, sometimes forming lines. 5 inches. Gulf of Manaar.

LITTLE VELVET FISH

Family APLOACTIDAE

Cottoid fishes with the dorsal fin originating above or slightly before the eye. Head bones with knob-like prominences. Scales absent and skin covered with velvety villi. No palatine teeth. Vomer toothed. Gill-membranes free. No slit behind last gill. No free pectoral rays. Anal spines absent or indistinct.

724. ANICULEROSA TAPROBANENSIS Whitley MANAAR VELVET FISH

Aniculerosa taprobanensis Whitley, Rec. Aust. Mus., xix (1), 1933, p. 101. *Aploactis aspersa* Johnstone, Rept. Pearl Oyster Fish., ii, Supplt. xv, Pl. i, fig. 5 (plate 48).

D. IV; 19. A. 10. P. 13. V. 4. Depth 3½. Cleft of mouth directed obliquely upwards. Eyes small, about 7 in head. Anterior nostrils tubular. Head with prominent preorbital processes. Cheeks and opercle with blunt spines and ridges. First 4 spines of dorsal separated from the rest, in which it is difficult to distinguish spines from rays. Ventrals reduced. Caudal rounded. Skin everywhere covered with blunt, soft tubercles; these are present also on fins and are especially large and rough on first part of dorsal. 11-13 large tubercles on lateral line. 2 inches. Adam's Bridge.

VELVET FISH

Family CARACANTHIDAE

Cottoid fishes with strongly compressed, suborbicular body which is scaleless or covered with minute villi or tubercles. The suborbital is articulated with the preopercle. Jaws with small teeth. Vomer and palatines without teeth. Gill-openings separated by a wide isthmus. Ventrals rudimentary. No opening behind the 4th gill.

725. CARACANTHUS ZEYLONICUS (Day) CEYLONESE VELVET FISH

Amphiprionichthys zeylonicus Day, Proc. Zool. Soc., 1869, p. 515. *Micropus zeylonicus* Day, Fish. India, 1875-78, Pl. xxxviii, fig. 6 (plate 48).

D. VII; 14. A. II, 11. P. 13. Ventrals rudimentary, I, 3. Depth 2-2½. Preopercle with 5 blunt spines. A slight notch between spinous and soft dorsal. Recognized by the orbicular body and villose skin. Bluish above and dirty brown below. Several rows of yellowish blotches along back and sides, separated by dark reticulation. Fins light. Caudal similar to body. 2½ inches. Harbours.

STONEFISH, GOBLINFISH, DEVILFISH

Family SYNANCEIDAE

Cottoid fishes with body and head spinate and rough externally. Gill-membranes fused with isthmus and openings restricted. 2-3 anal spines. Scales absent and skin often thick and warty but sometimes smooth with dermal appendages.

1 Body with small cycloid scales; body, head and fins with numerous dendritic dermal appendages .... *Peloropsis frondosus*
Body without scales; appendages few or absent 2
2. Dorsal fin with more spines (13) than rays (9) .... *Choridactylus multibarbis*
Dorsal fin with approximately same number of spines as rays .... *Minous monodactylus*

726. PELOROPSIS FRONDOSUS (Günther) LEAFY SEA GOBLIN

Scorpaena frondosa Günther, Proc. Zool. Soc., 1891, p. 482. *Peloropsis frondosus* Deraniyagala, Col. Atlas. Vert. Ceylon, i, 1952, Pl. xxxii (plate 48).

D. XI; I, 9. A. III, 5. P. 16. Small cycloid scales on body, none on head. Head and belly compressed. Third dorsal spine longest. Numerous large simple and tasselated dermal appendages on head and body. Similar appendages on fins. Pectoral not procumbent below. Red with darker mottling on body and fins. Indistinct bands and a black spot on soft dorsal. Pectoral with a semicircle of yellow spots. 10 inches. Pearl banks, living in coral and weeds.

727. CHORIDACTYLUS MULTIBARBIS Richardson
ORANGE-BANDED GOBLINFISH

Choridactylus multibarbis Richardson, Voy. Samarang, Fishes, 1848, p. 8; Day, Fish. India, 1875-78, Pl. xxxix, fig. 2 (plate 48).

D. XIII; 9. A. II, 8. P. 9+3. Depth 3½. Eyes with prominent orbits. Interorbital concave. A fleshy tentacle on eye and a few smaller ones on chin. Ventrals large, most of inner edge adnate to abdomen. Pectoral with 3 long detached lower rays. Brown with yellow shoulder mark and 2-3 vertical orange bands. Base of anal and ventrals with white spots. Fins blackish-brown; dorsal and caudal with lighter band. Pectoral with orange margin. 4 inches. Gulf of Manaar.

728. MINOUS MONODACTYLUS (Bloch and Schneider) GREY GOBLINFISH

Scorpaena monodactylus Bloch and Schneider, Syst. Ichth., 1801, p. 194. *Minous monodactylus* Day, Fish. India, 1875-78, Pl. xxxviii, fig. 7 (plate 48).

D. IX-XI; 9-12. A. 9-11. P. 10+1. Depth 4-4½. Opercle armed. Scales absent. Anal spines weak. First few spines of dorsal more or less free with membranes deeply cleft. Pectoral with a single detached basal ray. Head grey-brown. Sides and abdomen paler to flesh-coloured with dark blotches and marks. Dorsal light brown, margined with black. Pectoral black with detached ray white. Caudal buff with 3 vertical brown bars. 5 inches. Gulf of Manaar.

FLATHEAD

Family PLATYCEPHALIDAE

Cottoid fishes with elongate, subcylindrical body which tapers posteriorly. Head large, depressed, ridged and spinose. Body with rough adherent scales. Mouth large, lower jaw usually produced. Fine teeth in jaws and palate and some larger canines. Gill-rakers rudimentary. Ventrals directly below pectorals. 2 dorsal fins, spinous portion preceded by an isolated spine. Bottom dwelling carnivorous species.

1. More than 2 spines at angle of preoperculum .... *Rogadius asper*
   Only 2 spines at angle of preoperculum 2
2. Lateral line with spines throughout its length .... *Grammoplites scaber*
   Anterior third of lateral line with spines, the remainder smooth 3
   Lateral line without any spines 4
3. L.lat. 53-55; width of head 0·8 in its length .... *Suggrundus tuberculatus*
   L.lat. 75; width of head 0·5 in its length .... *Suggrundus macracanthus*
4. Interorbital ·25 of eye diameter 5
   Interorbital more than ·25 of eye diameter 6
5. Crests on head serrated .... *Suggrundus serratus*
   Crests on head spiniform .... *Suggrundus carbunculus*
6. Dorsal with 7 spines; eye 2 diameters apart; Tr. 12+24 .... *Thysanophrys indicus*
   Dorsal with 8 spines; eyes 0·5-0·7 diameters apart; Tr. 8+25 .... *Thysanophrys punctatus*

729. ROGADIUS ASPER (Cuvier) THORNY FLATHEAD

Platycephalus asper Cuvier and Valenciennes, Hist. Nat. Poiss., iv, 1829, p. 257. *Rogadius asper* Jordan and Richardson, Proc. U.S. Nat. Mus., xxxiii, 1908, fig. 1 (plate 48).

D. I, VIII; 11. A. 11. L. lat. 50. Space between eyes is an exceedingly narrow groove. Head 3 in length. Head ridges finely serrated, those on operculum smooth. 3 preopercular spines, the uppermost strongest. A 4th spine points downwards. Lateral line nearly smooth. Yellowish-brown with brown dots on the fins. Coastal waters.

730. GRAMMOPLITES SCABER (Linnaeus) ROUGH FLATHEAD

Cottus scaber Linnaeus, Syst. Nat., i, 1758, p. 451. *Insidiator scaber. Platycephalus scaber* Day, Fish. India, 1875-78, Pl. lx, fig. 4 (plate 48).

D. I, VII-VIII; 12. A. 12. P. 21. 105 scales in longitudinal series. L. lat. 55. Head ridged and spiny above. 2 spines on preopercle. Teeth small in jaw. Vomer with 2 oval patches subequal to eye. Palatines with elongate band. Eyes $4\frac{1}{2}$-$5\frac{1}{4}$ in head and $\frac{2}{3}$ diameter apart. Lateral line scales each with a strong backwardly directed spine. Brown above, white below. Fairly distinct vertical bars. First dorsal clouded with black, second dorsal spotted. Pectoral and ventral dark externally. Caudal grey distally. 9 inches. Coastal harbours, trawling grounds and Gulf of Manaar.

731. SUGGRUNDUS TUBERCULATUS (Cuvier) KNOBBY FLATHEAD

Platycephalus tuberculatus Cuvier and Valenciennes, Hist. Nat. Poiss., iv, 1829, p. 258; Day, Fish. India, 1875-78, Pl. lx, fig. 5 (plate 48).

D. I, VII-VIII; 11-12. A. 11-12. L. lat. 55, the front 15-20 spinose. Depth 6. Head ridged and spiny above. 2 preopercular spines. Eyes 3·6-4 in head, $\frac{1}{4}$ diameter apart. Teeth villiform in jaws, in 2 oval patches on vomer and an elongate band on palatines. Brownish above, white below. Dark mottling and vertical bands present. Spinous dorsal blackish-brown. Second dorsal spotted. Pectoral with brown spots in upper $\frac{2}{3}$, and a white margin and black submarginal band below. 20 inches. Coastal harbours, Gulf of Manaar and trawling grounds.

732. SUGGRUNDUS MACRACANTHUS (Bleeker) LARGE-SPINED FLATHEAD

Platycephalus macracanthus Bleeker, Versl. Meded. Akad. Wet. Amsterdam (2), iii, 1869, p. 253; Atlas Ichth. Ind. Neerl., ix, 1877, Pl. ccccxix, fig. 1 (plate 48).

D. I, VIII; 12. A. 12. P. 23. Lateral line with 75 scales above, 65 below in longitudinal series. Tr. 4+15. Depth $7\frac{1}{2}$. Eyes $4\frac{3}{4}$ in head, and $\frac{1}{2}$ diameter apart. Head ridged and spiny above. 2 preopercular spines, large. Teeth villiform in jaws, in 2 parallel patches on vomer and in narrow band on each palatine. First 15 scales of lateral line bearing spines. Brownish above, dull white below. Spinous dorsal and end of caudal blackish. Soft dorsal with brown spots on the rays. Gulf of Manaar.

733. SUGGRUNDUS SERRATUS (Cuvier) SERRATED FLATHEAD

Platycephalus serratus Cuvier and Valenciennes, Hist. Nat. Poiss., iv, 1829, p. 190. *Thysanophrys serratus.*

D. I, VIII; 11-12. A. 11. P. 19. Lateral line scales without spines. Head 4, crests and ridges without spines but serrated. Opercular ridge smooth. 2 preopercular spines, upper larger. Interorbital $\frac{1}{4}$ transverse eye diameter. Reddish-brown with 6-8 irregular brown bands. Abdomen white. Fins grey with black points. A black blotch on top of dorsal. Ventrals blue above, white below. 7 inches. Coastal harbours and trawling grounds. Mudhu valigouva (S).

734. SUGGRUNDUS CARBUNCULUS (Valenciennes)

CARBUNCLED FLATHEAD

Platycephalus carbunculus Cuvier and Valenciennes, Hist. Nat. Poiss., ix, 1833, p. 461.

D. I, VIII; 11-12. A. 11-12. L. lat. with 75-80 scales above and 65-70 below in longitudinal series. Tr. 6+26. Depth 6¾. Eyes 4 in head, ¼ diameter apart. Crests on head with spines. Preoperculum with 2 spines. Ventrals reach anal origin. Lateral line scales without spines. Brown with numerous brown spots and 3 vertical bands posteriorly. Opercle dark. Posterior half of spinous dorsal black. Soft dorsal spotted. Pectoral with several lines of spots and a dark basal mark. Caudal dark with 2 ill-defined vertical bands. 6½ inches. Trawling grounds.

735. THYSANOPHRYS INDICUS (Linnaeus) INDIAN FLATHEAD

Callionymus indicus Linnaeus, Syst. Nat., ed. 10, 1758, p. 250. *Platycephalus indicus. Platycephalus insidiator* Schlegel, Fauna Japonica, Pisces, 1842, Pl. xv, fig. 1 (plate 48).

D. I, VII; 13-14. A. 13. L. lat. 130. Tr. 12+24. Lateral line scales without spines. Head with only slight ridges above and flat spines on head low and feeble. Preoperculum with 2 spines. Depth 7. Eyes 6¾-7 in head, 2 diameters apart. Brown above, spotted with reddish-brown. Belly white. Dorsals pale, with indistinct spots. Pectorals and ventrals strongly spotted. Caudal yellow with white edged oblique bands across upper and lower lobes. 40 inches. Gulf of Manaar.
Mudhu valigouva (S).

736. THYSANOPHRYS CROCODILUS (Tilesius) SPOTTED FLATHEAD

Platycephalus crocodilus Tilesius, Reise Krusenstern, 1812. *Platycephalus punctatus. Thysanophrys crocodilus* Jordan and Richardson, Proc. U.S. Nat. Mus., xxxiii, 1908, fig. 4 (plate 48).

D. I, VII-VIII; 11-12. A. 11-12. P. 21. L. lat. 85, 100-110 in longitudinal series above and below. Tr. 8+25. Head ridged and spiny above. Eyes 5-5½ in head, ½-⅔ diameter apart. 2 preopercular spines. Lateral line entirely smooth. Brown, lighter below. 4-5 broad dark vertical bands. Head and body with numerous small black spots. First dorsal black. Soft dorsal and anal yellowish, first with brown spots. Caudal, ventrals and pectorals dark to black. 20 inches. Coastal waters and Gulf of Manaar.
Mal beluma (S). Eriyal (T).

CROCODILE FISH, SEA ROBINS

Family PERISTEDIIDAE

Body elongate, completely covered in bony plates, each with a lateral spine. Head completely encased in a bony shield. Each preorbital produced forward into a horn. 2 free lower pectoral rays. Lower jaw with barbels. Bottom living species.

1. D.VI; 22. A.21. Dorsal and pectorals with black markings .... *Peristedion rivers-andersoni*
   D.VII; 15. A.14-15. Dorsal and pectorals without black markings 2
2. L.lat. 34 .... *Peristedion halei*
   L.lat. 29 .... *Peristedion pothumaluva*

737. PERISTEDION RIVERS-ANDERSONI (Alcock)

ANDERSON'S SEA ROBIN

Peristethus Rivers-Andersoni Alcock, Jour. Asiatic. Soc. Bengal, lxiii (2), 1894, p. 121, Pl. vi, fig. 2 (plate 51).

D. VI; 22. A. 21. L. lat. 32. Preorbital processes bluntly pointed, depressed and hollow, their length 2 in distance from tip to anterior of eye. Preopercular ridge salient but not forming a spine. Opercular ridge with short bluntly rounded spine. Lower jaw thickly fringed with small tentacles. Strong recurved spines on each side of occiput and supraorbital margins. Body shields in 4 rows, each strongly carinated. Anterior ventral shields twice as long as wide. Flesh coloured. Pectorals with broad black band posteriorly and white tip. Spinous dorsal black superiorly. Soft dorsal with black edge. 3½ inches. 142-400 fathoms off Colombo.

738. PERISTEDION HALEI (Day)

HALY'S SEA ROBIN

Peristethus halei Day, Suppl. Fish. India, 1888, p. 791.

D. VII; 15. A. 15. L. lat. 34. Preorbital processes short, their length 3½ in distance from their tip to anterior of eye. A pair of spines on occiput, on either side of which is a low ridge terminating in a spine. Anterior vertical plates longer than broad. Opercular ridge forms a strong spine. Lower jaw with barbels. Uniform red. Deep water off coast.

739. PERISTEDION POTHUMALUVA Deraniyagala

YELLOW-FINNED SEA ROBIN

Peristedion pothumaluva Deraniyagala, Spol. Zeyl., xix, 1936, p. 221; Col. Atlas Vert. Ceylon, i, 1952, Pl. xxxiii (plate 51).

D. VII; 15. A. 14. L. lat. 29. Head 2. Eye 5·9 in head. Preorbital processes convergent. 4 rows of scutes on each side. Ventrally first 3 pairs of scutes enlarged, 2½ times as long as wide. Reddish. Dorsals, anal and caudal yellowish. Spinous dorsal with 2 rows and soft dorsal with 1 row of red spots. 14 inches. 46 fathoms. Coastal waters.

FLYING GURNARDS

Order DACTYLOPTERIFORMES

Family DACTYLOPTERIDAE

Robust angular body. Head blunt, completely encased in bony armour with strong spines and keels. A bony ridge across cheek to preopercle. Enlarged keel-like scales on lower part of body. Granular teeth in jaws. Pectorals enormous, with a few detached rays in front. Primarily bottom feeders but can leave the water using the pectorals for short flights.

740. DACTYLOPTENA ORIENTALIS (Cuvier) FLYING GURNARD

Dactylopterus orientalis Cuvier and Valenciennes, Hist. Nat. Poiss., iv, 1829, p. 134. *Dactyloptena volitans. Cephalacanthus orientalis* Jordan and Evermann, Bull. U.S. Fish. Comm., xxiii (1), 1905, fig. 208 (plate 51).

D. I; V-VI; 8. A. 6-7. P. 30. Body more or less quadrangular in cross section. First dorsal spine elongate and removed from the main part of the spinous dorsal which is lower. Scales bony, keeled, in 45 series along and 21 across. 4 enlarged keel-like scales along each side below. Depth 5½-6. Spine at preopercle angle ½ length of head. Pink, bluish along abdomen. Dorsal and caudal rays spotted. Pectoral grey-brown and spotted with red, yellow and blue. Young have also a blue-edged black ocellus near base of pectoral. 15 inches. Trawling grounds and Gulf of Manaar.

FLATFISHES, SOLES, FLOUNDERS

Order PLEURONECTIFORMES

Specialized fishes characterized by their asymmetry, primarily the eyes which lie on one side of the head, either to right or left of mouth. The two sides usually differ in colour, nature of scales and lateral line. Mostly bottom fishes resting on their blind side.

1. Dorsal fin not on head .... *Family* PSETTODIDAE, p. 256
   Dorsal fin on head also 2
2. Preopercular margin free 3
   Preopercular margin covered by skin 4
3. Eyes on right side .... *Family* PLEURONECTIDAE, p. 257
   Eyes on left side .... *Family* BOTHIDAE, p. 258
4. Eyes on right side .... *Family* SOLEIDAE, p. 262
   Eyes on left side .... *Family* CYNOGLOSSIDAE, p. 263

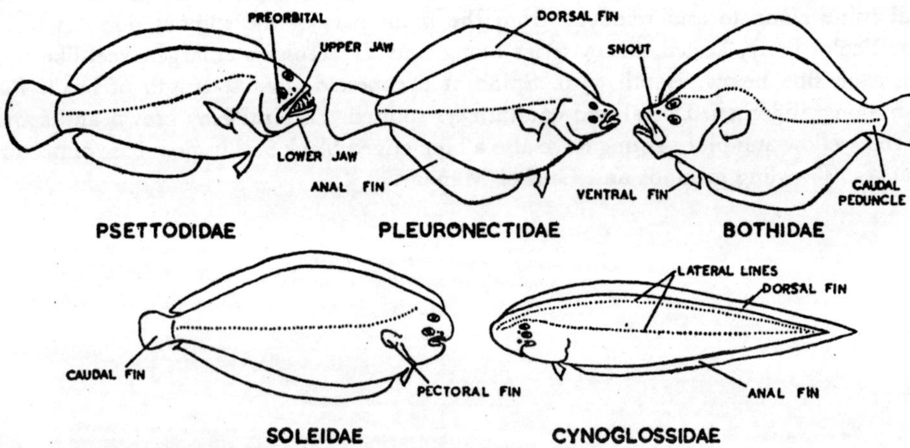

FIG. 18
Principal differences between the five families of Flat fishes, Order Pleuronectiformes.

HALIBUTS

Family PSETTODIDAE

Dorsal fin absent from head, front rays spinous. Mouth large, teeth well developed. Preopercular margin free. Migrated eye on top of head.

741. PSETTODES ERUMEI (Bloch) INDIAN HALIBUT
Pleuronectes erumei Bloch and Schneider, Syst. Ichth., 1801, p. 150. *Psettodes erumei* Day, Fish. India, 1875-78, Pl. xci, fig. 4 (plate 49).

D. 50-56. A. 34-43. L. lat. 68-76. Eye may be on either left or right. Dorsal fin does not extend on to head. Pectorals shorter than postorbital part of head. Caudal truncate. Uniform brown with 4 transverse dark bands in young specimens. 25 inches. Coastal down to 50 fathoms.
Pathamadiya (S).

RIGHT-HAND FLOUNDERS

Family PLEURONECTIDAE

Flatfishes with eyes on right hand. Dorsal extends on to head to above or beyond eye. Mouth terminal. No palatine teeth. Margin of preoperculum free.

1. Dorsal origin above eyes 2
   Dorsal origin in front of eyes 3
2. Depth about 3 in length; right pectoral with 8-10 simple rays .... *Poecilopsetta praelonga*
   Depth about 2 in length; right pectoral with 11-12 rays, middle rays branched ...... *Poecilopsetta colorata*
3. Some of anterior dorsal and right pelvic rays prolonged and filamentous; caudal rays simple .... *Samaris cristatus*
   Anterior dorsal and right pelvic rays not prolonged; middle caudal rays branched .... *Samariscus longimanus*

742. POECILOPSETTA PRAELONGA Alcock

ALCOCK'S CRESTED FLOUNDER

Poecilopsetta praelonga Alcock, J. Asiat. Soc. Bengal, lxiii, 1894, p. 139. *Boopsetta umbrarum* Alcock, Illust. Zool. Investigator, 1897, Pl. xvii, fig. 5 (plate 49).

D. 59-65. A. 50-54. L. lat. 90-95. Lateral line absent on blind side. Scales on ocular side weakly ctenoid in adults, cycloid in young; those of blind side cycloid. Pectorals with 8-10 simple rays; left pectoral about half head. Yellowish brown with large dark blotches on ocular side. Blind side white with traces of black spots. Young with 6-7 black blotches at upper and lower edges of body. 6½ inches. Deep waters, 180-250 fathoms.

743. POECILOPSETTA COLORATA Günther

LARGE-SPOT FLOUNDER

Poecilopsetta colorata Günther, Challenger Rept. Shore Fishes, p. 48. *Poecilopsetta maculosa* Alcock, Illust. Zool. Investigator, 1895, Pl. xv, fig. 1 (plate 49).

D. 56-61. A. 46-50. Eyes separated by narrow ridge. Right pectoral with 11-12 rays, the middle ones branched. Left pectoral 2¼ in head. Pale brown on ocular side; head and body with numerous blackish dots. Blind side white. Right pectoral with large black spot occupying most of fin. Caudal with pair of large dark spots. 6 inches. Deep water, 142-400 fathoms.

744. SAMARIS CRISTATUS Gray

GRAY'S CRESTED FLOUNDER

Samaris cristatus Gray, Zool. Miscell., 1831, p. 5; Alcock, Illust. Zool. Investigator, 1898, Pl. xxiii, fig. 2 (plate 49).

D. 77-79. A. 50-53. L. lat. 67-72. Scales of ocular side ctenoid; those of blind side cycloid or weakly ctenoid. First 13-14 dorsal rays greatly prolonged, 4-5 times length of head. Right pectoral with 4 rays. Brown mottled and spotted with paler and darker. Anterior dorsal rays white, rest of fins brown. $6\frac{1}{2}$ inches. Deep water, to 34 fathoms.

745. SAMARISCUS LONGIMANUS Norman — LONG-FINNED FLOUNDER

Samariscus longimanus Norman, Rec. Indian Mus., xxix, 1927, p. 46, Pl. vii (plate 49).

D. 66-71. A. 50-54. L. lat. 55-60. Scales ctenoid on both sides of body. Caudal with 16 rays, middle ones branched, pointed. Right pectoral with 5 rays, twice as long as head. No filamentous rays. Grey-brown with 5 black blotches on upper part of body and 3-4 on lower edge. Other blotches along lateral line. Pectoral black. 5 inches. Deep water, 102-105 fathoms.

LEFT-HAND FLOUNDERS

Family BOTHIDAE

Flatfishes with eyes on left side. Dorsal fin extends on head to above or beyond eye. Preopercular margin free. Lower jaw mostly prominent.

1. Pelvic fins equal, short based 2
   Left pelvic fin median with base much longer than that of the right 4
2. Anterior dorsal rays longer than those following; 23-25 gill-rakers on lower part of anterior arch; depth $1\frac{3}{5}$-$1\frac{3}{4}$ in length .... *Pseudorhombus triocellatus*
   Anterior dorsal rays not prolonged; 8-15 gill-rakers on lower part of anterior arch; depth $1\frac{3}{4}$-$2\frac{1}{8}$ in length 3
3. Maxillary extending to below posterior part of eye; 6-13 teeth on blind side of lower jaw; dorsal with 71-80 rays .... *Pseudorhombus arsius*
   Maxillary not extending beyond middle of eye; 23-32 teeth on blind side of lower jaw; dorsal with 67-72 rays .... *Pseudorhombus javanicus*
4. Mouth very large; maxillary more than $\frac{1}{2}$ head; lower jaw very prominent; body elongate .... *Chascanopsetta lugubris*
   Mouth small or moderate; maxillary $\frac{1}{2}$ or less than $\frac{1}{2}$ head; lower jaw not prominent 5
5. Lateral line on both sides of body .... *Grammatobothus polyophthalmus*
   Lateral line absent on blind side 6
6. Eyes separated by bony ridge .... *Arnoglossus macrolophus*
   Interorbital region concave, often broad 7
7. Scales small (80 or more); gill-opening ending a short distance above pectoral fin 8
   Scales moderate to large (38-60); gill-opening extending upwards to lateral line or slightly above pectoral when scales of head and body are continuous below lateral line 10
8. Anterior edge of upper eye opposite anterior part of lower; 83 dorsal rays .... *Bothus polylepis*
   Anterior edge of upper eye opposite middle or posterior part of lower; 86-100 dorsal rays 9
9. Depth $1\frac{5}{8}$ in length; eye $2\frac{3}{4}$-3 in head; maxilla $3\frac{1}{2}$-4 in head .... *Bothus ovalis*
   Depth $1\frac{3}{4}$-2 in length; eye $3\frac{1}{2}$-$6\frac{1}{4}$ in head; maxilla $2\frac{5}{8}$-3 in head .... *Bothus pantherinus*
10. Scales of ocular side weakly ctenoid .... *Engyprosopon grandisquama*
    Scales of ocular side strongly ctenoid 11
11. Anterior edge of upper eye opposite posterior part of lower; upper pectoral rays prolonged in males .... *Crossorhombus valde-rostratus*
    Anterior edge of upper eye opposite middle or anterior part of lower; upper pectoral rays not prolonged in males .... *Crossorhombus azureus*

746. PSEUDORHOMBUS TRIOCELLATUS (Bloch)

THREE-SPOT FLOUNDER

Pleuronectes triocellatus Bloch and Schneider, Syst. Ichth., 1801, p. 145. *Pseudorhombus triocellatus* Bleeker, Atlas Ichth. Ind. Neerl., vi, 1870, Pl. ccxxxix, fig. 1 (plate 49).

D. 65-70. A. 49-52. L. lat. 63-68. Tr. above 25-28. Depth $1\frac{1}{2}$-$1\frac{3}{4}$ in length. Dorsal profile a little notched in front of eyes. Scales on ocular side ctenoid; on blind side cycloid except around margins. Caudal round or double-truncate. Brown with indistinct darker spots and markings and white blotches. Three conspicuous dark ocelli arranged in a triangle. $5\frac{1}{2}$ inches. Coastal waters and pearl banks.

Kavuma, Thun pulli kavuma (S). Kuruvisa, Nellunthetti (T).

747. PSEUDORHOMBUS ARSIUS (Hamilton-Buchanan)

LARGE-TOOTHED FLOUNDER

Pleuronectes arsius Hamilton-Buchanan, Fish. Ganges, 1822, p. 128. *Pseudorhombus arsius* Day, Fish. India, 1875-78, Pl. xci, fig. 5 (plate 49).

D. 72-80. A. 56-61. L. lat. 67-78. Tr. above 26-34. Depth $1\frac{3}{4}$-$2\frac{1}{5}$ in length. Dorsal profile of head straight. Scales on ocular side ctenoid; on blind side cycloid. Anterior teeth large caniniform. Caudal double-truncate. Brown with darker spots and rings. Often a conspicuous dark spot surrounded by white dots in a ring at commencement of lateral line. Unpaired fins with brown spots and rings. 14 inches. Coastal waters, estuaries and pearl banks.

748. PSEUDORHOMBUS JAVANICUS (Bleeker) JAVANESE FLOUNDER

Rhombus javanicus Bleeker, Nat. Tijds. Ned. Ind., iv, 1853, p. 502. *Pseudorhombus javanicus* Bleeker, Atlas Ichth. Ind. Neerl., vi, 1870, Pl. ccxxxii, fig. 3 (plate 49).

D. 69-72. A. 54-55. L. lat. 68-72. Tr. above 27. Depth $2\frac{1}{10}$-$2\frac{1}{4}$ in length. Dorsal profile of head evenly convex. Scales on blind side cycloid; on ocular side ctenoid anteriorly and cycloid posteriorly. Anterior teeth not enlarged. Caudal double-truncate. Yellow-brown with or without small paler spots and ocelli which extend on to unpaired fins. A black blotch at bend in lateral line and another on straight portion. $8\frac{1}{2}$ inches. Trawling banks.

749. CHASCANOPSETTA LUGUBRIS Alcock GLOOMY FLOUNDER

Chascanopsetta lugubris Alcock, J. Asiat. Soc. Bengal, lxiii, 1894, p. 129; Illust. Zool. Investigator, 1895, Pl. xv, fig. 3 (plate 49).

D. 114-122. A. 81-85. L. lat. well developed on both sides of body, scales minute. Depth $3\frac{1}{4}$-4 in length. Maxillary extends well beyond eye. Scales on both sides of body cycloid. Single row of slender curved teeth in both jaws. Left pectoral with 14-17 simple rays. Caudal peduncle constricted. Grey or yellowish with or without numerous small dark spots. Fins dusky. 10 inches. Deep water, 143 fathoms.

750. GRAMMATOBOTHUS POLYOPHTHALMUS (Bleeker) MANY-EYED FLOUNDER

Platophrys polyophthalmus Bleeker, Ned. Tijds. Dierk., iii, 1866, p. 46. *Rhomboidichthys angustifrons. Platophrys polyophthalmus* Bleeker, Atlas Ichth. Ind. Neerl., vi, 1870, Pl. ccxxxiv, fig. 3 (plate 49).

D. 77-85. A. 61-68. L. lat. 77-83. Oval. Depth $1\frac{4}{5}$-$2\frac{1}{10}$ in length. Maxillary reaches front border of eye. Single row of conical teeth in both jaws. Caudal bluntly pointed. Scales ctenoid on ocular side, cycloid on blind side. Head totally scaled on both sides. Brownish with three large ocelli and many smaller bluish or brownish spots or ocelli. 6 inches. Coastal waters in 32 fathoms.

751. ARNOGLOSSUS MACROLOPHUS Alcock LARGE-CRESTED FLOUNDER

Arnoglossus macrolophus Alcock, J. Asiatic Soc. Bengal, lviii, 1889, p. 280. *Bothus (Arnoglossus) tapeinosoma. Arnoglossus macrolophus* Alcock, Illust. Zool. Investigator, 1898, Pl. xxiii, fig. 3 (plate 49).

D. 89-98. A. 67-72. L. lat. 48-55. Depth $2\frac{1}{2}$-3 in length. Head profile notched in front of eyes. Eyes separated by narrow bony ridge. Teeth small; close set and not enlarged anteriorly. Scales weakly ctenoid on ocular side, cycloid on blind side. Brownish with indistinct dark blotches along upper and lower edges of body. Dark blotch at bend in lateral line. Tips of ventrals black. $3\frac{1}{2}$ inches. Coastal waters, to 30 fathoms.

752. BOTHUS POLYLEPIS (Alcock) MANY-SCALED FLOUNDER

Arnoglossus polylepis Alcock, J. Asiatic. Soc. Bengal, lviii, 1889, p. 290. *Rhomboidichthys polylepis* Alcock, Illust. Zool. Investigator, 1898, Pl. xxiv, Fig. 4-5 (plate 49).

D. 83. A. 63-66. L. lat. 82-85. Depth $2\frac{1}{8}$ in length. Dorsal profile slightly notched at eye. Interorbital space narrow, concave and naked. Gill-opening close to commencement of lateral line, membrane connecting operculum with shoulder partly scaled. Scales of ocular side ctenoid, on blind side cycloid. Brown with series of dark rings at upper and lower edges of body. Two ocelli at bend in lateral line and another on straight portion. 5 inches. Off coast in 32-34 fathoms.

753. BOTHUS OVALIS (Regan) OVAL FLOUNDER

Platophrys ovalis Regan, Trans. Linn. Soc., xii, 1908, p. 232. *Psettylis ocellata. Platophrys ovalis* Regan, *loc. cit.*, Pl. xxvii, fig. 6 (plate 49).

D. 88-99. A. 66-73. L. lat. 95. Depth $1\frac{1}{4}$-$1\frac{2}{3}$ in length. Maxillary scarcely reaches anterior edge of eye. Teeth biserial in both jaws. Scales cycloid on both sides except along edges of ocular side. Left pectoral with 8-9 rays, one or more prolonged. Grey-brown clouded with darker. Large ocellus at commencement of straight part of lateral line. Head and body with numerous bi-coloured spots. Posterior part of caudal black. $4\frac{1}{2}$ inches. Coastal waters.

754. BOTHUS PELLUCIDA (Alcock) TRANSPARENT FLOUNDER

Psettylis pellucida Alcock, Ann. Mag. Nat. Hist., vi, 1890, p. 437, fig. 2 (plate 49).

A larval form of uncertain identity very similar to the above species but lacking pigmentation. Deep water off coast.

755. BOTHUS PANTHERINUS (Rüppell) LEOPARD FLOUNDER

Rhombus pantherinus Rüppell, Fische Roth. Meers, 1828, p. 121. *Platophrys pantherinus* Bleeker, Atlas Ichth. Ind. Neerl., vi, 1870, Pl. ccxxxiii, fig. 3 (plate 50).

D. 86-93. A. 65-70. L. lat. 82-87. Oval. Depth 2-2⅛ in length. Lower jaw slightly prominent. Maxillary reaches front border of eye. Teeth pointed, curved, in 2 series in both jaws. In males, 2nd, 3rd or 4th pectoral rays filamentous reaching caudal peduncle. Caudal obtusely pointed. Scales ctenoid on ocular side, cycloid on blind side. Brown or reddish with irregular dark blotches and ocelli. Conspicuous dark blotch on caudal half of lateral line. 9½ inches. Coastal waters.

756. ENGYPROSOPON GRANDISQUAMA (Schlegel)
LARGE-SCALED FLOUNDER

Rhombus grandisquama Schlegel, Fauna Japonica, Pisces, 1846, p. 183. *Arnoglossus spilurus. Rhomboidichthys spilurus* Günther, Challenger Repts., i, 1880, Pl. xxi, fig. A (plate 50).

D. 81-88. A. 59-68. L. lat. 38-43. Depth 1¾-2⅛ in length. Males with spine on snout. Teeth in single series in both jaws. Scales on ocular side weakly ctenoid, on blind side cycloid. Left pectoral with 10-12 simple rays, upper sometimes prolonged in male. Brown with irregular dark spots and markings. Unpaired fins with small brown spots. Two large blackish spots at upper and lower margins of caudal. 4 inches. Coastal waters and pearl banks.

757. CROSSORHOMBUS VALDE-ROSTRATUS (Alcock)
STRONG-SNOUT FLOUNDER

Rhomboidichthys valde-rostratus Alcock, Ann. Mag. Nat. Hist. (6), vi, 1890, p. 435; Alcock, Illust. Zool. Investigator, 1898, Pl. xxiv, fig. 2 (plate 50).

D. 85-88. A. 68-72. L. lat. 47-51. Depth 1⅘-2 in body length. Anterior profile notched in front of eyes. Males with strong spine on snout. Scales ctenoid on ocular side, cycloid on blind side. Left pectoral with 10-11 rays, upper prolonged in adult males, 1¼-1¾ times head. Right pectoral ½ head. Grey-brown with darker spots and blotches. 5 inches. Coastal waters in 32 fathoms.

758. CROSSORHOMBUS AZUREUS (Alcock) BLUE-SPOTTED FLOUNDER

Rhomboidichthys azureus Alcock, J. Asiatic Soc. Bengal, lviii, 1889, p. 283; Illust. Zool. Investigator, 1898, Pl. xxiv, fig. 3 (plate 50).

D. 84-90. A. 64-73. L. lat. 53-57. Depth 2 in body length. Left pectoral with 11-12 simple rays, uppermost not prolonged in males, 1½-1⅗ in head. Right pectoral ½ head. Brownish marked with pale and darker. One or two dark blotches on lateral line and males with 2 series of azure spots on head. Fins spotted and blotched with dark brown or black. 4 inches. Coastal waters, 16-32 fathoms.

SOLES

Family SOLEIDAE

Flatfishes with eyes on the right side. Dorsal fin extends on head to above eye or beyond. Preopercle edge covered with skin.

1. Pelvic fins markedly asymmetrical, that of right side median, elongate and joined to anal; snout prolonged into a hook, dorsal rays extending to its tip .... *Heteromycteris oculus*
   Pelvic fins symmetrical, that of right side free from anal or joined by a low membrane to base of first anal ray; snout not prolonged into a hook, dorsal rays not extending to its tip 2
2. Gill-opening on ocular side opposite lower edge or lower part of pectoral base; pectorals on both sides well developed, not connected to opercular membrane 3
   Gill-opening on ocular side ending opposite upper part of pectoral base; pectorals small, especially on blind side; opercular membrane joined to upper part of pectoral 5
3. Posterior rays of dorsal and anal short, not connected to caudal; urogenital papilla situated on right side of body some distance from anus .... *Solea elongata*
   Posterior rays of dorsal and anal elongate, united with rays of caudal; urogenital papilla situated between pelvic fins and close to anus 4
4. Form elongate; depth $3\frac{3}{4}$-$4\frac{1}{3}$ in length; a bony prominence on snout; 160 scales in longitudinal series .... *Synaptura commersoni*
   Form ovate; depth 2-$2\frac{2}{3}$ in length; no bony prominence on snout; 63-74 scales in longitudinal series .... *Brachirus orientalis*
5. Scales ctenoid; first dorsal ray not enlarged .... *Zebrias quagga*
   Scales cycloid; first dorsal ray enlarged, free .... *Aesopia cornuta*

759. HETEROMYCTERIS OCULUS (Alcock) EYED SOLE

Solea oculus Alcock, J. Asiat. Soc. Bengal, lviii (2), 1889, p. 285, Pl. xviii, fig. 3 (plate 50).

D. 90-103. A. 60-67. L. lat. 86-91. Pectoral fins asymmetrical, that of right side joined to anal. Snout hook-like with dorsal rays extending to its tip.

760. SOLEA ELONGATA Day ELONGATE SOLE

Solea elongata Day, Fish. India, 1875-78, p. 426, Pl. xc, fig. 4 (plate 50).

D. 72-77. A. 59-63. L. lat. 110. Snout obtusely pointed. No dilated nostril on blind side. Brownish or greyish, spotted or blotched with darker; markings more distinct near edges of body and often form irregular vertical bands. Black blotch on distal part of pectoral. $4\frac{1}{2}$ inches. Coastal waters. Handhalla (S).

761. SYNAPTURA COMMERSONI (Swainson) COMMERSON'S SOLE

Brcchirus commersoni Swainson, Nat. Hist. Fish., ii, 1839, p. 303. *Synaptura thomasoniana. Synaptura commersoniana* Day, Fish. India, 1875-78, Pl. xciv, fig. 1 (plate 50).

D. 70-81. A. 58-66. L. lat. 160. Bony prominence on anterior part of snout. Lower lip fringed. Two tubular nostrils on ocular side, in front of lower eye. Posterior rays of dorsal and anal completely joined to caudal which is obtusely pointed. Greyish-brown. Vertical fins blackish towards edges on both sides, and with conspicuous white margin. Right pectoral dusky. 13 inches. Shallow waters of pearl banks.

762. BRACHIRUS ORIENTALIS (Bloch) ORIENTAL SOLE

Pleuronectes orientalis Bloch and Schneider, Syst. Ichth., 1801, p. 157. *Synaptura cinerascens*. *Synaptura orientalis* Day, Fish. India, 1875-78, Pl. xciv, fig. 2 (plate 50).

D. 62-72. A. 47-57. L. lat. 63-74. Depth 2-2¼ in length. Lower lip feebly fringed. Two tubular nostrils on ocular side in front of lower eye. Posterior rays of dorsal and anal completely joined with caudal, which is rounded. Ocular side of body with numerous strong filamentous processes in irregular groups. Greyish or brownish, blotched or spotted with darker and a number of short dark vertical streaks crossing lateral line. Right pectoral dusky. 9 inches. Coastal waters.
Kalepu patha madiya, Kadol patha madiya, Maduwa (S).

763. ZEBRIAS QUAGGA (Kaup) ZEBRA SOLE

Aesopia quagga Kaup, Arch. Naturg., 1858, p. 98. *Synaptura quagga* or *zebra*. *Zebrias quagga* Norman, Rec. Indian Mus., xxx, 1928, Pl. vi (plate 50).

D. 66-73. A. 56-61. L. lat. 92-99. Depth 2½-2¾ in length. Eyes nearly contiguous. Nasal tube short. Posterior rays of dorsal and anal almost completely joined to caudal, which is scarcely distinct and rounded posteriorly. Upper rays of pectoral produced. Pale brown or grey, with a number of dark brown cross-bands, extending on to the vertical fins. Caudal irregularly marked with yellowish and black. 5½ inches. Coastal waters.
Vairan patha madiya (S).

764. AESOPIA CORNUTA Kaup HORNED SOLE

Aesopia cornuta Kaup, Arch. Naturg., 1858, p. 98. *Synaptura cornuta* Day, Fish. India, 1875-78, Pl. xciv, fig. 4 (plate 50).

D. 69-79. A. 61-66. L. lat. 87-98. Depth 3 in length. Eyes contiguous. Nasal tube short. Posterior rays of dorsal and anal completely joined to caudal. Right pectoral shorter than eye, upper rays not produced. Left pectoral about same size. Grey or brown with 13 dark brown cross-bands extending on to vertical fins. Caudal dark brown. 5 inches. Coastal waters.

TONGUE-SOLES

Family CYNOGLOSSIDAE

Flatfishes with eyes on left side. Dorsal originates on head. Preopercular margin beneath skin. Pectorals absent. Dorsal and anal confluent with caudal. Lateral line present, absent or multiple.

1. Two or three lateral lines on ocular side 2
   No lateral line on ocular side 10
2. Lips fringed .... *Paraplagusia bilineata*
   Lips not fringed 3
3. Three lateral lines on ocular side .... *Cynoglossus quinquelineatus*
   Two lateral lines on ocular side 4

4. Two lateral lines on blind side .... *Cynoglossus bilineatus*
Never more than one lateral line on blind side 5
5. Scales cycloid on blind side 6
Scales ctenoid on blind side 8
6. Scales between lateral lines 12-14 .... *Cynoglossus lingua*
Scales between lateral lines 7-10 7
7. Eye 15-17 in head; depth 5 in length .... *Cynoglossus arel*
Eye 8-12 in head; depth 4½ in length .... *Cynoglossus macrolepidotus*
8. Between lateral lines 16-21 scales .... *Cynoglossus puncticeps*
Between lateral lines 9-11 scales .... *Cynoglossus brachycephalus*
Between lateral lines 12-15 scales 9
9. Angle of mouth nearer end of snout than gill-opening; rostral hook not extending below eye .... *Cynoglossus cynoglossus*
Angle of mouth nearer gill-opening than end of snout; rostral hook extending to below lower eye .... *Cynoglossus lida*
10. Head less than 4 in length; 3-4 cross-bars .... *Symphurus trifasciatus*
Head more than 4½ in length; 6-7 cross-bars .... *Symphurus septemstriatus*

765. PARAPLAGUSIA BILINEATA (Bloch) TWO-LINED TONGUE-SOLE

Pleuronectes bilineatus Bloch, Nat. Ausländ. Fische, iii, 1785, p. 29. *Plagusia bilineata*, *Paraplagusia marmorata* or *bilineata*. *Plagusia marmorata* Day, Fish. India, 1875-78, Pl. xcv, fig. 1 (plate 51).

D. 106-118. A. 86-93. L. lat. 98-114. Two lateral lines on ocular side separated by 16-19 scales. No distinct lateral line on blind side. Rostral hook long. Fringes on lower lip strongly arborescent. Brownish, spotted and marbled with darker. 7½ inches. Coastal waters.
Kabara patha madiya (S).

766. CYNOGLOSSUS QUINQUELINEATUS Day FIVE-LINED TONGUE-SOLE

Cynoglossus quinquelineatus Day, Fishes India, 1875-78, p. 432, Pl. xcviii, fig. 1 (plate 50).

D. 109. A. 91. L. lat. 95. Three lateral lines on ocular side, the lower incomplete and extending only from middle of body to caudal base. 17 scales between upper and middle lateral lines. Two lateral lines on blind side. Scales ctenoid on ocular side and cycloid on blind side. Brown with irregular dark patch on operculum. 10 inches. Pearl banks.

767. CYNOGLOSSUS BILINEATUS (Lacépède) FOUR-LINED TONGUE-SOLE

Achirus bilineatus Lacépède, Hist. Nat. Poiss., iv, 1802, p. 659. *Cynoglossus sindensis* Day, Fish. India, 1875-78, Pl. xc, fig. 6 (plate 50).

D. 104-114. A. 84-95. L. lat. 84-96. Two lateral lines on ocular side separated by 14-19 scales. Two on blind side separated by 15-18 scales. Rostral hook short. Brownish with an irregular dark patch on the opercular region. 12 inches. Pearl banks.

768. CYNOGLOSSUS LINGUA Hamilton-Buchanan LONG TONGUE-SOLE

Cynoglossus lingua Hamilton-Buchanan, Fish. Ganges, 1822, p. 32. *Cynoglossus elongatus* Day, Fish. India, 1875-78, Pl. xc, fig. 5 (plate 51).

D. 130-175. A. 101-124. L. lat. 93-104. Two lateral lines on ocular side, separated by 12-14 scales. No distinct lateral line on blind side. Scales ctenoid on ocular side, cycloid on blind side. Rostral hook short. Brownish or greyish, with or without indistinct darker patches. Opercular region blackish. Fins darker posteriorly. 17 inches. Coastal waters and estuaries.

769. CYNOGLOSSUS AREL (Bloch) BROWN TONGUE-SOLE

Pleuronectes arel Bloch and Schneider, Syst. Ichth., 1801, p. 159. *Cynoglossus arel* Day, Fish. India, 1875-78, Pl. xcviii, fig. 2 (plate 51).

D. 122-138. A. 90-109. L. lat. 60-65. Two lateral lines on ocular side separated by 8-10 scales. No distinct lateral line on blind side. Scales ctenoid on ocular side, cycloid on blind side. Rostral hook short. Brownish with or without irregular blackish blotches. 11 inches. Coastal waters.

770. CYNOGLOSSUS MACROLEPIDOTUS (Bleeker)
LARGE-SCALED TONGUE-SOLE

Plagusia macrolepidota Bleeker, Nat. Tijds. Ned. Ind., i, 1850, p. 415. *Cynoglossus microlepidotus. Cynoglossus oligolepis* Bleeker, Atlas Ichth. Ind. Neerl., vi, 1870, Pl. ccxlii, fig. 3 (plate 51).

D. 105-130. A. 80-96. L. lat. 56-66. Two lateral lines on ocular side separated by 7-9 scales. No lateral line on blind side. Scales ctenoid on ocular side, cycloid on blind side. Rostral hook short. Uniform brownish, fins sometimes darker posteriorly. 14 inches. Coastal waters and pearl banks.

Adalla, Handhalla (S). Athal (T).

771. CYNOGLOSSUS PUNCTICEPS (Richardson) SPOTTED TONGUE-SOLE

Plagusia puncticeps Richardson, Rept. Brit. Assn. Adv. Sci., 1846, p. 280. *Cynoglossus brachyrhynchus. Cynoglossus puncticeps* Day, Fish. India, 1875-78, Pl. xcvii, fig. 1 (plate 51).

D. 92-107. A. 70-83. L. lat. 91-110. Two lateral lines on ocular side, separated by 16-21 scales. No lateral line on blind side. Scales ctenoid on both sides. Rostral hook short. Brownish with large dark blotches on head and body, often united to form irregular cross-bands. Longitudinal series of scales with fine dark stripes. Dark brown on some of vertical fin rays. 6½ inches. Pearl banks.

772. CYNOGLOSSUS BRACHYCEPHALUS Bleeker
SHORT-HEADED TONGUE-SOLE

Cynoglossus brachycephalus Bleeker, Atlas Ichth. Ind. Neerl., vi, 1866, p. 38. *Cynoglossus brevirostris. Cynoglossus brachycephalus* Bleeker, *loc. cit.*, Pl. ccxliv, fig. 6 (plate 51).

D. 102-112. A. 84-88. L. lat. 58-66. Two lateral lines on ocular side, separated by 9-10 scales. No lateral line on blind side. Scales ctenoid on both sides. Rostral hook short. Brownish with irregularly arranged darker spots and blotches. 5 inches. Pearl banks.

773. CYNOGLOSSUS CYNOGLOSSUS (Hamilton-Buchanan)

BENGAL TONGUE-SOLE

Achirus cynoglossus Hamilton-Buchanan, Fish. Ganges, 1822, p. 132. *Cynoglossus bengalensis* Day, Fish. India, 1875-78, Pl. xcvii, fig. 4 (plate 50).

D. 98-106. A. 78-83. L. lat. 80-89. Two lateral lines on ocular side separated by 13-15 scales. No lateral line on blind side. Scales ctenoid on both sides. Rostral hook short. Brownish or greyish, sometimes vaguely marbled with darker. 6 inches. Coastal waters.

774. CYNOGLOSSUS LIDA (Bleeker) SHOULDER-SPOT TONGUE-SOLE

Plagusia lida Bleeker, Nat. Tijds. Ned. Ind., i, 1851, p. 413. *Cynoglossus lida* Day, Fish. India, 1875-78, Pl. xcvii, fig. 3 (plate 51).

D. 99-112. A. 75-87. L. lat. 82-95. Two lateral lines on ocular side separated by 13-15 scales. No lateral line on blind side. Scales ctenoid on both sides. Rostral hook moderate. Brownish or greyish with sometimes a dark patch on opercular region. 7½ inches. Pearl banks.

775. SYMPHURUS TRIFASCIATUS (Alcock) THREE-BAND TONGUE-SOLE

Aphoristia trifasciata Alcock, J. Asiatic Soc. Bengal, lxiii, 1894, p. 132; Illust. Zool. Investigator, 1895, Pl. xv, fig. 5 (plate 51).

D. 87-89. A. 75-77. L. lat. 80-82. Tr. 38-40. No lateral line on ocular side. Head slightly less than body width, 4 in length without caudal. Mouth cleft reaching well beyond middle of lower eye. Longest rays of vertical fins more than ½ body depth. Olive brown with 3 broad black cross-bars. 4½ inches. Deeper waters off 100 fathom line.

776. SYMPHURUS SEPTEMSTRIATUS (Alcock)

SEVEN-BAND TONGUE-SOLE

Aphoristia septemstriata Alcock, Ann. Mag. Nat. Hist. (6), viii, 1891, p. 125; Illust. Zool. Investigator, 1892, Pl. ii, fig. 1 (plate 51).

D. 97. A. 80. L. lat. 92-94. Tr. 40. No lateral line on ocular side. Head less than width of body, about 5 in length without caudal. Mouth cleft hardly surpassing front edge of lower eye. Eyes almost in contact. Longest rays of vertical fins are ⅓-⅖ body width. Ocular side warm brown with 6-7 rather broad cross-bars. 4 inches. Coastal waters, in 142-200 fathoms.

SPINY EELS

Order MASTACEMBELIFORMES

Family MASTACEMBELIDAE

Fresh water eel-like spiny fishes. Dorsal, caudal and anal fins confluent. Some free spines before dorsal. Anal with 3 spines. No ventral fins. Pectorals present. Mouth bordered by pre-maxillaries only. Branchial openings small and ventral. Buccal cavity enlarged for air breathing.

Rostral groove dentigerous .... *Macrognathus aculeatus*
Rostral groove edentulous .... *Mastacembelus armatus*

777. MACROGNATHUS ACULEATUS (Bloch) — LESSER SPINY EEL

Ophidium aculeatum Bloch, Naturg. Ausländ. Fische, 1795, Pl. 159. *Rhynchobdella aculeata* Day, Fish. India, 1875-78, Pl. lxxii, fig. 1 (plate 52).

D. XIII-XVII; 50-54. A. II-III, 50-52. P. 22-25. Upper surface of head and snout without scales. 19-23 scales between lateral line and origin of soft dorsal. Spinous dorsal begins about 1 head length behind gill-opening; posterior spines longer than anterior ones. Bases of fins scaly. Head scales larger than body scales. Reddish-brown to olive above with median light stripe and yellow lateral band from eye to caudal above lateral line. An indistinct row of light spots between these bands. Sides greenish-yellow and belly salmon pink, yellow or white. Unpaired fins dusky brown; dorsal with 3-9 yellow or orange edged ocelli below a longitudinal yellow stripe. Anal uniform orange or yellow, sometimes with yellow line and one or 2 basal ocelli. Caudal with yellow spots and bars. Pectorals yellow. 10 inches. Fresh water streams, ponds, tanks in low-country.
Bata kola theliya (S).

778. MASTACEMBELUS ARMATUS (Lacépède) — SPINY EEL

Macrognathus armatus Lacépède, Hist. Nat. Poiss., ii, 1803, p. 286. *Mastocembelus armatus armatus* Deraniyagala, Col. Atlas Vert. Ceylon, i, 1952, Pl. xxxiv (plate 52).

D. XXXIV-XXXVII; 64-80. A. II-III, 64-78. P. 25. Top of snout naked, its side scaly at base. Single suborbital and 3 preopercular spines. 28-30 scales between lateral line and soft dorsal origin. Spinous dorsal originates above posterior third of pectoral. Fins scaly at base. Head scales equal to those on body. Brownish on back and sides, yellowish below. Body marked with darker blotches. Unpaired fins brown with a narrow basal light area. Pectoral yellowish with brown transverse stripes. 25 inches. Flowing and still fresh waters up to 4000 feet.
Theliya, Gan theliya, Oya theliya (S).

SUCKERFISH, REMORAS

Order ECHENEIFORMES

Family ECHENEIDAE

Body elongate with flattened head, on which the first dorsal fin is modified into a powerful laminated sucking disc. The disc is used to attach the fish to larger fish, turtles, ships, etc. No spines in second dorsal or anal.

1. Body slender .... *Echeneis naucrates*
   Body robust 2
2. Sucking disc with 12-13 laminae .... *Remora albescens*
   Sucking disc with 16-20 laminae .... *Remora remora*
   Sucking disc with 27 laminae .... *Remora scutata*

779. ECHENEIS NAUCRATES Linnaeus — SLENDER SUCKERFISH

Echeneis naucrates Linnaeus, Syst. Nat., ed. 10, 1758, p. 261. *Leptecheneis naucrates. Echeneis naucrates* Smith, Fish. Sth. Africa, 1949, fig. 949 (plate 52).

Sucking disc with 21-28 laminae, and more than twice as long as broad. D. 32-40. A. 32-37. Gill-rakers 9. Brownish-grey, lighter below. Sometimes with a brownish-black band along sides. External margins of caudal and anterior tips of dorsal and anal edged with white. Pectoral deep brown. 3 feet. Coastal waters, Gulf of Manaar and trawling grounds. Prefers sharks as host.

Kalu kalliya, Kalliya (S). — Ural, Ubbay, Karung ora (T).

780. REMORA ALBESCENS (Temminck and Schlegel) — WHITE REMORA

Echeneis albescens Temminck and Schlegel, Fauna Japonica, Pisces, 1843, p. 272, Pl. cxx, fig. 3 (plate 52).

Head rather broad and body robust. Sucking disc about 3 in length, $\frac{5}{8}$ as wide as long, with 12-13 laminae. D. 17-22. A. 20-23. Soft dorsal commences in 2nd third of distance between pectoral base and caudal base, opposite anal. Uniform brown, or more often colourless. 12 inches. Coastal waters. Prefers gill-chamber of large fish or devil rays.

781. REMORA REMORA (Linnaeus) — COMMON REMORA

Echeneis remora Linnaeus, Syst. Nat., ed. 10, 1758, p. 260; Day, Fish. Gt. Brit. and Ireland, 1880-84, Pl. xxxix, fig. 2 (plate 52).

Body moderately robust. Sucking disc $3\frac{1}{4}$ in length, $\frac{1}{2}$ as wide as long, with 16-20 laminae. D. 22-25. A. 23-25. G.R. 12. Soft dorsal commences mid-way between bases of pectoral and caudal, opposite anal. Usually brownish or dusky. 18 inches. Coastal waters and Gulf of Manaar.

782. REMORA SCUTATA (Günther) — CEYLONESE REMORA

Echeneis scutata Günther, Ann. Mag. Nat. Hist. (3), v, 1860, p. 401, Pl. x, fig. B (plate 52).

Body moderately robust. Sucking disc very long, $2\frac{1}{3}$ in total length, $\frac{1}{3}$ as wide as long with 27 laminae. D. 22. A. 21-23. Dorsal and anal not continued to the caudal. Brown. 20 inches. Coastal waters.

TRIGGER-FISH, FILE-FISH, LEATHER-JACKETS, BOX-FISH, GLOBE-FISH, PORCUPINE-FISH, SUN-FISH

Order TETRADONTIFORMES

Fishes without true scales, but with dermal spines, scutes, ossicles or complete covering of fused bony plates. Ventrals, when present, thoracic or subthoracic. Pelvic bones sometimes absent. No lower ribs. Gill-openings restricted.

1. Teeth not united into a beak 2
   Teeth united into a solid beak 3
2. Body compressed and usually fairly deep; scales absent or modified, sometimes marginally coalesced into hard scutes .... *Suborder* BALISTOIDEI, p. 269
   Body encased in an armour of hexagonal bony plates that are marginally fused to form a solid box .... *Suborder* OSTRACIOIDEI, p. 276
3. Small fishes with small or large spines embedded in skin. Caudal fin normal with peduncle .... *Suborder* TETRADONTOIDEI, p. 278
   Huge fishes with small plates embedded in a smooth thick skin. Caudal reduced, without peduncle .... *Suborder* MOLOIDEI, p. 284

Suborder BALISTOIDEI

Body mostly compressed and fairly deep. Scales absent or modified, frequently enlarged and marginally coalesced to form a hard covering. Premaxilla and maxilla coalesced. Teeth normal, not united. Gill-openings restricted. Fin spines mostly rudimentary or reduced, but usually a separate anterior dorsal with 1-3 spines, the first largest and often provided with a locking mechanism.

1. Skin fairly soft 2
   Skin hard, with large bony scales .... *Family* BALISTIDAE, p. 269
2. Pelvics each with a long stout locking spine .... *Family* TRIACANTHIDAE, p. 272
   Pelvics fused into a single spiny process, feeble or absent 3
3. Body moderately deep; no fleshy barbel on lower jaw .... *Family* ALUTERIDAE, p. 273
   Body very elongate; a fleshy barbel on lower jaw .... *Family* ANACANTHIDAE, p. 275

FILE-FISHES, TRIGGER-FISHES

Family BALISTIDAE

Deep compressed body, encased in heavy armour of large bony scales, hinder ones sometimes with tubercles or spines. Mouth small, teeth strong but not fused. Spinous dorsal separate, with 3 spines, the first largest and can be locked erect.

1. Cheeks normally scaly, no naked grooves 2
   Cheeks naked anteriorly, with small scales in separate elevated rows behind .... *Pseudobalistes fuscus*
2. Teeth red .... *Odonus niger*
   Teeth not red 3

3. A groove before eye 4
   No groove before eye 9
4. Caudal peduncle depressed .... *Abalistes stellaris*
   Caudal peduncle not depressed 5
5. No spines on caudal peduncle .... *Canthidermis rotundatus*
   Spines on peduncle at least 6
6. Soft dorsal and anal higher in front .... *Melichthys ringens*
   Soft dorsal and anal about uniform in height 7
7. Spines or tubercles not extending far forward on body .... *Balistoides viridescens*
   Spines or tubercles extend far forward on body 8
8. Cheek scales somewhat enlarged .... *Sufflamen capistratus*
   Cheek scales not enlarged .... *Hemibalistes chrysopterus*
9. Third dorsal spine minute .... *Rhinecanthus aculeatus*
   Third dorsal spine moderate .... *Balistapus undulatus*

783. PSEUDOBALISTES FUSCUS (Bloch) BROWN FILE-FISH

Balistes fuscus Bloch and Schneider, Syst. Ichth., 1801, p. 471; Günther, Jour. Mus. Godeffroy, vi (17), 1910, Pl. clxviii (plate 53).

D. III; 24-27. A. 22-24. About 30 scales between soft dorsal origin and anal. Depth 1·8. 45-55 scales in horizontal series. Cheeks with naked groove. Caudal rounded in young, emarginate with outer rays often prolonged in adult. Dull yellow with brown or orange spots, sometimes enclosing spaces. 3 dark narrow interorbital bands. Young with spots on fins, older fish with pale margin to fins. 20 inches. Coastal waters and reefs.

Mol kota (S). Anklathi, Vari klathi (T).

784. ODONUS NIGER (Rüppell) RED-TOOTHED FILE-FISH

Xenodon niger Rüppell, Neue Wirbelth., Fische, 1840, p. 53. *Odonus* or *Acanthurus erythrodon*. *Balistes erythrodon* Day, Fish. India, 1875-78, Pl. clxxv, fig. 4 (plate 53).

D. III; 34-36. A. 30. Depth 3. Teeth distinctly red. 33-35 scales in longitudinal series. Tr. 17. Cheeks entirely scaly, some enlarged scutes behind gill-opening. Scales along middle of body and caudal peduncle with an elevated tubercle, but no prominent spines. Deep brown with black stripe from mouth to pectoral base. Unpaired fins almost black. Caudal with a white margin. 20 inches. Trawling grounds.

Kalu pothypora, Valla (S). Karuppu klathi (T).

785. ABALISTES STELLARIS (Bloch) STARRY FILE-FISH

Balistes stellaris Bloch and Schneider, Syst. Ichth., 1801, p. 476. *Balistes stellatus. Leurus stellatus* Bleeker, Atlas Ichth. Ind. Neerl., v, 1865, Pl. ccxv (plate 53).

D. III; 26-27. A. 24-26. Depth 2·6-3. 42-46 scales in horizontal series. Tr. 25-26. A groove in front of eye. Caudal concave, ends of lobes produced in adults. Caudal peduncle depressed, wider than deep. Some enlarged scutes behind gill-opening. Grey, dashed with olive green. 3 large white blotches on back together with numerous small bluish spots. Ventral surface white. 3-4 yellow lines from mouth to pectoral base. Pectoral yellow with dark base. Fins banded with black, yellow and olive. 24 inches. Gulf of Manaar, trawling grounds and coastal reefs.

786. CANTHIDERMIS ROTUNDATUS (Proce) ROTUND FILE-FISH

Balistus rotundatus Proce, Bull. Soc. Philom. Paris, 1822, p. 130. *Balistes* (*Canthidermis*) *oculatus* Bleeker, Atlas Ichth. Ind. Neerl., v, 1865, Pl. ccxviii, fig. 2 (plate 53).

D. III; 25-26. A. 23-24. Depth 1¾ without caudal. Body well compressed and of rounded outline. A groove in front of eye. 42-57 scales in longitudinal series. Tr. 30-54 counted from dorsal origin to anal fin. Brown above, buff below. Undulating brown lines and fine bluish spots on sides. Fins brownish. Soft dorsal with obscure blotch on middle of highest rays. Caudal dark brown. 4½ inches. Coastal waters.
Dhivul pothupora (S).

787. MELICHTHYS RINGENS (Osbeck) BLACK FILE-FISH

Balistes ringens Osbeck, Voy. China, ii, 1771, p. 93. *Balistes buniva. Melichthys ringens* Bleeker, Atlas Ichth. Ind. Neerl., v, 1865, Pl. ccxx, fig. 2 (plate 53).

D. III; 31-33. A. 28-30. Depth 2. 53-65 scales in longitudinal series. Tr. 26. Enlarged osseous plates behind gill-openings. 7-8 raised lines of spines on caudal peduncle. Black with a white line along bases of soft dorsal and anal fins. 20 inches. Coastal waters.

788. BALISTOIDES VIRIDESCENS (Bloch) BLUE-FIN FILE-FISH

Balistes viridescens Bloch and Schneider, Syst. Ichth., 1801, p. 447; Day, Fish. India, 1875-78, Pl. clxxvii, fig. 2 (plate 53).

D. III; 24-25. A. 23-24. Depth 2½. 30-34 scales in longitudinal series. Tr. 17-18. A groove in front of eye. Osseous scutes behind gill-openings. 4½ rows curved spines on caudal peduncle. Brownish olive with a blue tinge. Lips black, separated by light band from black band on forehead. A black band from eye to pectoral base. Unpaired fins yellowish with dark margins and sinuous lines, or blotched with blue marginally. 24 inches. Gulf of Manaar and trawling grounds.

789. SUFFLAMEN CAPISTRATUS (Shaw) MASKED FILE-FISH

Balistes capistratus Shaw, Gen. Zool., v, 1804, p. 417. *Balistes mitis. Balistes* (*Balistapus*) *frenatus* Bleeker, Atlas Ichth. Ind. Neerl., v, 1865, Pl. ccxxiii, fig. 2 (plate 53).

D. III; 29-31. A. 26-28. Depth 2¼. 55-65 scales in longitudinal series. Tr. 34. Well-marked groove in front of eye. Some enlarged scutes behind gill-opening. Each scale in caudal region with a large tubercle, more distinct on middle 7-8 rows. Dull yellow-brown. Pectoral and ventral spines straw-coloured, rest of fins black with light bases. A yellow band passes around mouth towards pectoral base. 20 inches. Coastal reefs, Gulf of Manaar banks, trawling grounds.
Kukula maha (S).

790. HEMIBALISTES CHRYSOPTERUS (Bloch)
GOLDEN-FINNED FILE-FISH

Balistes chrysopterus Bloch and Schneider, Syst. Ichth., 1801, p. 466. *Balistes niger* Day, Fish. India, 1875-78, Pl. clxxxi, fig. 1 (plate 53).

D. III; 26-28. A. 23-25. Depth $2\frac{1}{2}$. 45-47 scales in longitudinal series. Tr. 25. A groove in front of eye. Caudal truncate or undulated. Some osseous scutes behind gill-opening. 6-8 rows of small recurved spines on side of tail. Rich brown. Lips and ring behind them yellow. Caudal dark with light upper and lower edge and white margin. Soft dorsal and anal yellow. 12 inches. Coastal reefs.

791. RHINECANTHUS ACULEATUS (Linnaeus) PRICKLY FILE-FISH

Balistes aculeatus Linnaeus, Syst. Nat., ed. 10, 1758, p. 238. *Balistapus aculeatus* Jordan and Evermann, Bull. U.S. Fish. Comm., xxiii (1), 1905, Pl. lxii (plate 53).

D. III; 24-25. A. 22. Depth 2·6. 38-40 scales in longitudinal series. Tr. 24. No groove in front of eye. Some osseous scutes behind gill-openings. $2\frac{1}{2}$ rows of recurved spines on caudal peduncle. Greyish with large dark blotch extending to anal. 4 dark blue interocular bands divided by 3 black ones. 3 blue lines from eye to pectoral base, first 2 separated by a black band. An orange band from angle of mouth. 4-5 oblique white bands from middle of body to anal base. Fins pale. 12 inches. Coastal reefs.
Ratoo potoobarah, Mol kotah (S).

792. BALISTAPUS UNDULATUS (Mungo Park) UNDULATED FILE-FISH

Balistes undulatus Mungo Park, Trans. Linn. Soc., iii, 1797, p. 37. *Balistes lineatus*. *Balistes undulatus* Day, Fish. India, 1875-78, Pl. clxxvii, fig. 4 (plate 53).

D. III; 26-27. A. 23-24. Depth 2·3. 40 scales in longitudinal series. Tr. 24. No groove in front of eye. A few enlarged scutes behind gill-opening. 4-6 strong spines in 2 rows on caudal peduncle. Brown with numerous curved yellow lines from eye and back to anal and caudal. 3-4 more from mouth to ventral and anal. A dark band along base of soft dorsal and anal. Black marginal mark between 1st and 2nd dorsal spines. Sometimes a black blotch on caudal base. 9 inches. Coastal waters.
Yakada pothypora (S).

793. "BALISTES" ELLIOTI Day ELLIOT'S FILE-FISH

Balistes ellioti Day, Fauna Brit. India, ii, 1889, p. 478.

An insufficiently characterized species of uncertain genus, not here included in the key to this family. Caudal peduncle not depressed. Teeth white. A groove in front of snout. Dorsal and anal high anteriorly. Caudal deeply lunate. Olive dashed with green, lighter below. 4 purple bands on cheek and numerous similar ones obliquely along length of body, the superior ones terminating on dorsal profile and inferior ones broken into spots. Unpaired fins bluish with dark base. Caudal light grey with carmine and yellow margin. Gulf of Manaar.

TRIPOD-FISH

Family TRIACANTHIDAE

Fairly compressed body with non-overlapping small spiny scales. Dorsal fin spines well-developed. Ventrals with one strong spine and one feeble soft ray; spine can be locked erect. Nasal pit present. No lateral line.

1. Tips of dorsal and ventral spines naked; second dorsal spine more than half length of first; spinous dorsal with high scaly sheath .... *Pseudotriacanthus strigilifer*
   Tips of dorsal and ventral spines covered with integument; second dorsal spine not more than half length of first; spinous dorsal without scaly sheath 2
2. Snout not produced, with upper profile nearly straight .... *Triacanthus brevirostris*
   Snout produced, with upper profile strongly concave .... *Triacanthus biaculeatus*

794. PSEUDOTRIACANTHUS STRIGILIFER (Cantor)

LONG-SPINED TRIPOD-FISH

Triacanthus strigilifer Cantor, Jour. Asiatic Soc. Bengal, xviii (2), 1849, p. 363 (1345); Day, Fish. India, 1875-78, Pl. clxxv, fig. 2 (plate 54).

D. V; 22. A. 15-16. Head 4-4¼. Depth 3¾. First dorsal spine very strong and longer than head. Second dorsal spine 2-3 times as long as third. Snout produced with upper profile concave. Scales larger than other species. Spinous dorsal without black spot. Body silvery with wide irregular yellow lines and blotches. 7½ inches. Gulf of Manaar.

795. TRIACANTHUS BREVIROSTRIS Schlegel

SHORT-NOSED TRIPOD-FISH

Triacanthus brevirostris Temminck and Schlegel, Fauna Japon., Pisces, xv, 1850, p. 294; Day, Fish. India, 1875-78, Pl. clxxv, fig. 1 (plate 54).

D. V; 22-25. A. 16-20. Head 4½-4⅜, depth 3¼-3½. First dorsal spine shorter than head in adults, as long as or longer in young. Snout not produced and upper profile straight. Scales minute. Silvery. A black spot on spinous dorsal, extending a short distance on to the back. Usually a dark supraorbital blotch. 9½ inches. Gulf of Manaar and trawling grounds.

Ankatilla, Thun katuva (S). Mullu klathi (T).

796. TRIACANTHUS BIACULEATUS (Bloch)

HOLLOW-SNOUTED TRIPOD-FISH

Balistes biaculeatus Bloch, Naturg. Ausländ. Fische, ii, 1786, p. 17; Bennett, Fish. Ceylon, 1828-30, Pl. xv (plate 54).

D. V; 22-25. A. 16-19. Depth 2½-3⅙ in length without caudal. Snout produced, with upper profile distinctly concave. First dorsal spine as long as, or longer than head. Second dorsal spine very short, not much longer than third. Scales minute. Spinous dorsal with or without a black spot, when present restricted to membrane between 1st and 2nd spines. 7 inches. Coastal waters.

LEATHER-JACKETS, TRIGGER-FISH, SANDPAPER-FISH

Family ALUTERIDAE

Highly compressed body, ovoid or angular. No normal scales, those present being minute and modified into rough or spinous projections. First dorsal represented by a single isolated large spine above or near eye, locking erect by a rudimentary second spine. Soft dorsal separate. A pelvic flap present with an internal fixed fused spine. It may or may not have a hinged movable extremity. Mouth small, with 6 outer teeth in each jaw.

1. Pelvic spine with freely movable portion 2
   Pelvic spine without movable portion 5
2. Dorsal spine originates before middle of eye .... *Pervagor tomentosus*
   Dorsal spine originates behind middle of eye 3
3. Dorsal spine with 2 series of spinules in front and 2 series behind .... *Laputa cingalensis*
   Dorsal spine with no spinules in front and 2 series behind 4
4. Scales minute, skin smooth .... *Paramonacanthus choirocephalus*
   Scales larger and distinct, skin rough .... *Stephanolepis diaspros*
5. Anal rays less than 40 .... *Amanses pardalis*
   Anal rays 40 or more 6
6. Depth 3 in length .... *Alutera monoceros*
   Depth 4 in length .... *Alutera scripta*

797. PERVAGOR TOMENTOSUS (Linnaeus) MATTED LEATHER-JACKET

Balistes tomentosus Linnaeus, Syst. Nat., ed. 10, 1758, p. 328. *Monacanthus tomentosus. Monacanthus hajam* Bleeker, Atlas Ichth. Ind. Neerl., v, 1865, Pl. ccxxx, fig. 3 (plate 53).

D. I; 26-29. A. 25-27. Depth 2½. Eyes 2 diameters from snout. Dorsal spine as long as head, with 2 rows of spinules behind. Scales with 4-5 marginal spines. Males with patch of long setiform spines on either side of caudal peduncle. Short fleshy tentacles on sides. Brown, spotted and marbled with black and a light band along anterior half of body. Caudal with 2 dark vertical bands. 5 inches. Harbours and Gulf of Manaar.

798. LAPUTA CINGALENSIS Fraser-Brunner SINHALESE LEATHER-JACKET

Laputa cingalensis Fraser-Brunner, Ann. Mag. Nat. Hist. (11), viii, 1941, p. 191, fig. 5 (plate 52).

D. I; 28. A. 28. Depth 2½. Eyes 2½ diameters from snout. Dorsal spine slender, ⅞ length of head, with 2 rows of 12 spinules anteriorly and 2 similar rows posteriorly. Movable pelvic spine, spinate at tip. Grey. Upper part of snout dark. 4 longitudinal bands, the two inner ones passing through eye anteriorly. Caudal with 2 dark transverse bands. 2¾ inches. Coastal waters.

799. PARAMONACANTHUS CHOIROCEPHALUS (Bleeker) PIG-FACED LEATHER-JACKET

Monacanthus choirocephalus Bleeker, Verh. Bat. Genootsch., xxiv, 1852, p. 19. *Monacanthus choerocephalus* Day, Fish. India, 1875-78, Pl. clxxix, fig. 3 (plate 53).

D. I; 28. A. 28-30. Depth 2⅝. Eyes 2¼ diameters from snout. Dorsal spine rough anteriorly and 2 rows of barbs posteriorly. Pelvic spine movable. Caudal rounded, sometimes with a setiform prolongation of one of upper rays. Scales indistinct, skin not very rough. Numerous small fleshy tentacles on body. Stone grey with irregular and rather large black blotches. 2 dark bands across dorsal. Gulf of Manaar.

Ang orava (S). Kombu ora (T).

800. STEPHANOLEPIS DIASPROS Fraser-Brunner
RETICULATED LEATHER-JACKET

Stephanolepis diaspros Fraser-Brunner, Ann. Mag. Nat. Hist. (11), v, 1940, p. 538. *Monacanthus* or *Alutera setifer*. *Stephanolepis diaspros* Fraser-Brunner, *loc. cit.*, fig. 4 (plate 52).

D. I; 32. A. 32. Eye 4-4½ in head. Dorsal spine slightly curved, stout and with 6-7 retrorse barbs in 2 series posteriorly. 2-3 in head length. Gill-opening wider than eye. Caudal angulate in male, rounded in female. Dark with pale network of fine lines. A triangular patch of spines on caudal peduncle. 9 inches. Gulf of Manaar and trawling grounds.

801. AMANSES PARDALIS (Rüppell) SPOTTED LEATHER-JACKET

Monacanthus pardalis Rüppell, Neue Wirbelth., Fische, 1855, p. 57. *Liomonacanthus pardalis* Bleeker, Atlas Ichth. Ind. Neerl., v, 1865, Pl. ccxxx, fig. 2 (plate 53).

D. I; 35-36. A. 30-31. Depth 1½ or more in length without caudal. Dorsal spine almost straight, shorter than head and without barbs, above anterior part of eye. Soft dorsal and anal low. Ventral spine not movable. 2 pairs of strong recurved spines on each side of caudal peduncle. Colour variable, often brown spots separated by fine lighter meshwork and a white spot behind last dorsal ray. 12 inches. Gulf of Manaar.

802. ALUTERA MONOCEROS (Linnaeus)
YELLOW-FINNED LEATHER-JACKET

Balistes monoceros Linnaeus, Syst. Nat., ed. 10, 1758, p. 327. *Monacanthus monoceros* Day, Fish. India, 1875-78, Pl. clxxix, fig. 2 (plate 52).

D. I; 46-48. A. 48-53. Depth 2¾-3⅓. Eyes small. Body oblong. Snout moderately produced, with convex profile above. Unpaired fins low. Dorsal spine weak and without barbs. Caudal truncate or slightly convex. Brown or black with yellow fins. 15 inches. Trawling grounds.

803. ALUTERA SCRIPTA (Osbeck) SCRIBBLED LEATHER-JACKET

Balistes scriptus Osbeck, Iter. China, 1765, p. 145. *Monacanthus scriptus*. *Anacanthus scriptus* Day, Fish. India, 1875-78, Pl. clxxvi, fig. 3 (plate 52).

D. I; 43-48. A. 46-52. Depth 4¼. Eyes moderate. Body oblong and snout produced with concave upper profile. Unpaired fins low. Dorsal spine weak, without barbs, and sometimes absent in large specimens. No ventral spine. Caudal wedge-shaped. Buff with small brown spots and blue lines. 40 inches. Gulf of Manaar and trawling grounds.

BARBELED LEATHER-JACKETS

Family ANACANTHIDAE

Body compressed and rather elongate. A fleshy barbel below symphysis of lower jaw. 2 rows incisor-like teeth in upper jaw and 1 row in lower. Spinous dorsal reduced to a single flexible spine. Soft dorsal and anal with many rays. Ventrals and ventral spines absent. Body covered with fine asperites.

804. ANACANTHUS BARBATUS (Gray) BARBELED LEATHER-JACKET

Balistes (*Anacanthus*) *barbatus* Gray, Illust. Indian Zool. Hardwicke, 1830-34, Pl. 84. *Psilocephalus barbatus*. *Anacanthus barbatus* Day, Fish. India, 1875-78, Pl. clxxix, fig. 1 (plate 52).

D. I; 49-52. A. 57-64. Head 3 in body without caudal. Body strongly compressed. A fleshy barbel below symphysis of lower jaw. Males with a skin-covered prolongation from throat, supported by pelvic bone and extending almost to anal. Caudal wedge-shaped. Dull brown or grey, fins yellow. Caudal with about 6 vertical or angular dark bands. 10 inches. Trawling grounds.

BOX-FISH, COW-FISH

Suborder OSTRACIOIDEI

Family OSTRACIONTIDAE

Body encased in an armour of hexagonal bony plates, marginally fused to form a solid box. Teeth not fused, about 10 in each jaw. No fin spines. Pelvics absent, replaced by external bony ridges.

1. Carapace three-angled .... *Tetrosomus gibbosus*
   Carapace four-angled (no ridge on back) 4
   Carapace five-angled (or four-angled with ridge on back) 2
2. Spines on carapace 3
   No spines on carapace 5
3. Large squat spine in middle of back; the horns in front of eyes and on posterior of ventral ridges robust and thorn-like .... *Lactoria fornasini*
   No spine in middle of back; the horns in front of the eyes and on posterior of ventral ridges long and slender .... *Lactoria cornuta*
4. Brown with small blue black-edged ocelli .... *Ostracion tuberculata*
   Reddish-brown with numerous small white spots .... *Ostracion lentiginosa*
5. A small projection directly above mouth followed by a long concave profile; ridge on back well-defined .... *Rhynchostracion nasus*
   A large lump occupying most of the snout; ridge on back poorly defined ... *Rhynchostracion rhinorhynchus*

805. TETROSOMUS GIBBOSUS (Linnaeus) THORN-BACKED BOX-FISH

Ostracion gibbosus Linnaeus, Syst. Nat., ed. 10, 1758, p. 409. *Ostracion turritus* Day, Fish. India, 1875-78, Pl. clxxxi, fig. 4 (plate 54).

D. 9. A. 9. Carapace three-ridged. A large sharp triangular spine on middle of dorsal ridge. This spine diminishes in size with age. Ventral ridges well-developed and each armed with 4 backward-directed strong spines. Olive brown with 3 ill-defined dark bands inferiorly. A light blue spot in centre of each scute. 12 inches. Gulf of Manaar and trawling grounds.

806. LACTORIA FORNASINI (Bianconi) SHORT-SPINED COW-FISH

Ostracion fornasini Bianconi, Nouv. Ann. Sci. Nat. Bologna (1), v, 1846, p. 113. *Lactoria fornasini* Smith, Fish. Sth. Africa, 1949, fig. 1180 (plate 54).

D. 9. A. 9. Carapace five-ridged. Ridge on back distinct with strong recurved spine. A pair of strong but not elongate antorbital spines. Ventral ridges terminating in a short strong claw-like spine, directed backwards. Interorbital deeply concave. Yellowish or brown with dark irregular spots and lines. 6 inches. Coastal waters in 34 fathoms.

807. LACTORIA CORNUTA (Linnaeus) LONG-HORNED COW-FISH

Ostracion cornutus Linnaeus, Syst. Nat., ed. 10, 1758, p. 331; Day, Fish. India, 1875-78, Pl. clxxvi, fig. 4 (plate 54).

D. 9. A. 9. Carapace five-ridged. Dorsal ridge not prominent and usually without a spine. A pair of slender, elongate antorbital spines directed forwards. Each ventral ridge ending in a similar long slender spine directed backwards. Caudal very elongate in adults. Brownish. Body and tail with bluish or blackish spots, about one to each scute. 20 inches. Gulf of Manaar and trawling grounds.

808. OSTRACION TUBERCULATA Linnaeus OCELLATED BOX-FISH

Ostracion tuberculatus Linnaeus, Syst. Nat., ed. 10, 1758, p. 409. *Ostracion tetragonus. Ostracion cubiceps* Day, Fish. India, 1875-78, Pl. clxxxi, fig. 3 (plate 54).

D. 9. A. 9. Body four-ridged with back gently rounded. Mouth opening in carapace greater than eye. No spines on carapace. Brown with a single blue black-edged ocellus in centre of most scutes. Spots frequently black on head; on abdomen may be dark or white instead of true ocelli. Fins yellow, with or without black spots. 18 inches. Gulf of Manaar and trawling grounds.
Dhara peyttheya (S).

809. OSTRACION LENTIGINOSA Bloch FRECKLED BOX-FISH

Ostracion lentiginosus Bloch and Schneider, Syst. Ichth., 1801, p. 501. *Tetrodon punctatus. Ostracion punctatus* Day, Fish. India, 1875-78, Pl. clxxxi, fig. 2 (plate 54).

D. 9. A. 9. Body four-ridged, ridges blunt and back slightly convex. Mouth opening in carapace less than eye. No spines on carapace. Reddish-brown to olive or green with white spots. Spots very numerous and in places confluent into lines. 9 inches. Gulf of Manaar.

810. RHYNCHOSTRACION NASUS (Bloch) SMALL-NOSED BOX-FISH

Ostracion nasus Bloch, Naturg. Ausländ. Fische, i, 1785, p. 118; Bleeker, Atlas Ichth. Ind. Neerl., v, 1865, Pl. cciv, fig. 2 (plate 54).

D. 9. A. 9. Body 5-ridged and spineless. Ridges sharp. Ridge along middle of back well-defined. A small hump above mouth followed by a long concavity to eyes. Greenish-yellow with irregularly scattered round black spots, or one on each scute. Black spots on free tail and caudal fin. Young are orange with black spots. 8½ inches. Coastal waters and Gulf of Manaar.

811. RHYNCHOSTRACION RHINORHYNCHUS (Bleeker)

HORN-NOSED BOX-FISH

Ostracion rhinorhynchus Bleeker, Verh. Bat. Gen., xxiv, 1852, p. 34; Atlas Ichth. Ind. Neerl., v, 1865, Pl. cci, fig. 1 (plate 54).

Body 5-ridged, but ridge along back is only poorly defined. No spines on carapace. Snout with a very large lump which occupies most of the snout, leaving only a slight concavity before the eyes. Back and tail, and sometimes the sides with numerous small brown dots; about 2-10 to each scute. Abdomen immaculate. 11 inches. Gulf of Manaar.

Suborder TETRADONTOIDEI

Teeth united into a solid beak. Body naked with small or large spines embedded in the skin. No pelvic fins and no fin spines. Lateral line present or absent, sometimes multiple. Most capable of inflating body.

1. Teeth in each jaw fused to one unit .... *Family* DIODONTIDAE, p. 278
   Fused teeth in each jaw with a distinct line of division in front 2
2. Body compressed; no lateral line .... *Family* CANTHIGASTERIDAE, p. 279
   Body rotund; lateral line present 3
3. 2 nostrils on each side .... *Family* LAGOCEPHALIDAE, p. 279
   1 nostril on each side, or nasal sac open, appearing as 2 tentacles .... *Family* TETRAODONTIDAE, p. 281

PORCUPINE-FISH

Family DIODONTIDAE

Teeth in each jaw fused to a single unit without suture in front, forming a powerful beak. Body covered with spines, short and stout, or longer and movable. Body inflatable almost to a sphere.

Black spots on fins .... *Diodon hystrix*
Fins not spotted .... *Diodon maculifer*

812. DIODON HYSTRIX Linnaeus SPOTTED PORCUPINE-FISH

Diodon hystrix Linnaeus, Syst. Nat., ed. 10, 1758, p. 335; Day, Fish. India, 1875-78, Pl. clxxix, fig. 4 (plate 54).

D. 12. A. 12. All spines with 2 roots and movable. 14-17 series of spines between snout and dorsal fin; frontal spines being shorter than the longest on the body, $1\frac{1}{2}$ times eye. Light brown above, pale below. Covered with blackish-blue or brownish spots. Fins yellowish. 36 inches. Gulf of Manaar and trawling grounds.
Katu peyttheya (S). Mullu peytthai (T).

813. DIODON MACULIFER Kaup BLOTCHED PORCUPINE-FISH

Diodon maculifer Kaup, Weigmans Arch. Nat., 1855, p. 229. *Diodon* or *Tetrodon maculatus*. *Diodon maculifer* Smith, Fish. Sth. Africa, 1949, fig. 1181 (plate 54).

D. 12. A. 12. All spines with 2 roots and movable. About 16 series of spines between snout and dorsal fin. No spines on peduncle. Dark above, light below and with large black yellow-edged blotches over body. The blotches are variable and may be reduced to black spots. Fins immaculate. 10 inches. Coastal waters and Gulf of Manaar.

TOBYS

Family CANTHIGASTERIDAE

Rather compressed body with rough skin, somewhat ridged on back. 2 large plates in each jaw, with suture in front forming a powerful beak. Nostrils single, usually inconspicuous. No lateral line.

814. CANTHIGASTER MARGARITATUS (Rüppell) OCELLATED TOBY

Tetraodon margaritatus Rüppell, Atlas Reise Nordl. Afrika, Fische, 1828, p. 66. *Tetrodon ocellatus. Canthigaster margaritatus* Smith, Fish. Sth. Africa, 1949, Pl. xciv, fig. 1210 (plate 54).

D. 9-10. A. 9. Skin rather rough. Colour variable, sometimes brilliant, often dull. Reddish above, creamy below. Fins yellow. Large black blue-edged ocellus at base of dorsal. Horizontal blue lines and dark-edged blue spots on head, cheeks and back. Caudal orange with yellow spots. 6 inches. Coastal waters and pearl banks.

PUFFER-FISH, BLOWFISH, TOAD-FISH

Family LAGOCEPHALIDAE

Two large plates in each jaw with anterior suture forming a powerful beak. 2 distinct nostrils on each side. 2 lateral lines, the lower sometimes restricted to the chin. Elongate fishes.

1. Lower lateral line without skin fold along tail .... *Amblyrhynchotes hypselogenion*
   Lower lateral line on a raised skin fold along tail and body 2
2. Caudal truncate or rounded .... *Torquigener oblongus*
   Caudal lunate 3
3. No spines on back .... *Lagocephalus inermis*
   Whole, or almost whole of back spiny 4
4. Body with spots superiorly .... *Gastrophysus scleratus*
   Body without spots .... *Gastrophysus lunaris*

815. AMBLYRHYNCHOTES HYPSELOGENION (Bleeker)
ORANGE-BANDED BLOWFISH

Tetraodon hypselogenion Bleeker, Nat. Tijds. Ned. Ind., iii, 1852, p. 300. *Tetrodon* or *Sphaeroides hypselogenion. Amblyrhynchotes hypselogenion* Smith, Fish. Sth. Africa, 1949, Pl. xciii, fig. 1199 (plate 55).

D. 9. A. 7-8. Depth $3\frac{1}{2}$ without caudal. Whole of belly and a band across body behind pectoral spiny. 2 nostrils. Caudal slightly emarginate. Back dark green to brownish, creamy white on belly, the two colours divided by a broad bright orange band. Back with light bluish dots, often with dark edges. 3-5 dark vertical bands under the eye. Fins yellowish, caudal sometimes banded. 7 inches. Coastal waters and Gulf of Manaar. Thora peyttheya (S).

816. TORQUIGENER OBLONGUS (Bloch) OBLONG BLOWFISH

Tetrodon oblongus Bloch, Naturg. Ausländ. Fische, ii, 1786, p. 6. *Torquinger oblongus* Smith, Fish. Sth. Africa, 1949, Pl. xciii, fig. 1197 (plate 55).

D. 12. A. 10-11. Depth 3-3½ without caudal. Flanks smooth but spines present on back and belly and across body at level of pectoral. 2 nostrils. Lower lateral line with a distinct fold or keel. Brownish olive above and yellowish on sides and belly. Darker brown bands from back down sides and enclosing pale spots. Fins stained with orange. 15 inches. Gulf of Manaar.

817. LAGOCEPHALUS INERMIS (Schlegel) SMOOTH-BACKED BLOWFISH

Tetrodon inermis Temminck and Schlegel, Fauna Japon., Pisces, 1844, p. 278; Day, Fish. India, 1875-78, Pl. clxxx, fig. 1 (plate 55).

D. 12. A. 11-12. Depth 4 without caudal. Body quite smooth but with some granules along belly. Caudal emarginate. Upper surface of head and back greenish or leaden, silvery white below. A broad dull brown band from eye towards tail. Gill-opening black internally. Dorsal dark superiorly. Caudal dark distally with white margin. 20 inches. Gulf of Manaar and Wadge Bank.

818. GASTROPHYSUS SCLERATUS (Gmelin)
SPOTTED ROUGH-BACKED BLOWFISH

Tetrodon scleratus Gmelin, Syst. Nat. Linnaeus, 1789, p. 1444. *Tetrodon argentatus* or *argyropleura*. *Pleuranacanthus scleratus* Whitley, C.S.I.R. (Aust.) Bull. 159, 1943, Pl. i, fig. A (plate 55).

D. 11-12. A. 10-11. Body elongate. Whole body spinose except the sides. Olive green above with numerous brown or black spots. Whitish around eye, laterally in a wide band, and on the belly. Sometimes a greenish narrow area between lateral band and belly. Fins pinkish. Gill-opening black internally. 30 inches. Coastal waters.

819. GASTROPHYSUS LUNARIS (Bloch)
GREEN ROUGH-BACKED BLOWFISH

Tetrodon lunaris Bloch and Schneider, Syst. Ichth., 1801, p. 505. *Tetrodon tepa*. *Gastrophysus lunaris* Smith, Fish. Sth. Africa, 1949, Pl. xciii, fig. 1195 (plate 55).

D. 11-12. A. 11-12. Depth 4 without caudal. Snout, sides and whole of tail smooth. Belly and whole of back spiny. Caudal emarginate. 2 nostrils. Greenish olive above, sides and abdomen satin white. A yellowish line from eye to middle of caudal. Fins yellowish. Caudal with light outer edge. 12 inches. Coastal reefs, Gulf of Manaar and trawling grounds.

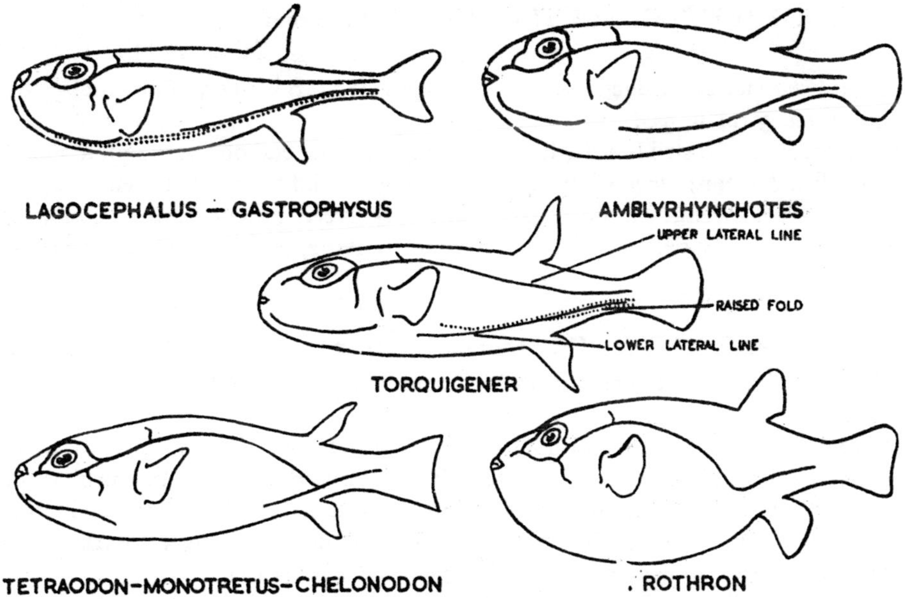

FIG. 19.

Variation in the development of lateral lines and raised skin folds in the genera of the families Lagocephalidae and Tetraodontidae (Puffer-fish).

PUFFER-FISH, BLOWFISH, TOAD-FISH

Family TETRAODONTIDAE

2 large plates in each jaw, with anterior suture and forming a powerful beak. A single nostril on each side or with an open nasal sac which may appear like 2 solid tentacles. 1 or 2 lateral lines. Robust bodied fishes.

1. Two lateral lines, the upper not reaching end of tail and meeting the lower above anal fin 2
   One lateral line reaching end of tail, although sharply bent above anal fin 4
2. A single nostril in the form of a moderately produced tube, which may have 2 lips ... *Monotretus cutcutia*
   Nasal sac open forming 2 skinny flaps or solid tentacles 3
3. Back brown or black with numerous white spots; no black blotches on sides .... *Chelonodon patoca*
   Back green; back and sides with large black blotches .... *Chelonodon fluviatilis*
4. Whole body spiny 5
   Back and abdomen spiny, sides spineless .... *Arothron leopardus*
5. Body without spots .... *Arothron immaculatus*
   Body always with spots 6
6. Spots on back pale 9
   Spots on back black 7
7. Black spots on belly; caudal without spots .... *Arothron nigropunctatus*
   No black spots on belly; caudal spotted 8
8. Dorsal fin spotted; belly without bands .... *Arothron stellatus*
   Dorsal fin without spots; belly with dark oblique narrow bands .... *Arothron aerostaticus*
9. Caudal without spots; sides with broad vertical bars .... *Arothron hispidus*
   Caudal spotted; belly with light and dark horizontal bands .... *Arothron reticularis*

820. MONOTRETUS CUTCUTIA (Hamilton-Buchanan)

OCELLATED BLOWFISH

Tetrodon cutcutia Hamilton-Buchanan, Fish. Ganges, 1822, p. 8; Day, Fish. India, 1875-78, Pl. clxxxii, fig. 5 (plate 55).

D. 10-11. A. 10. Nostril is a single orifice situated at the end of a very simple short tube. Fins rounded. Greenish-yellow above, white below. A light band between eyes. A large light-edged black ocellus on side. Back with dark green reticulations enclosing lighter spaces. Fins grey. Caudal tipped with carmine. A red spot on throat. 3½ inches. Fresh waters.

821. CHELONODON PATOCA (Hamilton-Buchanan) GANGETIC BLOWFISH

Tetrodon patoca Hamilton-Buchanan, Fish. Ganges, 1822, p. 7; Day, Fish. India, 1875-78, Pl. clxxxii, fig. 4 (plate 55).

D. 9-11. A. 8-10. Depth 3½ without caudal. Whole of back to dorsal origin, and belly spiny. Body otherwise smooth. Single nostril with raised margin. Interorbital broad. Brown or black above with numerous round or oval light spots. Sides and abdomen silvery. Caudal dark distally. Immature specimens sometimes with 3-4 dark cross-bars. 15 inches. Estuaries and lagoons, Gulf of Manaar and Wadge Bank.

822. CHELONODON FLUVIATILIS (Hamilton-Buchanan)

ESTUARINE BLOWFISH

Tetrodon fluviatilis Hamilton-Buchanan, Fish. Ganges, 1822, p. 6; Day, Fish. India, 1875-78, Pl. clxxxiii, fig. 1 (plate 55).

D. 14-16. A. 12-13. 2 nasal tentacles on each side. Interorbital broad. Spines on back from eyes to dorsal fin, on cheeks, and on abdomen to vent. Greenish olive above, white on sides and below. Back and sides with large black blotches. One or more irregular light cross-bands on back and another between eyes. Abdomen with round or angular black spots and blotches, much wider than interspaces. Fins yellowish. Caudal dark distally, sometimes with black spots. 6½ inches. Ascending coastal rivers and lagoons.

823. AROTHRON LEOPARDUS (Day) BANDED LEOPARD BLOWFISH

Tetrodon leopardus Day, Fish. India, 1875-78, p. 706, Pl. clxx, fig. 2 (plate 55).

D. 11. A. 8-9. 2 nasal tentacles on each side. Eyes moderate, 1½ diameters apart and similar distance from snout. Spines extend from front edge of eyes to dorsal; below they are closer together and reach vent. Olive above and on upper ⅔ of sides, with an interrupted network enclosing white spots. 3 black cross-bands, one across head. Reticulated narrow black interspaces between large white spots on caudal and dorsal. Coastal waters, Gulf of Manaar and trawling grounds, marine.

824. AROTHRON IMMACULATUS (Bloch) IMMACULATE BLOWFISH

Tetrodon immaculatus Bloch and Schneider, Syst. Ichth., xciii, 1801, p. 507. *Arothron immaculatus* Smith, Fish. Sth. Africa, 1949, Pl. xciv, fig. 1203 (plate 55).

D. 9-10. A. 9-10. Depth 3 without caudal. Whole body spiny except lips and caudal peduncle. 2 nasal tentacles separated by an impervious nasal fossa. Yellowish-olive above, lighter below and brownish along middle of sides. Upper and lower edges as well as margin of caudal black. Sometimes a white spot over eye. Sometimes several parallel greyish longitudinal bands on sides. 12 inches. Gulf of Manaar and trawling grounds.

825. AROTHRON NIGROPUNCTATUS (Bloch) BLACK-SPOTTED BLOWFISH

Tetrodon nigropunctatus Bloch and Schneider, Syst. Ichth., 1801, p. 507. *Arothron nigropunctatus* Smith, Fish. Sth. Africa, 1949, Pl. xciv, fig. 1204 (plate 55).

D. 9-10. A. 9-10. Body mostly spiny but spines small. 2 nasal tentacles on each side. Eyes small, 3 diameters apart and 2 from snout. Greenish or brown above, pale below. Some scattered black dots over body including belly. Vent in a black spot. Snout and around mouth black or brown. Body sometimes lemon. 10 inches. Gulf of Manaar.

826. AROTHRON STELLATUS (Bloch) STARRY BLOWFISH

Tetrodon stellatus Bloch and Schneider, Syst. Ichth., 1801, p. 503; Day, Fish. India, 1875-78, Pl. clxxxiii, fig. 3 (plate 55).

D. 10-11. A. 10-12. Heavy clumsy spiny body. 2 nasal tentacles on each side. Brownish-yellow on back and sides. Back and sides to below level of pectorals covered with brown spots. All fins similarly spotted. Pectoral yellowish with brown spots. Anus in a black ring. Lower lip dark. 36 inches. Gulf of Manaar and trawling grounds.

827. AROTHRON AEROSTATICUS (Jenyns) BLACK-LINED BLOWFISH

Tetrodon aerostaticus Jenyns, Voy. Beagle, Zool., 1842, p. 152. *Tetrodon lineatus. Arothron aerostaticus* Smith, Fish. Sth. Africa, 1949, Pl. xciv, fig. 1206 (plate 55).

D. 10-11. A. 10-11. Heavy flabby body. Spines over entire body. 2 nasal tentacles on each side. Variable, often brown, orange or green. Back with numerous small irregular black spots extending to level of pectoral. Below this on belly they form oblique black lines which run forwards and downwards. Caudal spotted. Anal sometimes with spots. Dorsal immaculate. 8 inches. Gulf of Manaar.

828. AROTHRON HISPIDUS (Lacépède) WHITE-SPOTTED BLOWFISH

Tetrodon hispidus Lacépède, Hist. Nat. Poiss., i, 1802, p. 487; Day, Fish. India, 1875-78, Pl. clxxxiii, fig. 2 (plate 55).

D. 10. A. 10-11. Body short and flabby. Whole body spiny. 2 nasal tentacles on each side. Eyes 3 diameters apart and 2½-3 from snout. Brown above with moderate number of bluish-white small spots and sometimes 1-2 bluish-white bands around orbit and gill-openings. Sometimes lines or spots on abdomen. Some with black blotches or broad black bars on sides. 20 inches. Coastal reefs, Gulf of Manaar and trawling grounds.
Bandi peyttheya, Pippa peyttheya (S).

829. AROTHRON RETICULARIS (Bloch) RETICULATED BLOWFISH

Tetrodon reticularis Bloch and Schneider, Syst. Ichth., 1801, p. 506; Day, Fish. India, 1875-78, Pl. clxxx, fig. 5 (plate 55).

D. 10-11. A. 10-11. 2 nasal tentacles on each side. Interorbital space broad. Small spines all over body except front of snout. Deep grey or brown above, white below. 8-10 longitudinal black stripes, which are curved under the eye, around mouth and pectoral fin. Back with round white spots with a reticulated black interspacing. 17 inches. Gulf of Manaar and trawling grounds.

Pulli peyttheya (S). Pulli peytthai (T).

SUNFISH, TRUNK-FISH

Suborder MOLOIDEI

Family MOLIDAE

No caudal fin or caudal peduncle. Teeth fused to a single unit, without suture in front, forming a powerful bony beak. Skin rough or smooth with small plates. Body firm, compressed and non-inflatable. Gigantic forms which idle at surface of the sea.

Form elongate; lips produced forward as a funnel .... *Ranzania laevis*
Form ovoid; lips not funnel-like .... *Mola mola*

830. RANZANIA LAEVIS (Pennant) LESSER TRUNK-FISH

Ostracion laevis Pennant, British Zool., ed. 4, iii, 1776, p. 129. *Orthagoriscus truncatus*. *Ranzania makua* Jordan and Evermann, Bull. U.S. Fish Comm., xxiii (1), 1905, fig. 194 (plate 52).

Body comparatively elongate, depth about 2. Lips produced forwards beyond teeth as a funnel, closing as a vertical slit. Gill-rakers free. 2 uppermost branchiostegal rays coalesced. Axis of pectoral fin above level of centre of eye. Anal fin $\frac{2}{3}$ length of head. 32 inches. Coastal waters.

831. MOLA MOLA (Linnaeus) GIANT SUNFISH

Tetrodon mola Linnaeus, Syst. Nat., ed. 10, i, 1758, p. 334. *Orthagoriscus mola* Schlegel, Fauna Japon., Pisces, 1850, Pl. cxxvii (plate 52).

Body comparatively short and ovoid, depth 1-1½. Lips not funnel-like. Gill-rakers concealed in thick skin. 6 distinct branchiostegal rays. Skin firm and rough, sometimes with patches of tubercles. Head and chin sometimes swollen. 10 feet. Coastal waters.

FROGFISH

Order BATRACHOIDIFORMES

Family BATRACHOIDIDAE

Elongate robust body with broad and depressed head. Scales absent, or if present concealed in thick mucus. Mouth large. Curved canines in jaws and on palate. Only 3 gills. Opercles with spines. 2 dorsals, first with 2-4 short spines. Ventrals I, 2-3, sometimes with a pocket behind axil. Grunt or croak when taken from the watèr.

832. BATRACHUS GRUNNIENS (Linnaeus) FROGFISH

Cottus grunniens Linnaeus, Syst. Nat., 1758, p. 1209. *Batrachus grunniens* Blegvad and Løppenthin, Fish. Iranian Gulf, 1944, Pl. xii, fig. 3 (plate 56).

D. III; 20-22. A. 15-18. P. 21. V. I, 2. Depth 5. Depressed and broad head. Opercles with 2 spines, subopercle with 2 more. Snout broad and depressed, surrounded by a wreath of tentacles, those above orbit very distinct. A round foramen in upper part of pectoral axilla. Several pointed canines in both jaws. Yellowish, marbled with darker. 8 inches. Coastal waters.
Munda (S).

ANGLER-FISH, MONKFISH, HANDFISH FISHING FROGFISH

Order LOPHIIFORMES

Abnormal specialized fishes, mostly deep water forms, with first ray of dorsal, when present, transformed into an illicium or line and bait, placed on head above large mouth. The illicium usually bears fringes, filaments or luminous appendages. Scales usually obsolete and skin often with flaps or cilia. Gill-openings 2-3, usually reduced. Pectoral powerful and with elbow-like bend, used in locomotion, etc.

1. Body depressed, head broad 2
   Body not depressed 3
2. Body soft, with a large mouth .... *Family* LOPHIIDAE, p. 286
   Body hard and spiny, with small mouth .... *Family* OGCOCEPHALIDAE, p. 286
3. Dorsal with 1 spine .... *Family* CHAUNACIDAE, p. 287
   Dorsal with 3 spines .... *Family* ANTENNARIIDAE, p. 287

MONKFISH, ANGLER-FISH

Family LOPHIIDAE

Large flabby fishes, with greatly depressed head and enormous mouth, fitted with large depressible canines. 3 isolated anterior filamentous dorsal spines, followed by 3 smaller ones in one fin. Pectoral only slightly bent.

833. LOPHIODES LUGUBRIS Alcock — CEYLONESE MONKFISH

Lophius lagubris Alcock, Jour. Asiatic Soc.- Bengal, lxiii (2), 1894, p. 118; Illust. Zool. Investigator, iii, 1895, Pl. xiv, fig. 1 (plate 56).

D. III, I, 7-8. A. 5-6. V. I, 5. Cephalic disc subcircular, width nearly ½ length without caudal, studded with scattered knobs and spines. Mouth cleft full width of disc. Depressible canines in 2 series in upper jaw and in 3 series in lower. Eyes 7 in head, 2 diameters apart. Skin loose and granular. Scattered fringe of tentacles around edge of disc and along tail. Second part of dorsal is a single filament, 2 times eye diameter. 4¼ inches. 142-400 fathoms off coast.

HANDFISH

Family OGCOCEPHALIDAE

Spiny depressed fishes with broad flat head and body and slender tapering tail. Mouth small and with feeble teeth. No true fin spines. One modified dorsal spine on snout, with short stalk and skinny flap (illicium). Pectorals muscular and limb-like. Ventrals jugular and also limb-like.

Disc almost circular .... *Halieutea stellata*
Disc somewhat triangular .... *Dibranchus micropus*

834. HALIEUTEA STELLATA (Vahl) STELLATE HANDFISH

Lophius stellatus Vahl, Skr. Nat. Kjob., iv, 1797, p. 214. *Halieutaea stellata* Day, Fish. India, 1875-78, Pl. liv, fig. 1 (plate 56).

D. 4. A. 4. V. I, 5. P. 13. Body flattened, disc-shaped with slender tapering tail. Illicium with trefoil extremity. Gill-openings above axilla in middle of total length. Dorsal in third fifth of total length. Body covered with distantly placed spines, each with 4 or more roots. Lateral spines largest. Pinkish. 8 inches. Coastal waters.

835. DIBRANCHUS MICROPUS Alcock LESSER HANDFISH

Dibranchus micropus Alcock, Ann. Mag. Nat. Hist. (6), viii, 1891, p. 25; Illust. Zool. Investigator, v, 1898, Pl. xx, fig. 1 (plate 56).

D. 5. A. 4. P. 15. Head and anterior part of body considerably flattened forming a triangular disc, nearly as broad as long and truncated anteriorly. Strong, sharp simple and bifid spines along the margin and a large trifid one at subopercular angle. Eyes small. Mouth cleft horizontal, 2⅓ in width of disc. Jaws with narrow band of minute teeth. Gill-cleft above axilla and hardly larger than nostril. Entire surface covered with fine short bristle-like spines with stellate bases. Ventrals very small. Uniform blue-black. 2½ inches. Gulf of Manaar in 900 fathoms.

ANGLER-FISH

Family CHAUNACIDAE

Body flabby and skin rough. Mouth large, almost vertical, and with small teeth. Gill-openings small, situated behind pectoral base. No true fin spines. An illicium on snout which can be withdrawn into a groove. Pelvics small. Belly inflatable.

836. CHAUNAX PICTUS Lowe PAINTED ANGLER-FISH

Chaunax pictus Lowe, Trans. Zool. Soc., iii, 1849, p. 339; Günther, Challenger Repts., Zool., xxii, 1887, Pl. x, fig. A (plate 56).

D. I, 11. A. 5-7. P. 11-12. Skin very rough. Illicium very short, pediculate. Muciferous canals very distinct, visible externally as chains of pits. Bright orange above, rosy on sides. Fins and lips vermilion. 18 inches. 142-400 fathoms, marine.

FISHING-FROGS

Family ANTENNARIIDAE

Compressed rather lumpy and misshapen body. Scales absent. Skin smooth or rough. Mouth oblique, with small teeth in jaws and palate. Gill-openings small, situated below pectoral base. Pectoral with long base and elbow-like bend. Dorsal with separate spines in front; first on snout slender and movable with fringed apex forming the illicium; 2nd and 3rd enveloped in thick skin.

1. Skin smooth .... *Histrio histrio*
   Skin rough 2
2. Body uniform black or brown .... *Antennarius commersoni*
   Body and fins with ocelli 3
   Body with marbling or reticulations 4
3. Illicium longer than 2nd spine, with fringed lobe .... *Antennarius leprosus*
   Illicium very short, with wedge-shaped tip .... *Antennarius nummifer*
4. Head and anterior part of back immaculate, rose-coloured, and remainder reticulated with black .... *Antennarius bigibbus*
   Light-coloured with darker variegations and marblings 5
5. Illicium moderately long, with lanceolate ciliated lobe; dorsal without spots .... *Antennarius hispidus*
   Illicium as long as second spine, with 3 lanceolate lobes; dorsal with round spots ... *Antennarius pinniceps*

837. HISTRIO HISTRIO (Linnaeus) SMOOTH FISHING-FROG

Lophius histrio Linnaeus, Syst. Nat., ed. 10, 1758, p. 237. *Antennarius marmoratus* Bleeker, Atlas Ichth. Ind. Neerl., v, 1865, Pl. cxcviii, fig. 4 (plate 56).

D. I, I, I, 12-14. A. 7-8. Anterior dorsal spine as long as second, with 2 long and thick cutaneous flaps. 3rd dorsal spine not continuous with soft dorsal. Dorsal spines, head, back and sides of the body with numerous cutaneous fringes, those on dorsal spines forming a dense cluster. Skin smooth or granular, without spines. Reddish or yellow marbled with brown. Eye with radiating brown streaks. Dorsal and anal with 3 series of brown spots. Sometimes more uniform in colour. 5 inches. Found in floating weeds. Coastal waters.

838. ANTENNARIUS COMMERSONI (Cuvier) BLACK FISHING-FROG

Chironectes commersonii Cuvier, Mem. Mus. Nat. Hist., iii, 1817, p. 431. *Antennarius commersonii* Günther, Jour. Mus. Godeffroy, iv (13), 1877, Pl. cv, fig. A (plate 56).

D. I, I, I, 13-14. A. 7-8. First dorsal spine much longer than the second, terminating in a single fringed lobe. Skin very rough and covered with small spines. Nearly uniform black or reddish-brown with dark mottling. Sometimes a few small white spots. 10 inches. Coastal waters.

839. ANTENNARIUS LEPROSUS Eydoux and Souleyet OCELLATED FISHING-FROG

Chironectes leprosus Eydoux and Souleyet, Zool. Voy. Bonite, i, p. 187. *Antennarius multiocellatus*. *Antennarius leprosus* Jordan and Evermann, Bull. U.S. Fish. Comm., xxiii (1), 1905, fig. 228 (plate 56).

D. I, I, I, 11-12. A. 6-7. Skin rather rough and velvety, covered with minute spines. First dorsal spine longer than 2nd, terminating in a fringed lobe. Without cutaneous fringes. Greyish, more or less marbled with dusky. A ragged brown blotch between 2nd and 3rd dorsal spine. Several large light-edged black ocelli on body and dorsal and anal fins. 5 inches. Coastal waters.

840. ANTENNARIUS NUMMIFER (Cuvier)
OCELLATED FRINGED FISHING-FROG

Chironectes nummifer Cuvier, Mem. Mus. Nat. Hist., iii, 1817, p. 430. *Antennarius nummifer* Bleeker, Atlas Ichth. Ind. Neerl., v, 1865, Pl. cxcviii, fig. 2 (plate 56).

D. I, I, I, 12. A. 7. Anterior dorsal spine shorter than second, with wedge-shaped extremity. Skin rough, covered with small spines. Cutaneous fringes not numerous. Grey-brown with a purplish tinge. Black ocellus with yellow edge behind and above end of pectoral base. Fins yellow. Dorsal with ocellus on 7th-8th rays. Fins also with some black bands and spots. 6½ inches. Pearl banks.

841. ANTENNARIUS BIGIBBUS (Lacépède) RETICULATED FISHING-FROG

Lophius bigibbus Lacépède, Hist. Nat. Poiss., i, 1802, p. 325. *Antennarius bigibbus* Günther, Jour. Mus. Godeffroy, iv (13), 1877, Pl. cv, fig. B (plate 56).

D. I, I, I, 12. A. 7. Anterior dorsal spine filiform, about 2 times as long as 2nd, with a simple flap. 3rd spine nearly entirely enveloped in skin. Skin rough, covered in minute spines but without cutaneous fringes. Head and anterior part of back red, rosy or orange, without markings. Remainder of body with dark reticulation. Dorsal also with similar reticulation. Caudal with broad black band. 8 inches. Coastal waters.

842. ANTENNARIUS HISPIDUS (Bloch) SHAGGY FISHING-FROG

Lophius hispidus Bloch and Schneider, Syst. Ichth., 1801, p. 142. *Antennarius hispidus* Day, Fish. India, 1875-78, Pl. lx, fig. 2 (plate 56).

D. I, I, I, 12. A. 7. Anterior dorsal spine of moderate length, terminating in a single lanceolate ciliated flap. Skin very rough, covered with small spines. Yellowish with brown spots and streaks and some radiating from eye. Marbling extends on to dorsal fin, which has no series of large round spots. 8 inches. Coastal waters.
Mudhu gemba (S). Kadal thavalai (T).

843. ANTENNARIUS PINNICEPS Valenciennes MARBLED FISHING-FROG

Antennarius pinniceps Cuvier and Valenciennes, Hist. Nat. Poiss., xii, 1837, p. 410; Bleeker, Atlas Ichth. Ind. Neerl., v, 1865, Pl. cxcvii, fig. 5 (plate 56).

D. I, I, I, 12-13. A. 7-8. Anterior dorsal spine as long as or longer than second and terminating in 3 lanceolate flaps. Skin rough, with small spines and some small cutaneous fringes. Yellowish, irregularly spotted and streaked with brown, some streaks radiating from eye. Dorsal with large elongate spots. Caudal and anal with round spots. 7 inches. Coastal waters.

DRAGON-FISH

Order PEGASIFORMES

Family PEGASIDAE

Degenerate specialized fishes superficially resembling those in the Syngnathiformes. Body armoured with bony rings which are fused into a solid case on the head and thorax, but flexible in the tail region. Mouth tiny, placed below a long flattened rostrum. No teeth. Pectoral fins horizontal, enlarged and with 10-18 unbranched rays. Ventrals I, 1-3.

1. Tail short, with 8 rings and quadrangular posteriorly .... *Pegasus draconis*
   Tail long, with 11-12 rings and flattened posteriorly 2
2. Back with 2 median ridges, each bearing 3 compressed tubercles .... *Parapegasus volans*
   Median ridges on back obtuse, without tubercles .... *Parapegasus natans*

844. PEGASUS DRACONIS Linnaeus — SHORT DRAGON-FISH

Pegasus draconis Linnaeus, Syst. Nat., ed. 10, i, 1758, p. 418. *Zalises amitengu* Jordan and Snyder, Proc. U.S. Nat. Mus., xxiv, 1901, Pl. i, fig. 2 (plate 56).

D. 5. A. 5. P. 11. V. 2. Anus midway between posterior margin of eye and caudal base. Tail with 8 rings. Main depressed part of trunk with cross-ridges, obtuse tubercles situated at junction of ridges. Snout prolonged with 4 denticulated ridges. Body finely reticulated with brown. Snout and 2 or 3 last caudal rings black. Pectorals with brown dots and white margin. 4 inches. Pearl banks, Gulf of Manaar.

845. PARAPEGASUS VOLANS (Linnaeus) — WINGED DRAGON-FISH

Pegasus volans Linnaeus, Syst. Nat., ed. 10, i, 1758, p. 418. *Pegasus volitans*. *Pegasus laternarius* Kaup, Cat. Lophobr., 1856, Pl. i, fig. 4 (plate 56).

D. 5. A. 5. P. 10. V. 2. Anus midway between anterior margin of eye and caudal base. Tail with 11 rings. Back of trunk with 2 median ridges, each with 3 compressed and more or less trenchant tubercles. Snout sometimes short, pointed, with or without spines, sometimes prolonged into a 5- or 6-ridged process, each ridge denticulated. Pectoral rays with brown spots. 2 broad bands across tail. Pearl banks.

846. PARAPEGASUS NATANS (Linnaeus) — FLOATING DRAGON-FISH

Pegasus natans Linnaeus, Syst. Nat., ed. 12, 1766, p. 418. *Parapegasus natans* McCulloch, Biol. Res. Endeavour, iii (3), 1915, fig. 2 (plate 56).

D. 5. A. 5. P. 11. V. 3. Tail without fin as long or longer than body to end of snout. Tail with 12 rings, tapering and much flattened behind. Ridges on upper side of body obtuse, without tubercles. Shields uniformly and finely granulated. Snout prolonged into a long flat sword-like process, with backward directed teeth along sides. Tail with broad brown cross-bands, the one below dorsal fin being more constant. Dorsal and pectoral fins with brown dots. 7 inches. Pearl banks.

LITERATURE ON CEYLONESE FISHES

The following bibliography includes only those published works which deal specifically with the fish fauna of Ceylon. All references herein listed contain either descriptions, distribution records or specific mention of fish specimens that have been collected either in Ceylon or its closely adjacent waters.

ALCOCK, A. W. (1892).—On the bathybial fishes collected during the season of 1891-92. *Ann. Mag. Nat. Hist.* (ser. 6), X: 345-465.

ALCOCK, A. W. (1896).—A supplementary list of the marine fishes of India, with descriptions of 2 new genera and 8 new species. *Jour. Asiatic Soc. Bengal,* LXV: 301-338.

ALCOCK, A. W. (1898).—A note on the deep-sea fishes, with descriptions of some genera and species, including another probably viviparous Ophidioid. *Ann. Mag. Nat. Hist.* (ser. 7), II: 136-156.

ALCOCK, A. W. (1899).—*Halimochirurgus centriscoides,* a new deep-sea fish from the Gulf of Manár. *Proc. Asiatic Soc. Bengal,* LXVIII: 78.

ALCOCK, A. W. (1899).—A descriptive catalogue of the Indian deep-sea fishes in the Indian Museum. collected by the Royal Indian marine survey ship "Investigator". Calcutta: 1-200.

AMIRTHALINGAM, C. (1939).—The nesting habits of the Gourami (*Osphronemus goramy*). *Jour. Bombay Nat. Hist. Soc.,* XLI (2): 436-437.

AMIRTHALINGAM, C. (1948).— The wealth of Ceylon waters. *Proc. 4th. Ann. Session Ceylon Assn. Sci.,* 1-10.

BENNETT, E. T. (1832).—Characters of new species of fishes from Ceylon. *Proc. Zool. Soc.,* 182-184.

BENNETT, J. W. (1828).—A selection of rare and curious fishes found upon the coast of Ceylon. London. Also editions 1834 and 1851.

BLANFORD, W. T. (1901).—The distribution of vertebrate animals in India, Ceylon and Burma. *Philos. Trans. Roy. Soc.,* CXCIV: 335-436.

BLEEKER, P. (1863).—Description de quelques espèces nouvelles de Cyprinoides de Ceylon. *Versl. Akad. Amsterdam,* XV: 239-253; *Ned. Tijds. Dierk.,* I: 373-380.

BLEEKER, P. (1863).—Description d'une nouvelle espèce du genre Nemacheilus. *Versl. Akad. Amsterdam,* XV: 254-256; *Ned. Tijds. Dierk.,* I: 380-381.

BLEEKER, P. (1864).—Description de quelques espèces de Cobitioides et de Cyprinoides de Ceylon. *Nat. Verh. Holl. Maatsch. Wetensch.* (ser. 2), XX: 1-23.

BLYTH, E. (1852).—Report on Ceylon mammals, birds, reptiles and fishes. In E. F. Kelaert, Prodromus faunae Zeylanicae. Ceylon.

BOAKE, B. (1866).—On the air-breathing fish of Ceylon. *Jour. Ceylon Br. Roy. Asiatic Soc.,* 128-142.

BORODIN, N. A. (1930).—Scientific results of the yacht "Ara" expedition during the years 1926-1930, while in command of William K. Vanderbilt. Fishes collected in 1929. *Bull. Vanderbilt Mar. Mus.,* I (2): 39-64.

BORODIN, N. A. (1932).—Scientific results of the yacht "Alva" world cruise, July 1931 to March 1932 in command of William K. Vanderbilt. Fishes. *Ibid.,* I (3): 65-101.

BRAUER, A. (1902).—Diagnosen von neuen Tiefseefischen welche von der Valdivia-Expedition gesammelt sind. *Zool. Anz.,* XXV: 277-298.

CHACKO, P. I. (1944).—Occurrence of a new variety of the skate, *Urogymnus asperrimus,* around Krusadai Island, Gulf of Manaar. *Current Sci., Bangalore,* XIII: 81.

CHACKO, P. I. (1949).—Food and feeding habits of the fishes of the Gulf of Manaar. *Proc. Ind. Acad. Sci.,* XXIX B: 83-97.

CUVIER, G. and VALENCIENNES, A. (1828-1850).—Histoire naturelle des poissons. 22 vols. Paris.

DAY, F. (1875-78).—The fishes of India; being a natural history of the fishes known to inhabit the seas and fresh waters of India, Burma and Ceylon. Text and Atlas in 4 pts. London.

DAY, F. (1888).—Supplement to the fishes of India. London.

DAY, F. (1889).—Fishes. In W. T. Blanford, The fauna of British India, including Ceylon and Burma. 2 vols. London and Calcutta.

De BEAUFORT, L. F. (1940).—Fishes of the Indo-Australian Archipelago (Weber and De Beaufort), VIII (Leiden).

De BEAUFORT, L. F. and CHAPMAN, W. M. (1951).—Fishes of the Indo-Australian Archipelago (Weber and De Beaufort), IX (Leiden).

DERANIYAGALA, P. E. P. (1929).—Some Anguilliform fishes of Ceylon. *Ceylon Jour. Sci.* (B), XV (1): 1-29.

DERANIYAGALA, P. E. P. (1929).—Ceylon sardines. *Ibid.*, XV (1): 31-41.

DERANIYAGALA, P. E. P. (1929).—Two new freshwater fishes. *Ibid.*, XV (2): 73-77.

DERANIYAGALA, P. E. P. (1929).—The Labyrinthica of Ceylon. *Ibid.*, XV (2): 79-111.

DERANIYAGALA, P. E. P. (1930).—The Eventognathi of Ceylon. *Ibid.*, XVI (1): 1-41.

DERANIYAGALA, P. E. P. (1930).—Notes on the breeding habits of the eel, *Leiuranus semicinctus*. *Ibid.*, XVI (1): 107.

DERANIYAGALA, P. E. P. (1930).—Some probable Ceylon larvivores. *Ceylon Jour. Sci.* (D), II: 191-201.

DERANIYAGALA, P. E. P. (1931).—Further notes on the Anguilliform fishes of Ceylon. *Ceylon Jour. Sci.* (B), XVI (2): 131-137.

DERANIYAGALA, P. E. P. (1932).—The Opisthomi of Ceylon. *Ibid.*, XVI (3): 265-269.

DERANIYAGALA, P. E. P. (1932).—The Nematognathoidea of Ceylon, Part I. *Ibid.*, XVI (3): 274-289.

DERANIYAGALA, P. E. P. (1932).—A curious association between *Ophichthys apicalis* and percoid fishes. *Ibid.*, XVI (3): 355-356.

DERANIYAGALA, P. E. P. (1932).—Ichthyological notes—The systematic position of the genus *Channa*; Some mineral spring fishes; A rain of fishes. *Ibid.*, XVII (1): 40-44.

DERANIYAGALA, P. E. P. (1932).—The fishes of Ceylon—A lecture. *Ceylon Economic Soc.*, IV: 35-44.

DERANIYAGALA, P. E. P. (1933).—Some names of fishes from Ceylon. *Ceylon Jour. Sci.* (C), V: 79-111.

DERANIYAGALA, P. E. P. (1933).—Cured marine products of Ceylon. *Ibid.*, V: 49-78.

DERANIYAGALA, P. E. P. (1933).—A new mountain stream fish. *Ceylon Jour. Sci.* (B), XVII (3): 227-229.

DERANIYAGALA, P. E. P. (1933).—Two shark monstrosities. *Ibid.*, XVII (3): 243-244.

DERANIYAGALA, P. E. P. (1933).—Some larger Rhegnopteri of Ceylon. *Ibid.*, XVIII (1): 37-60.

DERANIYAGALA, P. E. P. (1934).—The embryo of the dog-fish *Chiloscyllium indicum*. *Ibid.*, XVIII (2): 249-252.

DERANIYAGALA, P. E. P. (1934).—Some apodal larvae from Ceylon waters. *Ibid.*, XIX (1): 89-96.

DERANIYAGALA, P. E. P. (1936).—Two new fishes from Ceylon. *Ibid.*, XIX (3): 219-224.

DERANIYAGALA, P. E. P. (1936).—Two Xiphiiform fishes from Ceylon. *Ibid.*, XIX (3): 211-218.

DERANIYAGALA, P. E. P. (1937).—Two catadromus fishes new to Ceylon. *Ibid.*, XX (2): 181-184.

DERANIYAGALA, P. E. P. (1937).—The swordfish *Xiphias* of the Indian Ocean. *Ibid.*, XX (2): 347-349.

DERANIYAGALA, P. E. P. (1937).—*Malpulutta kretseri*—a new genus and species of fish from Ceylon. *Ibid.*, XX (3): 351-353.

DERANIYAGALA, P. E. P. (1938-40).—Administrative Report of the Acting Marine Biologist for the years 1937, 1939. *Ceylon Admin. Rept.*, IV (G): 1-16. *Ibid.*, IV (G): 1-9.

DERANIYAGALA, P. E. P. (1943).—A new cyprinoid fish from Ceylon. *Jour. Ceylon Br. Roy. Asiatic Soc.*, XXXV (96): 158-159.

DERANIYAGALA, P. E. P. (1944).—Some whale sharks and sunfishes captured off Ceylon. *Jour. Bombay Nat. Hist. Soc.*, XLIV (3): 426-430.

DERANIYAGALA, P. E. P. (1945).—A further embryonic phase of the Dog Fish, *Hemiscyllium colax*. *Ceylon Jour. Sci.* (B), XXIV (2): 91-92.

DERANIYAGALA, P. E. P. (1945).—A new subspecies among the Ophiocephalids of Ceylon and India. *Ibid.*, XXIV (2): 93.

DERANIYAGALA, P. E. P. (1948).—Some scientific results of two visits to Africa. *Ibid.*, XXV (2): 1-42.

DERANIYAGALA, P. E. P. (1949).—Some vertebrate animals of Ceylon. Vol. I, Ceylon: 1-119.

DERANIYAGALA, P. E. P. (1951).—The Istiophoridae and Xiphiidae of Ceylon. *Ceylon Jour. Sci.* (B), XXVI (2): 137-142.

DERANIYAGALA, P. E. P. (1952).—A coloured atlas of some vertebrates from Ceylon. Vol. I, Fishes. Colombo.

De ZYLVA, E. R. A. (1948).—Observations on the mouth breeding Cichlid fish, *Tilapia mossambica* (Peters). *Proc. 4th Meet. Ceylon Assn. Sci.*, II: 11.

De ZYLVA, E. R. A. (1951-3).—Administration Report of the Acting Director of Fisheries, 1940-1950, 1951, 1952. *Ceylon Admin. Rept.*, IV (1951): L 1-20. *Ibid.*, IV (1952): L 1-62, L 1-30. *Ibid.*, IV (1953): L 1-26,

DOFLEIN, F. J. (1906).—Ostasienfahrt; Ergebnisse und Beobachtungen eines Naturforschers in China. Japan und Ceylon. Leipzig: 1-511.

DRIEBERG, G. C. (1907).—Singing fish of Batticaloa. *Ceylon Jour. Sci.* (B), V (17): 67-68.

DRIEBERG, G. C. and WILLEY, A. (1910).—Crows as fishers. *Ibid.*, VII (25): 48.

DUNCKER, G. (1910).—On some syngnathids (pipe-fish) from Ceylon. *Ibid.*, VII (25): 25-34.

DUNCKER, G. (1912).—Die Süsswasserfische Ceylons. *Mitth. Naturhist. Mus. Hamburg*, XXIX (2): 241-272.

EGE, V. (1933).—On some new fishes of the Family Sudidae and Stomiatidae. *Vidensk. Medd. Naturh. Foren. Kjøb.*, XCIV: 223-236.

FOWLER, H. W. (1928).—Notes and descriptions of fishes from Ceylon. *Journ. Bombay Nat. Hist. Soc.*, 1928: 704-710.

FOWLER, H. W. (1931-1940).—The fishes of the Philippine and adjacent seas. *Bull. U.S. Nat. Mus.* 100; XI (1931), XII (1933), XIII (1940).

FOWLER, H. W. and BEAN, B. A. (1928-1930).—The Fishes of the Philippine and adjacent seas. *Ibid.*, 100, VII (1928), VIII (1929), X (1930).

FRASER-BRUNNER, A. (1933).—A revision of the Chaetodont fishes of the subfamily Pomacanthinae. *Proc. Zool. Soc.*: 543-599.

FRASER-BRUNNER, A. (1941).—Notes on the Plectognath fishes.—VI. A synopsis of the genera of the family Aluteridae, and descriptions of seven new species. *Ann. Mag. Nat. Hist.* (ser. 11.), VIII(45): 176-199.

GUDGER, E. W. (1933).—The whale sharks in the waters around Ceylon. *Nature*, CXXXI: 165.

GÜNTHER, A. (1859-70).—Catalogue of the fishes of the British Museum. 8 vols. London. I (1859), II (1860), III (1861), IV (1862), V (1864), VI (1866), VII (1868), VIII (1870).

GÜNTHER, A. (1860).—On the history of Echeneis. *Ann. Mag. Nat. Hist.* (ser. 3), V: 386-402.

GÜNTHER, A. (1886).—Description of a new siluroid fish from Ceylon (*Arius layardi*). *Ibid.*, (ser. 3), XVIII: 473-474.

HALY, A. (1883).—Occurrence of *Rhinodon typicus* Smith, on the west coast of Ceylon. *Ibid.*, (ser. 5), XII: 48-49.

HALY, A. (1884).—On *Rhinodon typicus* Smith. *Ceylon Admin. Rept.*, 1883 (D): 129-130.

HALY, A. (1886).—Notes on the collection of fish in the Colombo Museum. *Taprobanian*, I: 165-167.

HALY, A. (1887).—Fish new to Ceylon and India. *Ibid.*, II (1): 169.

HALY, A. (1890).—On *Rhinodon typicus* Smith. Fishes of India likely to be found on the coasts of Ceylon. *Ceylon Admin. Rept.*, 1889 (D): 14.

HENRY, G. M. (1912).—Fishes of Batticaloa, Trincomalee, and Jaffna. *Ceylon Jour. Sci.* (B), VIII: 62-64

HERRE, A. W. C. T. (1946).—Fishes collected in Ceylon in April, 1934. *Ibid.*, XXIV (3): 173-179.

HORA, S. L. (1934).—Siluroid fishes of India, Burma and Ceylon, I. *Rec. Ind. Mus.*, XXXV (4): 607-621.

HORA, S. L. (1936).—Siluroid fishes of India, Burma and Ceylon, II, III, IV and V. *Ibid.*, XXXVIII (2): 199-209. VI, VII and VIII. *Ibid.*, XXXVIII (3): 347-361.

HORA, S. L. (1941).—Siluroid fishes of India, Burma and Ceylon, IX and X. *Ibid.*, XLIII (1): 9-42. XI and XII. *Ibid.*, XLIII (2): 97-111.

HORA, S. L. (1949).—Siluroid fishes of India, Burma and Ceylon, XIII. *Ibid.*, XLVII (2): 183-202.

HORA, S. L. and MUKERJII, D. D. (1928).—Notes on fish in the Indian Museum, XVI.—On fishes of the genus *Esomus* Swainson. *Ibid.*, XXX: 41-56.

HORNELL, J. (1916).—Notes upon the two exploring cruises in search of trawl grounds off the Indian and Ceylon coasts. *Madras Fish. Bull.*, VIII: 23-41.

JOHNSTONE, J. (1904).—Report on the marine fishes collected by Professor Herdmann, at Ceylon, in 1902. *Rept. Govt. Ceylon Pearl Oyster Fish. Gulf of Manaar*, II, Suppl. Rept. XV: 201-222.

JONES, S. (1938).—External features in the embryonic development of a Ceylon mountain carp, *Garra ceylonensis ceylonensis* (Bleeker). *Ceylon Jour. Sci.* (C), VI: 91-98.

JONES, S. (1938).—On the breeding habits and development of a cyprinid, *Danio* (*Danio*) *malabaricus* (Jerdon) in Ceylon. *Ibid.*, VI: 79-89.

JONES, S. (1941).—Larval development of *Garra ceylonensis ceylonensis* (Blkr.). *Rec. Ind. Mus.*, XLIII (11): 117-120.

JORDAN, D. S. and STARKS, E. C. (1917).—Notes on a collection of fishes from Ceylon with descriptions of new species. *Ann. Carnegie Mus.*, XI: 430-460.

KÁROLI, J. (1882).—Prodromus piscium Asiae orientalis a Domino Joanne Xanthus annis 1868-70 collectorum. *Termeszet. Füzetek*, V: 147-187.

KIRTHISINGHE, P. (1932).—Notes on the course of the lateral line canals and the cranial nerves in the viviparous shark, *Scoliodon* (*Carcharias*). *Ceylon Jour. Sci.* (B), XVII (2): 103-108.

KNER, R. (1865).—Fische.—Reise der österreichischen Fregatte "Novara" um die Erde in den Jahren 1857-1859. Zool. 1-3. Vienna.

KOUMANS, F. P. (1941).—Gobioid fishes of India. *Mem. Ind. Mus.*, XIII (3): 205-329.

KOUMANS, F. P. (1953).—Fishes of the Indo-Australian Archipelago (Weber and De Beaufort), X. Leiden.

MALPAS, A. H. (1926).—The marine biological survey of the littoral waters of Ceylon. *Ceylon Jour. Sci.* (C), II: 13-165.

MALPAS, A. H. (1929-1939).—Administration report of the Acting Marine Biologist for 1928, 1934, 1935, 1936, 1938. *Ceylon Admin. Rept.*, IV (1929): G 1-24. *Ibid.*, IV (1935): G 1-7. *Ibid.*, IV (1936): G 1-9. *Ibid.*, IV (1937): G 1-12. *Ibid.*, (1939): G 1-11.

MISRA, K. S. (1947).—A check-list of the fishes of India, Burma and Ceylon—Pt. I. Elasmobranchii and Holocephali. *Rec. Ind. Mus.*, XLV (1): 1-5. Pt. II, Clupeiformes, Scopeliformes, Ateleopiformes. *Ibid.*, XLV (4): 377-431.

MISRA, K. S. (1952).—An aid to the identification of the fishes of India, Burma and Ceylon. I, Elasmobranchii and Holocephali. *Ibid.*, XLIX (1): 89-137.

NAIR, R. V. (1948).—Leptocephali of the Gulf of Manaar. *Proc. Ind. Acad. Sci.*, XXVII B: 87-91.

NORMAN, J. R. (1927-28).—The flatfishes (Heterosomata) of India, with a list of the species in the Indian Museum. Part I. *Rec. Ind. Mus.*, XXIX: 7-48. Part II. *Ibid.*, XXX: 173-215.

PEARSON, J. (1911).—Sun-fish caught near Jaffna. *Ceylon Jour. Sci.* (B), VII: 208.

PEARSON, J. (1911).—The Gourami. *Ibid.*, VII: 209-210.

PEARSON, J. (1912).—Biological survey of Trincomalee Harbour. *Ibid.*, VIII (29): 30.

PEARSON, J. (1912).—Crustaceans and fish in the sewage tanks at Madampithya (Ceylon). *Ibid.*, VIII: 152.

PEARSON, J. (1912).—The singing fish of Batticaloa. *Ibid.*, VIII: 150-151.

PEARSON, J. (1912-1933).—Administration report of the Government Marine Biologist for 1910-11, 1912-13, 1914, 1925, 1926, 1927, 1929, 1930, 1931, 1932. *Ceylon Admin. Rept.*, IV (1911): E 1-4. *Ibid.*, IV (1913): E 2-16. *Ibid.*, IV (1914): E 1-9. *Ibid.*, IV (1926): F 1-16. *Ibid.*, IV (1927): F 1-29. *Ibid.*, IV (1928): F 1-27. *Ibid.*, IV (1930): G 1-18. *Ibid.*, IV (1931): G 1-10. *Ibid.*, IV (1932): G 1-11. *Ibid.*, IV (1933): G 1-12.

PEARSON, J. and MALPAS, A. H. (1926).—A preliminary report on the possibilities of Commercial trawling in the sea around Ceylon. *Ceylon Jour. Sci.* (C), II: 1-12.

PERTWEE, A. H. (1911).—Proposed further attempt to introduce the Gourami (*Osphronemus olfax*) into Ceylon, with notes on a suitable locality. *Ceylon Jour. Sci.* (B), VII: 210-212.

PERTWEE, A. H. (1913).—Notes on the freshwater fishes of Ceylon. *Ibid.*, VIII: 243-250.

REGAN, C. T. (1909).—The Asiatic fishes of the family Anabantidae. *Proc. Zool. Soc.* (2): 767-787.

SHIPLEY, A. E. and HORNELL, J. (1906).—Cestode and Nematode parasites from the marine fishes of Ceylon. *Rept. Ceylon Govt. Pearl Oyster Fish. Gulf of Manaar*, V: 43.

SONNERAT, F. (1782).—Voyages aux Indes orientales et à la Chine, fait . . . depuis 1774 jusqu'en 1781: dans lequel on traite . . . des Indiens, des Chinois, des Pégouins et des Madégasses; suivi d'observations sur le cap de Bonne-Esperénce, les isles de France et de Bourbon, les Maldives, Ceylan, Malacca, les Philippines et les Moluques, et de recherches sur l'histoire naturelle de ces pays. 2 vols. Paris. Edition in German, 1783, 2 vols. Zürich.

SOUTHWELL, T. (1910).—A description of a large ray, *Taeniura melanospiles* Bleeker, from the Ceylon pearl banks. *Repts. Ceylon Marine Lab.*, I (4): 185-186.

SOUTHWELL, T. (1910).—On the relation of certain predatory fish to pearl oysters (*Margaritifera vulgaris*) on the Ceylon pearl banks. *Ibid.*, I (4): 175-178.

SOUTHWELL, T. (1910).—Description of nine new species of Cestode parasites including new genera from marine fishes of Ceylon. *Ibid.*, V (18): 216.

SOUTHWELL, T. (1910).—A descriptive note on the capture of a large sawfish (*Pristis cuspidatus*) containing intra-uterine embryos. *Ceylon Jour. Sci.* (B), VI (24): 137-139.

SOUTHWELL, T. (1911).—Some notes on the Ceylon pearl-inducing worm. *Ibid.*, VII (27): 124-134.

SOUTHWELL, T. (1911).—A description of ten new species of Cestode parasites from marine fishes of Ceylon, with notes on seven other species from the same region. *Rept. Ceylon Marine Lab.*, VI, (22): 259.

SOUTHWELL, T. (1913).—Notes on the fauna of the Ceylon pearl banks. *Ceylon Admin. Rept.* 1912-13, IV: E 32-49.

SOUTHWELL, T. (1913).—Notes on the food of certain marine fishes from Ceylon. *Ibid.*, IV: E 49-52.

SOUTHWELL, T. (1929).—A monograph of the cestodes of the Order Trypanorhyncha from Ceylon and India, Part I. *Ceylon Jour. Sci.* (B), XV (3): 169-312.

SUNDARA-RAJ, B. (1916).—Notes on the freshwater fish of Madras. *Rec. Ind. Mus.*, XII: 249-294.

TENNET, *Sir* J. E. (1861).—Sketches of the natural history of Ceylon; with narratives and anecdotes illustrative of the habits and instincts of the mammalia, birds, reptiles, fishes, insects, etc. London. Fishes: 323-368, including Günther's list on 359-362, and Huxley's memorandum on 364-366, 367-368.

THURSTON, E. (1894).—Inspection of Ceylon pearl banks. *Bull. Madras Govt. Mus.*, I: 36-38.

VASILIU, G. D. (1931).—Beschreibung der Ceylon Fische aus der Sammlung Plate. *Jena Z. Naturw.*, LXV: 319-360.

WEBER, M. and De BEAUFORT, L. F. (1913-1936).—Fishes of the Indo-Australian archipelago. II (1913), III (1916), IV (1922), V (1929), VI (1931), VII (1936). Leiden.

WILLEY, A. (1903).—On the more remarkable fresh water fishes. *Ceylon Jour. Sci.* (B), I: 5-9.

WILLEY, A. (1903).—The mahseer and murrel in Ceylon. *Ibid.*, I: 19-22.

WILLEY, A. (1903).—Constitution of the fauna of Ceylon. *Ibid.*, I: 1-13.

WILLEY, A. (1904).—Leaf-mimicry. *Ibid.*, II: 51-55.

WILLEY, A. (1906).—A guide to the collections in the Colombo Museum. (Fishes: 44-46). *Ibid.*, III: 1-66.

WILLEY, A. (1908).—Fishery observations. *Ibid.*, V (19): 144-152.

WILLEY, A. (1908).—*Rhynchobdella aculeata* in Ceylon. *Nature*, LXXVII: 345.

WILLEY, A. (1909).—The occurrence of *Solenostoma* off the coast of Ceylon. *Ceylon Jour. Sci.*, VI (23): 102-107.

WILLEY, A. (1909).—Observations on the nests, eggs and larvae of *Ophiocephalus striatus*. *Ibid.*, VI (23): 108-123.

WILLEY, A. (1910).—Notes on the freshwater fisheries of Ceylon. *Ibid.*, VII (26): 88-105.

WILLEY, A. (1910).—Report of the Marine Biologist for 1909-1910. *Ceylon Admin. Rept.*, IV: G 1-19.

INDEX

The following index contains scientific names of species and groups and vernacular names in the English, Sinhalese, and Tamil languages. Valid specific names and synonyms are in Italics and are listed under both generic and specific components. English vernaculars are indexed under the substantive only. Sinhalese and Tamil names are marked (S) and (T) respectively. Names of classes, sub-classes, orders, sub-orders, and families are in capitals.

Figures in boldface type are the serial numbers of the species according to their arrangement in the text. They are also the figure reference numbers for the illustrations on Plates 1-56.

Figures in ordinary type are the principal page references. In the case of group names the first page quoted refers to the appearance of that name in a key.